Protein Folding Handbook

Edited by J. Buchner and T. Kiefhaber

Further Titles of Interest

K. H. Nierhaus, D. N. Wilson (eds.)

Protein Biosynthesis and Ribosome Structure

ISBN 3-527-30638-2

R. J. Mayer, A. J. Ciechanover, M. Rechsteiner (eds.)

Protein Degradation

ISBN 3-527-30837-7 (Vol. 1)
ISBN 3-527-31130-0 (Vol. 2)

G. Cesareni, M. Gimona, M. Sudol, M. Yaffe (eds.)

Modular Protein Domains

ISBN 3-527-30813-X

S. Brakmann, A. Schwienhorst (eds.)

Evolutionary Methods in Biotechnology

ISBN 3-527-30799-0

Protein Folding Handbook

Edited by Johannes Buchner and Thomas Kiefhaber

WILEY-VCH Verlag GmbH & Co. KGaA

Editors

Prof. Dr. Johannes Buchner
Institut für Organische Chemie und Biochemie
Technische Universität München
Lichtenbergstrasse 4
85747 Garching
Germany
johannes.buchner@ch.tum.de

Prof. Dr. Thomas Kiefhaber
Biozentrum der Universität Basel
Division of Biophysical Chemistry
Klingelbergstrasse 70
4056 Basel
Switzerland
t.kiefhaber@unibas.ch

Cover
Artwork by Prof. Erich Gohl, Regensburg

Library of Congress Card No. applied for
A catalogue record for this book is available from the British Library.
Bibliographic information published by Die Deutsche Bibliothek
Die Deutsche Bibliothek lists this publication in the Deutsche Nationalbibliografie; detailed bibliographic data is available in the Internet at http://dnb.ddb.de

Printed in the Federal Republic of Germany.
Printed on acid-free paper.

Typesetting Asco Typesetters, Hong Kong
Printing betz-druck gmbh, Darmstadt
Bookbinding Litges & Dopf Buchbinderei GmbH, Heppenheim

ISBN-13 978-3-527-30784-5
ISBN-10 3-527-30784-2

Contents

Part I, Volume 1

Preface *LVIII*

Contributors of Part I *LX*

I/1 Principles of Protein Stability and Design *1*

1 Early Days of Studying the Mechanism of Protein Folding *3*
Robert L. Baldwin
1.1 Introduction *3*
1.2 Two-state Folding *4*
1.3 Levinthal's Paradox *5*
1.4 The Domain as a Unit of Folding *6*
1.5 Detection of Folding Intermediates and Initial Work on the Kinetic Mechanism of Folding *7*
1.6 Two Unfolded Forms of RNase A and Explanation by Proline Isomerization *9*
1.7 Covalent Intermediates in the Coupled Processes of Disulfide Bond Formation and Folding *11*
1.8 Early Stages of Folding Detected by Antibodies and by Hydrogen Exchange *12*
1.9 Molten Globule Folding Intermediates *14*
1.10 Structures of Peptide Models for Folding Intermediates *15*
Acknowledgments *16*
References *16*

2 Spectroscopic Techniques to Study Protein Folding and Stability *22*
Franz Schmid
2.1 Introduction *22*
2.2 Absorbance *23*
2.2.1 Absorbance of Proteins *23*
2.2.2 Practical Considerations for the Measurement of Protein Absorbance *27*

Protein Folding Handbook. Part II. Edited by J. Buchner and T. Kiefhaber

ISBN: 3-527-30784-2

2.2.3 Data Interpretation 29
2.3 Fluorescence 29
2.3.1 The Fluorescence of Proteins 30
2.3.2 Energy Transfer and Fluorescence Quenching in a Protein: Barnase 31
2.3.3 Protein Unfolding Monitored by Fluorescence 33
2.3.4 Environmental Effects on Tyrosine and Tryptophan Emission 36
2.3.5 Practical Considerations 37
2.4 Circular Dichroism 38
2.4.1 CD Spectra of Native and Unfolded Proteins 38
2.4.2 Measurement of Circular Dichroism 41
2.4.3 Evaluation of CD Data 42
References 43

3 Denaturation of Proteins by Urea and Guanidine Hydrochloride 45
C. Nick Pace, Gerald R. Grimsley, and J. Martin Scholtz
3.1 Historical Perspective 45
3.2 How Urea Denatures Proteins 45
3.3 Linear Extrapolation Method 48
3.4 $\Delta G(H_2O)$ 50
3.5 *m*-Values 55
3.6 Concluding Remarks 58
3.7 Experimental Protocols 59
3.7.1 How to Choose the Best Denaturant for your Study 59
3.7.2 How to Prepare Denaturant Solutions 59
3.7.3 How to Determine Solvent Denaturation Curves 60
3.7.3.1 Determining a Urea or GdmCl Denaturation Curve 62
3.7.3.2 How to Analyze Urea or GdmCl Denaturant Curves 63
3.7.4 Determining Differences in Stability 64
Acknowledgments 65
References 65

4 Thermal Unfolding of Proteins Studied by Calorimetry 70
George I. Makhatadze
4.1 Introduction 70
4.2 Two-state Unfolding 71
4.3 Cold Denaturation 76
4.4 Mechanisms of Thermostabilization 77
4.5 Thermodynamic Dissection of Forces Contributing to Protein Stability 79
4.5.1 Heat Capacity Changes, ΔC_p 81
4.5.2 Enthalpy of Unfolding, ΔH 81
4.5.3 Entropy of Unfolding, ΔS 83
4.6 Multistate Transitions 84
4.6.1 Two-state Dimeric Model 85

4.6.2 Two-state Multimeric Model *86*
4.6.3 Three-state Dimeric Model *86*
4.6.4 Two-state Model with Ligand Binding *88*
4.6.5 Four-state (Two-domain Protein) Model *90*
4.7 Experimental Protocols *92*
4.7.1 How to Prepare for DSC Experiments *92*
4.7.2 How to Choose Appropriate Conditions *94*
4.7.3 Critical Factors in Running DSC Experiments *94*
References *95*

5 Pressure–Temperature Phase Diagrams of Proteins *99*
Wolfgang Doster and Josef Friedrich
5.1 Introduction *99*
5.2 Basic Aspects of Phase Diagrams of Proteins and Early Experiments *100*
5.3 Thermodynamics of Pressure–Temperature Phase Diagrams *103*
5.4 Measuring Phase Stability Boundaries with Optical Techniques *110*
5.4.1 Fluorescence Experiments with Cytochrome *c* *110*
5.4.2 Results *112*
5.5 What Do We Learn from the Stability Diagram? *116*
5.5.1 Thermodynamics *116*
5.5.2 Determination of the Equilibrium Constant of Denaturation *117*
5.5.3 Microscopic Aspects *120*
5.5.4 Structural Features of the Pressure-denatured State *122*
5.6 Conclusions and Outlook *123*
Acknowledgment *124*
References *124*

6 Weak Interactions in Protein Folding: Hydrophobic Free Energy, van der Waals Interactions, Peptide Hydrogen Bonds, and Peptide Solvation *127*
Robert L. Baldwin
6.1 Introduction *127*
6.2 Hydrophobic Free Energy, Burial of Nonpolar Surface and van der Waals Interactions *128*
6.2.1 History *128*
6.2.2 Liquid–Liquid Transfer Model *128*
6.2.3 Relation between Hydrophobic Free Energy and Molecular Surface Area *130*
6.2.4 Quasi-experimental Estimates of the Work of Making a Cavity in Water or in Liquid Alkane *131*
6.2.5 Molecular Dynamics Simulations of the Work of Making Cavities in Water *133*
6.2.6 Dependence of Transfer Free Energy on the Volume of the Solute *134*
6.2.7 Molecular Nature of Hydrophobic Free Energy *136*

6.2.8 Simulation of Hydrophobic Clusters 137
6.2.9 ΔC_p and the Temperature-dependent Thermodynamics of Hydrophobic Free Energy 137
6.2.10 Modeling Formation of the Hydrophobic Core from Solvation Free Energy and van der Waals Interactions between Nonpolar Residues 142
6.2.11 Evidence Supporting a Role for van der Waals Interactions in Forming the Hydrophobic Core 144
6.3 Peptide Solvation and the Peptide Hydrogen Bond 145
6.3.1 History 145
6.3.2 Solvation Free Energies of Amides 147
6.3.3 Test of the Hydrogen-Bond Inventory 149
6.3.4 The Born Equation 150
6.3.5 Prediction of Solvation Free Energies of Polar Molecules by an Electrostatic Algorithm 150
6.3.6 Prediction of the Solvation Free Energies of Peptide Groups in Different Backbone Conformations 151
6.3.7 Predicted Desolvation Penalty for Burial of a Peptide H-bond 153
6.3.8 Gas–Liquid Transfer Model 154
Acknowledgments 156
References 156

7 Electrostatics of Proteins: Principles, Models and Applications *163*
Sonja Braun-Sand and Arieh Warshel
7.1 Introduction 163
7.2 Historical Perspectives 163
7.3 Electrostatic Models: From Microscopic to Macroscopic Models 166
7.3.1 All-Atom Models 166
7.3.2 Dipolar Lattice Models and the PDLD Approach 168
7.3.3 The PDLD/S-LRA Model 170
7.3.4 Continuum (Poisson-Boltzmann) and Related Approaches 171
7.3.5 Effective Dielectric Constant for Charge–Charge Interactions and the GB Model 172
7.4 The Meaning and Use of the Protein Dielectric Constant 173
7.5 Validation Studies 176
7.6 Systems Studied 178
7.6.1 Solvation Energies of Small Molecules 178
7.6.2 Calculation of pK_a Values of Ionizable Residues 179
7.6.3 Redox and Electron Transport Processes 180
7.6.4 Ligand Binding 181
7.6.5 Enzyme Catalysis 182
7.6.6 Ion Pairs 183
7.6.7 Protein–Protein Interactions 184
7.6.8 Ion Channels 185
7.6.9 Helix Macrodipoles versus Localized Molecular Dipoles 185
7.6.10 Folding and Stability 186
7.7 Concluding Remarks 189

Acknowledgments 190
References 190

8 **Protein Conformational Transitions as Seen from the Solvent: Magnetic Relaxation Dispersion Studies of Water, Co-solvent, and Denaturant Interactions with Nonnative Proteins** 201
Bertil Halle, Vladimir P. Denisov, Kristofer Modig, and Monika Davidovic
8.1 The Role of the Solvent in Protein Folding and Stability 201
8.2 Information Content of Magnetic Relaxation Dispersion 202
8.3 Thermal Perturbations 205
8.3.1 Heat Denaturation 205
8.3.2 Cold Denaturation 209
8.4 Electrostatic Perturbations 213
8.5 Solvent Perturbations 218
8.5.1 Denaturation Induced by Urea 219
8.5.2 Denaturation Induced by Guanidinium Chloride 225
8.5.3 Conformational Transitions Induced by Co-solvents 228
8.6 Outlook 233
8.7 Experimental Protocols and Data Analysis 233
8.7.1 Experimental Methodology 233
8.7.1.1 Multiple-field MRD 234
8.7.1.2 Field-cycling MRD 234
8.7.1.3 Choice of Nuclear Isotope 235
8.7.2 Data Analysis 236
8.7.2.1 Exchange Averaging 236
8.7.2.2 Spectral Density Function 237
8.7.2.3 Residence Time 239
8.7.2.4 ^{19}F Relaxation 240
8.7.2.5 Coexisting Protein Species 241
8.7.2.6 Preferential Solvation 241
References 242

9 **Stability and Design of α-Helices** 247
Andrew J. Doig, Neil Errington, and Teuku M. Iqbalsyah
9.1 Introduction 247
9.2 Structure of the α-Helix 247
9.2.1 Capping Motifs 248
9.2.2 Metal Binding 250
9.2.3 The 3_{10}-Helix 251
9.2.4 The π-Helix 251
9.3 Design of Peptide Helices 252
9.3.1 Host–Guest Studies 253
9.3.2 Helix Lengths 253
9.3.3 The Helix Dipole 253
9.3.4 Acetylation and Amidation 254
9.3.5 Side Chain Spacings 255

9.3.6 Solubility 256
9.3.7 Concentration Determination 257
9.3.8 Design of Peptides to Measure Helix Parameters 257
9.3.9 Helix Templates 259
9.3.10 Design of 3_{10}-Helices 259
9.3.11 Design of π-helices 261
9.4 Helix Coil Theory 261
9.4.1 Zimm-Bragg Model 261
9.4.2 Lifson-Roig Model 262
9.4.3 The Unfolded State and Polyproline II Helix 265
9.4.4 Single Sequence Approximation 265
9.4.5 N- and C-Caps 266
9.4.6 Capping Boxes 266
9.4.7 Side-chain Interactions 266
9.4.8 N1, N2, and N3 Preferences 267
9.4.9 Helix Dipole 267
9.4.10 3_{10}- and π-Helices 268
9.4.11 AGADIR 268
9.4.12 Lomize-Mosberg Model 269
9.4.13 Extension of the Zimm-Bragg Model 270
9.4.14 Availability of Helix/Coil Programs 270
9.5 Forces Affecting α-Helix Stability 270
9.5.1 Helix Interior 270
9.5.2 Caps 273
9.5.3 Phosphorylation 276
9.5.4 Noncovalent Side-chain Interactions 276
9.5.5 Covalent Side-chain interactions 277
9.5.6 Capping Motifs 277
9.5.7 Ionic Strength 279
9.5.8 Temperature 279
9.5.9 Trifluoroethanol 279
9.5.10 pK_a Values 280
9.5.11 Relevance to Proteins 281
9.6 Experimental Protocols and Strategies 281
9.6.1 Solid Phase Peptide Synthesis (SPPS) Based on the Fmoc Strategy 281
9.6.1.1 Equipment and Reagents 281
9.6.1.2 Fmoc Deprotection and Coupling 283
9.6.1.3 Kaiser Test 284
9.6.1.4 Acetylation and Cleavage 285
9.6.1.5 Peptide Precipitation 286
9.6.2 Peptide Purification 286
9.6.2.1 Equipment and Reagents 286
9.6.2.2 Method 286
9.6.3 Circular Dichroism 287
9.6.4 Acquisition of Spectra 288

9.6.4.1 Instrumental Considerations 288
9.6.5 Data Manipulation and Analysis 289
9.6.5.1 Protocol for CD Measurement of Helix Content 291
9.6.6 Aggregation Test for Helical Peptides 291
9.6.6.1 Equipment and Reagents 291
9.6.6.2 Method 292
9.6.7 Vibrational Circular Dichroism 292
9.6.8 NMR Spectroscopy 292
9.6.8.1 Nuclear Overhauser Effect 293
9.6.8.2 Amide Proton Exchange Rates 294
9.6.8.3 ^{13}C NMR 294
9.6.9 Fourier Transform Infrared Spectroscopy 295
9.6.9.1 Secondary Structure 295
9.6.10 Raman Spectroscopy and Raman Optical Activity 296
9.6.11 pH Titrations 298
9.6.11.1 Equipment and Reagents 298
9.6.11.2 Method 298
Acknowledgments 299
References 299

10 Design and Stability of Peptide β-Sheets 314
Mark S. Searle
10.1 Introduction 314
10.2 β-Hairpins Derived from Native Protein Sequences 315
10.3 Role of β-Turns in Nucleating β-Hairpin Folding 316
10.4 Intrinsic ϕ, ψ Propensities of Amino Acids 319
10.5 Side-chain Interactions and β-Hairpin Stability 321
10.5.1 Aromatic Clusters Stabilize β-Hairpins 322
10.5.2 Salt Bridges Enhance Hairpin Stability 325
10.6 Cooperative Interactions in β-Sheet Peptides: Kinetic Barriers to Folding 330
10.7 Quantitative Analysis of Peptide Folding 331
10.8 Thermodynamics of β-Hairpin Folding 332
10.9 Multistranded Antiparallel β-Sheet Peptides 334
10.10 Concluding Remarks: Weak Interactions and Stabilization of Peptide β-Sheets 339
References 340

11 Predicting Free Energy Changes of Mutations in Proteins 343
Raphael Guerois, Joaquim Mendes, and Luis Serrano
11.1 Physical Forces that Determine Protein Conformational Stability 343
11.1.1 Protein Conformational Stability [1] 343
11.1.2 Structures of the N and D States [2–6] 344
11.1.3 Studies Aimed at Understanding the Physical Forces that Determine Protein Conformational Stability [1, 2, 8, 19–26] 346
11.1.4 Forces Determining Conformational Stability [1, 2, 8, 19–27] 346

11.1.5 Intramolecular Interactions 347
11.1.5.1 van der Waals Interactions 347
11.1.5.2 Electrostatic Interactions 347
11.1.5.3 Conformational Strain 349
11.1.6 Solvation 350
11.1.7 Intramolecular Interactions and Solvation Taken Together 350
11.1.8 Entropy 351
11.1.9 Cavity Formation 352
11.1.10 Summary 353
11.2 Methods for the Prediction of the Effect of Point Mutations on in vitro Protein Stability 353
11.2.1 General Considerations on Protein Plasticity upon Mutation 353
11.2.2 Predictive Strategies 355
11.2.3 Methods 356
11.2.3.1 From Sequence and Multiple Sequence Alignment Analysis 356
11.2.3.2 Statistical Analysis of the Structure Databases 356
11.2.3.3 Helix/Coil Transition Model 357
11.2.3.4 Physicochemical Method Based on Protein Engineering Experiments 359
11.2.3.5 Methods Based only on the Basic Principles of Physics and Thermodynamics 364
11.3 Mutation Effects on in vivo Stability 366
11.3.1 The N-terminal Rule 366
11.3.2 The C-terminal Rule 367
11.3.3 PEST Signals 368
11.4 Mutation Effects on Aggregation 368
References 369

I/2 Dynamics and Mechanisms of Protein Folding Reactions *377*

12.1 Kinetic Mechanisms in Protein Folding *379*
Annett Bachmann and Thomas Kiefhaber
12.1.1 Introduction 379
12.1.2 Analysis of Protein Folding Reactions using Simple Kinetic Models 379
12.1.2.1 General Treatment of Kinetic Data 380
12.1.2.2 Two-state Protein Folding 380
12.1.2.3 Complex Folding Kinetics 384
12.1.2.3.1 Heterogeneity in the Unfolded State 384
12.1.2.3.2 Folding through Intermediates 388
12.1.2.3.3 Rapid Pre-equilibria 391
12.1.2.3.4 Folding through an On-pathway High-energy Intermediate 393
12.1.3 A Case Study: the Mechanism of Lysozyme Folding 394
12.1.3.1 Lysozyme Folding at pH 5.2 and Low Salt Concentrations 394
12.1.3.2 Lysozyme Folding at pH 9.2 or at High Salt Concentrations 398
12.1.4 Non-exponential Kinetics 401

12.1.5 Conclusions and Outlook 401
12.1.6 Protocols – Analytical Solutions of Three-state Protein Folding Models 402
12.1.6.1 Triangular Mechanism 402
12.1.6.2 On-pathway Intermediate 403
12.1.6.3 Off-pathway Mechanism 404
12.1.6.4 Folding Through an On-pathway High-Energy Intermediate 404
Acknowledgments 406
References 406

12.2 Characterization of Protein Folding Barriers with Rate Equilibrium Free Energy Relationships 411
Thomas Kiefhaber, Ignacio E. Sánchez, and Annett Bachmann
12.2.1 Introduction 411
12.2.2 Rate Equilibrium Free Energy Relationships 411
12.2.2.1 Linear Rate Equilibrium Free Energy Relationships in Protein Folding 414
12.2.2.2 Properties of Protein Folding Transition States Derived from Linear REFERs 418
12.2.3 Nonlinear Rate Equilibrium Free Energy Relationships in Protein Folding 420
12.2.3.1 Self-Interaction and Cross-Interaction Parameters 420
12.2.3.2 Hammond and Anti-Hammond Behavior 424
12.2.3.3 Sequential and Parallel Transition States 425
12.2.3.4 Ground State Effects 428
12.2.4 Experimental Results on the Shape of Free Energy Barriers in Protein Folding 432
12.2.4.1 Broadness of Free Energy Barriers 432
12.2.4.2 Parallel Pathways 437
12.2.5 Folding in the Absence of Enthalpy Barriers 438
12.2.6 Conclusions and Outlook 438
Acknowledgments 439
References 439

13 A Guide to Measuring and Interpreting ϕ-values 445
Nicholas R. Guydosh and Alan R. Fersht
13.1 Introduction 445
13.2 Basic Concept of ϕ-Value Analysis 445
13.3 Further Interpretation of ϕ 448
13.4 Techniques 450
13.5 Conclusions 452
References 452

14 Fast Relaxation Methods 454
Martin Gruebele
14.1 Introduction 454

14.2 Techniques 455
14.2.1 Fast Pressure-Jump Experiments 455
14.2.2 Fast Resistive Heating Experiments 456
14.2.3 Fast Laser-induced Relaxation Experiments 457
14.2.3.1 Laser Photolysis 457
14.2.3.2 Electrochemical Jumps 458
14.2.3.3 Laser-induced pH Jumps 458
14.2.3.4 Covalent Bond Dissociation 459
14.2.3.5 Chromophore Excitation 460
14.2.3.6 Laser Temperature Jumps 460
14.2.4 Multichannel Detection Techniques for Relaxation Studies 461
14.2.4.1 Small Angle X-ray Scattering or Light Scattering 462
14.2.4.2 Direct Absorption Techniques 463
14.2.4.3 Circular Dichroism and Optical Rotatory Dispersion 464
14.2.4.4 Raman and Resonance Raman Scattering 464
14.2.4.5 Intrinsic Fluorescence 465
14.2.4.6 Extrinsic Fluorescence 465
14.3 Protein Folding by Relaxation 466
14.3.1 Transition State Theory, Energy Landscapes, and Fast Folding 466
14.3.2 Viscosity Dependence of Folding Motions 470
14.3.3 Resolving Burst Phases 471
14.3.4 Fast Folding and Unfolded Proteins 472
14.3.5 Experiment and Simulation 472
14.4 Summary 474
14.5 Experimental Protocols 475
14.5.1 Design Criteria for Laser Temperature Jumps 475
14.5.2 Design Criteria for Fast Single-Shot Detection Systems 476
14.5.3 Designing Proteins for Fast Relaxation Experiments 477
14.5.4 Linear Kinetic, Nonlinear Kinetic, and Generalized Kinetic Analysis of Fast Relaxation 477
14.5.4.1 The Reaction $D \rightleftharpoons F$ in the Presence of a Barrier 477
14.5.4.2 The Reaction $2A \rightleftharpoons A_2$ in the Presence of a Barrier 478
14.5.4.3 The Reaction $D \rightleftharpoons F$ at Short Times or over Low Barriers 479
14.5.5 Relaxation Data Analysis by Linear Decomposition 480
14.5.5.1 Singular Value Decomposition (SVD) 480
14.5.5.2 χ-Analysis 481
Acknowledgments 481
References 482

15 Early Events in Protein Folding Explored by Rapid Mixing Methods 491
Heinrich Roder, Kosuke Maki, Ramil F. Latypov, Hong Cheng, and M. C. Ramachandra Shastry
15.1 Importance of Kinetics for Understanding Protein Folding 491
15.2 Burst-phase Signals in Stopped-flow Experiments 492
15.3 Turbulent Mixing 494

15.4 Detection Methods *495*
15.4.1 Tryptophan Fluorescence *495*
15.4.2 ANS Fluorescence *498*
15.4.3 FRET *499*
15.4.4 Continuous-flow Absorbance *501*
15.4.5 Other Detection Methods used in Ultrafast Folding Studies *502*
15.5 A Quenched-Flow Method for H-D Exchange Labeling Studies on the Microsecond Time Scale *502*
15.6 Evidence for Accumulation of Early Folding Intermediates in Small Proteins *505*
15.6.1 B1 Domain of Protein G *505*
15.6.2 Ubiquitin *508*
15.6.3 Cytochrome *c* *512*
15.7 Significance of Early Folding Events *515*
15.7.1 Barrier-limited Folding vs. Chain Diffusion *515*
15.7.2 Chain Compaction: Random Collapse vs. Specific Folding *516*
15.7.3 Kinetic Role of Early Folding Intermediates *517*
15.7.4 Broader Implications *520*
Appendix *521*
A1 Design and Calibration of Rapid Mixing Instruments *521*
A1.1 Stopped-flow Equipment *521*
A1.2 Continuous-flow Instrumentation *524*
Acknowledgments *528*
References *528*

16 Kinetic Protein Folding Studies using NMR Spectroscopy *536*
Markus Zeeb and Jochen Balbach
16.1 Introduction *536*
16.2 Following Slow Protein Folding Reactions in Real Time *538*
16.3 Two-dimensional Real-time NMR Spectroscopy *545*
16.4 Dynamic and Spin Relaxation NMR for Quantifying Microsecond-to-Millisecond Folding Rates *550*
16.5 Conclusions and Future Directions *555*
16.6 Experimental Protocols *556*
16.6.1 How to Record and Analyze 1D Real-time NMR Spectra *556*
16.6.1.1 Acquisition *556*
16.6.1.2 Processing *557*
16.6.1.3 Analysis *557*
16.6.1.4 Analysis of 1D Real-time Diffusion Experiments *558*
16.6.2 How to Extract Folding Rates from 1D Spectra by Line Shape Analysis *559*
16.6.2.1 Acquisition *560*
16.6.2.2 Processing *560*
16.6.2.3 Analysis *561*
16.6.3 How to Extract Folding Rates from 2D Real-time NMR Spectra *562*

16.6.3.1 Acquisition 563
16.6.3.2 Processing 563
16.6.3.3 Analysis 563
16.6.4 How to Analyze Heteronuclear NMR Relaxation and Exchange Data 565
16.6.4.1 Acquisition 566
16.6.4.2 Processing 567
16.6.4.3 Analysis 567
Acknowledgments 569
References 569

Part I, Volume 2

17 Fluorescence Resonance Energy Transfer (FRET) and Single Molecule Fluorescence Detection Studies of the Mechanism of Protein Folding and Unfolding 573
Elisha Haas
Abbreviations 573
17.1 Introduction 573
17.2 What are the Main Aspects of the Protein Folding Problem that can be Addressed by Methods Based on FRET Measurements? 574
17.2.1 The Three Protein Folding Problems 574
17.2.1.1 The Chain Entropy Problem 574
17.2.1.2 The Function Problem: Conformational Fluctuations 575
17.3 Theoretical Background 576
17.3.1 Nonradiative Excitation Energy Transfer 576
17.3.2 What is FRET? The Singlet–Singlet Excitation Transfer 577
17.3.3 Rate of Nonradiative Excitation Energy Transfer within a Donor–Acceptor Pair 578
17.3.4 The Orientation Factor 583
17.3.5 How to Determine and Control the Value of R_o? 584
17.3.6 Index of Refraction n 584
17.3.7 The Donor Quantum Yield Φ_D^o 586
17.3.8 The Spectral Overlap Integral J 586
17.4 Determination of Intramolecular Distances in Protein Molecules using FRET Measurements 586
17.4.1 Single Distance between Donor and Acceptor 587
17.4.1.1 Method 1: Steady State Determination of Decrease of Donor Emission 587
17.4.1.2 Method 2: Acceptor Excitation Spectroscopy 588
17.4.2 Time-resolved Methods 588
17.4.3 Determination of E from Donor Fluorescence Decay Rates 589
17.4.4 Determination of Acceptor Fluorescence Lifetime 589
17.4.5 Determination of Intramolecular Distance Distributions 590

17.4.6 Evaluation of the Effect of Fast Conformational Fluctuations and Determination of Intramolecular Diffusion Coefficients 592
17.5 Experimental Challenges in the Implementation of FRET Folding Experiments 594
17.5.1 Optimized Design and Preparation of Labeled Protein Samples for FRET Folding Experiments 594
17.5.2 Strategies for Site-specific Double Labeling of Proteins 595
17.5.3 Preparation of Double-labeled Mutants Using Engineered Cysteine Residues (strategy 4) 596
17.5.4 Possible Pitfalls Associated with the Preparation of Labeled Protein Samples for FRET Folding Experiments 599
17.6 Experimental Aspects of Folding Studies by Distance Determination Based on FRET Measurements 600
17.6.1 Steady State Determination of Transfer Efficiency 600
17.6.1.1 Donor Emission 600
17.6.1.2 Acceptor Excitation Spectroscopy 601
17.6.2 Time-resolved Measurements 601
17.7 Data Analysis 603
17.7.1 Rigorous Error Analysis 606
17.7.2 Elimination of Systematic Errors 606
17.8 Applications of trFRET for Characterization of Unfolded and Partially Folded Conformations of Globular Proteins under Equilibrium Conditions 607
17.8.1 Bovine Pancreatic Trypsin Inhibitor 607
17.8.2 The Loop Hypothesis 608
17.8.3 RNase A 609
17.8.4 Staphylococcal Nuclease 611
17.9 Unfolding Transition via Continuum of Native-like Forms 611
17.10 The Third Folding Problem: Domain Motions and Conformational Fluctuations of Enzyme Molecules 611
17.11 Single Molecule FRET-detected Folding Experiments 613
17.12 Principles of Applications of Single Molecule FRET Spectroscopy in Folding Studies 615
17.12.1 Design and Analysis of Single Molecule FRET Experiments 615
17.12.1.1 How is Single Molecule FRET Efficiency Determined? 615
17.12.1.2 The Challenge of Extending the Length of the Time Trajectories 617
17.12.2 Distance and Time Resolution of the Single Molecule FRET Folding Experiments 618
17.13 Folding Kinetics 619
17.13.1 Steady State and trFRET-detected Folding Kinetics Experiments 619
17.13.2 Steady State Detection 619
17.13.3 Time-resolved FRET Detection of Rapid Folding Kinetics: the "Double Kinetics" Experiment 621
17.13.4 Multiple Probes Analysis of the Folding Transition 622
17.14 Concluding Remarks 625

Acknowledgments 626
References 627

18 Application of Hydrogen Exchange Kinetics to Studies of Protein Folding 634
Kaare Teilum, Birthe B. Kragelund, and Flemming M. Poulsen
18.1 Introduction 634
18.2 The Hydrogen Exchange Reaction 638
18.2.1 Calculating the Intrinsic Hydrogen Exchange Rate Constant, k_{int} 638
18.3 Protein Dynamics by Hydrogen Exchange in Native and Denaturing Conditions 641
18.3.1 Mechanisms of Exchange 642
18.3.2 Local Opening and Closing Rates from Hydrogen Exchange Kinetics 642
18.3.2.1 The General Amide Exchange Rate Expression – the Linderstrøm-Lang Equation 643
18.3.2.2 Limits to the General Rate Expression – EX1 and EX2 644
18.3.2.3 The Range between the EX1 and EX2 Limits 646
18.3.2.4 Identification of Exchange Limit 646
18.3.2.5 Global Opening and Closing Rates and Protein Folding 647
18.3.3 The "Native State Hydrogen Exchange" Strategy 648
18.3.3.1 Localization of Partially Unfolded States, PUFs 650
18.4 Hydrogen Exchange as a Structural Probe in Kinetic Folding Experiments 651
18.4.1 Protein Folding/Hydrogen Exchange Competition 652
18.4.2 Hydrogen Exchange Pulse Labeling 656
18.4.3 Protection Factors in Folding Intermediates 657
18.4.4 Kinetic Intermediate Structures Characterized by Hydrogen Exchange 659
18.5 Experimental Protocols 661
18.5.1 How to Determine Hydrogen Exchange Kinetics at Equilibrium 661
18.5.1.1 Equilibrium Hydrogen Exchange Experiments 661
18.5.1.2 Determination of Segmental Opening and Closing Rates, k_{op} and k_{cl} 662
18.5.1.3 Determination of ΔG_{fluc}, m, and ΔG°_{unf} 662
18.5.2 Planning a Hydrogen Exchange Folding Experiment 662
18.5.2.1 Determine a Combination of t_{pulse} and pH_{pulse} 662
18.5.2.2 Setup Quench Flow Apparatus 662
18.5.2.3 Prepare Deuterated Protein and Chemicals 663
18.5.2.4 Prepare Buffers and Unfolded Protein 663
18.5.2.5 Check pH in the Mixing Steps 664
18.5.2.6 Sample Mixing and Preparation 664
18.5.3 Data Analysis 664
Acknowledgments 665
References 665

19 Studying Protein Folding and Aggregation by Laser Light Scattering *673*
Klaus Gast and Andreas J. Modler
19.1 Introduction *673*
19.2 Basic Principles of Laser Light Scattering *674*
19.2.1 Light Scattering by Macromolecular Solutions *674*
19.2.2 Molecular Parameters Obtained from Static Light Scattering (SLS) *676*
19.2.3 Molecular Parameters Obtained from Dynamic Light Scattering (DLS) *678*
19.2.4 Advantages of Combined SLS and DLS Experiments *680*
19.3 Laser Light Scattering of Proteins in Different Conformational States – Equilibrium Folding/Unfolding Transitions *680*
19.3.1 General Considerations, Hydrodynamic Dimensions in the Natively Folded State *680*
19.3.2 Changes in the Hydrodynamic Dimensions during Heat-induced Unfolding *682*
19.3.3 Changes in the Hydrodynamic Dimensions upon Cold Denaturation *683*
19.3.4 Denaturant-induced Changes of the Hydrodynamic Dimensions *684*
19.3.5 Acid-induced Changes of the Hydrodynamic Dimensions *685*
19.3.6 Dimensions in Partially Folded States – Molten Globules and Fluoroalcohol-induced States *686*
19.3.7 Comparison of the Dimensions of Proteins in Different Conformational States *687*
19.3.8 Scaling Laws for the Native and Highly Unfolded States, Hydrodynamic Modeling *687*
19.4 Studying Folding Kinetics by Laser Light Scattering *689*
19.4.1 General Considerations, Attainable Time Regions *689*
19.4.2 Hydrodynamic Dimensions of the Kinetic Molten Globule of Bovine α-Lactalbumin *690*
19.4.3 RNase A is Only Weakly Collapsed During the Burst Phase of Folding *691*
19.5 Misfolding and Aggregation Studied by Laser Light Scattering *692*
19.5.1 Overview: Some Typical Light Scattering Studies of Protein Aggregation *692*
19.5.2 Studying Misfolding and Amyloid Formation by Laser Light Scattering *693*
19.5.2.1 Overview: Initial States, Critical Oligomers, Protofibrils, Fibrils *693*
19.5.2.2 Aggregation Kinetics of Aβ Peptides *694*
19.5.2.3 Kinetics of Oligomer and Fibril Formation of PGK and Recombinant Hamster Prion Protein *695*
19.5.2.4 Mechanisms of Misfolding and Misassembly, Some General Remarks *698*
19.6 Experimental Protocols *698*
19.6.1 Laser Light Scattering Instrumentation *698*

19.6.1.1 Basic Experimental Set-up, General Requirements *698*
19.6.1.2 Supplementary Measurements and Useful Options *700*
19.6.1.3 Commercially Available Light Scattering Instrumentation *701*
19.6.2 Experimental Protocols for the Determination of Molecular Mass and Stokes Radius of a Protein in a Particular Conformational State *701*
Protocol 1 *702*
Protocol 2 *704*
Acknowledgments *704*
References *704*

20 Conformational Properties of Unfolded Proteins *710*
Patrick J. Fleming and George D. Rose
20.1 Introduction *710*
20.1.1 Unfolded vs. Denatured Proteins *710*
20.2 Early History *711*
20.3 The Random Coil *712*
20.3.1 The Random Coil – Theory *713*
20.3.1.1 The Random Coil Model Prompts Three Questions *716*
20.3.1.2 The Folding Funnel *716*
20.3.1.3 Transition State Theory *717*
20.3.1.4 Other Examples *717*
20.3.1.5 Implicit Assumptions from the Random Coil Model *718*
20.3.2 The Random Coil – Experiment *718*
20.3.2.1 Intrinsic Viscosity *719*
20.3.2.2 SAXS and SANS *720*
20.4 Questions about the Random Coil Model *721*
20.4.1 Questions from Theory *722*
20.4.1.1 The Flory Isolated-pair Hypothesis *722*
20.4.1.2 Structure vs. Energy Duality *724*
20.4.1.3 The "Rediscovery" of Polyproline II Conformation *724*
20.4.1.4 P_{II} in Unfolded Peptides and Proteins *726*
20.4.2 Questions from Experiment *727*
20.4.2.1 Residual Structure in Denatured Proteins and Peptides *727*
20.4.3 The Reconciliation Problem *728*
20.4.4 Organization in the Unfolded State – the Entropic Conjecture *728*
20.4.4.1 Steric Restrictions beyond the Dipeptide *729*
20.5 Future Directions *730*
Acknowledgments *731*
References *731*

21 Conformation and Dynamics of Nonnative States of Proteins studied by NMR Spectroscopy *737*
Julia Wirmer, Christian Schlörb, and Harald Schwalbe
21.1 Introduction *737*
21.1.1 Structural Diversity of Polypeptide Chains *737*

21.1.2 Intrinsically Unstructured and Natively Unfolded Proteins *739*
21.2 Prerequisites: NMR Resonance Assignment *740*
21.3 NMR Parameters *744*
21.3.1 Chemical shifts δ *745*
21.3.1.1 Conformational Dependence of Chemical Shifts *745*
21.3.1.2 Interpretation of Chemical Shifts in the Presence of Conformational Averaging *746*
21.3.2 *J* Coupling Constants *748*
21.3.2.1 Conformational Dependence of *J* Coupling Constants *748*
21.3.2.2 Interpretation of *J* Coupling Constants in the Presence of Conformational Averaging *750*
21.3.3 Relaxation: Homonuclear NOEs *750*
21.3.3.1 Distance Dependence of Homonuclear NOEs *750*
21.3.3.2 Interpretation of Homonuclear NOEs in the Presence of Conformational Averaging *754*
21.3.4 Heteronuclear Relaxation (^{15}N R_1, R_2, hetNOE) *757*
21.3.4.1 Correlation Time Dependence of Heteronuclear Relaxation Parameters *757*
21.3.4.2 Dependence on Internal Motions of Heteronuclear Relaxation Parameters *759*
21.3.5 Residual Dipolar Couplings *760*
21.3.5.1 Conformational Dependence of Residual Dipolar Couplings *760*
21.3.5.2 Interpretation of Residual Dipolar Couplings in the Presence of Conformational Averaging *763*
21.3.6 Diffusion *765*
21.3.7 Paramagnetic Spin Labels *766*
21.3.8 H/D Exchange *767*
21.3.9 Photo-CIDNP *767*
21.4 Model for the Random Coil State of a Protein *768*
21.5 Nonnative States of Proteins: Examples from Lysozyme, α-Lactalbumin, and Ubiquitin *771*
21.5.1 Backbone Conformation *772*
21.5.1.1 Interpretation of Chemical Shifts *772*
21.5.1.2 Interpretation of NOEs *774*
21.5.1.3 Interpretation of *J* Coupling Constants *780*
21.5.2 Side-chain Conformation *784*
21.5.2.1 Interpretation of *J* Coupling Constants *784*
21.5.3 Backbone Dynamics *786*
21.5.3.1 Interpretation of ^{15}N Relaxation Rates *786*
21.6 Summary and Outlook *793*
Acknowledgments *794*
References *794*

22 Dynamics of Unfolded Polypeptide Chains *809*
Beat Fierz and Thomas Kiefhaber
22.1 Introduction *809*

22.2 Equilibrium Properties of Chain Molecules 809
22.2.1 The Freely Jointed Chain 810
22.2.2 Chain Stiffness 810
22.2.3 Polypeptide Chains 811
22.2.4 Excluded Volume Effects 812
22.3 Theory of Polymer Dynamics 813
22.3.1 The Langevin Equation 813
22.3.2 Rouse Model and Zimm Model 814
22.3.3 Dynamics of Loop Closure and the Szabo-Schulten-Schulten Theory 815
22.4 Experimental Studies on the Dynamics in Unfolded Polypeptide Chains 816
22.4.1 Experimental Systems for the Study of Intrachain Diffusion 816
22.4.1.1 Early Experimental Studies 816
22.4.1.2 Triplet Transfer and Triplet Quenching Studies 821
22.4.1.3 Fluorescence Quenching 825
22.4.2 Experimental Results on Dynamic Properties of Unfolded Polypeptide Chains 825
22.4.2.1 Kinetics of Intrachain Diffusion 826
22.4.2.2 Effect of Loop Size on the Dynamics in Flexible Polypeptide Chains 826
22.4.2.3 Effect of Amino Acid Sequence on Chain Dynamics 829
22.4.2.4 Effect of the Solvent on Intrachain Diffusion 831
22.4.2.5 Effect of Solvent Viscosity on Intrachain Diffusion 833
22.4.2.6 End-to-end Diffusion vs. Intrachain Diffusion 834
22.4.2.7 Chain Diffusion in Natural Protein Sequences 834
22.5 Implications for Protein Folding Kinetics 837
22.5.1 Rate of Contact Formation during the Earliest Steps in Protein Folding 837
22.5.2 The Speed Limit of Protein Folding vs. the Pre-exponential Factor 839
22.5.3 Contributions of Chain Dynamics to Rate- and Equilibrium Constants for Protein Folding Reactions 840
22.6 Conclusions and Outlook 844
22.7 Experimental Protocols and Instrumentation 844
22.7.1 Properties of the Electron Transfer Probes and Treatment of the Transfer Kinetics 845
22.7.2 Test for Diffusion-controlled Reactions 847
22.7.2.1 Determination of Bimolecular Quenching or Transfer Rate Constants 847
22.7.2.2 Testing the Viscosity Dependence 848
22.7.2.3 Determination of Activation Energy 848
22.7.3 Instrumentation 849
Acknowledgments 849
References 849

23 **Equilibrium and Kinetically Observed Molten Globule States** *856*
Kosuke Maki, Kiyoto Kamagata, and Kunihiro Kuwajima
23.1 Introduction *856*
23.2 Equilibrium Molten Globule State *858*
23.2.1 Structural Characteristics of the Molten Globule State *858*
23.2.2 Typical Examples of the Equilibrium Molten Globule State *859*
23.2.3 Thermodynamic Properties of the Molten Globule State *860*
23.3 The Kinetically Observed Molten Globule State *862*
23.3.1 Observation and Identification of the Molten Globule State in Kinetic Refolding *862*
23.3.2 Kinetics of Formation of the Early Folding Intermediates *863*
23.3.3 Late Folding Intermediates and Structural Diversity *864*
23.3.4 Evidence for the On-pathway Folding Intermediate *865*
23.4 Two-stage Hierarchical Folding Funnel *866*
23.5 Unification of the Folding Mechanism between Non-two-state and Two-state Proteins *867*
23.5.1 Statistical Analysis of the Folding Data of Non-two-state and Two-state Proteins *868*
23.5.2 A Unified Mechanism of Protein Folding: Hierarchy *870*
23.5.3 Hidden Folding Intermediates in Two-state Proteins *871*
23.6 Practical Aspects of the Experimental Study of Molten Globules *872*
23.6.1 Observation of the Equilibrium Molten Globule State *872*
23.6.1.1 Two-state Unfolding Transition *872*
23.6.1.2 Multi-state (Three-state) Unfolding Transition *874*
23.6.2 Burst-phase Intermediate Accumulated during the Dead Time of Refolding Kinetics *876*
23.6.3 Testing the Identity of the Molten Globule State with the Burst-Phase Intermediate *877*
References *879*

24 **Alcohol- and Salt-induced Partially Folded Intermediates** *884*
Daizo Hamada and Yuji Goto
24.1 Introduction *884*
24.2 Alcohol-induced Intermediates of Proteins and Peptides *886*
24.2.1 Formation of Secondary Structures by Alcohols *888*
24.2.2 Alcohol-induced Denaturation of Proteins *888*
24.2.3 Formation of Compact Molten Globule States *889*
24.2.4 Example: β-Lactoglobulin *890*
24.3 Mechanism of Alcohol-induced Conformational Change *893*
24.4 Effects of Alcohols on Folding Kinetics *896*
24.5 Salt-induced Formation of the Intermediate States *899*
24.5.1 Acid-denatured Proteins *899*
24.5.2 Acid-induced Unfolding and Refolding Transitions *900*
24.6 Mechanism of Salt-induced Conformational Change *904*
24.7 Generality of the Salt Effects *906*

24.8 Conclusion 907
References 908

25 Prolyl Isomerization in Protein Folding 916
Franz Schmid
25.1 Introduction 916
25.2 Prolyl Peptide Bonds 917
25.3 Prolyl Isomerizations as Rate-determining Steps of Protein Folding 918
25.3.1 The Discovery of Fast and Slow Refolding Species 918
25.3.2 Detection of Proline-limited Folding Processes 919
25.3.3 Proline-limited Folding Reactions 921
25.3.4 Interrelation between Prolyl Isomerization and Conformational Folding 923
25.4 Examples of Proline-limited Folding Reactions 924
25.4.1 Ribonuclease A 924
25.4.2 Ribonuclease T1 926
25.4.3 The Structure of a Folding Intermediate with an Incorrect Prolyl Isomer 928
25.5 Native-state Prolyl Isomerizations 929
25.6 Nonprolyl Isomerizations in Protein Folding 930
25.7 Catalysis of Protein Folding by Prolyl Isomerases 932
25.7.1 Prolyl Isomerases as Tools for Identifying Proline-limited Folding Steps 932
25.7.2 Specificity of Prolyl Isomerases 933
25.7.3 The Trigger Factor 934
25.7.4 Catalysis of Prolyl Isomerization During de novo Protein Folding 935
25.8 Concluding Remarks 936
25.9 Experimental Protocols 936
25.9.1 Slow Refolding Assays (“Double Jumps”) to Measure Prolyl Isomerizations in an Unfolded Protein 936
25.9.1.1 Guidelines for the Design of Double Jump Experiments 937
25.9.1.2 Formation of U_S Species after Unfolding of RNase A 938
25.9.2 Slow Unfolding Assays for Detecting and Measuring Prolyl Isomerizations in Refolding 938
25.9.2.1 Practical Considerations 939
25.9.2.2 Kinetics of the Formation of Fully Folded IIHY-G3P* Molecules 939
References 939

26 Folding and Disulfide Formation 946
Margherita Ruoppolo, Piero Pucci, and Gennaro Marino
26.1 Chemistry of the Disulfide Bond 946
26.2 Trapping Protein Disulfides 947
26.3 Mass Spectrometric Analysis of Folding Intermediates 948
26.4 Mechanism(s) of Oxidative Folding so Far – Early and Late Folding Steps 949

26.5 Emerging Concepts from Mass Spectrometric Studies 950
26.5.1 Three-fingered Toxins 951
26.5.2 RNase A 953
26.5.3 Antibody Fragments 955
26.5.4 Human Nerve Growth Factor 956
26.6 Unanswered Questions 956
26.7 Concluding Remarks 957
26.8 Experimental Protocols 957
26.8.1 How to Prepare Folding Solutions 957
26.8.2 How to Carry Out Folding Reactions 958
26.8.3 How to Choose the Best Mass Spectrometric Equipment for Your Study 959
26.8.4 How to Perform Electrospray (ES)MS Analysis 959
26.8.5 How to Perform Matrix-assisted Laser Desorption Ionization (MALDI) MS Analysis 960
References 961

27 Concurrent Association and Folding of Small Oligomeric Proteins 965
Hans Rudolf Bosshard
27.1 Introduction 965
27.2 Experimental Methods Used to Follow the Folding of Oligomeric Proteins 966
27.2.1 Equilibrium Methods 966
27.2.2 Kinetic Methods 968
27.3 Dimeric Proteins 969
27.3.1 Two-state Folding of Dimeric Proteins 970
27.3.1.1 Examples of Dimeric Proteins Obeying Two-state Folding 971
27.3.2 Folding of Dimeric Proteins through Intermediate States 978
27.4 Trimeric and Tetrameric Proteins 983
27.5 Concluding Remarks 986
Appendix – Concurrent Association and Folding of Small Oligomeric Proteins 987
A1 Equilibrium Constants for Two-state Folding 988
A1.1 Homooligomeric Protein 988
A1.2 Heterooligomeric Protein 989
A2 Calculation of Thermodynamic Parameters from Equilibrium Constants 990
A2.1 Basic Thermodynamic Relationships 990
A2.2 Linear Extrapolation of Denaturant Unfolding Curves of Two-state Reaction 990
A2.3 Calculation of the van't Hoff Enthalpy Change from Thermal Unfolding Data 990
A2.4 Calculation of the van't Hoff Enthalpy Change from the Concentration-dependence of T_m 991
A2.5 Extrapolation of Thermodynamic Parameters to Different Temperatures: Gibbs-Helmholtz Equation 991

A3 Kinetics of Reversible Two-state Folding and Unfolding: Integrated Rate Equations 992
A3.1 Two-state Folding of Dimeric Protein 992
A3.2 Two-state Unfolding of Dimeric Protein 992
A3.3 Reversible Two-state Folding and Unfolding 993
A3.3.1 Homodimeric protein 993
A3.3.2 Heterodimeric protein 993
A4 Kinetics of Reversible Two-state Folding: Relaxation after Disturbance of a Pre-existing Equilibrium (Method of Bernasconi) 994
Acknowledgments 995
References 995

28 Folding of Membrane Proteins 998
Lukas K. Tamm and Heedeok Hong
28.1 Introduction 998
28.2 Thermodyamics of Residue Partitioning into Lipid Bilayers 1000
28.3 Stability of β-Barrel Proteins 1001
28.4 Stability of Helical Membrane Proteins 1009
28.5 Helix and Other Lateral Interactions in Membrane Proteins 1010
28.6 The Membrane Interface as an Important Contributor to Membrane Protein Folding 1012
28.7 Membrane Toxins as Models for Helical Membrane Protein Insertion 1013
28.8 Mechanisms of β-Barrel Membrane Protein Folding 1015
28.9 Experimental Protocols 1016
28.9.1 SDS Gel Shift Assay for Heat modifiable Membrane Proteins 1016
28.9.1.1 Reversible Folding and Unfolding Protocol Using OmpA as an Example 1016
28.9.2 Tryptophan Fluorescence and Time-resolved Distance Determination by Tryptophan Fluorescence Quenching 1018
28.9.2.1 TDFQ Protocol for Monitoring the Translocation of Tryptophans across Membranes 1019
28.9.3 Circular Dichroism Spectroscopy 1020
28.9.4 Fourier Transform Infrared Spectroscopy 1022
28.9.4.1 Protocol for Obtaining Conformation and Orientation of Membrane Proteins and Peptides by Polarized ATR-FTIR Spectroscopy 1023
Acknowledgments 1025
References 1025

29 Protein Folding Catalysis by Pro-domains 1032
Philip N. Bryan
29.1 Introduction 1032
29.2 Bimolecular Folding Mechanisms 1033
29.3 Structures of Reactants and Products 1033
29.3.1 Structure of Free SBT 1033

29.3.2 Structure of SBT/Pro-domain Complex 1036
29.3.3 Structure of Free ALP 1037
29.3.4 Structure of the ALP/Pro-domain Complex 1037
29.4 Stability of the Mature Protease 1039
29.4.1 Stability of ALP 1039
29.4.2 Stability of Subtilisin 1040
29.5 Analysis of Pro-domain Binding to the Folded Protease 1042
29.6 Analysis of Folding Steps 1043
29.7 Why are Pro-domains Required for Folding? 1046
29.8 What is the Origin of High Cooperativity? 1047
29.9 How Does the Pro-domain Accelerate Folding? 1048
29.10 Are High Kinetic Stability and Facile Folding Mutually Exclusive? 1049
29.11 Experimental Protocols for Studying SBT Folding 1049
29.11.1 Fermentation and Purification of Active Subtilisin 1049
29.11.2 Fermentation and Purification of Facile-folding Ala221 Subtilisin from *E. coli* 1050
29.11.3 Mutagenesis and Protein Expression of Pro-domain Mutants 1051
29.11.4 Purification of Pro-domain 1052
29.11.5 Kinetics of Pro-domain Binding to Native SBT 1052
29.11.6 Kinetic Analysis of Pro-domain Facilitated Subtilisin Folding 1052
29.11.6.1 Single Mixing 1052
29.11.6.2 Double Jump: Renaturation–Denaturation 1053
29.11.6.3 Double Jump: Denaturation–Renaturation 1053
29.11.6.4 Triple Jump: Denaturation–Renaturation–Denaturation 1054
References 1054

30 The Thermodynamics and Kinetics of Collagen Folding 1059
Hans Peter Bächinger and Jürgen Engel
30.1 Introduction 1059
30.1.1 The Collagen Family 1059
30.1.2 Biosynthesis of Collagens 1060
30.1.3 The Triple Helical Domain in Collagens and Other Proteins 1061
30.1.4 N- and C-Propeptide, Telopeptides, Flanking Coiled-Coil Domains 1061
30.1.5 Why is the Folding of the Triple Helix of Interest? 1061
30.2 Thermodynamics of Collagen Folding 1062
30.2.1 Stability of the Triple Helix 1062
30.2.2 The Role of Posttranslational Modifications 1063
30.2.3 Energies Involved in the Stability of the Triple Helix 1063
30.2.4 Model Peptides Forming the Collagen Triple Helix 1066
30.2.4.1 Type of Peptides 1066
30.2.4.2 The All-or-none Transition of Short Model Peptides 1066
30.2.4.3 Thermodynamic Parameters for Different Model Systems 1069
30.2.4.4 Contribution of Different Tripeptide Units to Stability 1075

30.2.4.5 Crystal and NMR Structures of Triple Helices *1076*
30.2.4.6 Conformation of the Randomly Coiled Chains *1077*
30.2.4.7 Model Studies with Isomers of Hydroxyproline and Fluoroproline *1078*
30.2.4.8 *Cis* ⇄ *trans* Equilibria of Peptide Bonds *1079*
30.2.4.9 Interpretations of Stabilities on a Molecular Level *1080*
30.3 Kinetics of Triple Helix Formation *1081*
30.3.1 Properties of Collagen Triple Helices that Influence Kinetics *1081*
30.3.2 Folding of Triple Helices from Single Chains *1082*
30.3.2.1 Early Work *1082*
30.3.2.2 Concentration Dependence of the Folding of $(PPG)_{10}$ and $(POG)_{10}$ *1082*
30.3.2.3 Model Mechanism of the Folding Kinetics *1085*
30.3.2.4 Rate Constants of Nucleation and Propagation *1087*
30.3.2.5 Host–guest Peptides and an Alternative Kinetics Model *1088*
30.3.3 Triple Helix Formation from Linked Chains *1089*
30.3.3.1 The Short N-terminal Triple Helix of Collagen III in Fragment Col1–3 *1089*
30.3.3.2 Folding of the Central Long Triple Helix of Collagen III *1090*
30.3.3.3 The Zipper Model *1092*
30.3.4 Designed Collagen Models with Chains Connected by a Disulfide Knot or by Trimerizing Domains *1097*
30.3.4.1 Disulfide-linked Model Peptides *1097*
30.3.4.2 Model Peptides Linked by a Foldon Domain *1098*
30.3.4.3 Collagen Triple Helix Formation can be Nucleated at either End *1098*
30.3.4.4 Hysteresis of Triple Helix Formation *1099*
30.3.5 Influence of *cis–trans* Isomerase and Chaperones *1100*
30.3.6 Mutations in Collagen Triple Helices Affect Proper Folding *1101*
References *1101*

31 Unfolding Induced by Mechanical Force *1111*
Jane Clarke and Phil M. Williams
31.1 Introduction *1111*
31.2 Experimental Basics *1112*
31.2.1 Instrumentation *1112*
31.2.2 Sample Preparation *1113*
31.2.3 Collecting Data *1114*
31.2.4 Anatomy of a Force Trace *1115*
31.2.5 Detecting Intermediates in a Force Trace *1115*
31.2.6 Analyzing the Force Trace *1116*
31.3 Analysis of Force Data *1117*
31.3.1 Basic Theory behind Dynamic Force Spectroscopy *1117*
31.3.2 The Ramp of Force Experiment *1119*
31.3.3 The Golden Equation of DFS *1121*
31.3.4 Nonlinear Loading *1122*

31.3.4.1 The Worm-line Chain (WLC) 1123
31.3.5 Experiments under Constant Force 1124
31.3.6 Effect of Tandem Repeats on Kinetics 1125
31.3.7 Determining the Modal Force 1126
31.3.8 Comparing Behavior 1127
31.3.9 Fitting the Data 1127
31.4 Use of Complementary Techniques 1129
31.4.1 Protein Engineering 1130
31.4.1.1 Choosing Mutants 1130
31.4.1.2 Determining $\Delta\Delta G_{D-N}$ 1131
31.4.1.3 Determining $\Delta\Delta G_{TS-N}$ 1131
31.4.1.4 Interpreting the Φ-values 1132
31.4.2 Computer Simulation 1133
31.5 Titin I27: A Case Study 1134
31.5.1 The Protein System 1134
31.5.2 The Unfolding Intermediate 1135
31.5.3 The Transition State 1136
31.5.4 The Relationship Between the Native and Transition States 1137
31.5.5 The Energy Landscape under Force 1139
31.6 Conclusions – the Future 1139
References 1139

32 Molecular Dynamics Simulations to Study Protein Folding and Unfolding 1143
Amedeo Caflisch and Emanuele Paci
32.1 Introduction 1143
32.2 Molecular Dynamics Simulations of Peptides and Proteins 1144
32.2.1 Folding of Structured Peptides 1144
32.2.1.1 Reversible Folding and Free Energy Surfaces 1144
32.2.1.2 Non-Arrhenius Temperature Dependence of the Folding Rate 1147
32.2.1.3 Denatured State and Levinthal Paradox 1148
32.2.1.4 Folding Events of Trp-cage 1149
32.2.2 Unfolding Simulations of Proteins 1150
32.2.2.1 High-temperature Simulations 1150
32.2.2.2 Biased Unfolding 1150
32.2.2.3 Forced Unfolding 1151
32.2.3 Determination of the Transition State Ensemble 1153
32.3 MD Techniques and Protocols 1155
32.3.1 Techniques to Improve Sampling 1155
32.3.1.1 Replica Exchange Molecular Dynamics 1155
32.3.1.2 Methods Based on Path Sampling 1157
32.3.2 MD with Restraints 1157
32.3.3 Distributed Computing Approach 1158
32.3.4 Implicit Solvent Models versus Explicit Water 1160
32.4 Conclusion 1162
References 1162

33 **Molecular Dynamics Simulations of Proteins and Peptides: Problems, Achievements, and Perspectives** *1170*
Paul Tavan, Heiko Carstens, and Gerald Mathias
33.1 Introduction *1170*
33.2 Basic Physics of Protein Structure and Dynamics *1171*
33.2.1 Protein Electrostatics *1172*
33.2.2 Relaxation Times and Spatial Scales *1172*
33.2.3 Solvent Environment *1173*
33.2.4 Water *1174*
33.2.5 Polarizability of the Peptide Groups and of Other Protein Components *1175*
33.3 State of the Art *1177*
33.3.1 Control of Thermodynamic Conditions *1177*
33.3.2 Long-range Electrostatics *1177*
33.3.3 Polarizability *1179*
33.3.4 Higher Multipole Moments of the Molecular Components *1180*
33.3.5 MM Models of Water *1181*
33.3.6 Complexity of Protein–Solvent Systems and Consequences for MM-MD *1182*
33.3.7 What about Successes of MD Methods? *1182*
33.3.8 Accessible Time Scales and Accuracy Issues *1184*
33.3.9 Continuum Solvent Models *1185*
33.3.10 Are there Further Problems beyond Electrostatics and Structure Prediction? *1187*
33.4 Conformational Dynamics of a Light-switchable Model Peptide *1187*
33.4.1 Computational Methods *1188*
33.4.2 Results and Discussion *1190*
Summary *1194*
Acknowledgments *1194*
References *1194*

Part II, Volume 1

Contributors of Part II *LVIII*

1 **Paradigm Changes from "Unboiling an Egg" to "Synthesizing a Rabbit"** *3*
Rainer Jaenicke
1.1 Protein Structure, Stability, and Self-organization *3*
1.2 Autonomous and Assisted Folding and Association *6*
1.3 Native, Intermediate, and Denatured States *11*
1.4 Folding and Merging of Domains – Association of Subunits *13*
1.5 Limits of Reconstitution *19*
1.6 In Vitro Denaturation-Renaturation vs. Folding in Vivo *21*

1.7 Perspectives 24
Acknowledgements 26
References 26

2 Folding and Association of Multi-domain and Oligomeric Proteins 32
Hauke Lilie and Robert Seckler
2.1 Introduction 32
2.2 Folding of Multi-domain Proteins 33
2.2.1 Domain Architecture 33
2.2.2 γ-Crystallin as a Model for a Two-domain Protein 35
2.2.3 The Giant Protein Titin 39
2.3 Folding and Association of Oligomeric Proteins 41
2.3.1 Why Oligomers? 41
2.3.2 Inter-subunit Interfaces 42
2.3.3 Domain Swapping 44
2.3.4 Stability of Oligomeric Proteins 45
2.3.5 Methods Probing Folding/Association 47
2.3.5.1 Chemical Cross-linking 47
2.3.5.2 Analytical Gel Filtration Chromatography 47
2.3.5.3 Scattering Methods 48
2.3.5.4 Fluorescence Resonance Energy Transfer 48
2.3.5.5 Hybrid Formation 48
2.3.6 Kinetics of Folding and Association 49
2.3.6.1 General Considerations 49
2.3.6.2 Reconstitution Intermediates 50
2.3.6.3 Rates of Association 52
2.3.6.4 Homo- Versus Heterodimerization 52
2.4 Renaturation versus Aggregation 54
2.5 Case Studies on Protein Folding and Association 54
2.5.1 Antibody Fragments 54
2.5.2 Trimeric Tail Spike Protein of Bacteriophage P22 59
2.6 Experimental Protocols 62
References 65

3 Studying Protein Folding in Vivo 73
I. Marije Liscaljet, Bertrand Kleizen, and Ineke Braakman
3.1 Introduction 73
3.2 General Features in Folding Proteins Amenable to in Vivo Study 73
3.2.1 Increasing Compactness 76
3.2.2 Decreasing Accessibility to Different Reagents 76
3.2.3 Changes in Conformation 77
3.2.4 Assistance During Folding 78
3.3 Location-specific Features in Protein Folding 79
3.3.1 Translocation and Signal Peptide Cleavage 79
3.3.2 Glycosylation 80

3.3.3 Disulfide Bond Formation in the ER *81*
3.3.4 Degradation *82*
3.3.5 Transport from ER to Golgi and Plasma Membrane *83*
3.4 How to Manipulate Protein Folding *84*
3.4.1 Pharmacological Intervention (Low-molecular-weight Reagents) *84*
3.4.1.1 Reducing and Oxidizing Agents *84*
3.4.1.2 Calcium Depletion *84*
3.4.1.3 ATP Depletion *85*
3.4.1.4 Cross-linking *85*
3.4.1.5 Glycosylation Inhibitors *85*
3.4.2 Genetic Modifications (High-molecular-weight Manipulations) *86*
3.4.2.1 Substrate Protein Mutants *86*
3.4.2.2 Changing the Concentration or Activity of Folding Enzymes and Chaperones *87*
3.5 Experimental Protocols *88*
3.5.1 Protein-labeling Protocols *88*
3.5.1.1 Basic Protocol Pulse Chase: Adherent Cells *88*
3.5.1.2 Pulse Chase in Suspension Cells *91*
3.5.2 (Co)-immunoprecipitation and Accessory Protocols *93*
3.5.2.1 Immunoprecipitation *93*
3.5.2.2 Co-precipitation with Calnexin ([84]; adapted from Ou et al. [85]) *94*
3.5.2.3 Co-immunoprecipitation with Other Chaperones *95*
3.5.2.4 Protease Resistance *95*
3.5.2.5 Endo H Resistance *96*
3.5.2.6 Cell Surface Expression Tested by Protease *96*
3.5.3 SDS-PAGE [13] *97*
Acknowledgements *98*
References *98*

4 Characterization of ATPase Cycles of Molecular Chaperones by Fluorescence and Transient Kinetic Methods *105*
Sandra Schlee and Jochen Reinstein
4.1 Introduction *105*
4.1.1 Characterization of ATPase Cycles of Energy-transducing Systems *105*
4.1.2 The Use of Fluorescent Nucleotide Analogues *106*
4.1.2.1 Fluorescent Modifications of Nucleotides *106*
4.1.2.2 How to Find a Suitable Analogue for a Specific Protein *108*
4.2 Characterization of ATPase Cycles of Molecular Chaperones *109*
4.2.1 Biased View *109*
4.2.2 The ATPase Cycle of DnaK *109*
4.2.3 The ATPase Cycle of the Chaperone Hsp90 *109*
4.2.4 The ATPase Cycle of the Chaperone ClpB *111*
4.2.4.1 ClpB, an Oligomeric ATPase With Two AAA Modules Per Protomer *111*

4.2.4.2 Nucleotide-binding Properties of NBD1 and NBD2 *111*
4.2.4.3 Cooperativity of ATP Hydrolysis and Interdomain Communication *114*
4.3 Experimental Protocols *116*
4.3.1 Synthesis of Fluorescent Nucleotide Analogues *116*
4.3.1.1 Synthesis and Characterization of (P_β)MABA-ADP and (P_γ)MABA-ATP *116*
4.3.1.2 Synthesis and Characterization of N8-MABA Nucleotides *119*
4.3.1.3 Synthesis of MANT Nucleotides *120*
4.3.2 Preparation of Nucleotides and Proteins *121*
4.3.2.1 Assessment of Quality of Nucleotide Stock Solution *121*
4.3.2.2 Determination of the Nucleotide Content of Proteins *122*
4.3.2.3 Nucleotide Depletion Methods *123*
4.3.3 Steady-state ATPase Assays *124*
4.3.3.1 Coupled Enzymatic Assay *124*
4.3.3.2 Assays Based on [α-^{32}P]-ATP and TLC *125*
4.3.3.3 Assays Based on Released P_i *125*
4.3.4 Single-turnover ATPase Assays *126*
4.3.4.1 Manual Mixing Procedures *126*
4.3.4.2 Quenched Flow *127*
4.3.5 Nucleotide-binding Measurements *127*
4.3.5.1 Isothermal Titration Calorimetry *127*
4.3.5.2 Equilibrium Dialysis *129*
4.3.5.3 Filter Binding *129*
4.3.5.4 Equilibrium Fluorescence Titration *130*
4.3.5.5 Competition Experiments *132*
4.3.6 Analytical Solutions of Equilibrium Systems *133*
4.3.6.1 Quadratic Equation *133*
4.3.6.2 Cubic Equation *134*
4.3.6.3 Iterative Solutions *138*
4.3.7 Time-resolved Binding Measurements *141*
4.3.7.1 Introduction *141*
4.3.7.2 One-step Irreversible Process *142*
4.3.7.3 One-step Reversible Process *143*
4.3.7.4 Reversible Second Order Reduced to Pseudo-first Order *144*
4.3.7.5 Two Simultaneous Irreversible Pathways – Partitioning *146*
4.3.7.6 Two-step Consecutive (Sequential) Reaction *148*
4.3.7.7 Two-step Binding Reactions *150*
References *152*

5 Analysis of Chaperone Function in Vitro *162*
Johannes Buchner and Stefan Walter
5.1 Introduction *162*
5.2 Basic Functional Principles of Molecular Chaperones *164*
5.2.1 Recognition of Nonnative Proteins *166*

5.2.2 Induction of Conformational Changes in the Substrate *167*
5.2.3 Energy Consumption and Regulation of Chaperone Function *169*
5.3 Limits and Extensions of the Chaperone Concept *170*
5.3.1 Co-chaperones *171*
5.3.2 Specific Chaperones *171*
5.4 Working with Molecular Chaperones *172*
5.4.1 Natural versus Artificial Substrate Proteins *172*
5.4.2 Stability of Chaperones *172*
5.5 Assays to Assess and Characterize Chaperone Function *174*
5.5.1 Generating Nonnative Conformations of Proteins *174*
5.5.2 Aggregation Assays *174*
5.5.3 Detection of Complexes Between Chaperone and Substrate *175*
5.5.4 Refolding of Denatured Substrates *175*
5.5.5 ATPase Activity and Effect of Substrate and Cofactors *176*
5.6 Experimental Protocols *176*
5.6.1 General Considerations *176*
5.6.1.1 Analysis of Chaperone Stability *176*
5.6.1.2 Generation of Nonnative Proteins *177*
5.6.1.3 Model Substrates for Chaperone Assays *177*
5.6.2 Suppression of Aggregation *179*
5.6.3 Complex Formation between Chaperones and Polypeptide Substrates *183*
5.6.4 Identification of Chaperone-binding Sites *184*
5.6.5 Chaperone-mediated Refolding of Test Proteins *186*
5.6.6 ATPase Activity *188*
Acknowledgments *188*
References *189*

6 Physical Methods for Studies of Fiber Formation and Structure *197*
Thomas Scheibel and Louise Serpell
6.1 Introduction *197*
6.2 Overview: Protein Fibers Formed in Vivo *198*
6.2.1 Amyloid Fibers *198*
6.2.2 Silks *199*
6.2.3 Collagens *199*
6.2.4 Actin, Myosin, and Tropomyosin Filaments *200*
6.2.5 Intermediate Filaments/Nuclear Lamina *202*
6.2.6 Fibrinogen/Fibrin *203*
6.2.7 Microtubules *203*
6.2.8 Elastic Fibers *204*
6.2.9 Flagella and Pili *204*
6.2.10 Filamentary Structures in Rod-like Viruses *205*
6.2.11 Protein Fibers Used by Viruses and Bacteriophages to Bind to Their Hosts *206*
6.3 Overview: Fiber Structures *206*

6.3.1 Study of the Structure of β-sheet-containing Proteins 207
6.3.1.1 Amyloid 207
6.3.1.2 Paired Helical Filaments 207
6.3.1.3 β-Silks 207
6.3.1.4 β-Sheet-containing Viral Fibers 208
6.3.2 α-Helix-containing Protein Fibers 209
6.3.2.1 Collagen 209
6.3.2.2 Tropomyosin 210
6.3.2.3 Intermediate Filaments 210
6.3.3 Protein Polymers Consisting of a Mixture of Secondary Structure 211
6.3.3.1 Tubulin 211
6.3.3.2 Actin and Myosin Filaments 212
6.4 Methods to Study Fiber Assembly 213
6.4.1 Circular Dichroism Measurements for Monitoring Structural Changes Upon Fiber Assembly 213
6.4.1.1 Theory of CD 213
6.4.1.2 Experimental Guide to Measure CD Spectra and Structural Transition Kinetics 214
6.4.2 Intrinsic Fluorescence Measurements to Analyze Structural Changes 215
6.4.2.1 Theory of Protein Fluorescence 215
6.4.2.2 Experimental Guide to Measure Trp Fluorescence 216
6.4.3 Covalent Fluorescent Labeling to Determine Structural Changes of Proteins with Environmentally Sensitive Fluorophores 217
6.4.3.1 Theory on Environmental Sensitivity of Fluorophores 217
6.4.3.2 Experimental Guide to Labeling Proteins With Fluorophores 218
6.4.4 1-Anilino-8-Naphthalensulfonate (ANS) Binding to Investigate Fiber Assembly 219
6.4.4.1 Theory on Using ANS Fluorescence for Detecting Conformational Changes in Proteins 219
6.4.4.2 Experimental Guide to Using ANS for Monitoring Protein Fiber Assembly 220
6.4.5 Light Scattering to Monitor Particle Growth 220
6.4.5.1 Theory of Classical Light Scattering 221
6.4.5.2 Theory of Dynamic Light Scattering 221
6.4.5.3 Experimental Guide to Analyzing Fiber Assembly Using DLS 222
6.4.6 Field-flow Fractionation to Monitor Particle Growth 222
6.4.6.1 Theory of FFF 222
6.4.6.2 Experimental Guide to Using FFF for Monitoring Fiber Assembly 223
6.4.7 Fiber Growth-rate Analysis Using Surface Plasmon Resonance 223
6.4.7.1 Theory of SPR 223
6.4.7.2 Experimental Guide to Using SPR for Fiber-growth Analysis 224
6.4.8 Single-fiber Growth Imaging Using Atomic Force Microscopy 225

6.4.8.1 Theory of Atomic Force Microscopy 225
6.4.8.2 Experimental Guide for Using AFM to Investigate Fiber Growth 225
6.4.9 Dyes Specific for Detecting Amyloid Fibers 226
6.4.9.1 Theory on Congo Red and Thioflavin T Binding to Amyloid 226
6.4.9.2 Experimental Guide to Detecting Amyloid Fibers with CR and Thioflavin Binding 227
6.5 Methods to Study Fiber Morphology and Structure 228
6.5.1 Scanning Electron Microscopy for Examining the Low-resolution Morphology of a Fiber Specimen 228
6.5.1.1 Theory of SEM 228
6.5.1.2 Experimental Guide to Examining Fibers by SEM 229
6.5.2 Transmission Electron Microscopy for Examining Fiber Morphology and Structure 230
6.5.2.1 Theory of TEM 230
6.5.2.2 Experimental Guide to Examining Fiber Samples by TEM 231
6.5.3 Cryo-electron Microscopy for Examination of the Structure of Fibrous Proteins 232
6.5.3.1 Theory of Cryo-electron Microscopy 232
6.5.3.2 Experimental Guide to Preparing Proteins for Cryo-electron Microscopy 233
6.5.3.3 Structural Analysis from Electron Micrographs 233
6.5.4 Atomic Force Microscopy for Examining the Structure and Morphology of Fibrous Proteins 234
6.5.4.1 Experimental Guide for Using AFM to Monitor Fiber Morphology 234
6.5.5 Use of X-ray Diffraction for Examining the Structure of Fibrous Proteins 236
6.5.5.1 Theory of X-Ray Fiber Diffraction 236
6.5.5.2 Experimental Guide to X-Ray Fiber Diffraction 237
6.5.6 Fourier Transformed Infrared Spectroscopy 239
6.5.6.1 Theory of FTIR 239
6.5.6.2 Experimental Guide to Determining Protein Conformation by FTIR 240
6.6 Concluding Remarks 241
Acknowledgements 242
References 242

7 Protein Unfolding in the Cell 254
Prakash Koodathingal, Neil E. Jaffe, and Andreas Matouschek
7.1 Introduction 254
7.2 Protein Translocation Across Membranes 254
7.2.1 Compartmentalization and Unfolding 254
7.2.2 Mitochondria Actively Unfold Precursor Proteins 256
7.2.3 The Protein Import Machinery of Mitochondria 257
7.2.4 Specificity of Unfolding 259

7.2.5 Protein Import into Other Cellular Compartments 259
7.3 Protein Unfolding and Degradation by ATP-dependent Proteases 260
7.3.1 Structural Considerations of Unfoldases Associated With Degradation 260
7.3.2 Unfolding Is Required for Degradation by ATP-dependent Proteases 261
7.3.3 The Role of ATP and Models of Protein Unfolding 262
7.3.4 Proteins Are Unfolded Sequentially and Processively 263
7.3.5 The Influence of Substrate Structure on the Degradation Process 264
7.3.6 Unfolding by Pulling 264
7.3.7 Specificity of Degradation 265
7.4 Conclusions 266
7.5 Experimental Protocols 266
7.5.1 Size of Import Channels in the Outer and Inner Membranes of Mitochondria 266
7.5.2 Structure of Precursor Proteins During Import into Mitochondria 266
7.5.3 Import of Barnase Mutants 267
7.5.4 Protein Degradation by ATP-dependent Proteases 267
7.5.5 Use of Multi-domain Substrates 268
7.5.6 Studies Using Circular Permutants 268
References 269

8 Natively Disordered Proteins 275
Gary W. Daughdrill, Gary J. Pielak, Vladimir N. Uversky, Marc S. Cortese, and A. Keith Dunker
8.1 Introduction 275
8.1.1 The Protein Structure-Function Paradigm 275
8.1.2 Natively Disordered Proteins 277
8.1.3 A New Protein Structure-Function Paradigm 280
8.2 Methods Used to Characterize Natively Disordered Proteins 281
8.2.1 NMR Spectroscopy 281
8.2.1.1 Chemical Shifts Measure the Presence of Transient Secondary Structure 282
8.2.1.2 Pulsed Field Gradient Methods to Measure Translational Diffusion 284
8.2.1.3 NMR Relaxation and Protein Flexibility 284
8.2.1.4 Using the Model-free Analysis of Relaxation Data to Estimate Internal Mobility and Rotational Correlation Time 285
8.2.1.5 Using Reduced Spectral Density Mapping to Assess the Amplitude and Frequencies of Intramolecular Motion 286
8.2.1.6 Characterization of the Dynamic Structures of Natively Disordered Proteins Using NMR 287
8.2.2 X-ray Crystallography 288
8.2.3 Small Angle X-ray Diffraction and Hydrodynamic Measurements 293

8.2.4 Circular Dichroism Spectropolarimetry 297
8.2.5 Infrared and Raman Spectroscopy 299
8.2.6 Fluorescence Methods 301
8.2.6.1 Intrinsic Fluorescence of Proteins 301
8.2.6.2 Dynamic Quenching of Fluorescence 302
8.2.6.3 Fluorescence Polarization and Anisotropy 303
8.2.6.4 Fluorescence Resonance Energy Transfer 303
8.2.6.5 ANS Fluorescence 305
8.2.7 Conformational Stability 308
8.2.7.1 Effect of Temperature on Proteins with Extended Disorder 309
8.2.7.2 Effect of pH on Proteins with Extended Disorder 309
8.2.8 Mass Spectrometry-based High-resolution Hydrogen-Deuterium Exchange 309
8.2.9 Protease Sensitivity 311
8.2.10 Prediction from Sequence 313
8.2.11 Advantage of Multiple Methods 314
8.3 Do Natively Disordered Proteins Exist Inside Cells? 315
8.3.1 Evolution of Ordered and Disordered Proteins Is Fundamentally Different 315
8.3.1.1 The Evolution of Natively Disordered Proteins 315
8.3.1.2 Adaptive Evolution and Protein Flexibility 317
8.3.1.3 Phylogeny Reconstruction and Protein Structure 318
8.3.2 Direct Measurement by NMR 320
8.4 Functional Repertoire 322
8.4.1 Molecular Recognition 322
8.4.1.1 The Coupling of Folding and Binding 322
8.4.1.2 Structural Plasticity for the Purpose of Functional Plasticity 323
8.4.1.3 Systems Where Disorder Increases Upon Binding 323
8.4.2 Assembly/Disassembly 325
8.4.3 Highly Entropic Chains 325
8.4.4 Protein Modification 327
8.5 Importance of Disorder for Protein Folding 328
8.6 Experimental Protocols 331
8.6.1 NMR Spectroscopy 331
8.6.1.1 General Requirements 331
8.6.1.2 Measuring Transient Secondary Structure in Secondary Chemical Shifts 332
8.6.1.3 Measuring the Translational Diffusion Coefficient Using Pulsed Field Gradient Diffusion Experiments 332
8.6.1.4 Relaxation Experiments 332
8.6.1.5 Relaxation Data Analysis Using Reduced Spectral Density Mapping 333
8.6.1.6 In-cell NMR 334
8.6.2 X-ray Crystallography 334
8.6.3 Circular Dichroism Spectropolarimetry 336

Acknowledgements 337
References 337

9 The Catalysis of Disulfide Bond Formation in Prokaryotes *358*
Jean-Francois Collet and James C. Bardwell
9.1 Introduction *358*
9.2 Disulfide Bond Formation in the *E. coli* Periplasm *358*
9.2.1 A Small Bond, a Big Effect *358*
9.2.2 Disulfide Bond Formation Is a Catalyzed Process *359*
9.2.3 DsbA, a Protein-folding Catalyst *359*
9.2.4 How is DsbA Re-oxidized? *361*
9.2.5 From Where Does the Oxidative Power of DsbB Originate? *361*
9.2.6 How Are Disulfide Bonds Transferred From DsbB to DsbA? *362*
9.2.7 How Can DsbB Generate Disulfide by Quinone Reduction? *364*
9.3 Disulfide Bond Isomerization *365*
9.3.1 The Protein Disulfide Isomerases DsbC and DsbG *365*
9.3.2 Dimerization of DsbC and DsbG Is Important for Isomerase and Chaperone Activity *366*
9.3.3 Dimerization Protects from DsbB Oxidation *367*
9.3.4 Import of Electrons from the Cytoplasm: DsbD *367*
9.3.5 Conclusions *369*
9.4 Experimental Protocols *369*
9.4.1 Oxidation-reduction of a Protein Sample *369*
9.4.2 Determination of the Free Thiol Content of a Protein *370*
9.4.3 Separation by HPLC *371*
9.4.4 Tryptophan Fluorescence *372*
9.4.5 Assay of Disulfide Oxidase Activity *372*
References *373*

10 Catalysis of Peptidyl-prolyl *cis/trans* Isomerization by Enzymes *377*
Gunter Fischer
10.1 Introduction *377*
10.2 Peptidyl-prolyl *cis/trans* Isomerization *379*
10.3 Monitoring Peptidyl-prolyl *cis/trans* Isomerase Activity *383*
10.4 Prototypical Peptidyl-prolyl *cis/trans* Isomerases *388*
10.4.1 General Considerations *388*
10.4.2 Prototypic Cyclophilins *390*
10.4.3 Prototypic FK506-binding Proteins *394*
10.4.4 Prototypic Parvulins *397*
10.5 Concluding Remarks *399*
10.6 Experimental Protocols *399*
10.6.1 PPIase Assays: Materials *399*
10.6.2 PPIase Assays: Equipment *400*
10.6.3 Assaying Procedure: Protease-coupled Spectrophotometric Assay *400*

10.6.4 Assaying Procedure: Protease-free Spectrophotometric Assay 401
References 401

11 Secondary Amide Peptide Bond *cis/trans* Isomerization in Polypeptide Backbone Restructuring: Implications for Catalysis 415
Cordelia Schiene-Fischer and Christian Lücke
11.1 Introduction 415
11.2 Monitoring Secondary Amide Peptide Bond *cis/trans* Isomerization 416
11.3 Kinetics and Thermodynamics of Secondary Amide Peptide Bond *cis/trans* Isomerization 418
11.4 Principles of DnaK Catalysis 420
11.5 Concluding Remarks 423
11.6 Experimental Protocols 424
11.6.1 Stopped-flow Measurements of Peptide Bond *cis/trans* Isomerization 424
11.6.2 Two-dimensional ^{1}H-NMR Exchange Experiments 425
References 426

12 Ribosome-associated Proteins Acting on Newly Synthesized Polypeptide Chains 429
Sabine Rospert, Matthias Gautschi, Magdalena Rakwalska, and Uta Raue
12.1 Introduction 429
12.2 Signal Recognition Particle, Nascent Polypeptide–associated Complex, and Trigger Factor 432
12.2.1 Signal Recognition Particle 432
12.2.2 An Interplay between Eukaryotic SRP and Nascent Polypeptide–associated Complex? 435
12.2.3 Interplay between Bacterial SRP and Trigger Factor? 435
12.2.4 Functional Redundancy: TF and the Bacterial Hsp70 Homologue DnaK 436
12.3 Chaperones Bound to the Eukaryotic Ribosome: Hsp70 and Hsp40 Systems 436
12.3.1 Sis1p and Ssa1p: an Hsp70/Hsp40 System Involved in Translation Initiation? 437
12.3.2 Ssb1/2p, an Hsp70 Homologue Distributed Between Ribosomes and Cytosol 438
12.3.3 Function of Ssb1/2p in Degradation and Protein Folding 439
12.3.4 Zuotin and Ssz1p: a Stable Chaperone Complex Bound to the Yeast Ribosome 440
12.3.5 A Functional Chaperone Triad Consisting of Ssb1/2p, Ssz1p, and Zuotin 440
12.3.6 Effects of Ribosome-bound Chaperones on the Yeast Prion [*PSI*$^+$] 442
12.4 Enzymes Acting on Nascent Polypeptide Chains 443

12.4.1 Methionine Aminopeptidases 443
12.4.2 N^{α}-acetyltransferases 444
12.5 A Complex Arrangement at the Yeast Ribosomal Tunnel Exit 445
12.6 Experimental Protocols 446
12.6.1 Purification of Ribosome-associated Protein Complexes from Yeast 446
12.6.2 Growth of Yeast and Preparation of Ribosome-associated Proteins by High-salt Treatment of Ribosomes 447
12.6.3 Purification of NAC and RAC 448
References 449

Part II, Volume 2

13 The Role of Trigger Factor in Folding of Newly Synthesized Proteins 459
Elke Deuerling, Thomas Rauch, Holger Patzelt, and Bernd Bukau
13.1 Introduction 459
13.2 In Vivo Function of Trigger Factor 459
13.2.1 Discovery 459
13.2.2 Trigger Factor Cooperates With the DnaK Chaperone in the Folding of Newly Synthesized Cytosolic Proteins 460
13.2.3 In Vivo Substrates of Trigger Factor and DnaK 461
13.2.4 Substrate Specificity of Trigger Factor 463
13.3 Structure–Function Analysis of Trigger Factor 465
13.3.1 Domain Structure and Conservation 465
13.3.2 Quaternary Structure 468
13.3.3 PPIase and Chaperone Activity of Trigger Factor 469
13.3.4 Importance of Ribosome Association 470
13.4 Models of the Trigger Factor Mechanism 471
13.5 Experimental Protocols 473
13.5.1 Trigger Factor Purification 473
13.5.2 GAPDH Trigger Factor Activity Assay 475
13.5.3 Modular Cell-free *E. coli* Transcription/Translation System 475
13.5.4 Isolation of Ribosomes and Add-back Experiments 483
13.5.5 Cross-linking Techniques 485
References 485

14 Cellular Functions of Hsp70 Chaperones 490
Elizabeth A. Craig and Peggy Huang
14.1 Introduction 490
14.2 "Soluble" Hsp70s/J-proteins Function in General Protein Folding 492
14.2.1 The Soluble Hsp70 of *E. coli*, DnaK 492
14.2.2 Soluble Hsp70s of Major Eukaryotic Cellular Compartments 493
14.2.2.1 Eukaryotic Cytosol 493
14.2.2.2 Matrix of Mitochondria 494
14.2.2.3 Lumen of the Endoplasmic Reticulum 494

14.3 "Tethered" Hsp70s/J-proteins: Roles in Protein Folding on the Ribosome and in Protein Translocation *495*
14.3.1 Membrane-tethered Hsp70/J-protein *495*
14.3.2 Ribosome-associated Hsp70/J-proteins *496*
14.4 Modulating of Protein Conformation by Hsp70s/J-proteins *498*
14.4.1 Assembly of Fe/S Centers *499*
14.4.2 Uncoating of Clathrin-coated Vesicles *500*
14.4.3 Regulation of the Heat Shock Response *501*
14.4.4 Regulation of Activity of DNA Replication-initiator Proteins *502*
14.5 Cases of a Single Hsp70 Functioning With Multiple J-Proteins *504*
14.6 Hsp70s/J-proteins – When an Hsp70 Maybe Isn't Really a Chaperone *504*
14.6.1 The Ribosome-associated "Hsp70" Ssz1 *505*
14.6.2 Mitochondrial Hsp70 as the Regulatory Subunit of an Endonuclease *506*
14.7 Emerging Concepts and Unanswered Questions *507*
References *507*

15 Regulation of Hsp70 Chaperones by Co-chaperones *516*
Matthias P. Mayer and Bernd Bukau
15.1 Introduction *516*
15.2 Hsp70 Proteins *517*
15.2.1 Structure and Conservation *517*
15.2.2 ATPase Cycle *519*
15.2.3 Structural Investigations *521*
15.2.4 Interactions With Substrates *522*
15.3 J-domain Protein Family *526*
15.3.1 Structure and Conservation *526*
15.3.2 Interaction With Hsp70s *530*
15.3.3 Interactions with Substrates *532*
15.4 Nucleotide Exchange Factors *534*
15.4.1 GrpE: Structure and Interaction with DnaK *534*
15.4.2 Nucleotide Exchange Reaction *535*
15.4.3 Bag Family: Structure and Interaction With Hsp70 *536*
15.4.4 Relevance of Regulated Nucleotide Exchange for Hsp70s *538*
15.5 TPR Motifs Containing Co-chaperones of Hsp70 *540*
15.5.1 Hip *541*
15.5.2 Hop *542*
15.5.3 Chip *543*
15.6 Concluding Remarks *544*
15.7 Experimental Protocols *544*
15.7.1 Hsp70s *544*
15.7.2 J-Domain Proteins *545*
15.7.3 GrpE *546*
15.7.4 Bag-1 *547*

15.7.5 Hip 548
15.7.6 Hop 549
15.7.7 Chip 549
References 550

16 Protein Folding in the Endoplasmic Reticulum Via the Hsp70 Family 563
Ying Shen, Kyung Tae Chung, and Linda M. Hendershot
16.1 Introduction 563
16.2 BiP Interactions with Unfolded Proteins 564
16.3 ER-localized DnaJ Homologues 567
16.4 ER-localized Nucleotide-exchange/releasing Factors 571
16.5 Organization and Relative Levels of Chaperones in the ER 572
16.6 Regulation of ER Chaperone Levels 573
16.7 Disposal of BiP-associated Proteins That Fail to Fold or Assemble 575
16.8 Other Roles of BiP in the ER 576
16.9 Concluding Comments 576
16.10 Experimental Protocols 577
16.10.1 Production of Recombinant ER Proteins 577
16.10.1.1 General Concerns 577
16.10.1.2 Bacterial Expression 578
16.10.1.3 Yeast Expression 580
16.10.1.4 Baculovirus 581
16.10.1.5 Mammalian Cells 583
16.10.2 Yeast Two-hybrid Screen for Identifying Interacting Partners of ER Proteins 586
16.10.3 Methods for Determining Subcellular Localization, Topology, and Orientation of Proteins 588
16.10.3.1 Sequence Predictions 588
16.10.3.2 Immunofluorescence Staining 589
16.10.3.3 Subcellular Fractionation 589
16.10.3.4 Determination of Topology 590
16.10.3.5 *N*-linked Glycosylation 592
16.10.4 Nucleotide Binding, Hydrolysis, and Exchange Assays 594
16.10.4.1 Nucleotide-binding Assays 594
16.10.4.2 ATP Hydrolysis Assays 596
16.10.4.3 Nucleotide Exchange Assays 597
16.10.5 Assays for Protein–Protein Interactions in Vitro/in Vivo 599
16.10.5.1 In Vitro GST Pull-down Assay 599
16.10.5.2 Co-immunoprecipitation 600
16.10.5.3 Chemical Cross-linking 600
16.10.5.4 Yeast Two-hybrid System 601
16.10.6 In Vivo Folding, Assembly, and Chaperone-binding Assays 601
16.10.6.1 Monitoring Oxidation of Intrachain Disulfide Bonds 601
16.10.6.2 Detection of Chaperone Binding 602

Acknowledgements 603
References 603

17 **Quality Control In Glycoprotein Folding** 617
E. Sergio Trombetta and Armando J. Parodi
17.1 Introduction 617
17.2 ER *N*-glycan Processing Reactions 617
17.3 The UDP-Glc:Glycoprotein Glucosyltransferase 619
17.4 Protein Folding in the ER 621
17.5 Unconventional Chaperones (Lectins) Are Present in the ER Lumen 621
17.6 In Vivo Glycoprotein-CNX/CRT Interaction 623
17.7 Effect of CNX/CRT Binding on Glycoprotein Folding and ER Retention 624
17.8 Glycoprotein-CNX/CRT Interaction Is Not Essential for Unicellular Organisms and Cells in Culture 627
17.9 Diversion of Misfolded Glycoproteins to Proteasomal Degradation 629
17.10 Unfolding Irreparably Misfolded Glycoproteins to Facilitate Proteasomal Degradation 632
17.11 Summary and Future Directions 633
17.12 Characterization of *N*-glycans from Glycoproteins 634
17.12.1 Characterization of *N*-glycans Present in Immunoprecipitated Samples 634
17.12.2 Analysis of Radio-labeled *N*-glycans 636
17.12.3 Extraction and Analysis of Protein-bound *N*-glycans 636
17.12.4 GII and GT Assays 637
17.12.4.1 Assay for GII 637
17.12.4.2 Assay for GT 638
17.12.5 Purification of GII and GT from Rat Liver 639
References 641

18 **Procollagen Biosynthesis in Mammalian Cells** 649
Mohammed Tasab and Neil J. Bulleid
18.1 Introduction 649
18.1.1 Variety and Complexity of Collagen Proteins 649
18.1.2 Fibrillar Procollagen 650
18.1.3 Expression of Fibrillar Collagens 650
18.2 The Procollagen Biosynthetic Process: An Overview 651
18.3 Disulfide Bonding in Procollagen Assembly 653
18.4 The Influence of Primary Amino Acid Sequence on Intracellular Procollagen Folding 654
18.4.1 Chain Recognition and Type-specific Assembly 654
18.4.2 Assembly of Multi-subunit Proteins 654
18.4.3 Coordination of Type-specific Procollagen Assembly and Chain Selection 655

18.4.4 Hypervariable Motifs: Components of a Recognition Mechanism That Distinguishes Between Procollagen Chains? *656*
18.4.5 Modeling the C-propeptide *657*
18.4.6 Chain Association *657*
18.5 Posttranslational Modifications That Affect Procollagen Folding *658*
18.5.1 Hydroxylation and Triple-helix Stability *658*
18.6 Procollagen Chaperones *658*
18.6.1 Prolyl 4-Hydroxylase *658*
18.6.2 Protein Disulfide Isomerase *659*
18.6.3 Hsp47 *660*
18.6.4 PPI and BiP *661*
18.7 Analysis of Procollagen Folding *662*
18.8 Experimental Part *663*
18.8.1 Materials Required *663*
18.8.2 Experimental Protocols *664*
References *668*

19 Redox Regulation of Chaperones *677*
Jörg H. Hoffmann and Ursula Jakob
19.1 Introduction *677*
19.2 Disulfide Bonds as Redox-Switches *677*
19.2.1 Functionality of Disulfide Bonds *677*
19.2.2 Regulatory Disulfide Bonds as Functional Switches *679*
19.2.3 Redox Regulation of Chaperone Activity *680*
19.3 Prokaryotic Hsp33: A Chaperone Activated by Oxidation *680*
19.3.1 Identification of a Redox-regulated Chaperone *680*
19.3.2 Activation Mechanism of Hsp33 *681*
19.3.3 The Crystal Structure of Active Hsp33 *682*
19.3.4 The Active Hsp33-Dimer: An Efficient Chaperone Holdase *683*
19.3.5 Hsp33 is Part of a Sophisticated Multi-chaperone Network *684*
19.4 Eukaryotic Protein Disulfide Isomerase (PDI): Redox Shuffling in the ER *685*
19.4.1 PDI, A Multifunctional Enzyme in Eukaryotes *685*
19.4.2 PDI and Redox Regulation *687*
19.5 Concluding Remarks and Outlook *688*
19.6 Appendix – Experimental Protocols *688*
19.6.1 How to Work With Redox-regulated Chaperones in Vitro *689*
19.6.1.1 Preparation of the Reduced Protein Species *689*
19.6.1.2 Preparation of the Oxidized Protein Species *690*
19.6.1.3 In Vitro Thiol Trapping to Monitor the Redox State of Proteins *691*
19.6.2 Thiol Coordinating Zinc Centers as Redox Switches *691*
19.6.2.1 PAR-PMPS Assay to Quantify Zinc *691*
19.6.2.2 Determination of Zinc-binding Constants *692*
19.6.3 Functional Analysis of Redox-regulated Chaperones in Vitro/in Vivo *693*
19.6.3.1 Chaperone Activity Assays *693*

19.6.3.2 Manipulating and Analyzing Redox Conditions in Vivo 694
Acknowledgements 694
References 694

20 The *E. coli* GroE Chaperone 699
Steven G. Burston and Stefan Walter
20.1 Introduction 699
20.2 The Structure of GroEL 699
20.3 The Structure of GroEL-ATP 700
20.4 The Structure of GroES and its Interaction with GroEL 701
20.5 The Interaction Between GroEL and Substrate Polypeptides 702
20.6 GroEL is a Complex Allosteric Macromolecule 703
20.7 The Reaction Cycle of the GroE Chaperone 705
20.8 The Effect of GroE on Protein-folding Pathways 708
20.9 Future Perspectives 710
20.10 Experimental Protocols 710
Acknowledgments 719
References 719

21 Structure and Function of the Cytosolic Chaperonin CCT 725
José M. Valpuesta, José L. Carrascosa, and Keith R. Willison
21.1 Introduction 725
21.2 Structure and Composition of CCT 726
21.3 Regulation of CCT Expression 729
21.4 Functional Cycle of CCT 730
21.5 Folding Mechanism of CCT 731
21.6 Substrates of CCT 735
21.7 Co-chaperones of CCT 739
21.8 Evolution of CCT 741
21.9 Concluding Remarks 743
21.10 Experimental Protocols 743
21.10.1 Purification 743
21.10.2 ATP Hydrolysis Measurements 744
21.10.3 CCT Substrate-binding and Folding Assays 744
21.10.4 Electron Microscopy and Image Processing 744
References 747

22 Structure and Function of GimC/Prefoldin 756
Katja Siegers, Andreas Bracher, and Ulrich Hartl
22.1 Introduction 756
22.2 Evolutionary Distribution of GimC/Prefoldin 757
22.3 Structure of the Archaeal GimC/Prefoldin 757
22.4 Complexity of the Eukaryotic/Archaeal GimC/Prefoldin 759
22.5 Functional Cooperation of GimC/Prefoldin With the Eukaryotic Chaperonin TRiC/CCT 761

22.6 Experimental Protocols *764*
22.6.1 Actin-folding Kinetics *764*
22.6.2 Prevention of Aggregation (Light-scattering) Assay *765*
22.6.3 Actin-binding Assay *765*
Acknowledgements *766*
References *766*

23 Hsp90: From Dispensable Heat Shock Protein to Global Player *768*
Klaus Richter, Birgit Meinlschmidt, and Johannes Buchner
23.1 Introduction *768*
23.2 The Hsp90 Family in Vivo *768*
23.2.1 Evolutionary Relationships within the Hsp90 Gene Family *768*
23.2.2 In Vivo Functions of Hsp90 *769*
23.2.3 Regulation of Hsp90 Expression and Posttranscriptional Activation *772*
23.2.4 Chemical Inhibition of Hsp90 *773*
23.2.5 Identification of Natural Hsp90 Substrates *774*
23.3 In Vitro Investigation of the Chaperone Hsp90 *775*
23.3.1 Hsp90: A Special Kind of ATPase *775*
23.3.2 The ATPase Cycle of Hsp90 *780*
23.3.3 Interaction of Hsp90 with Model Substrate Proteins *781*
23.3.4 Investigating Hsp90 Substrate Interactions Using Native Substrates *783*
23.4 Partner Proteins: Does Complexity Lead to Specificity? *784*
23.4.1 Hop, p23, and PPIases: The Chaperone Cycle of Hsp90 *784*
23.4.2 Hop/Sti1: Interactions Mediated by TPR Domains *787*
23.4.3 p23/Sba1: Nucleotide-specific Interaction with Hsp90 *789*
23.4.4 Large PPIases: Conferring Specificity to Substrate Localization? *790*
23.4.5 Pp5: Facilitating Dephosphorylation *791*
23.4.6 Cdc37: Building Complexes with Kinases *792*
23.4.7 Tom70: Chaperoning Mitochondrial Import *793*
23.4.8 CHIP and Sgt1: Multiple Connections to Protein Degradation *793*
23.4.9 Aha1 and Hch1: Just Stimulating the ATPase? *794*
23.4.10 Cns1, Sgt2, and Xap2: Is a TPR Enough to Become an Hsp90 Partner? *796*
23.5 Outlook *796*
23.6 Appendix – Experimental Protocols *797*
23.6.1 Calculation of Phylogenetic Trees Based on Protein Sequences *797*
23.6.2 Investigating the in Vivo Effect of Hsp90 Mutations in *S. cerevisiae* *797*
23.6.3 Well-characterized Hsp90 Mutants *798*
23.6.4 Investigating Activation of Heterologously Expressed Src Kinase in *S. cerevisiae* *800*
23.6.5 Investigation of Heterologously Expressed Glucocorticoid Receptor in *S. cerevisiae* *800*

23.6.6 Investigation of Chaperone Activity 801
23.6.7 Analysis of the ATPase Activity of Hsp90 802
23.6.8 Detecting Specific Influences on Hsp90 ATPase Activity 803
23.6.9 Investigation of the Quaternary Structure by SEC-HPLC 804
23.6.10 Investigation of Binding Events Using Changes of the Intrinsic Fluorescence 806
23.6.11 Investigation of Binding Events Using Isothermal Titration Calorimetry 807
23.6.12 Investigation of Protein-Protein Interactions Using Cross-linking 807
23.6.13 Investigation of Protein-Protein Interactions Using Surface Plasmon Resonance Spectroscopy 808
Acknowledgements 810
References 810

24 Small Heat Shock Proteins: Dynamic Players in the Folding Game *830*
Franz Narberhaus and Martin Haslbeck
24.1 Introduction 830
24.2 α-Crystallins and the Small Heat Shock Protein Family: Diverse Yet Similar 830
24.3 Cellular Functions of α-Hsps 831
24.3.1 Chaperone Activity in Vitro 831
24.3.2 Chaperone Function in Vivo 835
24.3.3 Other Functions 836
24.4 The Oligomeric Structure of α-Hsps 837
24.5 Dynamic Structures as Key to Chaperone Activity 839
24.6 Experimental Protocols 840
24.6.1 Purification of sHsps 840
24.6.2 Chaperone Assays 843
24.6.3 Monitoring Dynamics of sHsps 846
Acknowledgements 847
References 848

25 Alpha-crystallin: Its Involvement in Suppression of Protein Aggregation and Protein Folding *858*
Joseph Horwitz
25.1 Introduction 858
25.2 Distribution of Alpha-crystallin in the Various Tissues 858
25.3 Structure 859
25.4 Phosphorylation and Other Posttranslation Modification 860
25.5 Binding of Target Proteins to Alpha-crystallin 861
25.6 The Function of Alpha-crystallin 863
25.7 Experimental Protocols 863
25.7.1 Preparation of Alpha-crystallin 863
Acknowledgements 870
References 870

26 **Transmembrane Domains in Membrane Protein Folding, Oligomerization, and Function** *876*
Anja Ridder and Dieter Langosch
26.1 Introduction *876*
26.1.1 Structure of Transmembrane Domains *876*
26.1.2 The Biosynthetic Route towards Folded and Oligomeric Integral Membrane Proteins *877*
26.1.3 Structure and Stability of TMSs *878*
26.1.3.1 Amino Acid Composition of TMSs and Flanking Regions *878*
26.1.3.2 Stability of Transmembrane Helices *879*
26.2 The Nature of Transmembrane Helix-Helix Interactions *880*
26.2.1 General Considerations *880*
26.2.1.1 Attractive Forces within Lipid Bilayers *880*
26.2.1.2 Forces between Transmembrane Helices *881*
26.2.1.3 Entropic Factors Influencing Transmembrane Helix–Helix Interactions *882*
26.2.2 Lessons from Sequence Analyses and High-resolution Structures *883*
26.2.3 Lessons from Bitopic Membrane Proteins *886*
26.2.3.1 Transmembrane Segments Forming Right-handed Pairs *886*
26.2.3.2 Transmembrane Segments Forming Left-handed Assemblies *889*
26.2.4 Selection of Self-interacting TMSs from Combinatorial Libraries *892*
26.2.5 Role of Lipids in Packing/Assembly of Membrane Proteins *893*
26.3 Conformational Flexibility of Transmembrane Segments *895*
26.4 Experimental Protocols *897*
26.4.1 Biochemical and Biophysical Techniques *897*
26.4.1.1 Visualization of Oligomeric States by Electrophoretic Techniques *898*
26.4.1.2 Hydrodynamic Methods *899*
26.4.1.3 Fluorescence Resonance Transfer *900*
26.4.2 Genetic Assays *901*
26.4.2.1 The ToxR System *901*
26.4.2.2 Other Genetic Assays *902*
26.4.3 Identification of TMS-TMS Interfaces by Mutational Analysis *903*
References *904*

Part II, Volume 3

27 **SecB** *919*
Arnold J. M. Driessen, Janny de Wit, and Nico Nouwen
27.1 Introduction *919*
27.2 Selective Binding of Preproteins by SecB *920*
27.3 SecA-SecB Interaction *925*
27.4 Preprotein Transfer from SecB to SecA *928*
27.5 Concluding Remarks *929*
27.6 Experimental Protocols *930*
27.6.1 How to Analyze SecB-Preprotein Interactions *930*

27.6.2 How to Analyze SecB-SecA Interaction *931*
Acknowledgements *932*
References *933*

28 Protein Folding in the Periplasm and Outer Membrane of *E. coli* *938*
Michael Ehrmann
28.1 Introduction *938*
28.2 Individual Cellular Factors *940*
28.2.1 The Proline Isomerases FkpA, PpiA, SurA, and PpiD *941*
28.2.1.1 FkpA *942*
28.2.1.2 PpiA *942*
28.2.1.3 SurA *943*
28.2.1.4 PpiD *943*
28.2.2 Skp *944*
28.2.3 Proteases and Protease/Chaperone Machines *945*
28.2.3.1 The HtrA Family of Serine Proteases *946*
28.2.3.2 *E. coli* HtrAs *946*
28.2.3.3 DegP and DegQ *946*
28.2.3.4 DegS *947*
28.2.3.5 The Structure of HtrA *947*
28.2.3.6 Other Proteases *948*
28.3 Organization of Folding Factors into Pathways and Networks *950*
28.3.1 Synthetic Lethality and Extragenic High-copy Suppressors *950*
28.3.2 Reconstituted in Vitro Systems *951*
28.4 Regulation *951*
28.4.1 The Sigma E Pathway *951*
28.4.2 The Cpx Pathway *952*
28.4.3 The Bae Pathway *953*
28.5 Future Perspectives *953*
28.6 Experimental Protocols *954*
28.6.1 Pulse Chase Immunoprecipitation *954*
Acknowledgements *957*
References *957*

29 Formation of Adhesive Pili by the Chaperone-Usher Pathway *965*
Michael Vetsch and Rudi Glockshuber
29.1 Basic Properties of Bacterial, Adhesive Surface Organelles *965*
29.2 Structure and Function of Pilus Chaperones *970*
29.3 Structure and Folding of Pilus Subunits *971*
29.4 Structure and Function of Pilus Ushers *973*
29.5 Conclusions and Outlook *976*
29.6 Experimental Protocols *977*
29.6.1 Test for the Presence of Type 1 Piliated *E. coli* Cells *977*
29.6.2 Functional Expression of Pilus Subunits in the *E. coli* Periplasm *977*
29.6.3 Purification of Pilus Subunits from the *E. coli* Periplasm *978*

29.6.4 Preparation of Ushers *979*
Acknowledgements *979*
References *980*

30 Unfolding of Proteins During Import into Mitochondria *987*
Walter Neupert, Michael Brunner, and Kai Hell
30.1 Introduction *987*
30.2 Translocation Machineries and Pathways of the Mitochondrial Protein Import System *988*
30.2.1 Import of Proteins Destined for the Mitochondrial Matrix *990*
30.3 Import into Mitochondria Requires Protein Unfolding *993*
30.4 Mechanisms of Unfolding by the Mitochondrial Import Motor *995*
30.4.1 Targeted Brownian Ratchet *995*
30.4.2 Power-stroke Model *995*
30.5 Studies to Discriminate between the Models *996*
30.5.1 Studies on the Unfolding of Preproteins *996*
30.5.1.1 Comparison of the Import of Folded and Unfolded Proteins *996*
30.5.1.2 Import of Preproteins With Different Presequence Lengths *999*
30.5.1.3 Import of Titin Domains *1000*
30.5.1.4 Unfolding by the Mitochondrial Membrane Potential $\Delta\Psi$ *1000*
30.5.2 Mechanistic Studies of the Import Motor *1000*
30.5.2.1 Brownian Movement of the Polypeptide Within the Import Channel *1000*
30.5.2.2 Recruitment of mtHsp70 by Tim44 *1001*
30.5.2.3 Import Without Recruitment of mtHsp70 by Tim44 *1002*
30.5.2.4 MtHsp70 Function in the Import Motor *1003*
30.6 Discussion and Perspectives *1004*
30.7 Experimental Protocols *1006*
30.7.1 Protein Import Into Mitochondria in Vitro *1006*
30.7.2 Stabilization of the DHFR Domain by Methotrexate *1008*
30.7.3 Import of Precursor Proteins Unfolded With Urea *1009*
30.7.4 Kinetic Analysis of the Unfolding Reaction by Trapping of Intermediates *1009*
References *1011*

31 The Chaperone System of Mitochondria *1020*
Wolfgang Voos and Nikolaus Pfanner
31.1 Introduction *1020*
31.2 Membrane Translocation and the Hsp70 Import Motor *1020*
31.3 Folding of Newly Imported Proteins Catalyzed by the Hsp70 and Hsp60 Systems *1026*
31.4 Mitochondrial Protein Synthesis and the Assembly Problem *1030*
31.5 Aggregation versus Degradation: Chaperone Functions Under Stress Conditions *1033*
31.6 Experimental Protocols *1034*

31.6.1 Chaperone Functions Characterized With Yeast Mutants *1034*
31.6.2 Interaction of Imported Proteins With Matrix Chaperones *1036*
31.6.3 Folding of Imported Model Proteins *1037*
31.6.4 Assaying Mitochondrial Degradation of Imported Proteins *1038*
31.6.5 Aggregation of Proteins in the Mitochondrial Matrix *1038*
References *1039*

32 Chaperone Systems in Chloroplasts *1047*
Thomas Becker, Jürgen Soll, and Enrico Schleiff
32.1 Introduction *1047*
32.2 Chaperone Systems within Chloroplasts *1048*
32.2.1 The Hsp70 System of Chloroplasts *1048*
32.2.1.1 The Chloroplast Hsp70s *1049*
32.2.1.2 The Co-chaperones of Chloroplastic Hsp70s *1051*
32.2.2 The Chaperonins *1052*
32.2.3 The HSP100/Clp Protein Family in Chloroplasts *1056*
32.2.4 The Small Heat Shock Proteins *1058*
32.2.5 Hsp90 Proteins of Chloroplasts *1061*
32.2.6 Chaperone-like Proteins *1062*
32.2.6.1 The Protein Disulfide Isomerase (PDI) *1062*
32.2.6.2 The Peptidyl-prolyl *cis* Isomerase (PPIase) *1063*
32.3 The Functional Chaperone Pathways in Chloroplasts *1065*
32.3.1 Chaperones Involved in Protein Translocation *1065*
32.3.2 Protein Transport Inside of Plastids *1070*
32.3.3 Protein Folding and Complex Assembly Within Chloroplasts *1071*
32.3.4 Chloroplast Chaperones Involved in Proteolysis *1072*
32.3.5 Protein Storage Within Plastids *1073*
32.3.6 Protein Protection and Repair *1074*
32.4 Experimental Protocols *1075*
32.4.1 Characterization of Cpn60 Binding to the Large Subunit of Rubisco via Native PAGE (adopted from Ref. [6]) *1075*
32.4.2 Purification of Chloroplast Cpn60 From Young Pea Plants (adopted from Ref. [203]) *1076*
32.4.3 Purification of Chloroplast Hsp21 From Pea (*Pisum sativum*) (adopted from [90]) *1077*
32.4.4 Light-scattering Assays for Determination of the Chaperone Activity Using Citrate Synthase as Substrate (adopted from [196]) *1078*
32.4.5 The Use Of *Bis*-ANS to Assess Surface Exposure of Hydrophobic Domains of Hsp17 of *Synechocystis* (adopted from [202]) *1079*
32.4.6 Determination of Hsp17 Binding to Lipids (adopted from Refs. [204, 205]) *1079*
References *1081*

33 An Overview of Protein Misfolding Diseases *1093*
Christopher M. Dobson
33.1 Introduction *1093*

33.2 Protein Misfolding and Its Consequences for Disease *1094*
33.3 The Structure and Mechanism of Amyloid Formation *1097*
33.4 A Generic Description of Amyloid Formation *1101*
33.5 The Fundamental Origins of Amyloid Disease *1104*
33.6 Approaches to Therapeutic Intervention in Amyloid Disease *1106*
33.7 Concluding Remarks *1108*
Acknowledgements *1108*
References *1109*

34 Biochemistry and Structural Biology of Mammalian Prion Disease *1114*
Rudi Glockshuber
34.1 Introduction *1114*
34.1.1 Prions and the "Protein-Only" Hypothesis *1114*
34.1.2 Models of PrP^{Sc} Propagation *1115*
34.2 Properties of PrP^{C} and PrP^{Sc} *1117*
34.3 Three-dimensional Structure and Folding of Recombinant PrP *1120*
34.3.1 Expression of the Recombinant Prion Protein for Structural and Biophysical Studies *1120*
34.3.2 Three-dimensional Structures of Recombinant Prion Proteins from Different Species and Their Implications for the Species Barrier of Prion Transmission *1120*
34.3.2.1 Solution Structure of Murine PrP *1120*
34.3.2.2 Comparison of Mammalian Prion Protein Structures and the Species Barrier of Prion Transmission *1124*
34.3.3 Biophysical Characterization of the Recombinant Prion Protein *1125*
34.3.3.1 Folding and Stability of Recombinant PrP *1125*
34.3.3.2 Role of the Disulfide Bond in PrP *1127*
34.3.3.3 Influence of Point Mutations Linked With Inherited TSEs on the Stability of Recombinant PrP *1129*
34.4 Generation of Infectious Prions in Vitro: Principal Difficulties in Proving the Protein-Only Hypothesis *1131*
34.5 Understanding the Strain Phenomenon in the Context of the Protein-Only Hypothesis: Are Prions Crystals? *1132*
34.6 Conclusions and Outlook *1135*
34.7 Experimental Protocols *1136*
34.7.1 Protocol 1 [53, 55] *1136*
34.7.2 Protocol 2 [54] *1137*
References *1138*

35 Insights into the Nature of Yeast Prions *1144*
Lev Z. Osherovich and Jonathan S. Weissman
35.1 Introduction *1144*
35.2 Prions as Heritable Amyloidoses *1145*
35.3 Prion Strains and Species Barriers: Universal Features of Amyloid-based Prion Elements *1149*

35.4 Prediction and Identification of Novel Prion Elements *1151*
35.5 Requirements for Prion Inheritance beyond Amyloid-mediated Growth *1154*
35.6 Chaperones and Prion Replication *1157*
35.7 The Structure of Prion Particles *1158*
35.8 Prion-like Structures as Protein Interaction Modules *1159*
35.9 Experimental Protocols *1160*
35.9.1 Generation of Sup35 Amyloid Fibers in Vitro *1160*
35.9.2 Thioflavin T–based Amyloid Seeding Efficacy Assay (Adapted from Chien et al. 2003) *1161*
35.9.3 AFM-based Single-fiber Growth Assay *1162*
35.9.4 Prion Infection Protocol (Adapted from Tanaka et al. 2004) *1164*
35.9.5 Preparation of Lyticase *1165*
35.9.6 Protocol for Counting Heritable Prion Units (Adapted from Cox et al. 2003) *1166*
Acknowledgements *1167*
References *1168*

36 Polyglutamine Aggregates as a Model for Protein-misfolding Diseases *1175*
Soojin Kim, James F. Morley, Anat Ben-Zvi, and Richard I. Morimoto
36.1 Introduction *1175*
36.2 Polyglutamine Diseases *1175*
36.2.1 Genetics *1175*
36.2.2 Polyglutamine Diseases Involve a Toxic Gain of Function *1176*
36.3 Polyglutamine Aggregates *1176*
36.3.1 Presence of the Expanded Polyglutamine Is Sufficient to Induce Aggregation in Vivo *1176*
36.3.2 Length of the Polyglutamine Dictates the Rate of Aggregate Formation *1177*
36.3.3 Polyglutamine Aggregates Exhibit Features Characteristic of Amyloids *1179*
36.3.4 Characterization of Protein Aggregates in Vivo Using Dynamic Imaging Methods *1180*
36.4 A Role for Oligomeric Intermediates in Toxicity *1181*
36.5 Consequences of Misfolded Proteins and Aggregates on Protein Homeostasis *1181*
36.6 Modulators of Polyglutamine Aggregation and Toxicity *1184*
36.6.1 Protein Context *1184*
36.6.2 Molecular Chaperones *1185*
36.6.3 Proteasomes *1188*
36.6.4 The Protein-folding "Buffer" and Aging *1188*
36.6.5 Summary *1189*
36.7 Experimental Protocols *1190*
36.7.1 FRAP Analysis *1190*
References *1192*

37 Protein Folding and Aggregation in the Expanded Polyglutamine Repeat Diseases *1200*
Ronald Wetzel
37.1 Introduction *1200*
37.2 Key Features of the Polyglutamine Diseases *1201*
37.2.1 The Variety of Expanded PolyGln Diseases *1201*
37.2.2 Clinical Features *1201*
37.2.2.1 Repeat Expansions and Repeat Length *1202*
37.2.3 The Role of PolyGln and PolyGln Aggregates *1203*
37.3 PolyGln Peptides in Studies of the Molecular Basis of Expanded Polyglutamine Diseases *1205*
37.3.1 Conformational Studies *1205*
37.3.2 Preliminary in Vitro Aggregation Studies *1206*
37.3.3 In Vivo Aggregation Studies *1206*
37.4 Analyzing Polyglutamine Behavior With Synthetic Peptides: Practical Aspects *1207*
37.4.1 Disaggregation of Synthetic Polyglutamine Peptides *1209*
37.4.2 Growing and Manipulating Aggregates *1210*
37.4.2.1 Polyglutamine Aggregation by Freeze Concentration *1210*
37.4.2.2 Preparing Small Aggregates *1211*
37.5 In vitro Studies of PolyGln Aggregation *1212*
37.5.1 The Universe of Protein Aggregation Mechanisms *1212*
37.5.2 Basic Studies on Spontaneous Aggregation *1213*
37.5.3 Nucleation Kinetics of PolyGln *1215*
37.5.4 Elongation Kinetics *1218*
37.5.4.1 Microtiter Plate Assay for Elongation Kinetics *1219*
37.5.4.2 Repeat-length and Aggregate-size Dependence of Elongation Rates *1220*
37.6 The Structure of PolyGln Aggregates *1221*
37.6.1 Electron Microscopy Analysis *1222*
37.6.2 Analysis with Amyloid Dyes Thioflavin T and Congo Red *1222*
37.6.3 Circular Dichroism Analysis *1224*
37.6.4 Presence of a Generic Amyloid Epitope in PolyGln Aggregates *1225*
37.6.5 Proline Mutagenesis to Dissect the Polyglutamine Fold Within the Aggregate *1225*
37.7 Polyglutamine Aggregates and Cytotoxicity *1227*
37.7.1 Direct Cytotoxicity of PolyGln Aggregates *1228*
37.7.1.1 Delivery of Aggregates into Cells and Cellular Compartments *1229*
37.7.1.2 Cell Killing by Nuclear-targeted PolyGln Aggregates *1229*
37.7.2 Visualization of Functional, Recruitment-positive Aggregation Foci *1230*
37.8 Inhibitors of polyGln Aggregation *1231*
37.8.1 Designed Peptide Inhibitors *1231*
37.8.2 Screening for Inhibitors of PolyGln Elongation *1231*
37.9 Concluding Remarks *1232*
37.10 Experimental Protocols *1233*

37.10.1 Disaggregation of Synthetic PolyGln Peptides *1233*
37.10.2 Determining the Concentration of Low-molecular-weight PolyGln Peptides by HPLC *1235*
Acknowledgements *1237*
References *1238*

38 Production of Recombinant Proteins for Therapy, Diagnostics, and Industrial Research by in Vitro Folding *1245*
Christian Lange and Rainer Rudolph
38.1 Introduction *1245*
38.1.1 The Inclusion Body Problem *1245*
38.1.2 Cost and Scale Limitations in Industrial Protein Folding *1248*
38.2 Treatment of Inclusion Bodies *1250*
38.2.1 Isolation of Inclusion Bodies *1250*
38.2.2 Solubilization of Inclusion Bodies *1250*
38.3 Refolding in Solution *1252*
38.3.1 Protein Design Considerations *1252*
38.3.2 Oxidative Refolding With Disulfide Bond Formation *1253*
38.3.3 Transfer of the Unfolded Proteins Into Refolding Buffer *1255*
38.3.4 Refolding Additives *1257*
38.3.5 Cofactors in Protein Folding *1260*
38.3.6 Chaperones and Folding-helper Proteins *1261*
38.3.7 An Artificial Chaperone System *1261*
38.3.8 Pressure-induced Folding *1262*
38.3.9 Temperature-leap Techniques *1263*
38.3.10 Recycling of Aggregates *1264*
38.4 Alternative Refolding Techniques *1264*
38.4.1 Matrix-assisted Refolding *1264*
38.4.2 Folding by Gel Filtration *1266*
38.4.3 Direct Refolding of Inclusion Body Material *1267*
38.5 Conclusions *1268*
38.6 Experimental Protocols *1268*
38.6.1 Protocol 1: Isolation of Inclusion Bodies *1268*
38.6.2 Protocol 2: Solubilization of Inclusion Bodies *1269*
38.6.3 Protocol 3: Refolding of Proteins *1270*
Acknowledgements *1271*
References *1271*

39 Engineering Proteins for Stability and Efficient Folding *1281*
Bernhard Schimmele and Andreas Plückthun
39.1 Introduction *1281*
39.2 Kinetic and Thermodynamic Aspects of Natural Proteins *1281*
39.2.1 The Stability of Natural Proteins *1281*
39.2.2 Different Kinds of "Stability" *1282*
39.2.2.1 Thermodynamic Stability *1283*

39.2.2.2 Kinetic Stability *1285*
39.2.2.3 Folding Efficiency *1287*
39.3 The Engineering Approach *1288*
39.3.1 Consensus Strategies *1288*
39.3.1.1 Principles *1288*
39.3.1.2 Examples *1291*
39.3.2 Structure-based Engineering *1292*
39.3.2.1 Entropic Stabilization *1294*
39.3.2.2 Hydrophobic Core Packing *1296*
39.3.2.3 Charge Interactions *1297*
39.3.2.4 Hydrogen Bonding *1298*
39.3.2.5 Disallowed Phi-Psi Angles *1298*
39.3.2.6 Local Secondary Structure Propensities *1299*
39.3.2.7 Exposed Hydrophobic Side Chains *1299*
39.3.2.8 Inter-domain Interactions *1300*
39.3.3 Case Study: Combining Consensus Design and Rational Engineering to Yield Antibodies with Favorable Biophysical Properties *1300*
39.4 The Selection and Evolution Approach *1305*
39.4.1 Principles *1305*
39.4.2 Screening and Selection Technologies Available for Improving Biophysical Properties *1311*
39.4.2.1 In Vitro Display Technologies *1313*
39.4.2.2 Partial in Vitro Display Technologies *1314*
39.4.2.3 In Vivo Selection Technologies *1315*
39.4.3 Selection for Enhanced Biophysical Properties *1316*
39.4.3.1 Selection for Solubility *1316*
39.4.3.2 Selection for Protein Display Rates *1317*
39.4.3.3 Selection on the Basis of Cellular Quality Control *1318*
39.4.4 Selection for Increased Stability *1319*
39.4.4.1 General Strategies *1319*
39.4.4.2 Protein Destabilization *1319*
39.4.4.3 Selections Based on Elevated Temperature *1321*
39.4.4.4 Selections Based on Destabilizing Agents *1322*
39.4.4.5 Selection for Proteolytic Stability *1323*
39.5 Conclusions and Perspectives *1324*
Acknowledgements *1326*
References *1326*

Index *1334*

27
SecB

Arnold J. M. Driessen, Janny de Wit, and Nico Nouwen

27.1
Introduction

Most components of the protein export apparatus of *Escherichia coli* have been identified by genetic approaches that were conducted in the late 1980s. Since many of these components are essential for viability, genetic screens were aimed at the isolation of conditional lethal mutants with pleiotropic secretion defects. In 1983, Kumamoto and Beckwith [1] reported a gene that appeared nonessential for *E. coli* viability but that when mutated or even completely eliminated adversely affected the export of a subset of envelope proteins, including the periplasmic maltose-binding protein (MBP). This gene, termed *secB*, was found to map near min 81 on the *E. coli* chromosome [1]. Strains carrying SecB mutations, although unable to grow on Luria Bertani broth plates, were still viable on minimal media [2]. To define its function, it was necessary to biochemically analyze the SecB protein. Using a pulse-labeling approach, Gannon and Kumamoto [3] demonstrated that *secB* mutations caused a defect in co-translational processing of MBP. The preMBP that accumulated in cells of the *secB* mutant had folded into a protease-resistant conformation, and it was suggested that SecB stabilized the unfolded state of preproteins prior to their translocation. Direct evidence for an anti-folding activity of SecB was provided by in vitro experiments by the group of Phil Bassford [4]. In the absence of SecB, the rate of folding of in vitro synthesized wild-type preMBP was accelerated and was greatly retarded when excess SecB was present. This defined a biochemical function for SecB that complemented early work by Randall and Hardy, who correlated the folding of preMBP into a stable tertiary structure in the cytoplasm with a loss of export competence [5]. A further understanding of the function of the *secB* gene was obtained after overproduction and purification of the SecB protein in 1988 [6]. SecB was shown to be needed for the efficient in vitro translocation of preMBP into inverted inner-membrane vesicles. Moreover, SecB was found to associate and to retard the folding of preMBP and several other preproteins. In 1990, Hartl and Wickner [7] provided evidence for a targeting function of SecB. By means of biochemical assays, SecB and the binary SecB-preprotein complex were found to bind with high affinity to the membrane-

Protein Folding Handbook. Part II. Edited by J. Buchner and T. Kiefhaber

ISBN: 3-527-30784-2

Tab. 27.1. Chronology of some important findings in the SecB function.

Year	*Finding*	*References*
1983	Isolation of *secB* mutants	1
1988	Demonstration of an anti-folding activity of SecB	4
1988	Purification of SecB	6, 95, 96
	Demonstration that SecB stimulates preprotein translocation in vitro	
1990	Demonstration of a preprotein-targeting activity of SecB and evidence for a specific SecB-SecA interaction at the membrane	7
1994	Definition of the SecB-binding frame in the mature region of preproteins excluding the signal sequence	47
1995	Identification of the C-terminus of SecA as the SecB-binding domain	79
1997	Definition of the SecB-targeting cycle and preprotein transfer mechanism to SecA	8
1999	Identification of a critical zinc ion in the SecB-binding domain of SecA	80
2000	Elucidation of the crystal structure of SecB	9
2003	Elucidation of the structure of SecB in complex with the SecA C-terminal-binding domain	81

bound motor of the preprotein translocase, the SecA protein. This led to the understanding that SecB has a dual function in the cell, i.e., stabilizing the preprotein prior to translocation and transferring it to the membrane-bound SecA [8], a protein that catalyzes the next step in the translocation reaction. In 2000, the three-dimensional structure of the *Haemophilus influenzae* SecB protein was solved [9], which now provides a handle to correlate function to structure. A historical timeline is shown in Table 27.1.

27.2 Selective Binding of Preproteins by SecB

Folding of newly synthesized proteins inside the cell depends on a set of conserved proteins known as molecular chaperones. These molecular chaperones essentially prevent nonproductive interactions between nonnative polypeptides and typically increase the yield of a folding reaction. In *Escherichia coli* and other bacteria, export of newly synthesized proteins from the cytoplasm to the exterior involves at least one transport step across the inner membrane [10]. In order to pass the inner membrane, proteins need to be in an unfolded state [5]. Therefore, newly synthesized precursor proteins (preproteins) need to bypass the cellular folding apparatus and funnel into a pathway that directs them in an unfolded state to the preprotein translocase at the inner membrane. In *E. coli*, we can discriminate two independent targeting pathways for preproteins. One pathway involves a bacterial homologue of the eukaryotic signal recognition particle (SRP) and its membrane-associated receptor (FtsY), which will not be further discussed here. The other pathway involves the SecB protein. Both pathways converge at the translocase [11]. Translocase is a membrane-bound complex that consists of the SecA protein, a preprotein receptor and ATP-driven molecular motor device, and the SecYEG complex, a preprotein-conducting channel in the membrane [12].

SecB is a tetrameric chaperone in *E. coli* with a cellular function that is dedicated to preprotein translocation [13, 14]. It is important to note that SecB is found only in the subgroup of α-, β-, and γ-proteobacteria that includes the majority of known gram-negative bacteria of medical, industrial, and agricultural significance such as *Vibrio*, *Pseudomonas*, *Rickettsia*, *Neisseria*, and *Haemophilus* [15]. The *secB* gene is, however, not essential in Gram-negative bacteria [1] and is even absent in Gram-positive bacteria and archaea [16, 17]. Although *E. coli secB* null and mutant strains show pleiotropic defects in protein secretion, this growth defect is specific for rich media and has been attributed to a reduced expression of *gpsA* located downstream from the *secB* gene [18]. GpsA encodes an sn-glycerol-3-phosphate dehydrogenase that is involved in phospholipids biosynthesis. Its reduced expression affects the membrane lipid composition that indirectly influences the preprotein translocase, therefore leading to a secretion defect. One of the reasons that SecB is not essential relates to backup functions in the cell that ensure the proper translocation of essential proteins. In this respect, protein translocation in a *secB* null or mutant strain is supported by more general chaperones, in particular DnaK [19–21].

In vivo, SecB binds to a subset of nascent preproteins [22–24], while in vitro, it interacts with many unrelated proteins provided that they are in a nonnative state [25]. Studies with model protein substrates such as barstar revealed that SecB does not bind the folded or unfolded state but traps a near-native-like molten-globule state [26–28]. SecB has also been shown to bind partially folded states of α-lactalbumin, bovine pancreas trypsin inhibitor (BPTI), and ribonuclease A (RNase A) [29]. These model substrates are rather small, and in all cases, binding occurs with a stoichiometry of one polypeptide chain per SecB monomer [28, 30–32]. Calorimetric studies suggest that between 7 and 29 amino acid residues are buried upon substrate binding to SecB [30, 33]. Unfolded preprotein substrates, however, bind with a stoichiometry of one preprotein per SecB tetramer and a K_d that is in the nanomolar range [31, 34–36]. Short peptide substrates bind to SecB with a K_d value in the micromolar range [37]. SecB's key activity is to maintain preproteins in a so-called translocation-competent state [5, 38–40]. The exact nature of this state is unknown, but based on the studies with model substrates, it is generally assumed that this is a molten-globule state. Partially folded species such as molten globules have exposed hydrophobic surfaces, and hence nonproductive interactions readily lead to aggregation, a state that is not compatible with translocation. Indeed, with an authentic preprotein substrate such as the precursor of outer membrane protein A (proOmpA), SecB stabilizes a partially folded state with native-like secondary structure but incomplete tertiary structure [38]. The stoichiometric binding reaction prevents proOmpA, which is rich in β-sheet structure, from aggregation. With the precursor of maltose-binding protein (preMBP), the anti-folding activity of SecB seems to be of major importance. Unlike refolded preMBP, SecB-bound preMBP is particularly protease-sensitive, which is indicative of an unfolded conformation [5]. Moreover, various slowly folding mutants of preMBP exhibit a reduced SecB dependence for translocation [41]. Although SecB readily interacts with mature, unfolded MBP [42], it has been shown that the presence of a signal sequence retards the folding of the mature domain [43, 44], thereby allowing a greater time window for SecB interaction.

A range of biochemical methods has been used to understand the nature of the SecB-binding frame in preproteins [45–48]. An extensive screen of a peptide library for SecB binding indicates that a typical peptide substrate (SecB-binding motif) is approximately nine residues long and enriched in aromatic and basic residues; acidic residues are strongly disfavored. The majority of the binding energy for these peptides results from hydrophobic interactions [33]. The structures of *H. influenzae* (See Figure 27.1) [9] and *E. coli* SecB [49] have been solved, leading to some major insights as to the mode of polypeptide interaction. These highly homologous proteins show an identical structure. The SecB monomer has a simple $\alpha + \beta$ fold, but the overall rectangular-shaped structure is a tetramer organized as a dimer of dimers (Figure 27.1). This quaternary organization agrees well with the observed dynamic dimer-tetramer equilibrium of SecB in solution [36, 50] and with thermodynamic studies [51] that indicate that SecB is a stable, well-folded, and tightly packed tetramer. Although SecB has been crystallized without a peptide substrate, the structure suggests that substrate binding occurs at a surface site rather than at an interior cavity (Figure 27.2). A long surface-exposed channel is present on each side of the tetramer that has all the characteristics of a peptide-binding site. Each channel is proposed to contain two peptide-binding subsites

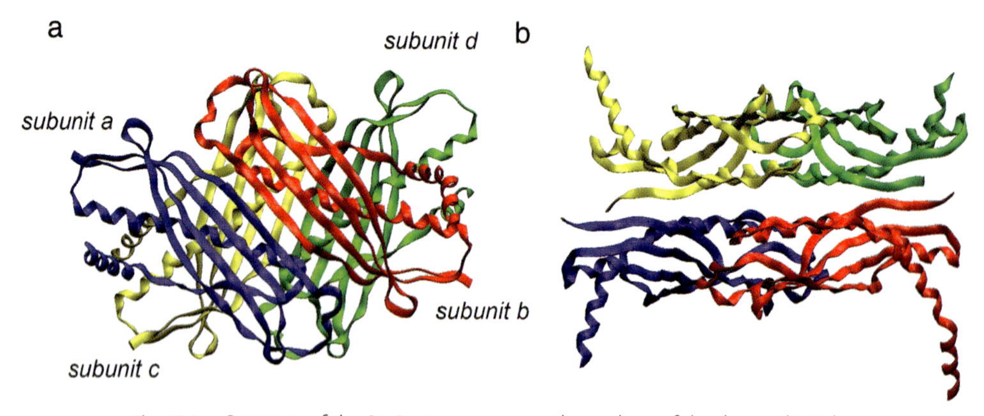

Fig. 27.1. Structure of the SecB tetramer. Ribbon drawing of the *H. influenzae* SecB tetramer (based on the coordinates deposited in the Protein Data Bank as 1FX3) [9] in two orthogonal views. (a) Front view, showing the four-stranded β-sheet of each monomer and the packing of the dimer. (b) Side view, showing the dimer-dimer interface formed by the α-helices. Each subunit in the tetramer is shown in a different color and is numbered from a to d.

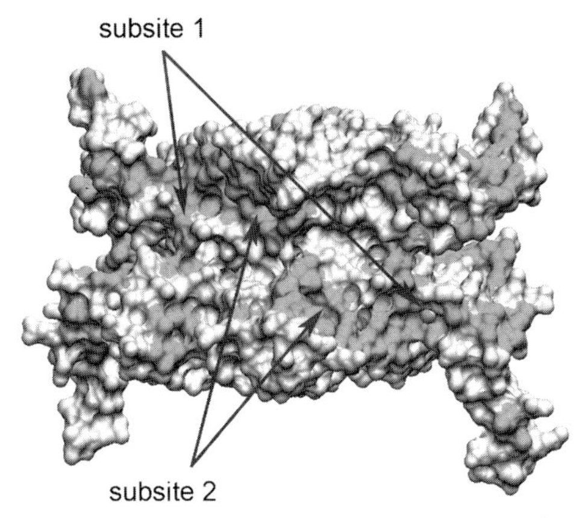

Fig. 27.2. The peptide-binding groove on SecB. Solvent-accessible surface of the proposed peptide-binding groove on one face of the *H. influenzae* SecB tetramer [9], shown in the same orientation as in Figure 27.1b. The hydrophobic residues are indicated in dark gray, and the positions of subsite 1 and the shallow subsite 2 are indicated by the arrows.

that might recognize distinct features of the preprotein [9]. Subsite 1 is a deep cleft located in a narrow constriction in the middle of the surface-exposed channel (Figure 27.2). It might recognize hydrophobic and aromatic regions of polypeptides, as most of the amino acids that line the surface are aromatic and conserved. The length of this cleft is sufficient to accommodate peptides of the size of the SecB-binding motif. Subsite 2 is a shallow, open groove with a hydrophobic surface (Figure 27.2). It lacks aromatic residues and might bind extended regions of polypeptides by forming regular main-chain hydrogen bonds. Negatively charged residues are positioned around the groove, and this could explain the selectivity for basic residues, which was particularly evident in studies with modified peptide substrates. Introduction of stretches of basic residues in preproteins that translocate in a SecB-independent manner renders these proteins SecB-dependent [52, 53]. Those studies, however, did not define a consensus SecB-binding motif. Moreover, since SecB is a highly acidic protein (see also Figure 27.4b) [9], it may more or less nonspecifically interact with basic protein stretches. The entry of subsite 1 contains a conserved tryptophan residue of SecB, and this residue has been used to monitor the binding of model substrates such as BPTI, which further support the notion that the structurally distinct grooves are indeed involved in substrate binding.

In vitro studies [25, 54, 55] have suggested that SecB discriminates between secretory and cytosolic proteins by the rate of folding. Cytosolic proteins would fold too fast and thereby escape the SecB binding. It is difficult to understand this "kinetic partitioning model" when it concerns the association of SecB with ribosome-bound nascent chains. Moreover, other studies indicate that SecB associates with unfolded proteins at a rate that is diffusion-limited [28, 31] and thus exceeds the rate of folding of most proteins. The kinetic partition model also does not explain how SecB differentiates between proteins in vivo [56], where binding seems to be dictated by a high affinity interaction [22, 23]. Although it has been suggested that SecB functions as a kind of cytosolic signal sequence recognition factor [57, 58], the signal sequence is not considered to be the prime target for SecB binding, as unfolded mature preprotein domains readily interact with SecB [42, 59, 60]. With some preproteins such as preLamB, the signal sequence showed some affinity for SecB, but it did not correspond to the major binding sites in preLamB [61]. SecB binds to long nascent chains of a subset of preproteins [23]. These ribosome-bound nascent proteins likely encounter a variety of chaperones before they are recognized by SecB. Stage- and site-specific cross-linking approaches suggest that the signal sequence of preproteins such as proOmpA, immediately after emerging from the ribosome, associates with Ffh, a subunit of the bacterial SRP, and with SecA, while remaining attached to the surface of the ribosome via protein L23 [62]. Hydrophobic signal sequences or transmembrane segments of nascent membrane proteins remain SRP-bound, and the ternary ribosome–nascent chain–SRP complex is subsequently targeted to the translocation sites at the membrane [11, 63]. With most preproteins, however, such proposed contacts with SRP will be only transient and lost upon further growth of the nascent chain. Experimental data indicate that the signal sequence then interacts with the chaperone trigger factor, a protein with peptidyl-prolyl *cis/trans* isomerase activity [63–65]. While the preprotein further emerges from the ribosome, trigger factor will scan the nascent chain, but, eventually, domains will become exposed that interact with SecB. It should be stressed that some of the cross-linking experiments [62] rely on cellular lysates that are supplemented with cross-linking partners. This may lead to a false bias of possible interactions. Nevertheless, in the cell, SecB is known to associate with ribosome-bound nascent chains after they have reached a length of ~150 residues [23, 24]. These chain lengths correspond with the location of SecB-binding sites mapped by proteolysis [45–47]. Because SecB recognizes peptide sequences that typically occur within core regions of folded proteins [48], it binds preferentially to the unfolded conformation of the mature part of preproteins. The long binding regions have been proposed to occupy multiple binding sites on SecB simultaneously [66], thereby allowing the high binding affinity of nonnative preproteins ($K_d = 5$–50 nM). To occupy the peptide-binding grooves on both sites, long, unstructured polypeptide segments presumably are wrapped around the rectangular tetrameric shape of SecB.

The SecB structure provides only a first glimpse of how preproteins might bind, but to reveal the molecular details of this interaction [9], a structure of SecB with bound polypeptide substrate is required. Specific mutations in *E. coli* SecB (C76Y, V78F, and Q80R) disrupt the interaction between SecB and preproteins [67]. In these mutants, which show only mild translocation defects, the tetramer-dimer equilibrium in solution is shifted towards the dimer [68]. In the SecB structure, these residues are located on a surface-exposed β-strand that on one face points towards the core of the molecule [9, 49]. The mutated residues are not part of the peptide-binding groove, nor do they directly contribute to the dimer-dimer interface. However, these residues appear crucial for tight packing that involves interactions between the β-sheet and long α-helices that are at the dimer-dimer interface (Figure 27.1b) [9, 49]. It is conceivable that the mutations destabilize this interface and thereby impair preprotein binding while leaving the SecB dimer intact. Although this process is induced by mutations, it may also have some functional significance for the SecB-mediated preprotein transfer reaction towards SecA, as will be discussed in the following section.

As to the physiological role of SecB in preventing aggregation of nonnative proteins, it is unclear to what extent the cell takes advantage of this general chaperone activity. During protein binding, there is a rapid equilibrium between bound and free states that allows partitioning of the polypeptide between folding and rebinding to SecB. As the signal sequence retards folding of the mature region of proteins, it will favor re-association with SecB and thus indirectly contribute to the specificity. However, the dedicated role of SecB in protein translocation more likely results from events downstream of the SecB-preprotein interaction, in particular the association of SecA with SecB and the exposed signal sequence [8]. In the past, suppressor mutants have been selected that restore the translocation of preproteins with a defective or even missing signal sequence. Such mutations reside in the core components of the preprotein translocase, in particular SecY (PrlA) and SecA (PrlD) [69]. Strikingly, translocation of such preprotein substrates in PrlA mutants is remarkably dependent on SecB, even though the original preprotein substrate may not have been SecB-dependent for translocation [70–73]. This phenomenon has been attributed to the targeting function of SecB rather than the unfolding activity [74]. The supportive targeting function may be more prominently required when the intrinsic targeting information in the signal sequence of preproteins is defective.

Although SecB fulfils its function as a secretion-specific chaperone that tightly interlinks with the Sec system, it is also involved in the secretion of HasA, a hemophore of the heme acquisition system in *Serratia marcescens* [75]. This protein is secreted by a specific ABC transporter. Inactivation of SecB abolishes HasA secretion, and this relates to the ability of SecB to maintain the translocation competence of HasA [76]. Point mutations in SecB that affect the interaction with SecA (see Section 27.3) are without effect, but mutations that disrupt the SecB-preprotein interaction also block HasA secretion [77]. HasA is a rapidly folding protein, and binding to SecB results in a nearly complete inhibition of folding. Slowly folding mutants of HasA are secreted essentially independently from SecB. This demonstrates that the anti-folding function of SecB is required for the HasA ABC secretion pathway. It is not known whether SecB also fulfils a function in the targeting of HasA, but so far, this appears as a unique requirement that is not observed with other polypeptide substrates of ABC transporters.

27.3 SecA-SecB Interaction

An essential aspect of the SecB function is its affinity for SecA, the ATPase motor domain of the preprotein translocase. SecB and SecA interact with low affinity in solution ($K_d = \sim 1.6$ μM) [78], but when SecA is bound to the protein-conducting SecYEG channel, it is primed for the high-affinity interaction with SecB ($K_d =$ 10–30 nM) [7, 8]. This high-affinity interaction directs the binary SecB-preprotein complex to the translocation site at the membrane, which is instrumental for the targeting function of SecB. A study with SecA deletion mutants and fusion proteins revealed that the site on SecA that binds SecB comprises only the highly conserved carboxyl-terminal 22-amino-acyl residues of SecA [8, 79]. This region is conserved in the SecA proteins of most bacteria, even though they lack a SecB protein. In bacteria such as *Streptomyces*, *Mycobacteria*, and *Mycoplasma*, the carboxyl-terminus consists of a non-conserved polypeptide stretch with a high abundance of positively charged amino acyl residues [16]. Strikingly, the conserved carboxyl-terminal domain of SecA is essential for viability [79] even though SecB is not! This apparent discrepancy likely relates to the reduced ability of SecB to transfer preproteins to SecA when the SecB-binding frame in the carboxyl-terminus is missing, whereas in the absence of SecB, proteins can rely on the signal sequence for targeting. The SecB-binding region of SecA is rich in arginyl and lysyl residues, resulting in a net positive charge. The carboxyl-terminus therefore also shows a high tendency to electrostatically associate with the lipid surface [79], although the functional significance of this phenomenon is not clear. In addition, the carboxyl-terminus of SecA contains conserved cysteine and histidine residues that coordinate a divalent zinc ion required for the functional interaction between SecB and SecA [80]. The crystal structure of SecB has been solved in complex with the carboxyl-terminal 27-amino-acyl residues from *H. influenzae* SecA (Figure 27.3) [81]. This SecA peptide is highly structured as a CCCH zinc-binding motif (Figure 27.3c). Mutagenesis of the cysteine residues, and thus likely the loss of zinc binding, results in defective SecB-mediated preprotein translocation [82], which has been attributed to the deficient interaction between SecA and SecB [8]. The high content of proline and glycine residues in this conserved stretch provides the conformational flexibility needed to coordinate the zinc ion. Importantly, the structure shows one SecB tetramer that is bound by two SecA peptides (Figure 27.3a) [81]. The positively charged SecA peptide interacts with an acidic surface on SecB that consists of the eight-stranded β-sheet formed by two monomers (Figure 27.3b). The interactions at the interface are predominantly electrostatic and involve extensive intermolecular salt bridges and hydrogen-bonding interactions. Four residues from the SecA peptide contribute to the high-affinity interaction: one neutral polar residue (Asn882) and three positively charged residues (Arg881, Lys891, and Lys893) (residue numbering according to the *E. coli* SecA) (Figure 27.3c). Site-directed mutagenesis of these residues into Ala disrupts the high affinity SecB-SecA peptide interaction but does not interfere with the ability of the SecA peptide to bind zinc [81]. On the other hand, the well-conserved amino acid residues Asp20, Glu24, Leu75, and Glu77 of the *E. coli* SecB are important for the high-affinity SecB-SecA interaction [3, 67]. Mutations of these residues cause disruption of the high-affinity binding of SecB to the membrane-bound SecA, but they do not interfere with preprotein binding [74]. The corresponding residues in *H. influenzae* SecB are spatially clustered on the flat solvent-exposed surface on both sides of the SecB tetramer [9]. The structural analysis confirms that this surface is negatively charged and that it electrostatically interacts with the SecA peptide that corresponds to the SecB-binding frame on SecA. The high degree of conservation of the interacting residues on both SecA and SecB supports the notion that the interaction between SecA and SecB is highly specific. In general, disruption of this high-

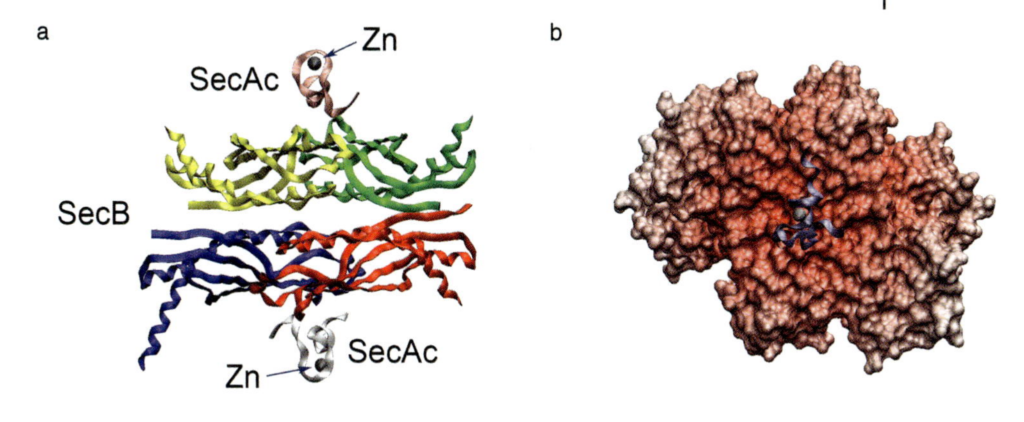

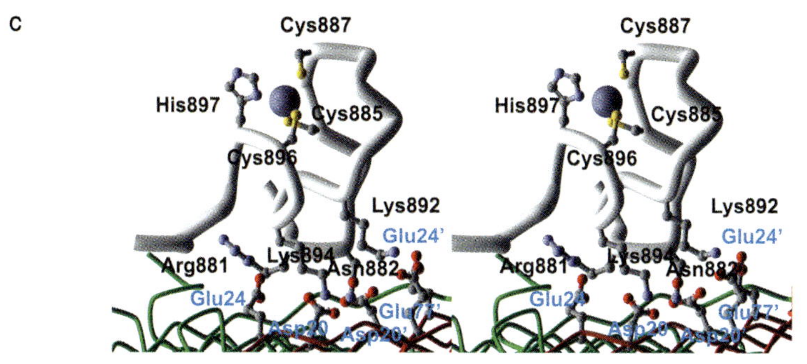

Fig. 27.3. Structural basis for the high-affinity SecB-SecA interaction. (a) Structure of *H. influenzae* SecB in complex with peptides that correspond to the carboxyl-terminal SecB-binding domain of SecA. SecB is shown in the same orientation as in Figure 27.1b, using identical coloring. The two SecA carboxyl-terminal peptides (residues 876–899, in white and pink) both contain a zinc ion, which is shown as a silver sphere. The figure was generated using the VMD software package [94] and is based on the coordinates deposited to the Protein Data Bank as 1OZB [81]. (b) Electrostatic interaction between SecB and the SecB-binding peptide of SecA. The solvent-accessible surface of the tetrameric SecB, shown as a front view that constitutes the SecA-binding site, is drawn and colored according to the surface potential (ranging from −30 to +30 kT); red: negative electrostatic potential; blue: positive electrostatic potential). The SecA peptide is show as a ribbon diagram with the central zinc ion. (c) A magnified stereo view of the coordination of the zinc ion by the SecA peptide and the specific interactions between SecB and the SecA carboxyl-terminal peptide. Residues of SecB and SecA involved in the interaction and zinc coordination are shown as ball and stick models. Residues of SecA are indicated in black, and residues of SecB are indicated in blue. Numbering is according to the *E. coli* SecA and SecB proteins.

affinity interaction leads to a loss of the targeting function of SecB and, consequently, a defect in preprotein translocation [74]. While wild-type SecB promotes preprotein translocation through its dual function, i.e., maintenance of the preprotein translocation-competent state and targeting to SecA, SecB mutants defective in

SecA interaction are inhibitory to translocation [74]. Although these mutants still maintain the preprotein in a translocatio-competent state, they are unable to release the preprotein to SecA (see Section 27.4).

The low-affinity SecB-SecA interaction in solution [78] requires neither the carboxyl terminus of SecA nor the negatively charged binding cluster on SecB [83]. At present, the functional significance of this low-affinity interaction for protein translocation is unclear. This will first require the selection of mutants specifically impaired in low-affinity binding. The 13-amino-acid carboxyl terminus of SecB has also been implicated in SecA binding [84]. Unfortunately, this solvent-exposed acidic region is barely resolved in the SecB structure, consistent with its being highly mobile. Interestingly, the (incomplete) α-helical carboxyl termini protrude out of the core structure as long arms, as if they embrace the carboxyl-terminal SecB-binding site on SecA. The crystals of the SecB-SecA peptide complex showed two SecB tetramers in the asymmetric unit, one with bound SecA peptide and one without [81]. This allowed comparison of two states; however, the structures revealed no significant conformational changes except for conformational variations in the amino- and carboxyl-terminal tails of SecB that extend away from the main SecB core and that are not in contact with the SecA peptide.

27.4
Preprotein Transfer from SecB to SecA

Another critical feature of the SecB function is that it mediates the transfer of the preprotein to SecA. In the presence of a preprotein, SecB more tightly interacts with the SecYEG-bound SecA [8]. This reaction can be mimicked with synthetic signal sequences, and it is believed that the increased SecA-SecB-binding affinity is triggered by the binding of the exposed signal sequence region of the SecB-bound preprotein to SecA. Due to the tighter interaction, the preprotein is dislocated from SecB and transferred to SecA [8]. The structural basis for this transfer reaction is unknown. Although the quaternary structure of SecB with and without the SecA peptide provides a nice solution to the dilemma of how the SecB tetramer binds the homodimeric SecA, it shows only a glimpse of how the preprotein may be transferred from SecB to SecA. Each SecA subunit within the dimer interacts with the negatively charged surface formed by one dimer of SecB [8, 81]. The stoichiometry of the SecA-SecB-preprotein ternary complex is 2:4:1. A docking model has been proposed based on the *H. influenzae* SecB [9] and *B. subtilis* SecA [85] structure, although with the latter the carboxyl-terminal residues are missing, including a region that is believed to function as a conformational flexible linker that connects the conserved SecB-binding domain with the SecA core. In the docking structure [81], the two conserved SecA zinc-binding carboxyl-termini clamp the SecB molecule such that the preprotein-binding channels are sandwiched between SecA and SecB. These channels are in juxtaposition close to the putative preprotein-binding domain on SecA. This spatial organization may facilitate the binding of the exposed preprotein signal sequence to SecA. Once this is achieved, the preprotein will be released from SecB, whereupon it is transferred to SecA. How this tighter SecB-SecA interaction is accomplished upon signal sequence binding by SecA is not known, but it may require additional interactions between SecA and SecB. Also, it is not known how the tighter binding elicits a reduced binding affinity of SecB for preproteins. It is of interest to note that some of the mutations that interfere with preprotein binding and the tetramer-dimer equilibrium [67, 68] map at different faces of a β-strand that contacts the long α-helices constituting the dimer-dimer interface of SecB [9, 49]. Hypothetically, the high-affinity binding of SecA to SecB could cause a reduction in the preprotein binding affinity, as may influencing the SecB dimer-dimer interfacial contact much akin to the mutations that shift the tetramer-dimer equilibrium. Loosing of the dimer-dimer interaction will result in a destabilization of the peptide-binding groove and thus cause the release the preprotein. In this respect, when bound with high affinity to SecA, SecB exhibits no affinity for the mature preprotein domain [8], which suggests that the peptide-binding groove indeed has an altered conformation in this interacting state. This mechanism also ensures that as long as SecB remains engaged with the carboxyl terminus of SecA, re-occupancy of the peptide-binding groove of SecB will not be possible. Release of SecB from SecA is directly coupled to translocation. Upon the binding of ATP to SecA, preprotein translocation is initiated and concomitantly results in the release of SecB from the membrane [8]. The structural basis of this event is also unknown, but it may simply reflect a reversal of the tightened SecA-SecB interaction that occurs when SecA releases the preprotein signal sequence domain to the SecYEG channel. It is noteworthy that the formation of SecA-SecB complexes results in a small stimulation of the SecA ATPase activity [86, 87], which may facilitate the SecB release and preprotein translocation initiation reaction. Figure 27.4 summarizes the ordered reactions that result in the targeting and transfer of preprotein from SecB to the SecYEG-bound SecA protein.

27.5 Concluding Remarks

SecB is a rather peculiar chaperone. It uses its entire molecular surface for its two key functions, i.e., the binding of unfolded preproteins and of SecA. However, for preprotein release and its own membrane release, SecB relies on the catalytic activity of SecA. This defines SecB as a translocation-specific molecular chaperone, and maybe even as a co-chaperone, as it has been suggested that SecA bears chaperone activity on its own [88]. The recent insights into the structure-function relationship of the SecB-preprotein and the SecB-SecA interaction have greatly increased our knowledge about the mechanism of chaperone-mediated protein targeting in bacteria. However, the exact mechanisms of preprotein recognition and selectivity have remained elusive, while the structural basis for the preprotein transfer reaction to SecA remains to be solved. These will be challenges for the future.

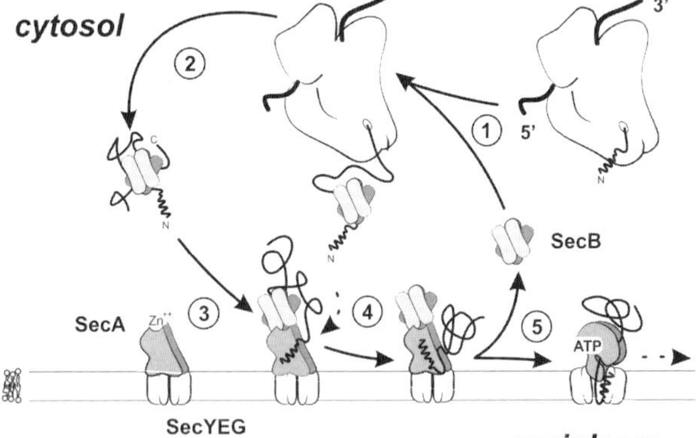

Fig. 27.4. Catalytic model SecB preprotein targeting cycle. Cytosolic SecB binds to the mature domain of a nascent preprotein (step 1) and stabilizes its unfolded state (step 2). The binary SecB-preprotein complex is targeted to the SecYEG-bound SecA (step 3) or, alternatively, first associates with cytosolic SecA through a low-affinity interaction (not shown) before it is targeted to the membrane to replace the SecYEG-bound SecA. Targeting requires the high-affinity binding of SecB to the carboxyl terminus of SecA (step 3). Binding of the signal sequence to SecA tightens the SecB-SecA interaction and elicits the release of the preprotein from the SecB-bound state with the concomitant transfer to SecA (step 4). Upon binding of ATP by SecA, preprotein translocation is initiated and SecB is released from the ternary complex (step 5) and can rebind a newly synthesized preprotein in the cytosol.

27.6 Experimental Protocols

27.6.1 How to Analyze SecB-Preprotein Interactions

The SecB-preprotein interaction is most conveniently assayed by utilizing a radiolabeled preprotein unfolded in urea or guanidine buffer, whereupon it is added to a suspension containing His-tagged SecB. The binary SecB-preprotein complex can than be collected by a batch-type of Ni-NTA affinity chromatography. Alternatively, SecB-preprotein complexes can be recovered by sucrose gradient centrifugation [39].

Materials

^{35}S-proOmpA is synthesized by an in vitro transcription/translation reaction using plasmid pET149 [89], affinity purified with OmpA antibodies [90], and solubilized in an unfolded state in 6 M urea and 50 mM TrisHCl, pH 7.8. His-tagged SecB is purified from *E. coli* MM152 (MC4100 *secB::Tn5 zhe::Tn10 mal Tcon*; [2]) transformed with plasmid pHKSB366 that contains the His-tagged *secB* gene under control of the *trc* promoter [74]. Cells are grown on Luria broth at 37 °C to an optical density of 1.0, induced with 1 mM IPTG, and grown for another 2 h. Next, the cells are harvested by centrifugation (10 min, 10 000 g) and passed twice through a French pressure cell (18 000 psi) in 50 mM NaP_i, pH 8.0. After removal of cell debris by another centrifugation step (15 min, 30 000 g), the supernatant is supplemented with 60 mM imidazole, pH 8.0, and applied onto a Ni-NTA column that was pretreated according to the manufacturer's instructions (Pharmacia, Biotech). The column is washed with several column volumes of 50 mM NaP_i and 60 mM imidazole, pH 8.0. Next, His-tagged SecB protein is eluted with a buffer containing 50 mM NaP_i, 500 mM imidazole, and 300 mM NaCl, pH 6.0. After overnight dialysis against 50 mM TrisHCl, pH 7.6, at 4 °C, the His-tagged SecB can be further purified by MonoQ FPLC anion-exchange chromatography [8].

Methods

^{35}S-proOmpA (20 nM) is 50-fold diluted into 50 μL of buffer A (50 mM HEPES-KOH, pH 7.5, 0.5 mg mL^{-1} BSA) containing His-tagged SecB (100 nM). The suspension is incubated on ice, and after 30 min 50 μL of prewashed Ni-NTA agarose slurry in buffer A is added. Incubation is continued on ice for 25 min with regular vortexing. Subsequently, the bead suspension is centrifuged (5 min, 3000 rpm) through a 200 μL sucrose cushion (0.2 M sucrose in buffer A) to separate the SecB-bound from free preprotein fraction. The pellet is washed once with 150 μL buffer A, and the pellet and combined supernatant fractions are collected and quantitated by liquid scintillation counting.

26.6.2
How to Analyze SecB-SecA Interaction

High-affinity interaction between SecB and SecA is observed only when SecA is bound to the SecYEG complex. Therefore, inner-membrane vesicles (IMVs) need to be used, preferentially containing overexpressed levels of SecYEG complex because of the greater number of binding sites that provide an increased sensitivity of the assay. These IMVs are prepared by French pressure treatment, which results in closed membranes vesicles with an inside-out orientation, i.e., the SecA-binding site of the SecYEG complex faces the outer surface of the vesicles. To these membranes, saturating amounts of SecA are added. Next, radio-labeled SecB is added and the membrane-bound and free SecB fractions are separated by centrifugation of the membranes through a sucrose cushion. Binding of SecB to the membranes is strictly SecA-dependent, and when an appropriate SecB concentration range is chosen, the binding reaction can be quantitated and analyzed by Scatchard analysis.

Materials

His-tagged SecB and SecA [91] are purified as described. Untagged SecB is purified from *E. coli* BL21 transformed with plasmid pJW25, which contains *secB* gene under control of the T7 promoter [6]. Cells are grown on Luria broth at 30 °C to an optical density of 1.0, induced with 1 mM IPTG, and grown for another 2 h. Next, the cells are harvested by centrifugation (10 min, 10 000 g), resuspended in 20 mM TrisHCl, pH 7.4, and passed twice through a French pressure cell (18 000 psi). After removal of cell debris by a centrifugation step (15 min, 30 000 g), membranes are removed by ultracentrifugation (1 h, 135 000 g). The cell-free extract is applied onto a Hi-Trap Q-Sepharose column equilibrated with 20 mM TrisHCl and 150 mM NaCl. The column is washed with seven volumes of 20 mM TrisHCl, pH 7.4, and 180 mM NaCl, and the bound SecB is eluted with 20 column volumes of a linear salt gradient (180–500 mM NaCl, 20 mM TrisHCl, pH 7.4). Fractions are analyzed on SDS-PAGE, pooled, and diluted with one volume of 20 mM TrisHCl, pH 7.4. To concentrate the sample, protein is applied onto a Hi-Trap Q-Sepharose column and step-eluted with 20 mM TrisHCl, pH 7.4, and 500 mM NaCl. Fractions containing concentrated SecB are pooled and applied onto a Sephacryl S300 gel filtration column (XK16/100) equilibrated with 20 mM TrisHCl, pH 7.4, and 150 mM NaCl. Fractions containing pure SecB are pooled and stored at −80 °C or concentrated by an additional anion-exchange step on a MonoQ 5/5 column.

IMVs are prepared from SF100 cells transformed with plasmid pET340, which allows for overproducing the SecYEG complex [92] by means of French pressure lysis and differential centrifugation, and treated with 6 M urea and 50 mM TrisHCl, pH 7.8 [93]. Purified SecB is labeled with carrier-free ^{125}I (Radiochemical Centre, Amersham, U.K.) to a specific activity of about 5×10^4 cpm pmol^{-1} according to the following procedure: SecB (100 μg) in buffer A (50 mM TrisHCl, pH 8.0, 50 mM KCl, 5 mM $MgCl_2$, and 10% glycerol) is transferred to a reaction vial coated with IODO-GEN Iodination Reagent (Pierce, Rockford, IL). The reaction is started by adding 2 μL of K^{125}I (200 μCi), incubated for 15 min at room temperature, and terminated by transferring the mixture into a new reaction vial containing dithiothreitol (DTT) at a final concentration of 10 mM. Free iodine is removed by chromatography on a PD-10 Sephadex column (Amersham Biotech AB, Uppsala, Sweden) which is prewashed with buffer A containing 1 mM DTT.

Methods

Binding of ^{125}I SecB to IMVs can be assayed essentially as described previously [7]. Urea-treated IMVs (10 μg of membrane protein) are suspended in 100 μL of buffer B (50 mM HEPES-KOH, pH 7.6, 30 mM KCl, 0.5 mg mL^{-1} BSA, 10 mM DDT, 2 mM Mg-acetate) and incubated for 20 min at room temperature with 0–50 nM ^{125}I-SecB in the presence or absence of 100 nM SecA. After 20 min, the solution containing the IMVs is layered on top of a 100-μL 0.2-M sucrose cushion in buffer B and spun with a Beckman Airfuge (30 psi, 10 min, 4 °C). The amount of ^{125}I-SecB in the supernatant and pellet is quantitated by liquid scintillation counting, and the data are analyzed by nonlinear regression assuming biphasic-binding kinetics. The biphasic kinetics results from a major high-affinity binding reaction that is SecA- and SecYEG-dependent and from a low, nonspecific component that is due to the non-saturable and SecA- and SecYEG-independent association of SecB with the membrane surface.

Acknowledgements

This work was supported by the Council for Chemical Sciences of the Netherlands Organization for Scientific Research and the Earth and Life Sciences Foundation, which are subsidized by the Dutch Organization for the Advancement of Scientific Research, the Royal Dutch Academy of Sciences (to N.N.) and by EC RTN contract HPRN-2000-00075. The authors wish to thank members of the protein secretion group for valuable discussion.

References

1 C. A. Kumamoto, J. Beckwith (1983). Mutations in a new gene, *secB*, cause defective protein localization in *Escherichia coli*. *J. Bacteriol.* **154**, 253–260.

2 C. A. Kumamoto, J. Beckwith (1985). Evidence for specificity at an early step in protein export in *Escherichia coli*. *J. Bacteriol.* **163**, 267–274.

3 P. M. Gannon, C. A. Kumamoto (1993). Mutations of the molecular chaperone protein SecB which alter the interaction between SecB and maltose-binding protein. *J. Biol. Chem.* **268**, 1590–1595.

4 D. N. Collier, V. A. Bankaitis, J. B. Weiss, P. J. Bassford, Jr. (1988). The antifolding activity of SecB promotes the export of the *E. coli* maltose-binding protein. *Cell* **53**, 273–283.

5 L. L. Randall, S. J. Hardy (1986). Correlation of competence for export with lack of tertiary structure of the mature species: a study *in vivo* of maltose-binding protein in *E. coli*. *Cell* **46**, 921–928.

6 J. B. Weiss, P. H. Ray, P. J. Bassford, Jr. (1988). Purified *secB* protein of *Escherichia coli* retards folding and promotes membrane translocation of the maltose-binding protein *in vitro*. *Proc. Natl. Acad. Sci. U.S.A.* **85**, 8978–8982.

7 F. U. Hartl, S. Lecker, E. Schiebel, J. P. Hendrick, W. Wickner (1990). The binding cascade of SecB to SecA to SecY/E mediates preprotein targeting to the *E. coli* plasma membrane. *Cell* **63**, 269–279.

8 P. Fekkes, C. van der Does, A. J. M. Driessen (1997). The molecular chaperone SecB is released from the carboxy-terminus of SecA during initiation of precursor protein translocation. *EMBO J.* **16**, 6105–6113.

9 Z. Xu, J. D. Knafels, K. Yoshino (2000). Crystal structure of the bacterial protein export chaperone secB. *Nat. Struct. Biol.* **7**, 1172–1177.

10 W. Wickner, A. J. M. Driessen, F. U. Hartl (1991). The enzymology of protein translocation across the *Escherichia coli* plasma membrane. *Annu. Rev. Biochem.* **60**, 101–124.

11 Q. A. Valent, P. A. Scotti, S. High, J. W. de Gier, G. von Heijne, G. Lentzen, W. Wintermeyer, B. Oudega, J. Luirink (1998). The *Escherichia coli* SRP and SecB targeting pathways converge at the translocon. *EMBO J.* **17**, 2504–2512.

12 L. Brundage, J. P. Hendrick, E. Schiebel, A. J. M. Driessen, W. Wickner (1990). The purified *E. coli* integral membrane protein SecY/E is sufficient for reconstitution of SecA-dependent precursor protein translocation. *Cell* **62**, 649–657.

13 A. J. M. Driessen (2001). SecB, a molecular chaperone with two faces. *Trends Microbiol.* **9**, 193–196.

14 L. L. Randall, S. J. Hardy (2002). SecB, one small chaperone in the complex milieu of the cell. *Cell Mol. Life Sci.* **59**, 1617–1623.

15 P. Fekkes, A. J. M. Driessen (1999). Protein targeting to the bacterial cytoplasmic membrane. *Microbiol. Mol. Biol. Rev.* **63**, 161–173.

16 K. H. M. van Wely, J. Swaving, M. Klein, R. Freudl, A. J. M. Driessen (2000). The carboxyl terminus of the Bacillus subtilis SecA is dispensable for protein secretion and viability. *Microbiology* **146**, 2573–2581.

17 K. H. M. van Wely, J. Swaving, R. Freudl, A. J. M. Driessen (2001). Translocation of proteins across the cell envelope of Gram-positive bacteria. *FEMS Microbiol. Rev.* **25**, 437–454.

18 H. Shimizu, K. Nishiyama, H. Tokuda (1997). Expression of *gpsA* encoding biosynthetic sn-glycerol 3-phosphate dehydrogenase suppresses both the LB-phenotype of a *secB* null mutant and the cold-sensitive phenotype of a *secG* null mutant. *Mol. Microbiol.* **26**, 1013–1021.

19 E. Altman, C. A. Kumamoto, S. D. Emr (1991). Heat-shock proteins can substitute for SecB function during protein export in *Escherichia coli*. *EMBO J.* **10**, 239–245.

20 J. Wild, E. Altman, T. Yura, C. A. Gross (1992). DnaK and DnaJ heat shock proteins participate in protein export in *Escherichia coli*. *Genes Dev.* **6**, 1165–1172.

21 J. Wild, P. Rossmeissl, W. A. Walter, C. A. Gross (1996). Involvement of the DnaK-DnaJ-GrpE chaperone team in protein secretion in *Escherichia coli*. *J. Bacteriol.* **178**, 3608–3613.

22 C. A. Kumamoto (1989). *Escherichia coli* SecB protein associates with exported protein precursors *in vivo*. *Proc. Natl. Acad. Sci. U.S.A.* **86**, 5320–5324.

23 C. A. Kumamoto, O. Francetic (1993). Highly selective binding of nascent polypeptides by an *Escherichia coli* chaperone protein *in vivo*. *J. Bacteriol.* **175**, 2184–2188.

24 L. L. Randall, T. B. Topping, S. J. Hardy, M. Y. Pavlov, D. V. Freistroffer, M. Ehrenberg (1997). Binding of SecB to ribosome-bound polypeptides has the same characteristics as binding to full-length, denatured proteins. *Proc. Natl. Acad. Sci. U.S.A.* **94**, 802–807.

25 S. J. Hardy, L. L. Randall (1991). A kinetic partitioning model of selective binding of nonnative proteins by the bacterial chaperone SecB. *Science* **251**, 439–443.

26 V. G. Panse, J. B. Udgaonkar, R. Varadarajan (1998). SecB binds only to a late native-like intermediate in the folding pathway of barstar and not to the unfolded state. *Biochemistry* **37**, 14477–14483.

27 R. Zahn, S. Perrett, A. R. Fersht (1996). Conformational states bound by the molecular chaperones GroEL and secB: a hidden unfolding (annealing) activity. *J. Mol. Biol.* **261**, *261*, 43–61.

28 G. Stenberg, A. R. Fersht (1997). Folding of barnase in the presence of the molecular chaperone SecB. *J. Mol. Biol.* **274**, *274*, 268–275.

29 R. Zahn, S. Perrett, G. Stenberg, A. R. Fersht (1996). Catalysis of amide proton exchange by the molecular chaperones GroEL and SecB. *Science* **271**, *271*, 642–645.

30 V. G. Panse, C. P. Swaminathan, A. Surolia, R. Varadarajan (2000). Thermodynamics of substrate binding to the chaperone SecB. *Biochemistry* **39**, 2420–2427.

31 P. Fekkes, T. den Blaauwen, A. J. M. Driessen (1995). Diffusion-limited interaction between unfolded polypeptides and the *Escherichia coli* chaperone SecB. *Biochemistry* **34**, 10078–10085.

32 N. L. Vekshin (1998). Protein sizes and stoichiometry in the chaperone SecB-RBPTI complex estimated by ANS fluorescence. *Biochemistry (Mosc)*. **63**, 485–488.

33 L. L. Randall, T. B. Topping, D. Suciu, S. J. Hardy (1998). Calorimetric analyses of the interaction between SecB and its ligands. *Protein Sci.* **7**, 1195–1200.

34 S. Lecker, D. Meyer, W. Wickner (1989). Export of prepro-alpha-factor from *Escherichia coli*. *J. Biol. Chem.* **264**, 1882–1886.

35 E. L. Powers, L. L. Randall (1995). Export of periplasmic galactose-binding protein in *Escherichia coli* depends on the chaperone SecB. *J. Bacteriol.* **177**, 1906–1907.

36 J. E. Bruce, V. F. Smith, C. Liu, L. L. Randall, R. D. Smith (1998). The observation of chaperone-ligand noncovalent complexes with electrospray ionization mass spectrometry. *Protein Sci.* **7**, 1180–1185.

37 L. L. Randall (1992). Peptide binding by chaperone SecB: implications for recognition of nonnative structure. *Science* **257**, 241–245.

38 S. H. Lecker, A. J. M. Driessen, W. Wickner (1990). ProOmpA contains secondary and tertiary structure prior to translocation and is shielded from aggregation by association with SecB protein. *EMBO J.* **9**, 2309–2314.

39 S. Lecker, R. Lill, T. Ziegelhoffer, C. Georgopoulos, P. J. Bassford, Jr., C. A. Kumamoto, W. Wickner (1989). Three pure chaperone proteins of *Escherichia coli*-SecB, trigger factor and GroEL-form soluble complexes with precursor proteins *in vitro*. *EMBO J.* **8**, 2703–2709.

40 R. Kusters, T. de Vrije, E.

Breukink, B. de Kruijff (1989). SecB protein stabilizes a translocation-competent state of purified prePhoE protein. *J. Biol. Chem.* **264**, 20827–20830.

41 D. L. Diamond, S. Strobel, S. Y. Chun, L. L. Randall (1995). Interaction of SecB with intermediates along the folding pathway of maltose-binding protein. *Protein Sci.* **4**, 1118–1123.

42 P. M. Gannon, P. Li, C. A. Kumamoto (1989). The mature portion of *Escherichia coli* maltose-binding protein (MBP) determines the dependence of MBP on SecB for export. *J. Bacteriol.* **171**, 813–818.

43 G. Liu, T. B. Topping, L. L. Randall (1989). Physiological role during export for the retardation of folding by the leader peptide of maltose-binding protein. *Proc. Natl. Acad. Sci. U.S.A.* **86**, 9213–9217.

44 G. P. Liu, T. B. Topping, W. H. Cover, L. L. Randall (1988). Retardation of folding as a possible means of suppression of a mutation in the leader sequence of an exported protein. *J. Biol. Chem.* **263**, 14790–14793.

45 V. F. Smith, S. J. Hardy, L. L. Randall (1997). Determination of the binding frame of the chaperone SecB within the physiological ligand oligopeptide-binding protein. *Protein Sci.* **6**, 1746–1755.

46 V. J. Khisty, G. R. Munske, L. L. Randall (1995). Mapping of the binding frame for the chaperone SecB within a natural ligand, galactose-binding protein. *J. Biol. Chem.* **270**, 25920–25927.

47 T. B. Topping, L. L. Randall (1994). Determination of the binding frame within a physiological ligand for the chaperone SecB. *Protein Sci.* **3**, 730–736.

48 N. T. Knoblauch, S. Rüdiger, H.-J. Schönfeld, A. J. M. Driessen, J. Schneider-Mergener, B. Bukau (1999). Substrate specificity of the SecB chaperone. *J. Biol. Chem.* **274**, 34219–34225.

49 C. Dekker, B. de Kruijff, P. Gros (2003). Crystal structure of SecB from *Escherichia coli*. *J. Struct. Biol.* **144**, 313–319.

50 T. B. Topping, R. L. Woodbury, D. L. Diamond, S. J. Hardy, L. L. Randall (2001). Direct demonstration that homotetrameric chaperone SecB undergoes a dynamic dimer-tetramer equilibrium. *J. Biol. Chem.* **276**, 7437–7441.

51 V. G. Panse, C. P. Swaminathan, J. J. Aloor, A. Surolia, R. Varadarajan (2000). Unfolding thermodynamics of the tetrameric chaperone, SecB. *Biochemistry* **39**, 2362–2369.

52 J. Kim, D. A. Kendall (1998). Identification of a sequence motif that confers SecB dependence on a SecB-independent secretory protein *in vivo*. *J. Bacteriol.* **180**, 1396–1401.

53 J. Kim, J. Luirink, D. A. Kendall (2000). SecB dependence of an exported protein is a continuum influenced by the characteristics of the signal peptide or early mature region. *J. Bacteriol.* **182**, 4108–4112.

54 D. L. Diamond, L. L. Randall (1997). Kinetic partitioning. Poising SecB to favor association with a rapidly folding ligand. *J. Biol. Chem.* **272**, 28994–28998.

55 T. B. Topping, L. L. Randall (1997). Chaperone SecB from *Escherichia coli* mediates kinetic partitioning via a dynamic equilibrium with its ligands. *J. Biol. Chem.* **272**, 19314–19318.

56 V. J. Khisty, L. L. Randall (1995). Mapping of the binding frame for the chaperone SecB within a natural ligand, galactose-binding protein. *J. Bacteriol.* **177**, 3277–3282.

57 M. Watanabe, G. Blobel (1995). High-affinity binding of *Escherichia coli* SecB to the signal sequence region of a presecretory protein. *Proc. Natl. Acad. Sci. U.S.A.* **92**, 10133–10136.

58 M. Watanabe, G. Blobel (1989). SecB functions as a cytosolic signal recognition factor for protein export in *E. coli*. *Cell* **58**, 695–705.

59 L. L. Randall, T. B. Topping, S. J. Hardy (1990). No specific recognition of leader peptide by SecB, a chaperone involved in protein export. *Science* **248**, 860–863.

60 S. Park, G. Liu, T. B. Topping, W. H.

Cover, L. L. Randall (1988). Modulation of folding pathways of exported proteins by the leader sequence. *Science* **239**, 1033–1035.

61 E. Altman, S. D. Emr, C. A. Kumamoto (1990). The presence of both the signal sequence and a region of mature LamB protein is required for the interaction of LamB with the export factor SecB. *J. Biol. Chem.* **265**, 18154–18160.

62 G. Eisner, H. G. Koch, K. Beck, J. Brunner, M. Muller (2003). Ligand crowding at a nascent signal sequence. *J. Cell Biol.* **163**, 35–44.

63 K. Beck, L. F. Wu, J. Brunner, M. Muller (2000). Discrimination between SRP- and SecA/SecB-dependent substrates involves selective recognition of nascent chains by SRP and trigger factor. *EMBO J.* **19**, 134–143.

64 Q. A. Valent, D. A. Kendall, S. High, R. Kusters, B. Oudega, J. Luirink (1995). Early events in preprotein recognition in *E. coli*: interaction of SRP and trigger factor with nascent polypeptides. *EMBO J.* **14**, 5494–5505.

65 T. Hesterkamp, S. Hauser, H. Lutcke, B. Bukau (1996). *Escherichia coli* trigger factor is a prolyl isomerase that associates with nascent polypeptide chains. *Proc. Natl. Acad. Sci. U.S.A.* **93**, 4437–4441.

66 L. L. Randall, S. J. Hardy, T. B. Topping, V. F. Smith, J. E. Bruce, R. D. Smith (1998). The interaction between the chaperone SecB and its ligands: evidence for multiple subsites for binding. *Protein Sci.* **7**, 2384–2390.

67 H. H. Kimsey, M. D. Dagarag, C. A. Kumamoto (1995). Diverse effects of mutation on the activity of the *Escherichia coli* export chaperone SecB. *J. Biol. Chem.* **270**, 22831–22835.

68 E. M. Muren, D. Suciu, T. B. Topping, C. A. Kumamoto, L. L. Randall (1999). Mutational alterations in the homotetrameric chaperone SecB that implicate the structure as dimer of dimers. *J. Biol. Chem.* **274**, 19397–19402.

69 P. N. Danese, T. J. Silhavy (1998). Targeting and assembly of periplasmic and outer-membrane proteins in *Escherichia coli*. *Annu. Rev. Genet.* **32**, 59–94.

70 O. Francetic, M. P. Hanson, C. A. Kumamoto (1993). *prlA* suppression of defective export of maltose-binding protein in *secB* mutants of *Escherichia coli*. *J. Bacteriol.* **175**, 4036–4044.

71 O. Francetic, C. A. Kumamoto (1996). *Escherichia coli* SecB stimulates export without maintaining export competence of ribose-binding protein signal sequence mutants. *J. Bacteriol.* **178**, 5954–5959.

72 N. J. Trun, J. Stader, A. Lupas, C. Kumamoto, T. J. Silhavy (1988). *prlA* suppression of defective export of maltose-binding protein in *secB* mutants of *Escherichia coli*. *J. Bacteriol.* **170**, 5928–5930.

73 A. I. Derman, J. W. Puziss, P. J. Bassford, Jr., J. Beckwith (1993). A signal sequence is not required for protein export in *prlA* mutants of *Escherichia coli*. *EMBO J.* **12**, 879–888.

74 P. Fekkes, J. G. de Wit, J. P. van der Wolk, H. H. Kimsey, C. A. Kumamoto, A. J. M. Driessen (1998). Preprotein transfer to the *Escherichia coli* translocase requires the co-operative binding of SecB and the signal sequence to SecA. *Mol. Microbiol.* **29**, 1179–1190.

75 P. Delepelaire, C. Wandersman (1998). The SecB chaperone is involved in the secretion of the *Serratia marcescens* HasA protein through an ABC transporter. *EMBO J.* **17**, 936–944.

76 L. Debarbieux, C. Wandersman (2001). Folded HasA inhibits its own secretion through its ABC exporter. *EMBO J.* **20**, 4657–4663.

77 G. Sapriel, C. Wandersman, P. Delepelaire (2002). The N terminus of the HasA protein and the SecB chaperone cooperate in the efficient targeting and secretion of HasA via the ATP-binding cassette transporter. *J. Biol. Chem.* **277**, 6726–6732.

78 T. den Blaauwen, E. Terpetschnig, J. R. Lakowicz, A. J. M. Driessen (1997). Interaction of SecB with soluble SecA. *FEBS Lett.* **416**, 35–38.

79 E. Breukink, N. Nouwen, A. van

Raalte, S. Mizushima, J. Tommassen, B. de Kruijff (1995). The C terminus of SecA is involved in both lipid binding and SecB binding. *J. Biol. Chem.* **270**, 7902–7907.

80 P. Fekkes, J. G. de Wit, A. Boorsma, R. H. Friesen, A. J. M. Driessen (1999). Zinc stabilizes the SecB binding site of SecA. *Biochemistry* **38**, 5111–5116.

81 J. Zhou, Z. Xu (2003). Structural determinants of SecB recognition by SecA in bacterial protein translocation. *Nat. Struct. Biol.* **10**, 942–947.

82 T. Rajapandi, D. Oliver (1994). Carboxy-terminal region of *Escherichia coli* SecA ATPase is important to promote its protein translocation activity *in vivo*. *Biochem. Biophys. Res. Commun.* **200**, 1477–1483.

83 R. L. Woodbury, T. B. Topping, D. L. Diamond, D. Suciu, C. A. Kumamoto, S. J. Hardy, L. L. Randall (2000). Complexes between protein export chaperone SecB and SecA. Evidence for separate sites on SecA providing binding energy and regulatory interactions. *J. Biol. Chem.* **275**, 24191–24198.

84 T. L. Volkert, J. D. Baleja, C. A. Kumamoto (1999). A highly mobile C-terminal tail of the *Escherichia coli* protein export chaperone SecB. *Biochem. Biophys. Res. Commun.* **264**, 949–954.

85 J. F. Hunt, S. Weinkauf, L. Henry, J. J. Fak, P. McNicholas, D. B. Oliver, J. Deisenhofer (2002). Nucleotide control of interdomain interactions in the conformational reaction cycle of SecA. *Science* **297**, 2018–2026.

86 J. Kim, A. Miller, L. Wang, J. P. Muller, D. A. Kendall (2001). Evidence that SecB enhances the activity of SecA. *Biochemistry* **40**, 3674–3680.

87 A. Miller, L. Wang, D. A. Kendall (2002). SecB modulates the nucleotide-bound state of SecA and stimulates ATPase activity. *Biochemistry* **41**, 5325–5332.

88 M. Eser, M. Ehrmann (2003). SecA-dependent quality control of intracellular protein localization. *Proc. Natl. Acad. Sci. U.S.A.* **100**, 13231–13234.

89 J. P. van der Wolk, P. Fekkes, A. Boorsma, J. L. Huie, T. J. Silhavy, A. J. M. Driessen (1998). PrlA4 prevents the rejection of signal sequence defective preproteins by stabilizing the SecA-SecY interaction during the initiation of translocation. *EMBO J.* **17**, 3631–3639.

90 E. Crooke, W. Wickner (1987). Trigger factor: a soluble protein that folds pro-OmpA into a membrane-assembly-competent form. *Proc. Natl. Acad. Sci. U.S.A.* **84**, 5216–5220.

91 R. J. Cabelli, L. Chen, P. C. Tai, D. B. Oliver (1988). SecA protein is required for secretory protein translocation into *E. coli* membrane vesicles. *Cell* **55**, 683–692.

92 C. van der Does, E. H. Manting, A. Kaufmann, M. Lutz, A. J. M. Driessen (1998). Interaction between SecA and SecYEG in micellar solution and formation of the membrane-inserted state. *Biochemistry* **37**, 201–210.

93 K. Cunningham, R. Lill, E. Crooke, M. Rice, K. Moore, W. Wickner, D. Oliver (1989). SecA protein, a peripheral protein of the *Escherichia coli* plasma membrane, is essential for the functional binding and translocation of proOmpA. *EMBO J.* **8**, 955–959.

94 W. Humphrey, A. Dalke, K. Schulten (1996). VMD: visual molecular dynamics. *J. Mol. Graph.* **14**, 33–38.

95 J. B. Weiss, P. J. Bassford, Jr. (1990). The folding properties of the *Escherichia coli* maltose-binding protein influence its interaction with SecB *in vitro*. *J. Bacteriol.* **172**, 3023–3029.

96 C. A. Kumamoto, L. Chen, J. Fandl, P. C. Tai (1989). Purification of the *Escherichia coli secB* gene product and demonstration of its activity in an *in vitro* protein translocation system. *J. Biol. Chem.* **264**, 2242–2249.

28
Protein Folding in the Periplasm and Outer Membrane of *E. coli*

Michael Ehrmann

28.1
Introduction

The cell envelope of Gram-negative bacteria such as *E. coli* is comprised of the periplasm and the outer membrane. The periplasm contains the proteoglycan layer providing cell shape and rigidity against osmotic stresses that often occur in the natural habitat of these organisms. The periplasm is thought to occupy about 30% of the cell volume and to have a width of around 50 nm [1]. However, these values can change according to the osmolarity of the growth medium. In addition, the periplasm is believed to be rather viscous and it is thought to have a higher concentration of protein than does the cytoplasm [2]. The outer membrane is, in contrast to the cytoplasmic membrane, asymmetric. It is composed of an inner leaflet of phospholipids and an outer leaflet containing the glycolipid lipopolysaccharide. Lipopolysaccharide comprises three components: a lipid A moiety containing sugar and multiple fatty acyl chains, a core oligosaccharide, and the O-antigen. The O-antigen is a complex polysaccharide consisting of repeats of oligosaccharide units. The cell envelope is responsible for communication with the environment, including, for example, nutrient uptake; exclusion of toxic compounds; interaction with host cells, other bacteria, or bacteriophages; and cell division. These functions are mediated by proteins whose folding, functionality, and abundance must be continuously monitored for optimal performance of the entire organism. Given the vast number of proteins and their different properties and abundance, protein quality control, repair, and degradation are challenging tasks. These issues have been elegantly solved by nature and are the focus of intensive research. This chapter describes the known cellular factors involved, the most commonly studied substrates, and the regulation of the unfolded protein response that coordinates important elements involved in the regulation of protein composition. The cell envelope of the gram-negative bacterium *E. coli* represents an experimental system that is well suited to address the important question of mechanism of protein quality control. The fact that most cellular factors involved in this process are nonessential and ATP independent is a major advantage, as both features simplify experimental approaches. Nonessential factors allow us to work with rather healthy

Protein Folding Handbook. Part II. Edited by J. Buchner and T. Kiefhaber

ISBN: 3-527-30784-2

mutants and stable genetic backgrounds. The absence of ATPase domains reduces the number of issues to consider and thus simplifies mutant construction and the interpretation of data. Another important property of the cell envelope is that it is oxidizing. Therefore, many proteins contain disulfide (S–S) bonds, providing additional structural stability.

Protein composition in *E. coli* can be derived from its genome sequence [3], indicating that about two-thirds of all proteins are localized in the cytoplasm, that 20% of all proteins are in the cytoplasmic membrane, and that roughly 10% of all proteins are sorted to the cell envelope. We estimate that about 150 different proteins are localized in the periplasm and about 320 in outer membrane. A comprehensive list of periplasmic and outer membrane proteins can be accessed via the ECCE database (http://www.cf.ac.uk/biosi/staff/ehrmann/tools/ecce/ecce.htm). The largest class of proteins in the periplasm is the substrate-binding proteins of ABC transporters. There are also enzymes that are involved in metabolism and murein biogenesis, proteins that are involved in regulation, and those that are part of large protein complexes involved in, for example, motility, cell division, and virulence. In addition there are quite a number of proteases, chaperones, and folding catalysts, such as proline isomerases and proteins involved in disulfide bond formation. The outer membrane has proteins that are involved in transport (the porins); enzymes such as proteases, lipases, and glucosidases; and components of export machines involved in detoxification and protein secretion. A common structural feature of outer membrane proteins is that their membrane-spanning segments are β-sheets, which is in contrast to cytoplasmic membrane proteins whose transmembrane segments adopt an α-helical secondary structure. Most outer membrane proteins contain a β-barrel domain that is integral to the membrane, and some proteins have additional domains either in the periplasm or on the cell surface, which can be quite extensive in size (for review, see Refs. [4–6]).

A common feature of periplasmic and outer membrane proteins is that they have to be initially translocated across the cytoplasmic membrane. There are two main routes for this purpose: via the classical secretion machinery (Sec) or via the twin arginine pathway (TAT) (For review, see Refs. [7–9]). About 420 proteins probably use the Sec route, while it is believed that the TAT pathway has about 30 substrates. These estimates were derived from inspection of the 452 signal sequences found in *E. coli* proteins for TAT motifs (see ECCE database). While substrates of the Sec system need to be unfolded during translocation, TAT substrates fold in the cytoplasm and are believed to be fully functional after translocation is completed. In fact, folding is a prerequisite for TAT-dependent translocation [10]. The post-translocational folding of proteins that are delivered by the Sec pathway may be considered as reminiscent of folding of nascent polypeptide chains in the cytoplasm after or during emergence from the ribosome. Thus, we should expect that similar principles would apply for the folding of polypeptide chains emerging from the translocon or the ribosome. After proteins have folded, each has its lifespan, which can be affected by certain types of environmental conditions that interfere with the folded and functional state. These conditions include osmo, redox, heat, radiation, pH, and chemical stresses. Because the outer membrane is "leaky" for compounds smaller than 600 Da, the periplasm is less protected than the cytoplasm, which is surrounded by the semipermeable cytoplasmic membrane. The fate of damaged proteins is determined by three pathways: repair, degradation, and, if the latter fail, aggregation. Aggregated proteins can impose a severe threat to cells because they can cause lethal damage, e.g., when an aggregated protein has an essential function or when the aggregates physically interfere with the function of other important proteins or the integrity of cells.

It is of interest to study these processes on the mechanistic level to understand the basic physiology of the cell and its regulatory principles, but also for practical aspects including recombinant protein production and vaccine or drug development. Initially, and as has been often the case with general scientific insights obtained from studying bacteria, knowledge was gained from genetic evidence. Because genetic evidence can be considered as indirect, in vitro biochemical evidence using purified components has helped to verify genetic data. So far we have learned that there are four proline isomerases, several proteins dealing with formation and isomerization of disulfide bonds, one general chaperone, several specific chaperones, and a large number of proteases. Another important aspect that has become apparent is that several proteins involved in quality control can be bifunctional. For example, some proline isomerases can have chaperone function that is independent of their enzymatic activity, which is also true for at least one protease (see Section 28.2.3.3).

28.2 Individual Cellular Factors

Molecular chaperones recognize and selectively bind proteins that are present in a nonnative state, which may occur during translocation across membranes, folding, assembly of oligomeric complexes, and various environmental stresses. Chaperones recognize surface-exposed hydrophobicity, thereby discriminating against native proteins. In contrast to the cytoplasm, where the major chaperone systems have been identified and are well understood [11], not much is known about general chaperones in the cell envelope. The only protein that is believed to function exclusively as a general chaperone is Skp, which has so far been attributed to the folding and assembly of outer membrane proteins and to improvement of the folding of recombinant antibody fragments (see below).

Protein folding catalysts carry out similar functions, but, in contrast to chaperones, they improve slow steps in protein folding by catalyzing covalent modifications such as S–S bonds or the *cis-trans* isomerization of proline residues. The redox machinery is genetically and biochemically well studied. The main players belong to the Dsb family, which are involved in the formation, reduction, and isomerization of S–S bonds (see Chapter 9). Native substrates of the Dsb system include dozens of proteins containing S–S bonds, such as alkaline phosphatase,

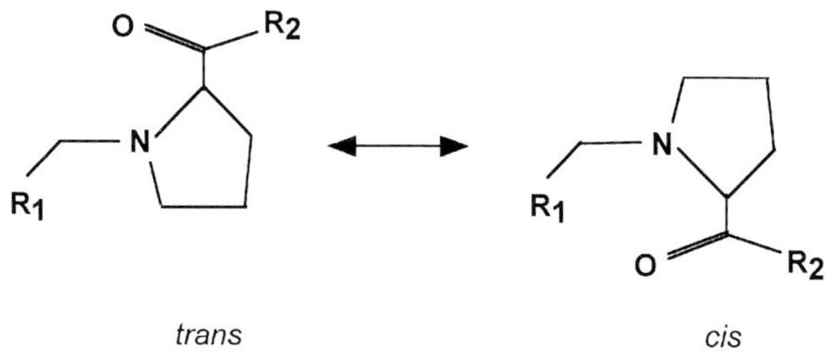

trans cis

Fig. 28.1. Proline isomerization. The *trans* and *cis* conformations of a proline residue are shown.

OmpA, heat-stable enterotoxins, the maltoporin LamB, periplasmic binding proteins, pilin proteins, and components of various secretion apparatuses involved in virulence. There are also four proline isomerases in the cell envelope. Even though knockout mutations are available, proline isomerase activity could be demonstrated in vitro, and involvement in outer membrane biogenesis was suggested, their entire physiological implications remain to be determined [12].

28.2.1 The Proline Isomerases FkpA, PpiA, SurA, and PpiD

One major rate-limiting step in protein folding is the *cis-trans* isomerization of peptidyl bonds occurring N-terminally to each proline residue (Figure 28.1). In general, the *trans* peptide bond is energetically most favorable, and this configuration is represented in almost all peptide bonds. However, *cis* and *trans* peptide bonds involving proline have similar stabilities and both are found in native proteins [13]. Peptidyl-proline bonds are mostly present in *trans* conformation, while about 10% are present in *cis* conformation [14], and the isomeric state of each proline is characteristic of a specific protein. Because the *cis-trans* isomerization is a slow process with high activation energies of about 20 kcal mol^{-1} [15], this step is catalyzed by peptidyl prolyl isomerases that are widely conserved in nature. The *E. coli* cell envelope has four enzymes that can be classified into three families, i.e., the cyclophilin PpiA, the FK506-binding protein (FKBP) FkpA, and the parvulins SurA and PpiD. Proline isomerases have been initially characterized by in vitro methods. These tests use the fact that proteases exclusively cleave a peptide bond that is present in a *trans* conformation, thus having the potential to change the *cis-trans* equilibrium of peptidyl-prolyl isomers. A classical setup uses a tetrameric peptide containing one Pro residue as well as a chromogenic group at the C-terminus and, e.g., α-chymotrypsin as a protease. After mixing these components, a rapid cleavage of the *trans* isomers is observed, which is followed by a slow cleavage after *cis-trans* isomerization. Since these tests were established, a number of more sensitive assays have been developed, including dynamic NMR, direct spectroscopic methods that are independent of an added protease, and refolding reactions using chemically denatured protein such as Rnase T1 (for review, see Ref. [14]). The latter provides direct evidence that proline isomerases are indeed involved in protein-folding reactions.

28.2.1.1 **FkpA**

FkpA was identified as a member of the FKBP family because its carboxyl domain contains the consensus FK506-binding motif, including those amino acids at the active site that form hydrogen bonds with the drug [16]. In addition, high-affinity binding to such drugs has been verified experimentally [17]. FkpA has two domains that can be expressed separately. In addition to its C-terminal PPIase domain, FkpA has a second domain, located between residues 1 and 122, that has been implicated in dimerization but not in chaperone activity. When both domains were expressed separately, and when using RNAse T1 and citrate synthase as substrates, it was found that the FKPA domain had full chaperone but reduced PPIase activity [17]. Chaperone activity of FkpA was also detected for single chain Fv (scFv) antibody fragments in vivo and in vitro [18, 19]. As FkpA-dependent refolding was observed in scFv fragments that do not contain *cis*-prolines, it was concluded that chaperone function of FkpA could be independent of its proline isomerase activity. Also, co-expression of the periplasmic PPIases PpiA and SurA did not increase the level of functional scFv fragments, indicating at least some specificity of the PPIases tested [20]. Similar results were obtained when the aggregation of MBP mutant 31 was analyzed [21]. It was shown that FkpA, but not PpiA or SurA, was able to suppress aggregation of MBP31. Again, after disrupting active site residues, the chaperone activity of FkpA remained functional. These data suggest that the capacity of binding denatured polypeptides might be sufficient to stimulate refolding of chemically denatured proteins and to prevent aggregation. It is also interesting to note that only small changes to the overall structure are sufficient to cause aggregation and inclusion body formation of MBP31. This was recognized when the protein, containing the two point mutations $Gly_{32}Asp$ and $Ile_{33}Pro$, was crystallized. The structure of MBP31 closely resembled the wild type. Only a small dislocation of the changed residues that are part of the loop connecting the first α-helix to the first β-strand of MBP was observed [22].

28.2.1.2 **PpiA**

In *E. coli*, PpiA (also called Rotamase A or CypA) was the first proline isomerase that was cloned, purified, and crystallized [23–25]. It belongs to the cyclophilin family that is inhibited by the immunosuppressant cyclosporin A. The apparent Ki is in the micromolar range, which is about several orders of magnitude higher than for the eukaryotic enzymes [26, 27]. Compton et al. have also shown catalyzed refolding of denatured type III collagen, but there are currently no native substrates of PpiA known. The only indirect evidence available that PpiA is important for general protein-folding reactions can be derived from the fact that its gene is regulated by the Cpx stress pathway ([28]; see also Section 28.4.2).

28.2.1.3 SurA

The *surA* gene has been identified to be essential for survival in the stationary phase [29], and it was shown later that the elevated pH of the growth medium and the absence of the stationary-phase sigma factor S were responsible for this phenotype [30]. Another striking phenotype of *surA* mutants is the marked sensitivity to toxic compounds such as detergents, dyes, chelators, and antibiotics [31] that may be attributed to a defective outer membrane. Consistent with this idea, it was found that in vivo, SurA is required for folding of the three outer membrane proteins OmpA, OmpF, and LamB [31–33]. Co-immunoprecipitation experiments suggested a preference for binding of outer membrane proteins such as PhoE, LamB, and OmpF versus periplasmic or cytoplasmic proteins such as MBP, beta-lactamase, or chloramphenicol transacetylase [33]. Studies with LamB revealed that SurA could convert unfolded monomers into assembly-competent folded monomers [33]. Therefore, these studies support existing models for the assembly of porins, which can be divided into at least three stages: release from the translocon, folding of monomers in the periplasm, followed by the final assembly into functional trimers (for review, see Ref. [8]).

SurA can be divided into four domains: an N-terminal domain, two PPIase domains, and a C-terminal domain of about 40 residues. A functional analysis of each of these domains indicated that the C-terminal PPIase domain is catalytically active, while the N-terminal domain PPIase domain is not [34]. The chaperone function is localized in the non-catalytic parts of SurA [34], and the crystal structure (Figure 28.2) showed that the chaperone and the PPIase domains indeed are distinct structural units, the catalytic domain being tethered via linkers to the rest of the molecule comprised of the N-terminus, the inactive PPIase domain, and the C-terminal domain [35]. The fortunes of crystal packing allowed identification of a possible substrate-binding site because nine residues of an adjacent SurA monomer bound to a 50 Å long channel formed by the non-catalytic domain. One part of the N-terminal domain forms a nonpolar flap facing the interior of a crevice that also has a hydrophobic binding pocket. This part actually binds to a Leu residue of the peptide. This finding is in agreement with earlier results showing that most binding sites of molecular chaperones have hydrophobic patches that are ideally suited to bind exposed hydrophobicity of unfolded substrates. The crystal structure, however, provides little information about the catalysis of peptidyl-prolyl isomerization, and the question of whether a peptide bound to the chaperone domain can undergo proline isomerization is not yet clear.

28.2.1.4 PpiD

Like SurA, PpiD belongs to the parvulin family. It has one N-terminal transmembrane segment, and residues 37–623 are localized in the periplasm. In the middle of this rather large protein, a PPIase domain can be recognized comprising residues 266–355. This domain has PPIase activity, and several mutations in this region have been shown to lead to a loss of catalytic activity. *ppiD* knockouts have phenotypes that are comparable to *surA* mutants, i.e., reduced levels of outer mem-

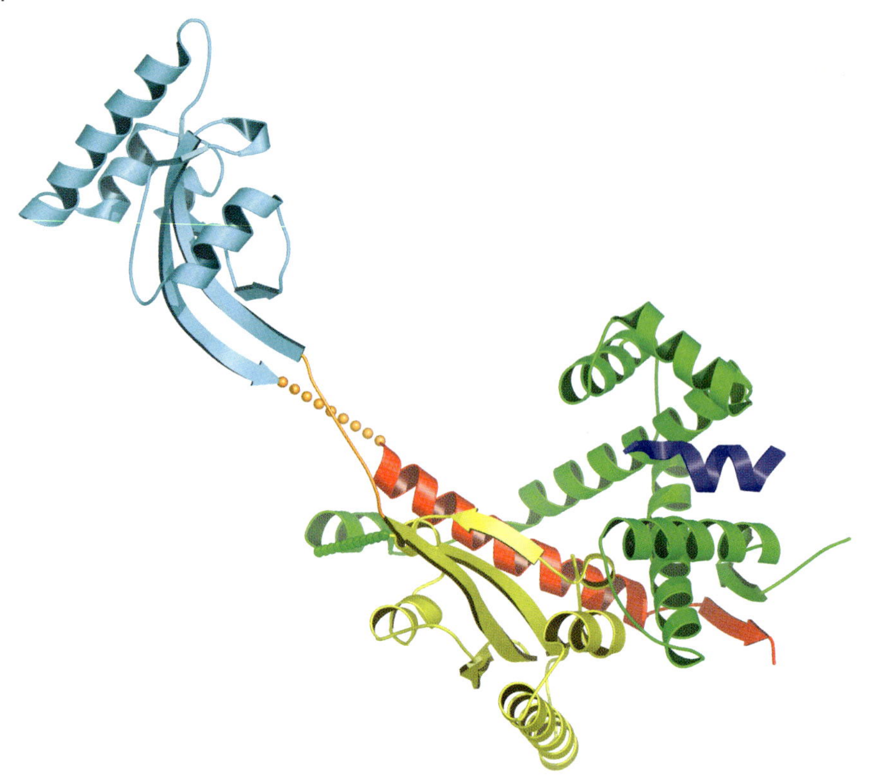

Fig. 28.2. Structure of SurA. Ribbon diagram of the SurA monomer with bound peptide from an adjacent molecule. The N-terminal domain is green, the PPIase domain 1 is gold, the PPIase domain 2 is light blue, the C-terminus is red, and the bound peptide is dark blue [35].

brane proteins OmpA, OmpF, OmpC, and LamB as well as increased sensitivity to SDS and antibiotics. Because *ppiD* was isolated as a high-copy suppressor of *surA* mutants, it can be speculated that SurA and PpiD are functionally redundant [36].

28.2.2 Skp

Skp is a small periplasmic protein of 17 kDa that exists as a homotetramer, and several lines of evidence suggest a general chaperone function. (1) Skp was shown to bind to an affinity column with Sepharose-bound outer membrane porin OmpF. It also binds to outer membrane proteins such as OmpA, OmpC, and LamB [37]. (2) Skp was found to interact in vitro with nascent chains of OmpA precursor and in spheroplasts with trapped translocation intermediates of PhoE. These data indicate that Skp can interact with folding intermediates of outer membrane proteins

[38, 39]. While *skp* knockout mutants are viable [37, 38], *skp degP* double mutants are lethal at 37 °C and accumulate protein aggregates in the periplasm [38]. (3) *skp surA* double mutants also exhibit a synthetic lethal phenotype, and a dramatic decrease in levels of LamB, MBP, and OmpA proteins was detected as early as after 6 h of depletion of SurA [40]. (4) Direct evidence for the participation of Skp in the folding of OmpA was obtained from in vitro refolding experiments. However, folding of OmpA required the additional presence of lipopolysaccharide. Also, it appeared that OmpA bound to Skp and lipopolysaccharide was kept in an unfolded state in solution. The fact that unfolded OmpA in complex with Skp and lipopolysaccharide folded faster into phospholipid bilayers than did urea-unfolded OmpA provides evidence that Skp assists in folding of outer membrane proteins [41].

From these data it could be speculated that Skp mainly functions in outer membrane protein biogenesis. However, several publications report beneficial effects of Skp in the production of functional antibody fragments [42–46]. These data support the model that Skp might have general molecular chaperone function. In fact, Skp is the only protein known to date that could exclusively function as a general chaperone in the cell envelope. All other proteins that were shown to have chaperone function normally either are involved in protein degradation, redox reactions, and proline isomerization or are specific chaperones. The large family of specific chaperones includes some that are well studied and that are involved in pilus assembly (for review, see Refs. [47, 48] and Chapter 29) and the periplasmic chaperone LolA, which is essential for viability and, together with other Lol proteins, is involved in transfer, insertion, and assembly of outer membrane lipoproteins [49–51].

The genomic localization of the *skp* gene is conserved in many gram-negative bacteria. Skp is part of a large gene cluster whose members mediate lipopolysaccharide-, phospholipid-, and fatty-acid synthesis. Directly upstream of *skp*, two other genes that have been shown to be essential for viability and important for cell envelope protein biogenesis, *yaeL* and *yaeT* (o810 or Omp85), are localized. The YaeL protease is involved in the unfolded protein response (Section 28.4.1), and, from work carried out in *Neisseria meningitidis*, YaeT has been postulated to be involved in lipid export [52]. After prolonged depletion, *yaeT* mutants were found to be defective in outer membrane protein assembly [53]. However, direct evidence from biochemical in vitro experiments using purified components is required to identify the precise function of this important gene.

28.2.3 Proteases and Protease/Chaperone Machines

Proteases are in charge of protein turnover by catalyzing the cleavage of peptide bonds. They serve essential housecleaning functions by degrading damaged proteins and signal peptides. Other important functions include regulation via, e.g., activation or inactivation of signaling proteins. There are at least 24 different peptidases and proteases in the cell envelope of *E. coli*. While a few are well studied, many have been identified only via bioinformatics (see the EPD database at www.cf.ac.uk/biosi/staff/ehrmann/tools/proteases.index.html). Among these, several genes are, at least in prokaryotes, widely conserved (*degP*, *degQ*, *degS*, *nlpC*, *ptrA*, *tsp*, *ydcP*, *ydgD*, *ydhO*, *yebA*, *yhbU*, *yhjJ*, *nlpD*, *ompT*, *sohB*, and *yfbL*), indicating an important physiological function.

28.2.3.1 **The HtrA Family of Serine Proteases**

The defining feature of the over 180 family members is the combination of a catalytic serine protease domain with one or more C-terminal PDZ domains. PDZ domains are protein modules that mediate specific protein-protein interactions and bind preferentially to the C-terminal 3–4 residues of the target protein. Some, but not all, family members are classic heat shock proteins. They are typically localized in extracytoplasmic compartments such as the periplasm of gram-negative bacteria, the ER or mitochondria of eukaryotes, the chloroplast of plants, or in the extracellular space. Prokaryotic HtrA has been attributed to the tolerance against various folding stresses as well as to pathogenicity. Human homologues are believed to be involved in arthritis, cell growth, unfolded stress response, apoptosis, muscular dystrophy, cancer, and aging. One family member, DegP of *E. coli*, has both chaperone and protease functions. These functions switch in a temperature-dependent manner, the protease activity being most apparent at elevated temperatures (for review, see Ref. [54]). There are other factors in the cell envelope that are believed to have more than one function. For example, chaperone activity has been demonstrated for the proline isomerases FkpA and SurA (Sections 28.2.1.1 and 28.2.1.3) and the redox proteins DsbC and DsbG [55, 56].

28.2.3.2 ***E. coli* HtrAs**

Three HtrA proteins, DegP, DegQ, and DegS, have been identified in *E. coli*. Originally, DegP was purified from cell extracts and was called protease Do. It was shown to be inhibited by the serine protease inhibitor diisopropyl fluorophosphate (DFP) and to be oligomeric [57]. Subsequently, the *degP* gene was identified as a heat shock gene [58] and *degP* mutants were shown to stabilize membrane protein-PhoA fusions [59].

degQ and *degS* (also termed *hhoA* and *hhoB*) were identified as multicopy suppressors of conditional lethal *prc* (tail specific protease) null mutants [60] and as genes located downstream of the *mdh* gene [61]. Both *degQ* and *degS* are located directly next to each other, but are not co-regulated, and neither are induced by heat shock.

28.2.3.3 **DegP and DegQ**

DegP has two functions in the regulation of protein composition in the cell envelope. As a rather unspecific protease, it is a key player in the turnover of damaged and nonnative cell envelope proteins, including misfolded periplasmic *E. coli* proteins such as MBP [62], PhoA [63], or MalS [64]; the misfolded outer membrane protein LamB [65]; mislocalized cytoplasmic proteins such as TreF [66] and β-Gal [67]; proteins that have failed to assemble into hetero-oligomeric complexes, such as HflKC of *E. coli* [68] and HMW1 adhesin of *Haemophilus influenzae* [69]; and hybrid proteins [59, 70] and recombinant proteins [71, 72]. In addition, DegP is required for maturation of colicin A [73]. It is interesting that the proteolytic function dominates at elevated temperatures, where protein repair is less effective compared with low temperatures [64].

At low temperature, DegP has a protein repair function that was first recognized by its complementation of a *dsbA* mutant in the folding process of the periplasmic amylase MalS. Additional evidence for the chaperone function of DegP was obtained from in vitro refolding assays [64] and from genetic studies [40, 74, 75].

Not much is known about DegQ, but because it can complement the temperature sensitivity of *degP* null mutants, it is thought to have functions similar to those of DegP [61].

28.2.3.4 DegS

DegS is homologous to DegP and DegQ but differs in substrate specificity, oligomeric state, and function. DegS has an N-terminal transmembrane segment, is a trimer, and has a regulatory function [76, 77]. It is highly specific [61] and is required for viability because it is essential for the sigma E pathway ([78]; see also Section 28.4.1). It is also required for virulence, although the precise function of DegS in pathogenesis is not understood [79]. The proteolytic activity of DegS seems to be modulated by its PDZ domains. Activation of DegS function is mediated by binding of peptides corresponding to the C-terminus of outer membrane proteins [80]. These peptides are thought to be related to the identified YQF or YYF motifs. These sequences must provide an appropriate fit to the specific substrate-binding pocket in the PDZ domain. It has been speculated that in the absence of activating peptide, the PDZ domain sterically blocks entry of substrates to the active site of the protease domain. Upon binding of peptide, the PDZ domain would then rotate away from the active site to make it accessible for substrates [80, 81]. However, other models are equally likely; for example, activating peptide could simply induce a conformational change in the catalytic triad of the protease without rotation of the PDZ domain. This model is based on the fact that the active sites of DegP and human HtrA2 were found in an inactive conformation in the recently determined crystal structures [54].

28.2.3.5 The Structure of HtrA

The functional unit of HtrA appears to be a trimer that is stabilized exclusively by residues of the protease domains [82–84]. The basic trimer has a funnel-like shape, with the protease domains located at its top and the PDZ domains protruding to the outside. While the protease domains form the rigid part of the funnel, the PDZ domains are highly mobile elements. This construction is reminiscent of a "molecular anemone," with the PDZ domains swinging around to capture substrates. After binding, substrates have to be delivered into the interior of the funnel and to the proteolytic sites.

The recently solved crystal structure of *E. coli* DegP indicates that it exists as a hexamer of stacked trimers and that the protease domains are localized in an inner cavity. Access to this cavity is controlled by 12 PDZ domains, which form the mobile sidewalls of the particle [82] (Figure 28.3). Therefore, DegP represents a novel type of self-compartmentalizing protease. Self-compartmentalization is an ingenious architectural feature by which the accessibility of active sites becomes exclusive, in this case for proteins that are at least partially unfolded. This strategy prevents folded proteins from being substrates. This mechanism is widely used in nature and is also found in other proteases such as ClpP (Hsp100) [85], in the proteasome [86] and the tricorn [87], and in the Hsp60 (GroEL) chaperone [88].

Compared to other proteases of known structure, the DegP particle has no rigid sidewalls. The PDZ domains that could fulfill this function are highly flexible and are able to adopt different conformations within the hexamer. The PDZ domains represent side doors, which may either swing open or close the complex. In addition, because short peptides binding to PDZ domains have been found to activate the proteolytic activity of DegP and DegS [80, 89], it is likely that the conformation or position of the PDZ domains could be involved in regulation of protease activity. It should be noted in this context that the protease activity can be reversibly switched on and off. This remarkable reversibility is mediated by the geometric arrangement of the active site. Perhaps binding of substrate to the PDZ domains could have knock-on effects leading to conformational changes in the active site i.e., the proper positioning of the catalytic triad and the formation of the oxyanion hole.

28.2.3.6 **Other Proteases**

Prc or tsp (tail specific protease) is a conserved periplasmic serine protease that has one PDZ domain that precedes the active site. It is a member of the S41 family according to the MEROPS classification and is therefore related to the tricorn protease. Prc is a rather large protein of 660 residues that recognizes preferentially hydrophobic C-termini [90]. The specificity of Prc seems to be mediated by its PDZ domain, which is capable of binding a nonpolar peptide with a dissociation constant of 1.9 μM [91]. It was also found that Prc has a preference for at least partially unfolded substrates [92] or substrates that are only partially synthesized [93]. Given this preference for nonnative conformations, it might be speculated that Prc could be involved in protection against protein folding stresses.

Protease III is a large periplasmic metalloprotease, and its sequence and size are highly conserved in evolution. Homologues can be found in most organisms, including higher plants and humans. In *E. coli* it was shown to be involved in the degradation of the misfolded MBP31 mutant [62], but it also cleaves insulin [94] and levels of beta-lactamase were found to be increased in protease III mutants [71]. Like Prc, protease III is expected to play a role in degradation of damaged proteins, but its precise contributions remain to be determined.

OmpT is an integral outer membrane protein that functions as a protease. The *ompT* gene is often deleted in strains that are used for the production of recombinant proteins. Among the several genes that are upregulated upon induction of overexpression of recombinant proteins are *degP* and *ompT* [95, 96]; OmpT was found to be associated with aggregates of yeast alpha-glucosidases when produced in *E. coli* [97] and it was also implicated in virulence [98]. Taken together, these

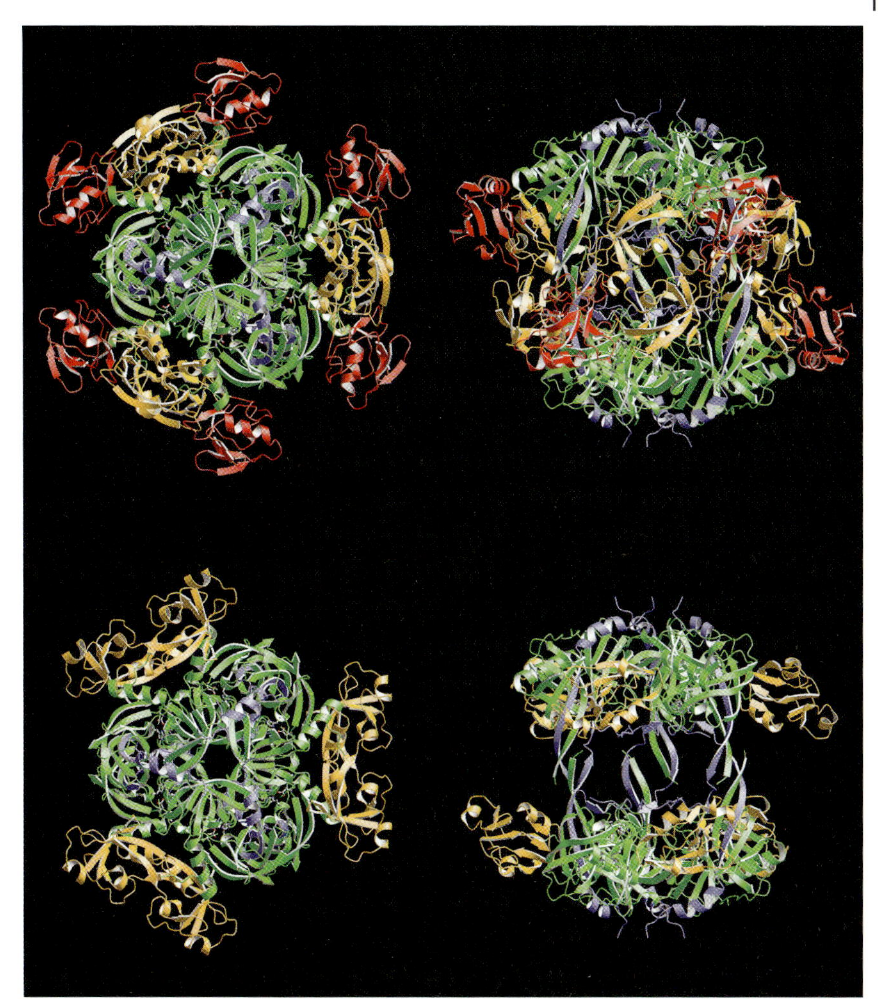

Fig. 28.3. Structure of DegP. DegP is shown in its hexameric conformation where two trimers are stacked upon each other. In the crystals, DegP was present in the closed (upper part) as well as in the open (lower part) conformation. While the structures in the top view (left-hand side) are similar, the side views show that in the open conformation the PDZ domains are rotated away from the protease domains and allow access to the inner cavity. On the top and bottom of the cavity, the active sites of the protease domains are localized. The pillars of the particle are formed by N-terminal extensions that reach from one monomer into the active site of another monomer in the opposite trimer (for review, see Refs. [82] and [54]). It should be noted that in the open conformation, the structure of only one of the two PDZ domains was solved. However, the density of the second PDZ domain is found directly adjacent to the first PDZ domain. The protease domains are shown in green, the N-terminal extensions in blue, and the PDZ domains in brown and red.

data suggest that OmpT is involved in protein quality control in the cell envelope and, given its cellular localization, perhaps particularly in the outer membrane.

28.3
Organization of Folding Factors into Pathways and Networks

Although direct evidence has been obtained for several individual folding catalysts and proteases that they are involved in protein folding and degradation, it is obvious that these factors do not work in isolation. Based on evidence from other biological systems, we can expect that several pathways dealing with protein quality control and protein metabolism exist. We can also expect that these pathways are connected and organized into networks. However, the knowledge of these pathways and networks is very limited. Because they are critical for the basic physiology of bacterial cells, as well as for commercial exploitation, they are well worth being studied experimentally.

28.3.1
Synthetic Lethality and Extragenic High-copy Suppressors

To distinguish the various pathways involved in protein folding, quality control, and protein turnover and to recognize functional redundancies, bacterial genetics is a useful tool for initial analysis. An experimentally simple and straightforward approach is to look for synthetic lethal phenotypes of combinations of individual knockout mutations of nonessential genes. Synthetic lethality occurs when the combination of two mutations that alone are not lethal makes the double mutant inviable. Such synthetic lethal mutations can indicate that the two mutations affect (1) a single function or (2) a pathway. Also, the complementation of essential genes or of synthetic lethality by extragenic multicopy suppressors can indicate functional redundancy.

As an example of (1) above, both Skp and DegP have chaperone activity, and the absence of both chaperone activities in an *skp degP* double null mutant is lethal. Also, expression of a catalytically inactive *degP* mutant that still has chaperone activity but no longer has protease activity complements the *skp degP* double mutant [38, 40]. These phenotypes indicate a functional redundancy of Skp and DegP in vivo.

As an example of (2) above, it was found that combining *degP* with other null mutations could cause a lethal phenotype in more than one case. For example, *degP* and *surA* as well as *degP* and *skp* double mutants are synthetic lethals. Because the protease activity of DegP is required to complement the *degP surA* double mutant but the chaperone activity of DegP complements the *degP skp* double mutant, it can be postulated that two pathways exist, one involving Skp and another involving SurA [40].

There is one potential problem that should not be ignored when carrying out such analyses: not every lethal combination represents a true synthetic lethal phenotype. It is possible that an observed lethality is due to two "unhealthy" single mutations, the combination of which is just too much. Even in light of such limitations, initial studies clearly indicate that the construction and careful analysis of synthetic lethal double mutants are very useful approaches to identifying and studying protein-folding and degradation pathways.

28.3.2 Reconstituted in Vitro Systems

To further verify and explain the observed phenotypes, in vitro refolding and proteolysis experiments can be informative. Such in vitro systems can provide direct quantitative evidence for the quality-control pathways identified. For this purpose, refolding and, if applicable, degradation of model substrates are studied in the presence and absence of folding factors and proteases that are thought to be part of a specific pathway. Interesting questions to address include: is folding or quality control different (faster or more efficient) if the entire network of cellular factors is present in the reconstituted system?

For example, a mini-chaperone derived from GroEL, together with DsbA and a prolyl isomerase, exhibited synergistic assistance during refolding of a scorpion toxin [99]. Also, the combination of a chaperone and protein disulfide isomerase increased the production of functional antibody in a cell-free system [100]. Although these experiments did not completely mimic the in vivo situation (because factors originally residing in different cellular compartments were used), they do provide initial evidence that such approaches could yield important information.

28.4 Regulation

E. coli has developed compartment-specific responses to react to the presence of misfolded proteins. In the cell envelope, three systems have been identified: sigma E, Cpx, and Bae. Because stress occurs in the cell envelope while the regulation of gene expression is carried out in the cytoplasm, these systems must detect misfolded cell envelope proteins, transmit signals across the cytoplasmic membrane, and stimulate transcription of protein folding and degradation factors in the cytoplasm. It is generally accepted that misfolded proteins, protein fragments, and mislocalized outer membrane proteins are activators of these stress response pathways. Such activation occurs under any kind of stress situation that causes protein-folding problems, including heat shock, pH, and osmo-stress, and also when proteins are overproduced from nonnative expression systems.

28.4.1 The Sigma E Pathway

The key component of the essential sigma E pathway is the alternative sigma factor RpoE. Sigma factors interact with RNA polymerase and stimulate binding to and transcription of promoters that contain specific binding sites for these sigma factors. Because sigma E is a cytoplasmic protein, there are additional factors in the cell envelope that regulate its activity. Three proteins are known to regulate sigma E function: the anti-sigma factor RseA, the RseB protein that modulates the stability of the sigma E–RseA complex, and RseC, whose function remains to be determined. RseA is a membrane protein comprised of a cytoplasmic domain, one transmembrane segment, and a periplasmic domain. Sigma E binds to the cytoplasmic segment of RseA and is inactive in this conformation. The anti-sigma activity of RseA is regulated on the level of protein stability. After proteolytic inactivation of RseA, sigma E is free to activate certain promoters (for review, see Ref. [12]). Recent evidence suggests that two proteases, DegS and YaeL, are involved in the proteolytic inactivation of RseA [77, 101]. YaeL is a metalloprotease and an integral cytoplasmic membrane protein that belongs to the RIP family (regulated intramembrane proteolysis) [102]. DegS is a member of the HtrA family of serine proteases (Section 28.2.3.4), and it was shown that peptides corresponding to the C-terminus of outer membrane proteins activate protease activity of DegS by binding to its PDZ domains [80]. Activated DegS initiates cleavage of RseA and the sigma E cascade. It is therefore believed that when outer membrane proteins accumulate in the periplasm, they serve as signals for folding stress and thus activate the sigma E pathway. Substrates of the sigma E pathway include the *degP* and *fkpA* promoters; *rpoH*, which encodes the cytoplasmic heat shock sigma factor; and *rpoE* and *rseABC* themselves (for review, see Ref. [12]).

28.4.2
The Cpx Pathway

Cpx belongs to the family of two-component signal transduction systems [103]. Two proteins, CpxR and CpxA, carry out signal transduction, and CpxP is also involved in regulation of Cpx function. The cytoplasmic membrane protein CpxA is the sensor kinase, and cytoplasmic CpxR is the response regulator and a DNA-binding protein. CpxA modulates the phosphorylation state of CpxR. The signal transduction cascade is initiated by a conformational change in CpxA that triggers autophosphorylation of a conserved His residue. Subsequently, the phosphorylation is transferred to a conserved Asp residue in CpxR. Phosphorylated CpxR then acts as a transcription factor. In the absence of signals, CpxA dephosphorylates CpxR, providing a tight regulation of CpxR activity.

CpxP is a periplasmic protein that inhibits CpxA. Because overproduction of CpxP increases inhibition of CpxA, it is the current hypothesis that CpxP interacts with CpxA. Release of CpxP induces conformational change and activates the Cpx cascade (for review, see Ref. [12]). A recent analysis indicates, however, that the mode of regulation is more sophisticated. This might be the case, because a rather large number of stresses induce the Cpx pathway. For example, typical envelope stress conditions induce Cpx via CpxA but not via NlpE, while the mechanisms of induction via adhesion and growth phase appear to be distinct from envelope stress. The response to adhesion requires the outer membrane lipoprotein NlpE but not CpxP, and the induction by growth phase seems to be independent of CpxA, indicating that induction by growth phase is distinct from adhesion and envelope stress [104].

Besides its role in the unfolded protein response, the Cpx pathway is also involved in regulation of chemotaxis [105], biofilms [106], and pili biogenesis [107] as well as in attachment of bacteria to surfaces [108]. However, these implications might represent only the tip of the iceberg, as a recent study suggests that Cpx may be directly involved in regulation of about 100 operons. Via its interaction with other signal transduction pathways, its effects can even be greater [109]. Among the known protein quality-control factors that are regulated by Cpx are the promoters of the *degP*, *dsbA*, *ppiD*, and *ppiA* genes [12] and also *htpX*, which encodes a heat shock protease that is an integral cytoplasmic membrane protein with a cytoplasmic active site [110]. Perhaps surprisingly, it was also shown that phosphorylated CpxR protein directly interacts with one promoter of the *rpoErseABC* operon, indicating that the sigma E pathway may be negatively controlled by the Cpx pathway [109].

28.4.3
The Bae Pathway

Spy (spheroplast protein Y), a protein of unknown function that has been identified to be induced by spheroplast formation [111], is also induced by the Cpx pathway, and its deletion leads to upregulation of the sigma E pathway [112]. However, since induction via spheroplast formation is only partially Cpx-dependent, a novel two-component signal transduction system, Bae, has been identified that regulates *spy* expression [113]. BaeS is the sensor kinase, while BaeR is the response regulator. It was also shown that, in analogy to Cpx, the Bae system responds to cell envelope stress, and because Spy is homologous to CpxP, it was suggested that the Bae system represents a third system that senses and responds to protein-folding stress in the cell envelope.

28.5
Future Perspectives

The study of protein folding in the cell envelope not only is adding to the increasing perception of basic cellular functions but also is leading to new methods for protein production as well as for diagnosis, therapy, and prevention of infectious diseases caused by pathogenic bacteria. Although several key cellular factors have been identified and studied in detail, much work remains to be done to complete our understanding and to eventually allow us to model a virtual cell and to simulate the cellular response to various important environmental factors such as heat, drought, antimicrobial drugs, and the immune system among many others.

It is now important to study the action of individual cellular factors not only in isolation but also in conjunction with each other. Perhaps only then it will be recognized that, for example, Skp promotes folding and not only holding (preventing aggregation), as is seen in some in vitro systems. Also, it may well be that the activity of individual factors is modulated by other factors, either directly via protein-protein interaction or indirectly when, for example, one factor acting at an early stage of folding produces a folding intermediate that is required so that a second factor can catalyze the folding process to completion. Therefore, future studies will address (1) the identification and characterization of pathways involved in protein folding, repair, and degradation; (2) their organization into networks of pathways; and (3) the identification of key players in this network that, for example, function as bottlenecks by catalyzing rate-limiting reactions or by connecting individual pathways.

Much remains to be learned about outer membrane proteins. For example, because in vitro the refolding and membrane insertion reactions are very slow (on the order of one generation time of the whole organism), it is obvious that cellular factors must exist that catalyze these processes (for review, see Ref. [114]). In this respect, it is surprising that despite intense efforts no cellular machine has been identified that catalyzes membrane insertion. In addition, a quality-control system for properly inserted, native outer membrane proteins has not yet been described or systematically studied. In the same context, the cell envelope has many hetero-oligomeric protein complexes such as pili, flagella, and multi-drug resistance pumps that can be very large and their architecture complex. The functional assembly and quality control of these complexes are not well understood and provide a challenge for future studies. The cell envelope hosts a great number of proteins of unknown function, many of which are conserved and several of which will turn out to be important for protein biogenesis and quality control.

Finally, additional regulatory systems and mechanisms remain to be discovered. For example, a recent study identified that ribonuclease E is involved in regulation of *dsbC* expression, and here a new protein, RraA (regulator of ribonuclease activity A), was shown to inhibit RNase E endonuclease by binding [115]. Clearly, posttranscriptional regulatory events must play an important role in cell envelope protein biogenesis, and exciting developments can be expected in the near future.

28.6 Experimental Protocols

28.6.1 Pulse Chase Immunoprecipitation

Rationale

Pulse-chase immunoprecipitation allows determination of the stability of proteins and thus provides in vivo evidence on the folded state of proteins and, if applicable, the rates of its degradation. In principle, all proteins are labeled with ^{35}S-methionine for a short period of time. After labeling, excess of unlabelled methionine is added to stop the labeling reaction. At various time points, further samples are taken. A specific antibody is used to isolate one target protein, the level and stability of which is investigated. This experiment allows us to follow the fate of a particular protein because it is labeled during synthesis and its stability is monitored over time.

Such an approach is useful if, for example, the effects of a mutation in a folding factor or of protease on a particular substrate are to be examined. Comparing the levels and thus stability of the target protein in wild-type and mutant background provides evidence for the involvement of a chosen cellular factor in the folding and stability of a substrate.

Protocol

1. Grow cells overnight in minimal medium M9, 0.2% glycerol supplemented with 18 amino acids and 0.01% thiamine.
2. Dilute cells 1/25 in the same medium.
3. Grow to $OD_{600} = 0.3–0.5$ (if required, induce the promoter at $OD \approx 0.2$).

Pulse

1. Place 4.5 mL of cells in a prewarmed plastic vial (do all steps at 37 °C with vigorous aeration).
2. Add 10–50 μCi ^{35}S-methionine.
3. Label for 1–2 min.
4. Transfer 1 mL of cells into Eppendorf tube containing ice (time point 1).

Chase

1. To the remaining 3.5 mL of cells, add 0.05% cold methionine (or 20 mM final) –640 μL (0.2 M methionine stock solution).
2. After 1, 5, and 15 min, transfer 1 mL of cells into Eppendorf tubes containing ice (time points 2–4).

Immunoprecipitation

1. Spin all samples for 1 min in an Eppendorf centrifuge.
2. Resuspend pellet in 1% SDS, 1 mM EDTA, 50 mM Tris HCl pH 7.5.
3. Heat 10 min at 65 °C (or 3 min at 95 °C).
4. Add 1 mL of HS buffer.
5. Spin 10 min.
6. Take 970 μL supernatant.
7. Add 10 μL Sephacryl protein A.
8. Incubate on ice for 30 min.
9. Spin 1 min.
10. Take 900 μL supernatant.
11. Add antibody (2–10 μL) directed against the target protein.

12. Keep on ice for 4 h or overnight.
13. Add 10 μL Sephacryl protein A.
14. Keep 30 min on ice.
15. Invert tubes occasionally.
16. Wash 1× in 1 mL HS buffer.
17. Wash 1× in 1 mL Li-buffer.
18. Wash 1× in 1 mL 50 mM Tris HCl pH 7.5
19. Resuspend in 50 μL sample buffer
20. Heat 10 min 65 °C (or 3–5 min 95 °C).
21. Spin 5 min to remove Sephacryl protein A.
22. Take supernatant.
23. Load half on SDS-PAGE.
24. Dry gel and place into phosphorimager.

Notes

Various proteins have different requirements for heating. Soluble proteins are easy: just boil for 2 min. Cytoplasmic membrane proteins: incubate 10 min at RT or 42 °C. Boiling can cause aggregation in some cases. Outer membrane proteins tend to form stable trimers, and monomers do not easily denature: 10–30 min of boiling is required.

Solutions and Buffers

HS buffer: 50 mM Tris HCl pH = 8
1 M NaCl
1% Triton × 100
1 mM EDTA

Li-buffer: 50 mM Tris HCl pH = 7.5
0.1% SDS
0.5 M LiCl

Minimal Medium 9 (10×):

$Na_2HPO_4 \cdot 2H_2O$: 76.54 g
KH_2PO_4: 30 g
NaCl: 5 g
NH_4Cl: 10 g
Autoclave, let cool and add:
MgSO4: 1 mM
$CaCl_2$: 0.1 mM (use 100 mM stock, 1 M precipitates immediately)

Liquid Culture (5 mL)

4.5 mL H_2O
0.5 mL M9 10×

0.2–0.4% carbon source
1× 18 amino acid mix
0.01% thiamine

Hydrophobic Amino Acids (50×)

1. Add 250 mg of each Ile, Phe, Trp, and Tyr to 100 mL H_2O.
2. Boil and adjust pH to 8.0 using $NaCO_2$.
3. Filter sterilize while hot.
4. Make aliquots and store at −70 °C.

Hydrophilic Amino Acids (100×)

1. Add 250 mg of each Gly, Ala, Val, Leu, Ser, Thr, Asp, Asn, Glu, Gln, Lys, Arg, His, and Pro to 50 mL H_2O.
2. Filter sterilize.
3. Make aliquots and store at −70 °C.

Note

Bacteria grown in minimal media tolerate a lower concentration of antibiotics compared to growth in rich media. Concentrations for minimal media (μg mL^{-1}):

Ampicillin: 100
Kanamycin: 50
Tetracycline: 5
Chloramphenicol: 12

Acknowledgements

I would like to thank Dave McKay for a picture of the SurA structure, Tim Clausen for a picture of the DegP structure, and Tracy Raivio, George Georgiou, and Susan Marqusee for communicating unpublished results.

References

1 Wülfing, C. & Plückthun, A. (1994). Protein folding in the periplasm of *E. coli*. *Mol. Microbiol.* 12, 685–692.

2 Brass, J. M., Higgins, C. F., Foley, M., Rugman, P. A., Birmingham, J. & Garland, P. B. (1986). Lateral diffusion of proteins in the periplasm of *Escherichia coli*. *J. Bacteriol.* 165, 787–795.

3 Blattner, F. R., Plunkett, G., 3rd, Bloch, C. A., Perna, N. T., Burland, V., Riley, M., Collado-Vides, J., Glasner, J. D., Rode, C. K., Mayhew, G. F., Gregor, J., Davis, N. W., Kirkpatrick, H. A., Goeden, M. A.,

Rose, D. J., Mau, B. & Shao, Y. (1997). The complete genome sequence of *Escherichia coli K-12. Science* 277, 1453–1474.
4 Schulz, G. E. (2002). The structure of bacterial outer membrane proteins. *Biochim. Biophys. Acta* 1565, 308–317.
5 Koronakis, V., Andersen, C. & Hughes, C. (2001). Channel-tunnels. *Curr. Opin. Struct. Biol.* 11, 403–407.
6 Andersen, C., Hughes, C. & Koronakis, V. (2001). Protein export and drug efflux through bacterial channel-tunnels. *Curr. Opin. Cell Biol.* 13, 412–416.
7 van Wely, K. H., Swaving, J., Freudl, R. & Driessen, A. J. (2001). Translocation of proteins across the cell envelope of Gram-positive bacteria. *FEMS Microbiol Rev.* 25, 437–454.
8 Danese, P. N. & Silhavy, T. J. (1998). Targeting and assembly of periplasmic and outer-membrane proteins in *Escherichia coli. Annu. Rev. Genet.* 32, 59–94.
9 Palmer, T. & Berks, B. C. (2003). Moving folded proteins across the bacterial cell membrane. *Microbiology* 149, 547–556.
10 DeLisa, M. P., Tullman, D. & Georgiou, G. (2003). Folding quality control in the export of proteins by the bacterial twin-arginine translocation pathway. *Proc. Natl. Acad. Sci. USA* 100, 6115–6120.
11 Bukau, B. & Horwich, A. (1998). The Hsp70 and Hsp60 chaperone machines. *Cell* 92, 351–366.
12 Raivio, T. L. & Silhavy, T. J. (2001). Periplasmic stress and ECF sigma factors. *Annu. Rev. Microbiol.* 55, 591–624.
13 Ramachandran, G. N. & Mitra, A. K. (1976). An explanation for the rare occurrence of cis peptide units in proteins and polypeptides. *J. Mol. Biol.* 107, 85–92.
14 Gothel, S. F. & Marahiel, M. A. (1999). Peptidyl-prolyl cis-trans isomerases, a superfamily of ubiquitous folding catalysts. *Cell. Mol. Life Sci.* 55, 423–436.
15 Schmid, F. X. & Baldwin, R. L. (1978). Acid catalysis of the formation of the slow-folding species of RNase A: evidence that the reaction is proline isomerization. *Proc. Natl. Acad. Sci. USA* 75, 4764–4768.
16 Horne, S. & Young, K. (1995). *Escherichia coli* and other species of the Enterobacteriaceae encode a protein similar to the family of Mip-like FK506-binding proteins. *Arch. Microbiol.* 163, 357–365.
17 Ramm, K. & Pluckthun, A. (2001). High enzymatic activity and chaperone function are mechanistically related features of the dimeric *E. coli* peptidyl-prolyl-isomerase FkpA. *J. Mol. Biol.* 310, 485–498.
18 Bothmann, H. & Pluckthun, A. (2000). The periplasmic *Escherichia coli* peptidylprolyl cis,trans-isomerase FkpA. I. Increased functional expression of antibody fragments with and without cis-prolines. *J. Biol. Chem.* 275, 17100–17105.
19 Ramm, K. & Pluckthun, A. (2000). The periplasmic *Escherichia coli* peptidylprolyl cis,trans-isomerase FkpA. II. Isomerase-independent chaperone activity in vitro. *J. Biol. Chem.* 275, 17106–17113.
20 Balbach, J., Steegborn, C., Schindler, T. & Schmid, F. X. (1999). A protein folding intermediate of ribonuclease T1 characterized at high resolution by 1D and 2D real-time NMR spectroscopy. *J. Mol. Biol.* 285, 829–842.
21 Arie, J. P., Sassoon, N. & Betton, J. M. (2001). Chaperone function of FkpA, a heat shock prolyl isomerase, in the periplasm of *Escherichia coli. Mol. Microbiol.* 39, 199–210.
22 Saul, F. A., Mourez, M., Vulliez-Le Normand, B., Sassoon, N., Bentley, G. A. & Betton, J. M. (2003). Crystal structure of a defective folding protein. *Protein Sci.* 12, 577–585.
23 Liu, J. & Walsh, C. T. (1990). Peptidyl-prolyl *cis-trans*-isomerase from *Escherichia coli*: a periplasmic homolog of cyclophilin that is not inhibited by cyclosporin A. *Proc. Natl. Acad. Sci. USA* 87, 4028–4032.

24 Hayano, T., Takahashi, N., Kato, S., Maki, N. & Suzuki, M. (1991). Two distinct forms of peptidylprolyl-*cis-trans*-isomerase are expressed separately in periplasmic and cytoplasmic compartments of *Escherichia coli* cells. *Biochemistry* 30, 3041–3048.

25 Clubb, R. T., Ferguson, S. B., Walsh, C. T. & Wagner, G. (1994). Three-dimensional solution structure of *Escherichia coli* periplasmic cyclophilin. *Biochemistry* 33, 2761–2772.

26 Liu, J., Chen, C. M. & Walsh, C. T. (1991). Human and *Escherichia coli* cyclophilins: sensitivity to inhibition by the immunosuppressant cyclosporin A correlates with a specific tryptophan residue. *Biochemistry* 30, 2306–23010.

27 Compton, L. A., Davis, J. M., Macdonald, J. R. & Bachinger, H. P. (1992). Structural and functional characterization of *Escherichia coli* peptidyl-prolyl *cis-trans* isomerases. *Eur. J. Biochem.* 206, 927–934.

28 Pogliano, J., Lynch, A., Belin, D., Lin, E. & Beckwith, J. (1997). Regulation of *Escherichia coli* cell envelope proteins involved in protein folding and degradation by the Cpx two-component system. *Genes and Dev.* 11, 1169–1182.

29 Tormo, A., Almiron, M. & Kolter, R. (1990). *surA*, an *Escherichia coli* gene essential for survival in stationary phase. *J. Bacteriol.* 172, 4339–4347.

30 Lazar, S. W., Almiron, M., Tormo, A. & Kolter, R. (1998). Role of the *Escherichia coli* SurA protein in stationary-phase survival. *J. Bacteriol.* 180, 5704–5711.

31 Missiakas, D., Betton, J. M. & Raina, S. (1996). New components of protein folding in extracytoplasmic compartments of *Escherichia coli* SurA, FkpA and Skp/OmpH. *Mol. Microbiol.* 21, 871–884.

32 Lazar, S. W. & Kolter, R. (1996). SurA assists the folding of *Escherichia coli* outer membrane proteins. *J. Bacteriol.* 178, 1770–1773.

33 Rouviere, P. E. & Gross, C. A. (1996). SurA, a periplasmic protein with peptidyl-prolyl isomerase activity, participates in the assembly of outer membrane porins. *Genes Dev.* 10, 3170–3182.

34 Behrens, S., Maier, R., de Cock, H., Schmid, F. X. & Gross, C. A. (2001). The SurA periplasmic PPIase lacking its parvulin domains functions in vivo and has chaperone activity. *EMBO J.* 20, 285–294.

35 Bitto, E. & McKay, D. B. (2002). Crystallographic structure of SurA, a molecular chaperone that facilitates folding of outer membrane porins. *Structure* 10, 1489–1498.

36 Dartigalongue, C. & Raina, S. (1998). A new heat-shock gene, *ppiD*, encodes a peptidyl-prolyl isomerase required for folding of outer membrane proteins in *Escherichia coli*. *Embo J.* 17, 3968–3980.

37 Chen, R. & Henning, U. (1996). A periplasmic protein (Skp) of *Escherichia coli* selectively binds a class of outer membrane proteins. *Mol. Microbiol.* 19, 1287–1294.

38 Schafer, U., Beck, K. & Muller, M. (1999). Skp, a molecular chaperone of gram-negative bacteria, is required for the formation of soluble periplasmic intermediates of outer membrane proteins. *J. Biol. Chem.* 274, 24567–24574.

39 Harms, N., Koningstein, G., Dontje, W., Muller, M., Oudega, B., Luirink, J. & de Cock, H. (2001). The early interaction of the outer membrane protein PhoE with the periplasmic chaperone Skp occurs at the cytoplasmic membrane. *J. Biol. Chem.* 276, 18804–18811.

40 Rizzitello, A. E., Harper, J. R. & Silhavy, T. J. (2001). Genetic evidence for parallel pathways of chaperone activity in the periplasm of *Escherichia coli*. *J. Bacteriol.* 183, 6794–6800.

41 Bulieris, P. V., Behrens, S., Holst, O. & Kleinschmidt, J. H. (2003). Folding and insertion of the outer membrane protein OmpA is assisted by the chaperone Skp and by

lipopolysaccharide. *J. Biol. Chem.* 278, 9092–9099.

42 Bothmann, H. & Pluckthun, A. (1998). Selection for a periplasmic factor improving phage display and functional periplasmic expression. *Nat. Biotechnol.* 16, 376–380.

43 Hayhurst, A. & Harris, W. J. (1999). *Escherichia coli skp* chaperone coexpression improves solubility and phage display of single-chain antibody fragments. *Protein Expr. Purif.* 15, 336–343.

44 Strachan, G., Williams, S., Moyle, S. P., Harris, W. J. & Porter, A. J. (1999). Reduced toxicity of expression, in *Escherichia coli*, of antipollutant antibody fragments and their use as sensitive diagnostic molecules. *J. Appl. Microbiol.* 87, 410–417.

45 Mavrangelos, C., Thiel, M., Adamson, P. J., Millard, D. J., Nobbs, S., Zola, H. & Nicholson, I. C. (2001). Increased yield and activity of soluble single-chain antibody fragments by combining high-level expression and the Skp periplasmic chaperonin. *Protein Expr. Purif.* 23, 289–295.

46 Levy, R., Weiss, R., Chen, G., Iverson, B. L. & Georgiou, G. (2001). Production of correctly folded Fab antibody fragment in the cytoplasm of *Escherichia coli trxB gor* mutants via the coexpression of molecular chaperones. *Protein Expr. Purif.* 23, 338–347.

47 Sauer, F. G., Barnhart, M., Choudhury, D., Knight, S. D., Waksman, G. & Hultgren, S. J. (2000). Chaperone-assisted pilus assembly and bacterial attachment. *Curr. Opin. Struct. Biol.* 10, 548–556.

48 Knight, S. D., Berglund, J. & Choudhury, D. (2000). Bacterial adhesins: structural studies reveal chaperone function and pilus biogenesis. *Curr. Opin. Chem. Biol.* 4, 653–660.

49 Takeda, K., Miyatake, H., Yokota, N., Matsuyama, S., Tokuda, H. & Miki, K. (2003). Crystal structures of bacterial lipoprotein localization factors, LolA and LolB. *Embo J.* 22, 3199–3209.

50 Yokota, N., Kuroda, T., Matsuyama, S. & Tokuda, H. (1999). Characterization of the LolA-LolB system as the general lipoprotein localization mechanism of *Escherichia coli*. *J. Biol. Chem.* 274, 30995–30999.

51 Matsuyama, S., Yokota, N. & Tokuda, H. (1997). A novel outer membrane lipoprotein, LolB (HemM), involved in the LolA (p20)-dependent localization of lipoproteins to the outer membrane of *Escherichia coli*. *Embo J.* 16, 6947–6955.

52 Genevrois, S., Steeghs, L., Roholl, P., Letesson, J. J. & van der Ley, P. (2003). The Omp85 protein of Neisseria meningitidis is required for lipid export to the outer membrane. *Embo J.* 22, 1780–1789.

53 Voulhoux, R., Bos, M. P., Geurtsen, J., Mols, M. & Tommassen, J. (2003). Role of a highly conserved bacterial protein in outer membrane protein assembly. *Science* 299, 262–265.

54 Clausen, T., Southan, C. & Ehrmann, M. (2002). The HtrA Family of Proteases. Implications for Protein Composition and Cell Fate. *Mol. Cell* 10, 443–455.

55 Chen, J., Song, J. L., Zhang, S., Wang, Y., Cui, D. F. & Wang, C. C. (1999). Chaperone activity of DsbC. *J. Biol. Chem.* 274, 19601–19605.

56 Shao, F., Bader, M. W., Jakob, U. & Bardwell, J. C. (2000). DsbG, a protein disulfide isomerase with chaperone activity. *J. Biol. Chem.* 275, 13349–13352.

57 Swamy, K. H., Chung, C. H. & Goldberg, A. L. (1983). Isolation and characterization of protease do from *Escherichia coli*, a large serine protease containing multiple subunits. *Arch. Biochem. Biophys.* 224, 543–554.

58 Lipinska, B., Sharma, S. & Georgopoulos, C. (1988). Sequence analysis and regulation of the *htrA* gene of *Escherichia coli*: a sigma 32-independent mechanism of heat-inducible transcription. *Nucleic Acids Res.* 16, 10053–10067.

59 Strauch, K. L. & Beckwith, J. (1988).

An *Escherichia coli* mutation preventing degradation of abnormal periplasmic proteins. *Proc. Natl. Acad. Sci. USA* 85, 1576–1580.

60 Bass, S., Gu, Q. & Christen, A. (1996). Multicopy suppressors of *prc* mutant *Escherichia coli* include two HtrA (DegP) protease homologs (HhoAB), DksA, and a truncated R1pA. *J. Bacteriol.* 178, 1154–1161.

61 Waller, P. & Sauer, R. (1996). Characterization of *degQ* and *degS*, *Escherichia coli* genes encoding homologs of the DegP protease. *J. Bacteriol.* 178, 1146–1153.

62 Betton, J.-M., Sassoon, N., Hofnung, M. & Laurent, M. (1998). Degradation versus aggregation of misfolded maltose-binding protein in the periplasm of *Escherichia coli*. *J. Biol. Chem.* 273, 8897–8902.

63 Sone, M., Kishigami, S., Yoshihisa, T. & Ito, K. (1997). Roles of disulfide bonds in bacterial alkaline phosphatase. *J. Biol. Chem.* 272, 6174–6178.

64 Spiess, C., Beil, A. & Ehrmann, M. (1999). A temperature-dependent switch from chaperone to protease in a widely conserved heat shock protein. *Cell* 97, 339–347.

65 Misra, R., Peterson, A., Ferenci, T. & Silhavy, T. J. (1991). A genetic approach for analyzing the pathway of LamB assembly into the outer membrane of *Escherichia coli*. *J. Biol. Chem.* 266, 13592–13597.

66 Uhland, K., Mondigler, M., Spiess, C., Prinz, W. & Ehrmann, M. (2000). Determinants of translocation and folding of TreF, a trehalase of *Escherichia coli*. *J. Biol. Chem.* 275, 23439–23445.

67 Snyder, W. & Silhavy, T. (1995). Beta-galactosidase is inactivated by intermolecular disulfide bonds and is toxic when secreted to the periplasm of *Escherichia coli*. *J. Bacteriol.* 177, 953–963.

68 Kihara, A. & Ito, K. (1998). Translocation, folding, and stability of the HflKC complex with signal anchor topogenic sequences. *J. Biol. Chem.* 273, 29770–29775.

69 St Geme, J. & Grass, S. (1998). Secretion of the *Haemophilus influenzae* HMW1 and HMW2 adhesins involves a periplasmic intermediate and requires the HMWB and HMWC proteins. *Mol. Microbiol.* 27, 617–630.

70 Guigueno, A., Belin, P. & Boquet, P. (1997). Defective export in *Escherichia coli* caused by DsbA′-PhoA hybrid proteins whose DsbA′ domain cannot fold into a conformation resistant to periplasmic proteases. *J. Bacteriol.* 179, 3260–3269.

71 Baneyx, F. & Georgiou, G. (1991). Construction and characterization of *Escherichia coli* strains deficient in multiple secreted proteases: protease III degrades high-molecular-weight substrates in vivo. *J. Bacteriol.* 173, 2696–2703.

72 Arslan, E., Schulz, H., Zufferey, R., Kunzler, P. & Thony-Meyer, L. (1998). Overproduction of the *Bradyrhizobium japonicum* c-type cytochrome subunits of the cbb3 oxidase in *Escherichia coli*. *Biochem. Biophys. Res. Commun.* 251, 744–777.

73 Cavard, D. (1995). Role of DegP protease on levels of various forms of colicin A lysis protein. *FEMS Microbiol. Lett.* 125, 173–178.

74 Misra, R., CastilloKeller, M. & Deng, M. (2000). Overexpression of protease-deficient DegP(S210A) rescues the lethal phenotype of *Escherichia coli* OmpF assembly mutants in a *degP* background. *J. Bacteriol.* 182, 4882–4888.

75 CastilloKeller, M. & Misra, R. (2003). Protease-deficient DegP suppresses lethal effects of a mutant OmpC protein by its capture. *J. Bacteriol.* 185, 148–154.

76 Ades, S. E., Connolly, L. E., Alba, B. M. & Gross, C. A. (1999). The *Escherichia coli* sigma(E)-dependent extracytoplasmic stress response is controlled by the regulated proteolysis of an anti-sigma factor. *Genes Dev.* 13, 2449–2461.

77 Alba, B. M., Leeds, J. A., Onufryk, C., Lu, C. Z. & Gross, C. A. (2002).

DegS and YaeL participate sequentially in the cleavage of RseA to activate the sigma(E)-dependent extracytoplasmic stress response. *Genes Dev.* 16, 2156–2168.

78 Alba, B. M., Zhong, H. J., Pelayo, J. C. & Gross, C. A. (2001). *degS* (*hhoB*) is an essential *Escherichia coli* gene whose indispensable function is to provide sigma activity. *Mol. Microbiol.* 40, 1323–1333.

79 Redford, P., Roesch, P. L. & Welch, R. A. (2003). DegS is necessary for virulence and is among extraintestinal *Escherichia coli* genes induced in murine peritonitis. *Infect. Immun.* 71, 3088–3096.

80 Walsh, N. P., Alba, B. M., Bose, B., Gross, C. A. & Sauer, R. T. (2003). OMP peptide signals initiate the envelope-stress response by activating DegS protease via relief of inhibition mediated by its PDZ domain. *Cell* 113, 61–71.

81 Young, J. C. & Hartl, F. U. (2003). A stress sensor for the bacterial periplasm. *Cell* 113, 1–2.

82 Krojer, T., Garrido-Franco, M., Huber, R., Ehrmann, M. & Clausen, T. (2002). Crystal structure of DegP (HtrA) reveals a new protease-chaperone machine. *Nature* 416, 455–459.

83 Li, W., Srinivasula, S. M., Chai, J., Li, P., Wu, J. W., Zhang, Z., Alnemri, E. S. & Shi, Y. (2002). Structural insights into the pro-apoptotic function of mitochondrial serine protease HtrA2/Omi. *Nat. Struct. Biol.* 9, 436–441.

84 Kim, D. Y., Kim, D. R., Ha, S. C., Lokanath, N. K., Lee, C. J., Hwang, H. Y. & Kim, K. K. (2003). Crystal structure of the protease domain of a heat-shock protein HtrA from *Thermotoga maritima*. *J. Biol. Chem.* 278, 6543–6551.

85 Wang, J., Hartling, J. A. & Flanagan, J. M. (1997). The structure of ClpP at 2.3 A resolution suggests a model for ATP-dependent proteolysis. *Cell* 91, 447–456.

86 Groll, M., Ditzel, L., Lowe, J., Stock, D., Bochtler, M., Bartunik, H. D. & Huber, R. (1997). Structure of 20S proteasome from yeast at 2.4 A resolution. *Nature* 386, 463–471.

87 Brandstetter, H., Kim, J. S., Groll, M. & Huber, R. (2001). Crystal structure of the tricorn protease reveals a protein disassembly line. *Nature* 414, 466–470.

88 Braig, K., Otwinowski, Z., Hegde, R., Boisvert, D., Joachimiak, A., Horwich, A. & Sigler, P. (1994). The crystal structure of the bacterial chaperonin GroEL at 2.8 A. *Nature* 371, 578–586.

89 Jones, C. H., Dexter, P., Evans, A. K., Liu, C., Hultgren, S. J. & Hruby, D. E. (2002). *Escherichia coli* DegP protease cleaves between paired hydrophobic residues in a natural substrate: the PapA pilin. *J. Bacteriol.* 184, 5762–5771.

90 Silber, K. R., Keiler, K. C. & Sauer, R. T. (1992). Tsp: a tail-specific protease that selectively degrades proteins with nonpolar C termini. *Proc. Natl. Acad. Sci. USA* 89, 295–299.

91 Beebe, K. D., Shin, J., Peng, J., Chaudhury, C., Khera, J. & Pei, D. (2000). Substrate recognition through a PDZ domain in tail-specific protease. *Biochemistry* 39, 3149–3155.

92 Keiler, K. C., Silber, K. R., Downard, K. M., Papayannopoulos, I. A., Biemann, K. & Sauer, R. T. (1995). C-terminal specific protein degradation: activity and substrate specificity of the Tsp protease. *Protein Sci.* 4, 1507–1515.

93 Keiler, K. C., Waller, P. R. & Sauer, R. T. (1996). Role of a peptide tagging system in degradation of proteins synthesized from damaged messenger RNA. *Science* 271, 990–993.

94 Ding, L., Becker, A. B., Suzuki, A. & Roth, R. A. (1992). Comparison of the enzymatic and biochemical properties of human insulin-degrading enzyme and *Escherichia coli* protease III. *J. Biol. Chem.* 267, 2414–2420.

95 Gill, R. T., Valdes, J. J. & Bentley, W. E. (2000). A comparative study of global stress gene regulation in response to overexpression of

recombinant proteins in *Escherichia coli*. *Metab. Eng.* 2, 178–189.

96 Gill, R. T., DeLisa, M. P., Shiloach, M., Holoman, T. R. & Bentley, W. E. (2000). OmpT expression and activity increase in response to recombinant chloramphenicol acetyltransferase over-expression and heat shock in *E. coli*. *J Mol Microbiol Biotechnol* 2, 283–289.

97 Jurgen, B., Lin, H. Y., Riemschneider, S., Scharf, C., Neubauer, P., Schmid, R., Hecker, M. & Schweder, T. (2000). Monitoring of genes that respond to overproduction of an insoluble recombinant protein in *Escherichia coli* glucose-limited fed-batch fermentations. *Biotechnol. Bioeng.* 70, 217–224.

98 Stathopoulos, C. (1998). Structural features, physiological roles, and biotechnological applications of the membrane proteases of the OmpT bacterial endopeptidase family: a micro-review. *Membr. Cell. Biol.* 12, 1–8.

99 Altamirano, M. M., Garcia, C., Possani, L. D. & Fersht, A. R. (1999). Oxidative refolding chromatography: folding of the scorpion toxin Cn5. *Nat. Biotechnol.* 17, 187–191.

100 Ryabova, L. A., Desplancq, D., Spirin, A. S. & Pluckthun, A. (1997). Functional antibody production using cell-free translation: effects of protein disulfide isomerase and chaperones. *Nat. Biotechnol.* 15, 79–84.

101 Kanehara, K., Ito, K. & Akiyama, Y. (2002). YaeL (EcfE) activates the sigma(E) pathway of stress response through a site-2 cleavage of anti-sigma(E), RseA. *Genes Dev.* 16, 2147–2155.

102 Brown, M. S., Ye, J., Rawson, R. B. & Goldstein, J. L. (2000). Regulated intramembrane proteolysis: a control mechanism conserved from bacteria to humans. *Cell* 100, 391–398.

103 Hellingwerf, K. J., Postma, P. W., Tommassen, J. & Westerhoff, H. V. (1995). Signal transduction in bacteria: phospho-neural network(s) in *Escherichia coli*? *FEMS Microbiol. Rev.* 16, 309–321.

104 DiGiuseppe, P. A. & Silhavy, T. J. (2003). Signal detection and target gene induction by the CpxRA two-component system. *J. Bacteriol.* 185, 2432–2440.

105 De Wulf, P., Kwon, O. & Lin, E. C. (1999). The CpxRA signal transduction system of *Escherichia coli*: growth-related autoactivation and control of unanticipated target operons. *J. Bacteriol.* 181, 6772–6778.

106 Dorel, C., Vidal, O., Prigent-Combaret, C., Vallet, I. & Lejeune, P. (1999). Involvement of the Cpx signal transduction pathway of *E. coli* in biofilm formation. *FEMS Microbiol. Lett.* 178, 169–175.

107 Hung, D. L., Raivio, T. L., Jones, C. H., Silhavy, T. J. & Hultgren, S. J. (2001). Cpx signaling pathway monitors biogenesis and affects assembly and expression of P pili. *Embo J.* 20, 1508–1518.

108 Otto, K. & Silhavy, T. J. (2002). Surface sensing and adhesion of *Escherichia coli* controlled by the Cpx-signaling pathway. *Proc Natl Acad Sci USA* 99, 2287–2292.

109 De Wulf, P., McGuire, A. M., Liu, X. & Lin, E. C. (2002). Genome-wide profiling of promoter recognition by the two-component response regulator CpxR-P in *Escherichia coli*. *J. Biol. Chem.* 277, 26652–26661.

110 Shimohata, N., Chiba, S., Saikawa, N., Ito, K. & Akiyama, Y. (2002). The Cpx stress response system of *Escherichia coli* senses plasma membrane proteins and controls HtpX, a membrane protease with a cytosolic active site. *Genes Cells* 7, 653–662.

111 Hagenmaier, S., Stierhof, Y. D. & Henning, U. (1997). A new periplasmic protein of *Escherichia coli* which is synthesized in spheroplasts but not in intact cells. *J. Bacteriol.* 179, 2073–2076.

112 Raivio, T. L., Laird, M. W., Joly, J. C. & Silhavy, T. J. (2000). Tethering of CpxP to the inner membrane prevents spheroplast induction of the cpx envelope stress response. *Mol. Microbiol.* 37, 1186–1197.

113 Raffa, R. G. & Raivio, T. L. (2002). A third envelope stress signal transduc-

tion pathway in *Escherichia coli*. *Mol. Microbiol.* 45, 1599–1611.

114 Tamm, L. K., Arora, A. & Kleinschmidt, J. H. (2001). Structure and assembly of beta-barrel membrane proteins. *J. Biol. Chem.* 276, 32399–32402.

115 Lee, K., Zhan, X., Gao, J., Qiu, J., Feng, Y., Meganathan, R., Cohen, S. & Georgiou, G. (2003). RraA: a Protein Inhibitor of RNase E Activity that Globally Modulates RNA Abundance in *E. coli*. *Cell* 114, 623–634.

29
Formation of Adhesive Pili by the Chaperone-Usher Pathway

Michael Vetsch and Rudi Glockshuber

29.1
Basic Properties of Bacterial, Adhesive Surface Organelles

Adhesive surface organelles of gram-negative bacteria are classified into four distinct categories based on their assembly pathway. A first pathway, found in a wide variety of bacteria, requires a specialized pair of a periplasmic chaperone and an outer-membrane assembly platform, termed usher [1]. This pathway is the main topic of this article. A second pathway is employed for the assembly of type 4 or bundle-forming pili. Type 4 pilus formation requires a minimum of 14 assembly components that are homologous to many accessory proteins of the type II secretion pathway in gram-negative bacteria and to the flagellum biogenesis machinery of archaea [2]. A third pathway is used for the formation of curli, which form amorphous matrices surrounding certain bacteria. The subunits of curli are thought to be secreted as monomers to the cell surface, where accessory proteins promote their polymerization into fibers [3]. A fourth pathway is proposed for the biogenesis of CS1, CS2, and CFA/I fimbriae from enterotoxic *Escherichia coli*. Like the chaperone-usher pathway, this mechanism appears to involve a dedicated periplasmic assembly factor and a specialized outer-membrane assembly platform, but these proteins share no sequence similarity with the components of the chaperone-usher pathway [4].

The chaperone-usher pathway is the common assembly mechanism of at least 37 different bacterial surface filaments found in pathogenic bacterial strains. Moreover, there are at least an additional 23 pairs of chaperones and ushers in protein databases that are thought to promote the assembly of yet unknown fibers (Table 29.1). Organelles assembled by the chaperone-usher pathway are divided into pilus and non-pilus fibers [5]. Nevertheless, the fiber subunits from both classes have the same fold, and the same principles underlie subunit-subunit interactions and interactions of subunits with chaperones and ushers. The best-studied member of non-pilus fibers is the capsular F1 antigen from the plague pathogen *Yersinia pestis* [6–8].

In contrast to the non-pilus fibers, pili have a complex architecture. They are composite structures made of a rigid helical rod and a thin, flexible tip fibrillum

Protein Folding Handbook. Part II. Edited by J. Buchner and T. Kiefhaber

ISBN: 3-527-30784-2

Tab. 29.1. Overview of adhesive organelles assembled by the chaperone-usher pathway

Organelle	***Organism***	***Chaperone***	***Usher***	***Disease***	***References***
FGS class					
AF/R1 pilus	*Escherichia coli*	AfrC	AfrB	Diarrhea in rabbits	61
Atf pilus	*Proteus mirabilis*	AtfB	AtfC	UTI	62
CS18	*E. coli*	FotB	FotD	Diarrhea	63
CS31A pilus	*E. coli*	ClpE	ClpD	Diarrhea	64, 65
F1C pilus	*E. coli*	FocC	FocD	Cystitis?	66, 67
F17 pilus	*E. coli*	F17D	F17papC	Diarrhea in piglets	68, 69
F18	*E. coli*	FedB	FedC	Diarrhea in piglets	70
Haf pilus	*Haemophilus influenzae* biogroup *aegyptus*	HafB	HafC	Brazilian purpuric fever	71
Hif pilus	*H. influenzae*	HifB	HifC	Otitis media, meningitis	72
K88 pilus	*E. coli*	FaeE	FaeD	Neonatal diarrhea in piglets	73–75
K99 pilus	*E. coli*	FanE	FanD	Diarrhea in lambs, calves, piglets	76, 75
Lpf pilus	*Salmonella typhimurium*, *E. coli*	LpfB	LpfC	Gastroenteritis, salmonellosis?	77
MRF	*Photorhabdus temperata*	MrfD	MrfC	Entomopathogenicity	78
MR/K (type3) pilus	*Klebsiella pneumoniae*	MrkB	MrkC	Pneumonia	79
MR/P pilus	*P. mirabilis*	MrpD	MrpC	Nosocomial UTI	80, 81
P pilus	*E. coli*	PapD	PapC	Pyelonephritis, cystitis	33, 82
Pef pilus	*S. typhimurium*	PefD	PefC	Gastroenteritis, salmonellosis	83
Pix pilus	*E. coli*	PixD	PixC	UTI	84
PMF pilus	*P. mirabilis*	PmfD	PmfC	Nosocomial UTI	85, 86
S pilus	*E. coli*	SfaE	SfaF	UTI, NBM	87
Sef	*S. typhi*	SefB	SefC	Gastroenteritis, salmonellosis	88, 89
Sef	*Serratia entomophila*	SefD	SefC	Amber disease	90
Sfp	*E. coli*	SfpD	SfpC	Diarrhea, hemolytic-uremic syndrome	91
Type 1 pilus	*E. coli*, *S. typhi*, *Pseudomonas putida*, *K. pneumoniae*	FimC	FimD	Cystitis	34
Type 2 and 3 pili	*Bordetella pertussis*	FimB (FhaD)	FimC (FhaA)	Whooping cough	92, 93
987P pilus	*E. coli*	FasB	FasD	Diarrhea in piglets	94, 95
	E. coli	CshC	CshB		*
	E. coli	EcpD	HtrE		96
	E. coli	RalE	RalD	Diarrhea in rabbits	97
	E. coli	SfmC	SfmD		*
	E. coli, *Shigella flexineri*	YbgP	YbgQ		*
	E. coli	YcbR	YcbS		*
	E. coli	YehC	YehB		*
	E. coli	YqiH	YqiG		*
	E. coli	YraI	YraJ		*
	Edwardsiella tarda	EtfB	EtfC		98, *
	Haemophilus spp	GhfB	GhfC	Neonatal genital tract infections	99

Tab. 29.1. *(continued)*

Organelle	*Organism*	*Chaperone*	*Usher*	*Disease*	*References*
	Photorhabdus luminescens	PhfC	PhfD		*
	S. typhimurium	BcfB,G?	BcfC	Gastroenteritis?	100
	S. typhi	StaB	StaC	Gastroenteritis?	100
	S. typhi, S. typhimurium	StbB	StbC	Gastroenteritis?	100
	S. typhi	StcB	StcC	Gastroenteritis?	100
	S. typhi	StdC	StdB	Gastroenteritis?	100
	S. typhi	SteC	SteB	Gastroenteritis?	100
	S. typhimurium	StfD	StfC	Gastroenteritis	101
	S. typhimurium	SthA	SthB	Gastroenteritis?	100
	S. typhi	StiB	StiC	Gastroenteritis?	102
	S. flexineri	YcbF	YcbS		*
FGL class					
AAF/I	*E. coli*	AggD	AggC	Diarrhea	103
AAF/II	*E. coli*	AafD	AafC	Infantile diarrhea	104
Afa-1	*E. coli*	AfaB	AfaC	UTI, diarrhea	105, 106
CS3	*E. coli*	Cs3-1	Cs3-2	Traveler's diarrhea	107, 108
CS6 pilus	*E. coli*	CssC	CssD	Diarrhea	109, 110
Dr/Afa-111	*E. coli*	DraB	DraC	UTI, diarrhea	111, *
F1 antigen	*Yersinia pestis*	Caf1M	Caf1A	Plague	112, 113
MFA1-6 family	*E. coli*	NfaE	NfaC	UTI, NBM	114, 115
Myf	*Y. enteritidis*	MyfB	MyfC	Enterocolitis	116
PH6 antigen	*Y. pestis, Y. pseudotuberculosis*	PsaB	PsaC	Plague	117
Saf	*S. typhimurium*	SafB	SafC	Gastroenteritis, enteric fever	118

* Entry in SwissProt or TrEMBL.

joined to the distal end of the rod. Type 1 pili and P pili from enteropathogenic *E. coli* are two prominent and well-characterized members of this class. The rod of type 1 pili consists of 500–3000 FimA subunits, has a diameter of 6.9 nm, and is up to 2 μm long. The tip fibrillum is approximately 15 nm long and is formed by a linear array of the adaptor subunits FimF and FimG and the adhesive subunit FimH at the end of the pilus (Figures 29.1 and 29.2) [9–11]. The role of another potential subunit, FimI, presently remains unclear [12, 13].

Compared to type 1 pili, P pili have a longer tip fibrillum but otherwise similar dimensions. The rod of P pili is formed by the main structural subunit PapA, and the tip fibrillum consists mainly of the subunit PapE. The minor subunit PapK links the tip fibrillum to the rod, while PapF serves as an adaptor between the tip fibrillum and the adhesin PapG [14, 15]. Whereas non-pilus fibers generally do not contain a specialized adhesin, binding of pili to specific target molecules is accomplished by dedicated adhesin subunits. The adhesins, in contrast to all other subunits, which are single-domain proteins, usually contain two domains with different functions. The N-terminal lectin domains mediate binding to sugar moieties in glycoprotein receptors, whereas the C-terminal pilin domains are required for incorporation of the adhesins into the pilus structure [16–18].

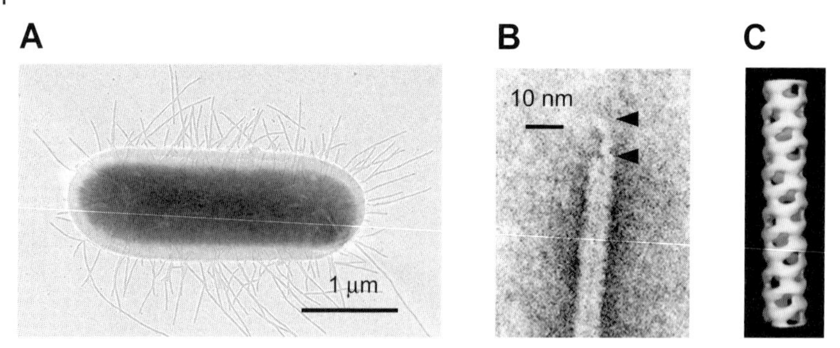

Fig. 29.1. Morphology of type 1 pili, which mediate binding of bacteria to the bladder epithelium. (A) Electron micrograph of an *Escherichia coli* cell bearing type 1 pili. A single *E. coli* cell may bear several hundred type 1 pili. (B) At the distal end of the rigid pilus rod, a thin, flexible tip fibrillum (between the arrowheads) is observed that contains the mannose-binding adhesin. (C) 3D reconstruction from electron microscopy data showing a type 1 pilus rod segment comprising 40 FimA subunits.

Pili are often associated with the capability of bacterial strains to cause infections (cf. Table 29.1). For example, the vast majority of urinary tract infections are caused by uropathogenic *E. coli* strains expressing a number of virulence factors that enable them to colonize the urinary tract and to persist despite robust host defense mechanisms [19]. The interplay between uropathogenic bacteria and the defense mechanisms in the bladder is reviewed thoroughly in Ref. [20]. Importantly, the initial step in establishing pathogenesis critically relies on the ability of the bacteria to attach to host tissue through specific adhesins. Among uropathogenic *E. coli* strains, type 1 pili are the most widely distributed adhesive organelles [21]. The adhesin of type 1 pili, FimH, mediates mannose-sensitive binding to the glycoprotein uroplakin Ia, an abundant integral membrane protein of the bladder epithelium [22]. Moreover, FimH triggers the internalization of bacteria into bladder cells [23]. Within the host cell, bacteria are protected from a variety of host defense mechanisms as well as antibiotics. Internalized bacteria not only persist but in some cases even replicate [24] and thus constitute a reservoir of pathogenic bacteria, which may cause the recurrent infections that are seen in a large portion of women with urinary tract infections [25]. P pili from uropathogenic *E. coli* strains enable the bacteria to bind to galα1-4gal units from glycolipids in the human kidney and thus to gain an initial foothold in the kidney [26], where they cause pyelonephritis [27].

Figure 29.2 illustrates important principles in the formation of fibers by the chaperone-usher pathway, using type 1 pili from *E. coli* as example. The formation of each type of fiber involves two dedicated proteins: a periplasmic chaperone (FimC in Figure 29.2) and an assembly platform, termed usher, in the outer membrane (FimD in Figure 29.2). Both components are essential for the assembly of fibers [28, 29]. The subunit polypeptides that eventually assemble into the fiber (FimA, FimF, FimG, and FimH in Figure 29.2) have N-terminal signal sequences and enter the periplasm through the SecYEG translocon. During translocation of the subunits or shortly thereafter, signal peptidase cleaves off the signal sequence

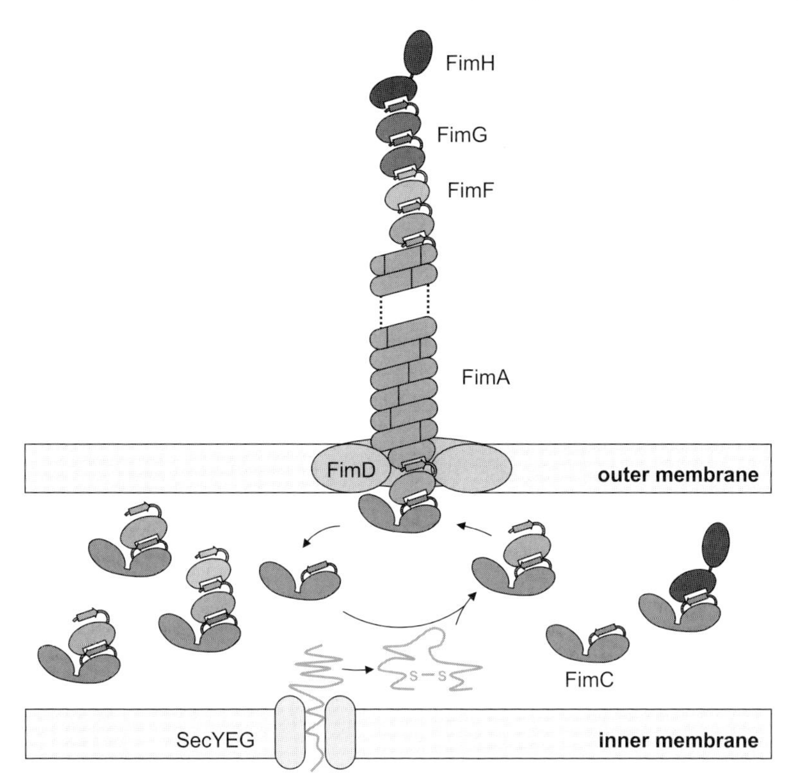

Fig. 29.2. Schematic presentation of the current view of type 1 pilus assembly via the chaperone-usher pathway. When newly synthesized structural pilus subunits (FimA, FimF, FimG, and FimH) enter the periplasm, their signal sequence is cleaved off and a single disulfide bridge is introduced into each pilin polypeptide. The subunits then fold and form heterodimers with the type 1 pilus chaperone FimC. These complexes bind to the usher FimD in the outer membrane, which triggers dissociation of the chaperone-subunit complexes. While the chaperone is released to the periplasm, the subunit is incorporated into the growing pilus. Translocation of subunits occurs via FimD, which is a ring-shaped oligomer with a central pore that allows translocation of folded subunits to the cell surface. In the quaternary structure of the pilus, every subunit donates an N-terminal extension, termed donor strand, to the preceding subunit, thereby complementing the fold of the preceding subunit.

and DsbA, the periplasmic dithiol oxidase of *E. coli* [30], introduces a single disulfide bond, which is invariant in all subunits. This disulfide bond most likely enhances the stability of the subunits but appears to be dispensable for the assembly process in the case of type 1 pili, as *E. coli* strains lacking DsbA retain the ability to produce functional type 1 fibers [31]. The fact that DsbA is not required for type 1 pilus assembly could be due to the extraordinary stability of the quaternary structure of type 1 pili against dissociation and denaturation, which may compensate for a lower intrinsic stability of non-oxidized subunits. In the case of P pili, however, DsbA is required [32], possibly because the single disulfide bond in the P pilus chaperone PapD is required for PapD's structural integrity.

After signal sequence cleavage and disulfide bond formation, the subunits fold and form soluble complexes with the periplasmic pilus chaperones [33, 34]. The chaperone-subunit complexes diffuse to the outer membrane, where they dissociate upon contact with the usher. The free pilus chaperone is released to the periplasm for another round of subunit binding and subunit transport to the usher. Thus, the chaperone is a true assembly factor and not part of the final pilus structure [33]. The usher itself is a ring-shaped oligomer with a central pore, which most likely allows translocation of subunits in their folded state [12, 35]. The thermodynamic drive for pilus assembly might come from quaternary structure formation in the extracellular space. Pilus formation after subunit secretion to the periplasm would thus be a spontaneous process that does not require ATP or other energy sources [36].

29.2
Structure and Function of Pilus Chaperones

To date, the structures of four bacterial pilus chaperones are known. The structures of PapD (P pilus system of *E. coli*), FimC (type 1 pilus system of *E. coli*), SfaE (S pilus system of *E. coli*), and Caf1M (capsular F1 antigen of *Y. pestis*) were solved by X-ray crystallography or NMR [7, 16, 37–41]. The tertiary structures of all these chaperones are very similar. Each pilus chaperone comprises two immunoglobulin (Ig)-like domains joined at approximately right angles. The two last strands of the N-terminal domain (the F1- and G1-strands) are connected by a long loop (F1-G1 loop) protruding away from the domain (Figure 29.3). Structure-based sequence alignments of all known pilus chaperones showed that the length of the F1-G1 loop is a criterion to distinguish between two closely related but different classes of pilus chaperones. FGS chaperones (e.g., PapD, FimC, or SfaE) have a short F1-G1 loop and are required for the assembly of rigid pili. In contrast, FGL chaperones (e.g., Caf1M) have a long F1-G1 loop and assist in the formation of non-pilus fibers [5]. The F1-G1 loop comprises a motif of alternating hydrophobic and hydrophilic residues, which play a critical role in subunit binding (see Section 29.3). The motif contains three hydrophobic residues in FGS chaperones and four to five hydrophobic residues in most FGL chaperones [6]. A comparison of monomeric chaperones and chaperones with bound pilus subunit shows that upon binding of subunits, the conformation of the F1-G1 loop alters significantly while the overall structure of the chaperone structure is not affected [16, 38]. In addition to the variable loop length, a second difference between the two classes is observed. FGL chaperones contain a conserved disulfide bond connecting the F- and the G-strand of the N-terminal domain [8], while FGS chaperones lack this disulfide bond. The disulfide bond that is found in some FGS chaperones (e.g., PapD) is close to the C-terminus and connects the two C-terminal β-strands. Formation of this disulfide is critical for the in vivo function of PapD [32].

Periplasmic pilus chaperones are multifunctional proteins. They prevent both premature assembly of pilus subunits [42, 43] and nonspecific aggregation of subunits in the periplasm [42, 44, 45]. But unlike classical chaperones, the pilus chaperones form stable complexes with folded substrate proteins. The kinetic stability

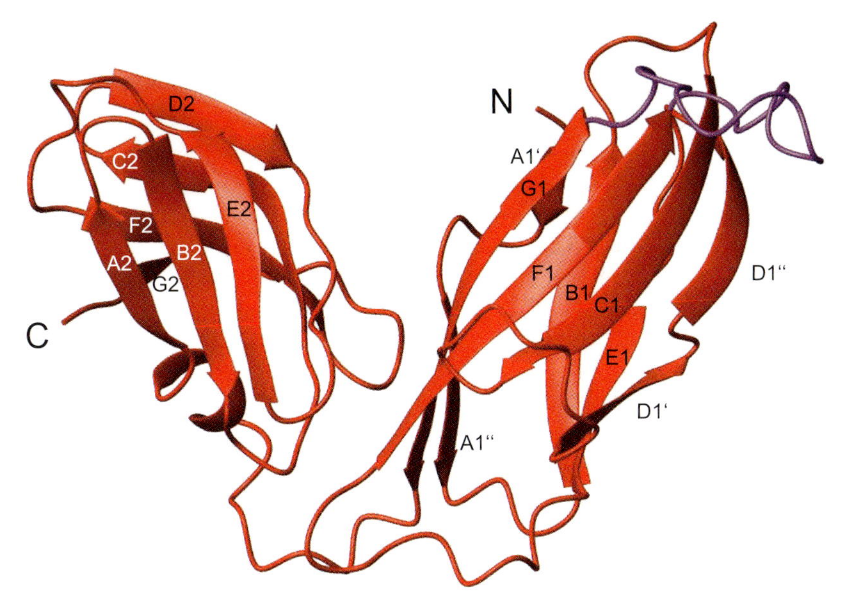

Fig. 29.3. Ribbon diagram of the type 1 pilus chaperone FimC [40]. The loop connecting the F1- and G1-strands is colored purple. Based on the length of the F1-G1 loop, pilus chaperones are classified into FGL chaperones (long loops) and FGS chaperones (short loops), of which FimC is a member. FimC (23 kDa) is a β-sheet protein composed of two domains (residues 1–115 and 131–205), connected by a 15-amino-acid linker.

of such complexes against dissociation most likely prevents formation of subunit-subunit complexes in the periplasm. This model is supported by the finding that pilus chaperones cap a region of the subunits that is required for the interaction with a neighboring subunit in the pilus [8, 39], as originally proposed by Soto et al. [43]. Despite the relatively strong interactions between pilus chaperones and their target subunits, bacterial pili assemble within a few minutes in vivo [36]. This suggests that chaperone-subunit complexes dissociate rapidly only when they are in contact with the outer-membrane usher. In addition, pilus chaperones bind to peptides corresponding to the C-terminal β-strand of target subunits, suggesting that pilus chaperones may recognize unfolded subunits and assist subunit folding [43, 46]. Another function of pilus chaperones, discussed in more detail below, is targeting of chaperone-subunit complexes to the ushers, which have a periplasmically oriented binding domain that exclusively recognizes chaperone-subunit complexes [47].

29.3
Structure and Folding of Pilus Subunits

Three recently determined crystal structures of chaperone-subunit complexes have contributed enormously to our understanding of the pilus assembly process [8, 16, 38]. The common, most striking finding was that the subunits have an atypical Ig-like fold that lacks the C-terminal G-strand. This creates a large hydrophobic groove on the subunit surface. In the chaperone-subunit complex, this groove is occupied by an extrinsic β-strand, provided by the chaperone [8, 16, 38]. This remarkably tight interaction is referred to as "donor strand complementation", and the extrinsic β-strand is called "donor strand". Specifically, a part of the G1-strand of the chaperone that includes the motif with alternating hydrophobic residues in the F1-G1 loop is inserted between the A″- and the F-strand of the subunit (Figures 29.4B and 29.5B, see p. 974/975). The conserved hydrophobic residues in the F1-G1 loop of the chaperone, which are solvent-exposed in the free chaperone, reach deep into the hydrophobic core of the bound subunit. The complemented structure of the subunit has an atypical Ig-like fold because the donated strand runs parallel to the F-strand of the subunit rather than antiparallel (Figures 29.4B and 29.5B).

The second important result from the structures of chaperone-subunit complexes, modeling studies, and biochemical experiments is that donor strand complementation also might govern subunit-subunit interactions in the quaternary structure of pili [16, 38]. Sequence alignment of structural pilus subunits shows that they are structurally homologous, single-domain β proteins in which the N-terminal ~14 amino acids are not part of the tertiary structure of the subunits. These N-terminal segments serve as donor strands for a preceding subunit in the pilus and replace the donor strand of the chaperone in this subunit. In contrast to the donor strand of the chaperone, however, the donor strand provided by a neighboring pilus subunit is assumed to insert antiparallel to the F-strand of the acceptor subunit (Figure 29.5C), creating a canonical Ig-like fold [16, 38]. The concept proposing that pilus subunits are both β-strand donors and β-strand acceptors also nicely explains why pilus adhesins are located only at the tip of the pilus [11]. Unlike other subunits, the adhesins lack the N-terminal donor strand, which is replaced by a lectin domain (e.g., PapG in the case of P pili and FimH in the case of type 1 pili). Thus, adhesins are only donor stand acceptors and act as pilus caps: they are either incorporated as the first subunit at the tip of a pilus or not at all (cf. also Figure 29.2).

The structural proof of donor strand complementation between pilus subunits came from the X-ray structure of the ternary complex between the chaperone Caf1M and two subunits of Caf1, which forms the homo-oligomeric fiber of the *Yersinia pestis* F1 capsular antigen [8]. In this structure, the first Caf1 subunit is bound to the chaperone via parallel donor strand complementation, while the second Caf1 subunit interacts with the first subunit via antiparallel donor strand complementation. Comparison of the structures of both Caf1 subunits revealed that the two β-sheets of the Ig-like fold in the chaperone-bound subunit have low shape correlation statistics [48], reflecting a loosely packed hydrophobic core. Meanwhile, the second Caf1 subunit that is complexed with the first Caf1 has a tightly packed hydrophobic core, as is typically observed for β-barrels consisting of two β-sheets. This, in more general terms, suggests a collapse of the hydrophobic core of a subunit upon release from the chaperone and assembly into the fiber oligomer. This conformational change could result in a gain in free folding energy. Zavaliov et al. [8] propose that this potential difference in free energy might constitute a driving force for fiber assembly, in addition to stabilizing intermolecular contacts in the quaternary structure. It remains to be tested whether subunit-subunit interactions are indeed thermodynamically more stable than chaperone-subunit interactions and, if this proves true, how periplasmic pilus chaperones prevent spontaneous formation of more stable subunit-subunit interactions in the periplasm.

Removal of an extrinsic donor strand from a pilus subunit causes a significant exposure of the subunit's hydrophobic core to the solvent. This led to the hypothesis that pilus subunits lack essential steric information to fold autonomously to a defined tertiary structure and that they strictly require assistance of the chaperone to fold [16, 38]. Support for this model came from the observation that high-level expression of pilus subunits is possible only when the corresponding chaperone is co-expressed [44] and from unsuccessful attempts to refold chemically denatured subunits in the absence of chaperone in vitro [49]. More recent experiments, however, challenge this hypothesis. It was found that the pilin domain of FimH folds autonomously and with a cooperativity that is characteristic for a protein of the size of the pilin domain [45]. However, the thermodynamic stability of the pilin domain is very low ($\Delta G^{\circ}_{fold} = -10$ kJ mol^{-1}), such that a small but significant fraction of molecules is unfolded even under native conditions. This explains the high tendency of the pilin domain to aggregate nonspecifically in the absence of stoichiometric amounts of the chaperone FimC. The simplest conclusion from these data is that pilin domains can fold autonomously but are intrinsically very unstable without the donor strand. Thermodynamic stabilization by donor strand complementation through the chaperone thus most likely prevents nonspecific aggregation of pilus subunits. The role of pilus chaperones in the subunit folding reactions is presently unknown. On the one hand, there is evidence that subunits can fold autonomously prior to binding to the chaperone in vitro [50, 51]. On the other hand, refolding of pilus subunits in the presence of the chaperone increases the yield of chaperone-subunit complexes compared to refolding of the subunit alone and subsequent addition of the chaperone. This may indicate a more "active" role of pilus chaperones in pilus subunit folding [49]. This view is supported by the observation that spontaneous folding of the FimH pilin domain is very slow [45], especially compared to the fast incorporation of subunits into pili in vivo [36].

29.4
Structure and Function of Pilus Ushers

The only specialized membrane protein involved in pilus assembly is the pilus usher, which is located in the outer bacterial membrane. In the absence of the usher, chaperone-subunit complexes accumulate in the periplasm and no fibers are assembled [7, 29, 33]. PapC (86 kDa per subunit) and FimD (91 kDa per subunit) are the ushers required for formation of P pili and type 1 pili, respectively. Electron microscopy revealed that they comprise 6–12 subunits forming ring-shaped oligomers with an outer diameter of 15 nm and a pore that is about 2 nm wide [12, 35]. Proteoliposome swelling assays with PapC and FimD showed that the ushers indeed form pores [35]. A pore with a diameter of 2 nm is too small to accommodate pilus rods, which are about 7 nm wide [11, 52]. It has therefore been

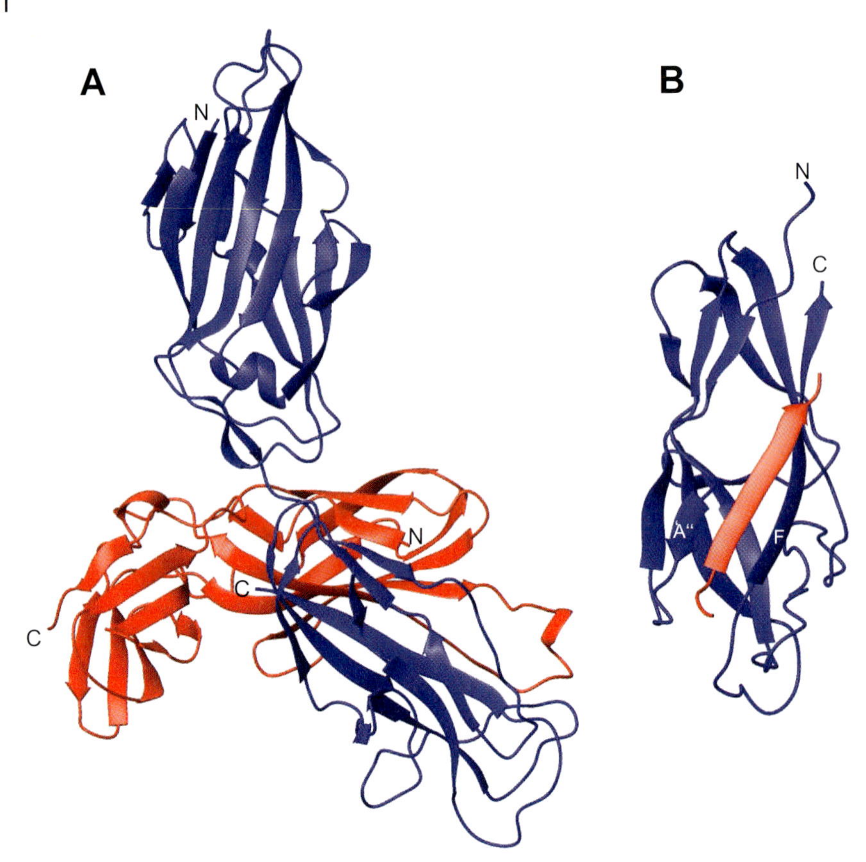

Fig. 29.4. Donor strand complementation in chaperone-subunit complexes [16]. (A) Ribbon diagram of the X-ray structure of the FimC-FimH complex. The type 1 pilus adhesin FimH (blue) has two domains with different roles. The N-terminal lectin domain binds to mannosylated uroplakin Ia on the bladder epithelium, while the C-terminal pilin domain mediates the association with the subunit FimG in the intact pilus. FimC (red) almost exclusively interacts with the pilin domain of FimH via its N-terminal domain. There is no interaction between the lectin domain of FimH and FimC. (B) View of the pilin domain of FimH in the context of the FimC-FimH complex. The pilin domain (blue) has an incomplete Ig-like β-fold lacking one β-strand and is structurally homologous to the other pilus subunits FimA, FimF, and FimG. In the chaperone-subunit complex, the missing strand, termed donor strand, is provided by the chaperone and inserted between the N-terminal A″-strand and the C-terminal F-strand of the pilin domain. Because the orientation of the donor strand of FimC (red) is parallel relative to the F-strand of the pilin domain, an atypical Ig-like fold is created.

proposed that assembled subunits leave the periplasm as a linear array of subunits and that formation of the helical rod occurs on the bacterial surface [35]. This model is supported by the observation that subunits interact head-to-tail with each other [8, 39]. Hence, it is conceivable that subunits are translocated through the central pore of the ushers as an extended, linear oligomer, including subunits like PapA and FimA, which eventually form helical pilus rods.

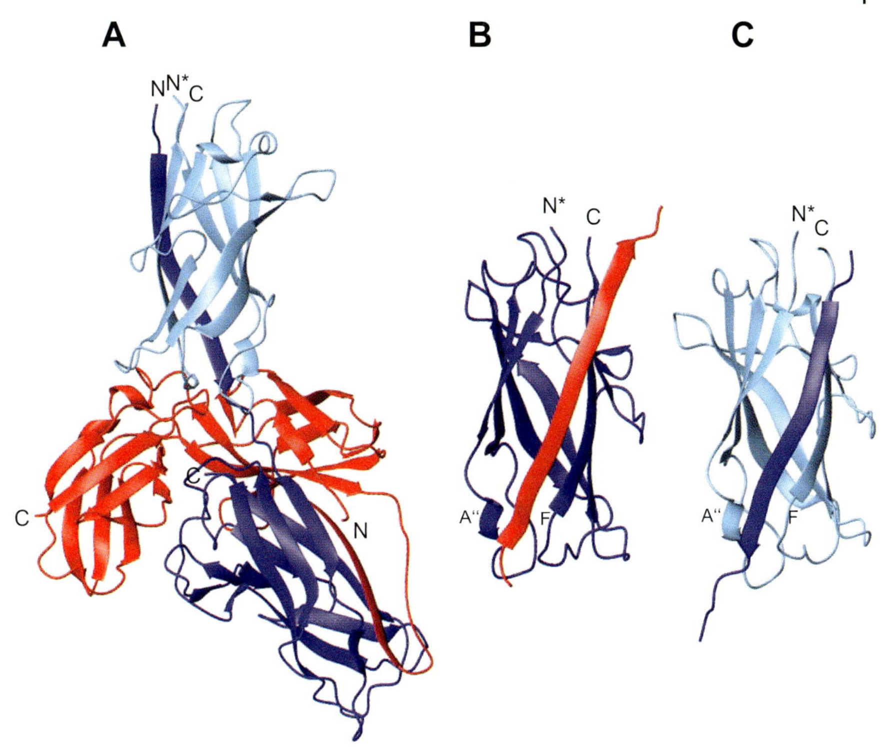

Fig. 29.5. Structural basis of donor strand complementation in subunit-subunit interactions. The F1 capsular antigen of *Yersinia pestis* is a linear, homo-oligomeric fiber of Caf1 subunits. The structure of the ternary complex between two Caf1 subunits and the assembly chaperone Caf1M was solved by X-ray crystallography [8]. (A) Ribbon diagram of the Caf1M-Caf1-Caf1 complex. The chaperone Caf1M (red) interacts with the first Caf1 subunit (dark blue), to which the second Caf1 subunit (light blue) is bound. (B) A close-up of the chaperone-bound Caf1 subunit reveals that the Ig-like fold of the subunit is complemented by a donor strand from the chaperone (red) running parallel to the C-terminal F-strand of the subunit. Exactly the same observation was made for the pilin domain of FimH in the FimC-FimH complex [16] (cf. Figure 4B) and for PapK in the PapD-PapK complex [38]. The asterisk indicates that the N-terminal 15-residue donor strand was omitted for clarity. (C) Inspection of the second Caf1 subunit shows that the donor strand provided by the first subunit also inserts between the A″- and F-strands of the acceptor subunit, but in the opposite, antiparallel orientation. Due to the antiparallel insertion of the extrinsic donor strand relative to the F-strand of the subunit, a canonical Ig-like fold is obtained.

The ushers of type 1 and P pili (FimD and PapC, respectively) were reported to bind their target chaperone-subunit complexes with affinities in the nanomolar to micromolar range [53]. In both pilus systems, the affinities of the ushers for the different chaperone-subunit complexes differ by several orders of magnitude. Strikingly, the chaperone-subunit complexes containing the adhesin (FimH and PapG, respectively) have the highest affinity for their cognate usher [53]. Because the adhesins are thought to be the first subunits to become incorporated into the pilus, this observation has led to the hypothesis that the relative affinities of the usher for the various chaperone-subunit complexes determine the order of the subunits within the pilus [53, 54]. At the least, this mechanism would ensure that a substantial fraction of pili bear an adhesin at their tip. Besides the specific affinities of the ushers for the various chaperone-subunit complexes, the relative affinities of the subunits towards each other are most likely another critical factor determining the final quaternary structure of pili.

Little is known about the molecular details of the binding of chaperone-subunit complexes to the ushers. It is established that isolated pilus chaperones are not bound by ushers [53]. This could mean that pilus subunits contain sufficient structural information for being recognized by the ushers. The finding that different chaperone-subunit complexes containing the same chaperone have different affinities for the target usher supports this view, together with the report that the pilin domain of the type 1 pilus subunit FimH has the capability to bind to the usher FimD even in the absence of the chaperone FimC [55].

Primary structure analysis of pilus ushers indicates that these proteins share a common topology. Computational sequence analysis and biochemical topology analysis of the usher FaeD from K88 fimbriae suggest a central transmembrane domain flanked by N- and C-terminal periplasmically oriented domains [56, 57]. Recent experiments with the P pilus usher PapC indicate that the ability of ushers to bind chaperone-subunit complexes resides in their N-terminal regions, while central and C-terminal parts appear to be required for subunit assembly or translocation [58]. Progress towards a more detailed understanding of the interactions between chaperone-subunit complexes and ushers has been made by Nishiyama et al. [47], who discovered that the N-terminal, periplasmic domain of the usher FimD ($FimD_N$), comprising the N-terminal 139 FimD residues, can be expressed as a soluble, monomeric protein that folds autonomously. In addition, $FimD_N$ specifically binds chaperone-subunit complexes and, like full-length FimD, does not bind the isolated chaperone FimC. Unilke FimD, however, $FimD_N$ fails to recognize isolated subunits [47, 55]. In addition, the affinity of $FimD_N$ for chaperone-subunit complexes is lower compared to that reported for full-length FimD. A possible reason for these lower affinities may be the monomeric state of $FimD_N$. Ring-shaped FimD oligomers are supposed to have multiple $FimD_N$ binding sites in close proximity, which may increase the apparent affinity of the oligomer for pilus subunits. Moreover, important segments of FimD involved in recognition of chaperone-subunit complexes may not be contained in $FimD_N$. Despite these uncertainties, the availability of $FimD_N$ opens the possibility of structure determination of ternary complexes among $FimD_N$, FimC, and different type 1 pilus subunits. These ternary complexes are predicted intermediates in pilus assembly.

29.5 Conclusions and Outlook

In the past few years, our understanding of the assembly of bacterial pili has tremendously improved through the determination of the three-dimensional structures of chaperone-subunit complexes from different pilus systems. These structures have revealed the mechanism of donor strand complementation as the common principle underlying pilus subunit stabilization and subunit binding by pilus chaperones, as well as subunit-subunit interactions in the context of the quaternary structure of pili. Despite these important fundamental insights, there are still many intriguing questions to be answered to gain a complete mechanistic understanding of bacterial pilus assembly: Do pilus chaperones influence or even catalyze folding of pilus subunits? Why are pili extremely stable macromolecular assemblies that resist unfolding by acid or high concentrations of denaturants, while individual pilus subunits are comparably unstable? Which factors determine the average pilus length distribution and average number of pili per cell? What are the three-dimensional structures and detailed mechanisms of the ushers? Ushers are predicted not only to bind specifically to chaperone-subunit complexes but also to catalyze pilus assembly by accelerating the dissociation of chaperone-subunit complexes and promoting subunit-subunit interactions. No functional in vitro assay for a purified, intact assembly platform complex could be established so far. In principle, catalytic amounts of a functional usher should immediately trigger pilus formation in vitro when added to a mixture of purified chaperone-subunit complexes. The latter experiment is most likely the key towards the complete in vitro reconstitution of macromolecular pilus assembly.

29.6 Experimental Protocols

29.6.1 Test for the Presence of Type 1 Piliated *E. coli* Cells

The presence of adhesive pili on the bacterial surface can often be tested easily by agglutination with other cells bearing the target molecule of the pilus adhesin. In the case of type 1 pili from *E. coli*, agglutination with yeast cells is used (Figure 29.6). Bacteria are grown in LB medium without shaking for two days, harvested by centrifugation, and resuspended in PBS buffer, pH 7.5, to a final optical density at 600 nm of 1. A 500-µl aliquot is mixed with 20 µl of a 10% (w/v) suspension of dry baker's yeast in PBS. The extent of agglutination is examined 5 min after mixing. Baker's yeast cells bear multiple mannose units on their surface and therefore agglutinate with bacterial strains expressing functional type 1 pili to three-dimensional networks. This leads to the formation of large, visible clots. In the absence of type 1 pili, no agglutination is observed.

29.6.2 Functional Expression of Pilus Subunits in the *E. coli* Periplasm

Both functional expression of pilus subunits in the periplasm of *E. coli* and purification of pilus subunits critically depend on the presence of stoichiometric amounts of chaperone. High expression yields are achieved with dicistronic operons ensuring a 1:1 ratio of both genes at the mRNA level. To minimize degradation of pilus

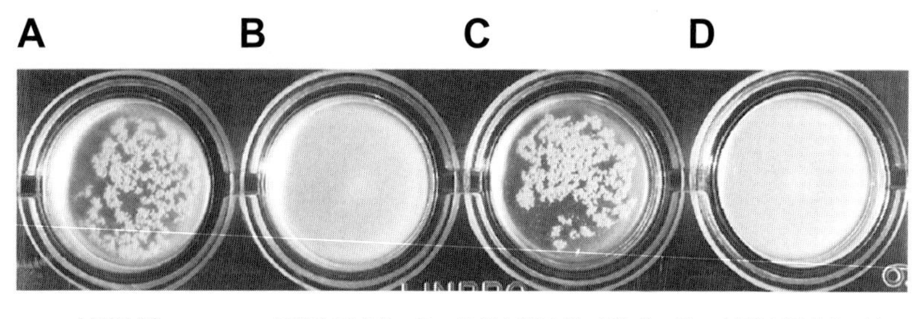

Fig. 29.6. Agglutination of yeast cells by *E. coli* cells bearing type 1 pili. The presence of functionnal type 1 pili can be tested by agglutination of the bacteria with bakers yeast. (A) The adhesin FimH at the tip of type 1 pili of *E. coli* strain W3110 binds to mannose units on the surface of yeast cells. As a consequence, visible clots are formed consisting of bacteria and yeast cells. (B) An isogenic W3110 strain lacking the gene encoding the pilus chaperone FimC (W3110Δ*fimC*) is unable to assemble pili. No agglutination is observed. (C) The ability of W3110Δ*fimC* to agglutinate with yeast cells is restored when the strain carries the plasmid pfimC for expression of FimC. (D) The strain W3110Δ*fimH* expresses type 1 pili, but these lack the adhesin FimH at their tip. Thus, no agglutination occurs.

subunits, *E. coli* strains lacking periplasmic proteases, e.g., strain HM 125 [59], can be employed. Expression yields strongly depend on the individual subunit. For example, comparison of the expression yields of the complexes FimC-FimH and FimC-FimA using analogous, dicistronic operons shows that the chaperone FimC is expressed at identical levels in both cases, while the subunit FimH is expressed in much higher amounts than the subunit FimA (M. Vetsch, unpublished results). Similar amounts of FimH and FimA are probably produced, but FimH possibly has a higher affinity for FimC than does FimA. Subunits that dissociate from the chaperone are subject to kinetic competition between rebinding to the chaperone on the one hand and nonspecific aggregation or self-polymerization on the other hand. The latter side reactions are most likely responsible for the difficulty in expressing soluble, monomeric subunits in the periplasm. In addition, isolated subunits are certainly also prone to proteolytic degradation, due to their low thermodynamic stability.

29.6.3
Purification of Pilus Subunits from the *E. coli* Periplasm

Chaperone-subunit complexes can be purified from periplasmic extracts by ion-exchange and hydrophobic chromatography. Alternatively, a hexahistidine tag may be fused to the C-terminus of the chaperone. This extension does not affect the interaction with subunits, which essentially interact with the N-terminal domain of the chaperone. When the chaperone is tagged, immobilized metal-affinity chromatography (IMAC) can be used to isolate the chaperone with the bound subunit from periplasmic extracts. Because co-expression of the chaperone and a subunit generally results in accumulation of both chaperone-subunit complexes and an excess of monomeric chaperone, an additional purification step is necessary to obtain a homogenous preparation of chaperone-subunit complex. Usually, ion-exchange chromatography is used for this purpose. Chaperone-subunit complexes are in most cases very stable and can therefore be stored for prolonged periods of time. To isolate the subunits, chaperone-subunit complexes are unfolded and dissociated with chemical denaturants such as 5 M urea or 2.5 M guanidinium chloride (GdmCl). The subunits can then be separated from the chaperone by ion-exchange chromatography in the presence of urea or by IMAC in the presence of GdmCl or urea (if the chaperone has a hexahistidine tag). Due to the high tendency of subunits to aggregate in the absence of chaperone, subunits should always be freshly prepared directly before further biochemical experiments [45, 47, 51].

29.6.4 Preparation of Ushers

Preparation of pure and functional oligomeric ushers is a prerequisite for biochemical studies aimed at a detailed understanding of usher function. So far, only the ushers PapC and FimD, both containing a hexahistidine tag at the C-terminus, have been purified successfully [35, 53]. The C-terminal tag does not interfere with the biological activity of the ushers, as the tagged usher variants complement usher deficiency in strains with chromosomal usher deletions. On the other hand, high-level expression of ushers is often detrimental for cell growth. Consequently, tightly regulated promoters such as the arabinose-inducible P_{BAD} promoter [60] and short-term induction of protein expression are used for high-level expression of ushers. After disruption of the cells, differential centrifugation is used to enrich the membranes. Subsequently, proteins of the inner membrane are selectively solubilized with the detergent sarcosyl (1%) to enrich outer-membrane proteins. Finally, the detergents Zwittergent 3–14 (1%) or Elugent (1%) are used to solubilize the outer-membrane ushers, which are then purified by IMAC [35, 53]. Ushers isolated according to this procedure were shown to retain the ability to bind chaperone-subunit complexes in the presence of Zwittergent 3–14 (0.1%) or Elugent (0.1%) [35, 53]. As this function lies in the N-terminal chaperone-subunit–binding domain of the ushers, however, this is not proof for an entirely functional usher molecule with an intact transmembrane region. Further functional assays for purified ushers have not been reported so far.

Acknowledgements

We would like to thank Chasper Pourger for preparing Table 29.1 and Ulla Grauschopf for critically reading the manuscript.

References

1 Sauer, F. G., Barnhart, M., Choudhury, D., Knight, S. D., Waksman, G. & Hultgren, S. J. (2000). Chaperone-assisted pilus assembly and bacterial attachment. *Curr Opin Struct Biol* **10**, 548–56.

2 Peabody, C. R., Chung, Y. J., Yen, M. R., Vidal-Ingigliardi, D., Pugsley, A. P. & Saier, M. H., Jr. (2003). Type II protein secretion and its relationship to bacterial type IV pili and archaeal flagella. *Microbiology* **149**, 3051–72.

3 Chapman, M. R., Robinson, L. S., Pinkner, J. S., Roth, R., Heuser, J., Hammar, M., Normark, S. & Hultgren, S. J. (2002). Role of Escherichia coli curli operons in directing amyloid fiber formation. *Science* **295**, 851–5.

4 Smyth, C. J., Marron, M. B., Twohig, J. M. & Smith, S. G. (1996). Fimbrial adhesins: similarities and variations in structure and biogenesis. *FEMS Immunol Med Microbiol* **16**, 127–39.

5 Hung, D. L., Knight, S. D., Woods, R. M., Pinkner, J. S. & Hultgren, S. J. (1996). Molecular basis of two subfamilies of immunoglobulin-like chaperones. *Embo J* **15**, 3792–805.

6 MacIntyre, S., Zyrianova, I. M., Chernovskaya, T. V., Leonard, M., Rudenko, E. G., Zav'Yalov, V. P. & Chapman, D. A. (2001). An extended hydrophobic interactive surface of Yersinia pestis Caf1M chaperone is essential for subunit binding and F1 capsule assembly. *Mol Microbiol* **39**, 12–25.

7 Zavialov, A. V., Kersley, J., Korpela, T., Zav'yalov, V. P., MacIntyre, S. & Knight, S. D. (2002). Donor strand complementation mechanism in the biogenesis of non-pilus systems. *Mol Microbiol* **45**, 983–95.

8 Zavialov, A. V., Berglund, J., Pudney, A. F., Fooks, L. J., Ibrahim, T. M., MacIntyre, S. & Knight, S. D. (2003). Structure and biogenesis of the capsular F1 antigen from Yersinia pestis: preserved folding energy drives fiber formation. *Cell* **113**, 587–96.

9 Klemm, P. & Christiansen, G. (1987). Three fim genes required for the regulation of length and mediation of adhesion of Escherichia coli type 1 fimbriae. *Mol Gen Genet* **208**, 439–45.

10 Jones, C. H., Pinkner, J. S., Roth, R., Heuser, J., Nicholes, A. V., Abraham, S. N. & Hultgren, S. J. (1995). FimH adhesin of type 1 pili is assembled into a fibrillar tip structure in the Enterobacteriaceae. *Proc Natl Acad Sci USA* **92**, 2081–5.

11 Hahn, E., Wild, P., Hermanns, U., Sebbel, P., Glockshuber, R., Haner, M., Taschner, N., Burkhard, P., Aebi, U. & Muller, S. A. (2002). Exploring the 3D molecular architecture of Escherichia coli type 1 pili. *J Mol Biol* **323**, 845–57.

12 Saulino, E. T., Bullitt, E. & Hultgren, S. J. (2000). Snapshots of usher-mediated protein secretion and ordered pilus assembly. *Proc Natl Acad Sci USA* **97**, 9240–5.

13 Valenski, M. L., Harris, S. L., Spears, P. A., Horton, J. R. & Orndorff, P. E. (2003). The Product of the fimI gene is necessary for Escherichia coli type 1 pilus biosynthesis. *J Bacteriol* **185**, 5007–11.

14 Lindberg, F., Lund, B., Johansson, L. & Normark, S. (1987). Localization of the receptor-binding protein adhesin at the tip of the bacterial pilus. *Nature* **328**, 84–7.

15 Jacob-Dubuisson, F., Heuser, J., Dodson, K., Normark, S. & Hultgren, S. (1993). Initiation of assembly and association of the structural elements of a bacterial pilus depend on two specialized tip proteins. *Embo J* **12**, 837–47.

16 Choudhury, D., Thompson, A., Stojanoff, V., Langermann, S., Pinkner, J., Hultgren, S. J. & Knight, S. D. (1999). X-ray structure of the FimC-FimH chaperone-adhesin complex from uropathogenic Escherichia coli. *Science* **285**, 1061–6.

17 Dodson, K. W., Pinkner, J. S., Rose, T., Magnusson, G., Hultgren, S. J. & Waksman, G. (2001). Structural basis of the interaction of the pyelone-

phritic E. coli adhesin to its human kidney receptor. *Cell* **105**, 733–43.
18 Merckel, M. C., Tanskanen, J., Edelman, S., Westerlund-Wikstrom, B., Korhonen, T. K. & Goldman, A. (2003). The structural basis of receptor-binding by Escherichia coli associated with diarrhea and septicemia. *J Mol Biol* **331**, 897–905.
19 Hooton, T. M. & Stamm, W. E. (1997). Diagnosis and treatment of uncomplicated urinary tract infection. *Infect Dis Clin North Am* **11**, 551–81.
20 Mulvey, M. A., Schilling, J. D., Martinez, J. J. & Hultgren, S. J. (2000). Bad bugs and beleaguered bladders: interplay between uropathogenic Escherichia coli and innate host defenses. *Proc Natl Acad Sci USA* **97**, 8829–35.
21 Langermann, S., Palaszynski, S., Barnhart, M., Auguste, G., Pinkner, J. S., Burlein, J., Barren, P., Koenig, S., Leath, S., Jones, C. H. & Hultgren, S. J. (1997). Prevention of mucosal Escherichia coli infection by FimH-adhesin-based systemic vaccination. *Science* **276**, 607–11.
22 Zhou, G., Mo, W. J., Sebbel, P., Min, G., Neubert, T. A., Glockshuber, R., Wu, X. R., Sun, T. T. & Kong, X. P. (2001). Uroplakin Ia is the urothelial receptor for uropathogenic Escherichia coli: evidence from in vitro FimH binding. *J Cell Sci* **114**, 4095–103.
23 Martinez, J. J., Mulvey, M. A., Schilling, J. D., Pinkner, J. S. & Hultgren, S. J. (2000). Type 1 pilus-mediated bacterial invasion of bladder epithelial cells. *Embo J* **19**, 2803–12.
24 Mulvey, M. A., Schilling, J. D. & Hultgren, S. J. (2001). Establishment of a persistent Escherichia coli reservoir during the acute phase of a bladder infection. *Infect Immun* **69**, 4572–9.
25 Schilling, J. D., Lorenz, R. G. & Hultgren, S. J. (2002). Effect of trimethoprim-sulfamethoxazole on recurrent bacteriuria and bacterial persistence in mice infected with uropathogenic Escherichia coli. *Infect Immun* **70**, 7042–9.
26 Lund, B., Lindberg, F., Marklund, B. I. & Normark, S. (1987). The PapG protein is the alpha-D-galactopyranosyl-(1–4)-beta-D-galactopyranose-binding adhesin of uropathogenic Escherichia coli. *Proc Natl Acad Sci USA* **84**, 5898–902.
27 Roberts, J. A., Marklund, B. I., Ilver, D., Haslam, D., Kaack, M. B., Baskin, G., Louis, M., Mollby, R., Winberg, J. & Normark, S. (1994). The Gal(alpha 1–4)Gal-specific tip adhesin of Escherichia coli P-fimbriae is needed for pyelonephritis to occur in the normal urinary tract. *Proc Natl Acad Sci USA* **91**, 11889–93.
28 Norgren, M., Baga, M., Tennent, J. M. & Normark, S. (1987). Nucleotide sequence, regulation and functional analysis of the papC gene required for cell surface localization of Pap pili of uropathogenic Escherichia coli. *Mol Microbiol* **1**, 169–78.
29 Klemm, P. & Christiansen, G. (1990). The fimD gene required for cell surface localization of Escherichia coli type 1 fimbriae. *Mol Gen Genet* **220**, 334–8.
30 Bardwell, J. C., McGovern, K. & Beckwith, J. (1991). Identification of a protein required for disulfide bond formation in vivo. *Cell* **67**, 581–9.
31 Hultgren, S. J., Jones, C. H. & Normark, S. (1996). Bacterial adhesins and their assebly. In *Escherichia coli and Salmonella* (Neidhardt, F. C., ed.), Vol. 2, pp. 2730–2756. 2 vols. ASM Press, Washington, DC.
32 Jacob-Dubuisson, F., Pinkner, J. S., Xu, Z., Striker, R. T., Padmanhban, A. & Hultgren, S. J. (1994a). PapD chaperone function in pilus biogenesis depends on oxidant and chaperone-like activities of DsbA. *Proc Natl Acad Sci USA* **91**, 11552–11556.
33 Lindberg, F., Tennent, J. M., Hultgren, S. J., Lund, B. & Normark, S. (1989). PapD, a periplasmic transport protein in P-pilus biogenesis. *J Bacteriol* **171**, 6052–8.
34 Jones, C. H., Pinkner, J. S., Nicholes, A. V., Slonim, L. N., Abraham, S. N. & Hultgren, S. J. (1993). FimC is a periplasmic PapD-like chaperone that directs assembly of

type 1 pili in bacteria. *Proc Natl Acad Sci USA* **90**, 8397–401.

35 Thanassi, D. G., Saulino, E. T., Lombardo, M. J., Roth, R., Heuser, J. & Hultgren, S. J. (1998). The PapC usher forms an oligomeric channel: implications for pilus biogenesis across the outer membrane. *Proc Natl Acad Sci USA* **95**, 3146–51.

36 Jacob-Dubuisson, F., Striker, R. T. & Hultgren, S. J. (1994b). Chaperone-assisted self-assembly of pili independent of cellular energy. *J. Biol. Chem.* **269**, 12447–12455.

37 Holmgren, A. & Branden, C. I. (1989). Crystal structure of chaperone protein PapD reveals an immunoglobulin fold. *Nature* **342**, 248–51.

38 Sauer, F. G., Futterer, K., Pinkner, J. S., Dodson, K. W., Hultgren, S. J. & Waksman, G. (1999). Structural basis of chaperone function and pilus biogenesis. *Science* **285**, 1058–61.

39 Sauer, F. G., Pinkner, J. S., Waksman, G. & Hultgren, S. J. (2002). Chaperone priming of pilus subunits facilitates a topological transition that drives fiber formation. *Cell* **111**, 543–51.

40 Pellecchia, M., Guntert, P., Glockshuber, R. & Wuthrich, K. (1998). NMR solution structure of the periplasmic chaperone FimC. *Nat Struct Biol* **5**, 885–90.

41 Knight, S. D., Choudhury, D., Hultgren, S., Pinkner, J., Stojanoff, V. & Thompson, A. (2002). Structure of the S pilus periplasmic chaperone SfaE at 2.2 A resolution. *Acta Crystallogr D Biol Crystallogr* **58**, 1016–22.

42 Kuehn, M. J., Normark, S. & Hultgren, S. J. (1991). Immunoglobulin-like PapD chaperone caps and uncaps interactive surfaces of nascently translocated pilus subunits. *Proc Natl Acad Sci USA* **88**, 10586–90.

43 Soto, G. E., Dodson, K. W., Ogg, D., Liu, C., Heuser, J., Knight, S., Kihlberg, J., Jones, C. H. & Hultgren, S. J. (1998). Periplasmic chaperone recognition motif of subunits mediates quaternary interactions in the pilus. *Embo J* **17**, 6155–67.

44 Jones, C. H., Danese, P. N., Pinkner, J. S., Silhavy, T. J. & Hultgren, S. J. (1997). The chaperone-assisted membrane release and folding pathway is sensed by two signal transduction systems. *Embo J* **16**, 6394–406.

45 Vetsch, M., Sebbel, P. & Glockshuber, R. (2002). Chaperone-independent folding of type 1 pilus domains. *J Mol Biol* **322**, 827–40.

46 Kuehn, M. J., Ogg, D. J., Kihlberg, J., Slonim, L. N., Flemmer, K., Bergfors, T. & Hultgren, S. J. (1993). Structural basis of pilus subunit recognition by the PapD chaperone. *Science* **262**, 1234–41.

47 Nishiyama, M., Vetsch, M., Puorger, C., Jelesarov, I. & Glockshuber, R. (2003). Identification and Characterization of the Chaperone-Subunit Complex-binding Domain from the Type 1 Pilus Assembly Platform FimD. *J Mol Biol* **330**, 513–25.

48 Lawrence, M. C. & Colman, P. M. (1993). Shape complementarity at protein/protein interfaces. *J Mol Biol* **234**, 946–50.

49 Barnhart, M. M., Pinkner, J. S., Soto, G. E., Sauer, F. G., Langermann, S., Waksman, G., Frieden, C. & Hultgren, S. J. (2000). From the cover: PapD-like chaperones provide the missing information for folding of pilin proteins [see comments]. *Proc Natl Acad Sci USA* **97**, 7709–14.

50 Hermanns, U., Sebbel, P., Eggli, V. & Glockshuber, R. (2000). Characterization of FimC, a Periplasmic Assembly Factor for Biogenesis of Type 1 Pili in Escherichia coli. *Biochemistry* **39**, 11564–11570.

51 Pellecchia, M., Sebbel, P., Hermanns, U., Wuthrich, K. & Glockshuber, R. (1999). Pilus chaperone FimC-adhesin FimH interactions mapped by TROSY-NMR. *Nat Struct Biol* **6**, 336–9.

52 Bullitt, E. & Makowski, L. (1995). Structural polymorphism of bacterial adhesion pili. *Nature* **373**, 164–7.

53 Saulino, E. T., Thanassi, D. G., Pinkner, J. S. & Hultgren, S. J. (1998). Ramifications of kinetic

partitioning on usher-mediated pilus biogenesis. *Embo J* **17**, 2177–85.

54 Dodson, K. W., Jacob-Dubuisson, F., Striker, R. T. & Hultgren, S. J. (1993). Outer-membrane PapC molecular usher discriminately recognizes periplasmic chaperone-pilus subunit complexes. *Proc Natl Acad Sci USA* **90**, 3670–4.

55 Barnhart, M. M., Sauer, F. G., Pinkner, J. S. & Hultgren, S. J. (2003). Chaperone-subunit-usher interactions required for donor strand exchange during bacterial pilus assembly. *J Bacteriol* **185**, 2723–30.

56 Valent, Q. A., Zaal, J., de Graaf, F. K. & Oudega, B. (1995). Subcellular localization and topology of the K88 usher FaeD in Escherichia coli. *Mol Microbiol* **16**, 1243–57.

57 Harms, N., Oudhuis, W. C., Eppens, E. A., Valent, Q. A., Koster, M., Luirink, J. & Oudega, B. (1999). Epitope tagging analysis of the outer membrane folding of the molecular usher FaeD involved in K88 fimbriae biosynthesis in Escherichia coli. *J Mol Microbiol Biotechnol* **1**, 319–25.

58 Thanassi, D. G., Stathopoulos, C., Dodson, K., Geiger, D. & Hultgren, S. J. (2002). Bacterial outer membrane ushers contain distinct targeting and assembly domains for pilus biogenesis. *J Bacteriol* **184**, 6260–9.

59 Meerman, H. J. & Georgiou, G. (1994). Construction and characterization of a set of E. coli strains deficient in all known loci affecting the proteolytic stability of secreted recombinant proteins. *Biotechnology (NY)* **12**, 1107–10.

60 Guzman, L. M., Belin, D., Carson, M. J. & Beckwith, J. (1995). Tight regulation, modulation, and high-level expression by vectors containing the arabinose PBAD promoter. *J Bacteriol* **177**, 4121–30.

61 Cantey, J. R., Blake, R. K., Williford, J. R. & Moseley, S. L. (1999). Characterization of the Escherichia coli AF/R1 pilus operon: novel genes necessary for transcriptional regulation and for pilus-mediated adherence. *Infect Immun* **67**, 2292–8.

62 Massad, G., Fulkerson, J. F., Jr., Watson, D. C. & Mobley, H. L. (1996). Proteus mirabilis ambient-temperature fimbriae: cloning and nucleotide sequence of the aft gene cluster. *Infect Immun* **64**, 4390–5.

63 Honarvar, S., Choi, B. K. & Schifferli, D. M. (2003). Phase variation of the 987P-like CS18 fimbriae of human enterotoxigenic Escherichia coli is regulated by site-specific recombinases. *Mol Microbiol* **48**, 157–71.

64 Girardeau, J. P., Der Vartanian, M., Ollier, J. L. & Contrepois, M. (1988). CS31A, a new K88-related fimbrial antigen on bovine enterotoxigenic and septicemic Escherichia coli strains. *Infect Immun* **56**, 2180–8.

65 Bertin, Y., Girardeau, J. P., Der Vartanian, M. & Martin, C. (1993). The ClpE protein involved in biogenesis of the CS31A capsule-like antigen is a member of a periplasmic chaperone family in gram-negative bacteria. *FEMS Microbiol Lett* **108**, 59–67.

66 Klemm, P., Christiansen, G., Kreft, B., Marre, R. & Bergmans, H. (1994). Reciprocal exchange of minor components of type 1 and F1C fimbriae results in hybrid organelles with changed receptor specificities. *J Bacteriol* **176**, 2227–34.

67 Klemm, P., Jorgensen, B. J., Kreft, B. & Christiansen, G. (1995). The export systems of type 1 and F1C fimbriae are interchangeable but work in parental pairs. *J Bacteriol* **177**, 621–7.

68 Lintermans, P. F., Pohl, P., Bertels, A., Charlier, G., Vandekerckhove, J., Van Damme, J., Schoup, J., Schlicker, C., Korhonen, T., De Greve, H. & et al. (1988). Characterization and purification of the F17 adhesin on the surface of bovine enteropathogenic and septicemic Escherichia coli. *Am J Vet Res* **49**, 1794–9.

69 Holmgren, A., Kuehn, M. J., Branden, C. I. & Hultgren, S. J. (1992). Conserved immunoglobulin-like features in a family of periplasmic pilus chaperones in bacteria. *Embo J* **11**, 1617–22.

tor is associated with the translocase at the *trans* side of the inner membrane and completes the translocation of the precursor proteins across the mitochondrial membranes.

Proteins are translocated into mitochondria in a largely unfolded state. In this respect, their import into mitochondria differs from their translocation into peroxisomes or their transport by the TAT translocase in bacteria and chloroplasts [46, 114], but it is similar to most translocation pathways in the endoplasmatic reticulum (ER) and chloroplasts and to the Sec translocase in bacteria [23, 51, 54, 56, 75]. Unfolding of proteins during translocation across the mitochondrial outer membrane (OM) and inner membrane (IM) is mediated by the mitochondrial import machinery [76, 91]. Protein unfolding by the mitochondrial import motor has become a paradigm for protein unfolding, a physiological process occurring in virtually every cell. Unfolding of proteins is also essential for other cellular processes, in particular protein degradation. Here, we discuss the import of proteins into mitochondria and the unfolding of proteins by the mitochondrial import motor during the translocation process.

30.2
Translocation Machineries and Pathways of the Mitochondrial Protein Import System

Three translocation machineries that mediate the import of precursor proteins into mitochondria are known (see Figure 30.1). The translocase of the outer membrane (TOM complex) functions in the recognition of precursor proteins, their insertion into the outer membrane, and in translocation across the outer membrane (for details, see [98, 101, 104]). Chaperones have been reported to interact with the precursor proteins in the cytosol to keep them in a translocation-competent state and to support their transport to the TOM complex [21, 41, 88, 158].

The process of translocation across the TOM complex is followed by sorting proteins either directly to the intermembrane space or to two translocases of the inner membrane (TIM): the TIM23 complex and the TIM22 complex [3, 52, 58, 113]. The TIM23 complex transports proteins with typical mitochondrial targeting sequences across or into the mitochondrial inner membrane, while the TIM22 complex inserts a subclass of inner-membrane proteins with internal targeting signals into the membrane. This subclass includes the proteins of the carrier family and components of the inner-membrane translocases such as Tim22 and Tim23.

An interesting case is the sorting of β-barrel proteins of the outer membrane. They first use the TOM complex and are then inserted and assembled in the membrane with the help of the recently discovered TOB/SAM complex [36, 60, 99, 152].

We later focus on the translocation pathway of proteins destined for the mitochondrial matrix. These proteins are translocated across the inner membrane by the TIM23 complex.

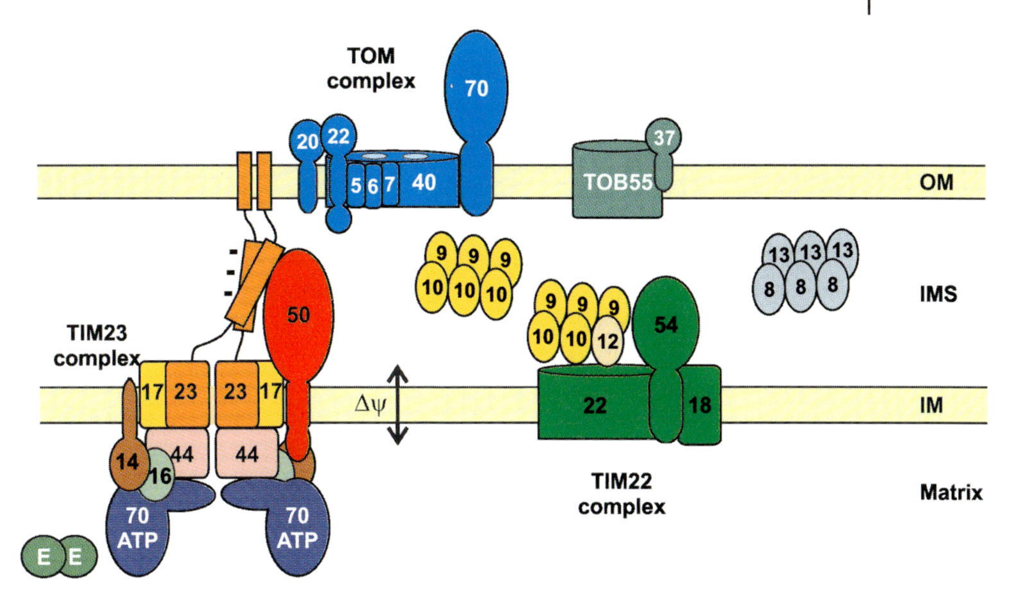

Fig. 30.1. Translocation machineries of the mitochondrial protein import system. The translocase of the outer membrane (TOM complex) mediates the translocation of proteins across and into the outer membrane. All proteins analyzed so far enter mitochondria via the TOM complex. The TOM complex consists of the receptor subunits Tom20, Tom70, and Tom22 and the general import pore. The general import pore is made up of the core subunit Tom40, the integral part of the Tom22 receptor and of the small TOM proteins Tom5, Tom6, and Tom7. There are two translocases of the inner membrane, the TIM23 and TIM22 complexes. The TIM22 complex inserts polytopic proteins with internal targeting signals into the inner membrane. It is composed of two subcomplexes. The peripherally attached Tim9/10/12 subcomplex is associated with the membrane-integrated components Tim22, Tim54, and Tim18. The soluble Tim9/10 complex and the Tim8/13 complex in the intermembrane space cooperate with the Tim22 complex in the transport of carrier proteins and of Tim23, respectively. The TIM23 complex mediates the translocation of proteins with N-terminal targeting sequences across or into the inner membrane. Tim50 and the N-terminal domains of Tim23 guide the precursor proteins to the translocation channel in the inner membrane, which is formed by the integral C-terminal segments of Tim23 and Tim17. Translocation is completed by the import motor of the TIM23 translocase. Tim44 recruits the Tim14/Tim16 (Pam18/Pam16) subcomplex and mtHsp70 to the translocation channel. The co-chaperone Mge1 interacts with the mtHsp70 chaperone. The TOB complex (topogenesis of outer-membrane β-barrel proteins) inserts and assembles β-barrel proteins into the outer membrane. Two subunits have been identified: the main pore-forming component Tob55 (Sam50) and Mas37.

30.2.1
Import of Proteins Destined for the Mitochondrial Matrix

Matrix-targeted proteins in most cases have amino-terminal presequences that have the potential to form an amphipathic α-helix [66, 146]. These targeting signals direct the proteins to mitochondria by binding to the surface-exposed receptor proteins of the TOM complex on the cytoplasmic side of the outer membrane, the *cis* side [28]. The receptor protein Tom20 preferentially recognizes presequence proteins, whereas the Tom70 receptor protein binds mostly proteins with internal targeting signals [28]. From the initial receptor proteins, the preproteins are transferred to the general import pore (GIP) via the receptor domain of Tom22 [47, 57]. GIP is made up of five other membrane-embedded components of the TOM complex: Tom40, Tom22, and the three small subunits Tom7, Tom6, and Tom5 [2, 20]. The only essential component of the TOM complex in *Saccharomyces cerevisiae*, Tom40, is able to form a conducting pore in the membrane and is the main constituent of the protein-conducting channel [1, 45].

Following translocation through this channel, the preprotein binds to the "*trans*-binding site" on the inner face of the outer membrane [79, 106]. It has been suggested that the *trans*-binding site has a higher affinity for preproteins than do the preprotein-binding sites of the TOM complex at the surface of the outer membrane. This would drive translocation of the presequence across the outer membrane [8, 79, 105, 120].

Next, the presequences interact with the TIM23 complex of the inner membrane. The TIM23 complex translocates the presequences across the inner membrane in a membrane potential–dependent ($\Delta\Psi$) and ATP-dependent manner. Three functions have been assigned to the $\Delta\Psi$. First, it triggers dimerization of the Tim23 presequence receptor [4]. Second, it exerts an electrophoretic effect and drives the translocation of the presequence across the inner membrane [73]. Third, it can support unfolding of protein domains on the surface of the mitochondria [48]. Once the presequence has reached the matrix side of the inner membrane, complete translocation of proteins into the matrix is driven by ATP hydrolysis [44, 78, 90, 102]. Inner-membrane proteins containing a matrix-targeting signal followed by one hydrophobic transmembrane segment are also imported via the TIM23 pathway. They are not completely translocated across the inner membrane but instead are laterally inserted into the inner membrane [82, 115]. If the hydrophobic segment of these proteins is located directly after the matrix-targeting signal, their import is independent of matrix ATP [34, 115].

The TIM23 complex can be structurally and functionally divided into three parts: the hydrophilic domains in the intermembrane space, the membrane-embedded part, and the mitochondrial import motor on the matrix side of the inner membrane. The amino-terminal domain of Tim23 spans the mitochondrial outer membrane and links the outer membrane with the inner membrane [22]. The conserved intermembrane space domain of the single transmembrane protein Tim50 facilitates the transfer of the preprotein from the TOM complex to the TIM23 complex, probably to the receptor domain of Tim23 in the intermembrane space [37, 84,

157]. The integral part of the complex is made up by the two proteins with four transmembrane domains, Tim17 and Tim23, which most likely form the protein-conducting channel [5, 18, 19, 27, 71, 81, 118, 119, 137].

The mitochondrial import motor consists of at least five components: Tim44, mtHsp70, Tim14/Pam18, Tim16/Pam16, and Mge1 (see Figure 30.1) [24, 32, 59, 85, 90, 138]. Tim44 is crucial for recruiting mtHsp70 to the outlet of the protein-conducting channel [62, 85, 109, 123]. The chaperone mtHsp70 is the central component of the import motor. Like all DnaK-type chaperones, it consists of an N-terminal adenine nucleotide-binding domain, a peptide-binding domain, and a C-terminal "lid" domain [11, 61, 86, 134]. When ATP is bound to mtHsp70, the peptide-binding domain is in an open conformation [11, 40]. MtHsp70 in the ATP form binds to the incoming polypeptide chain emerging from the channel [69, 85, 123]. Conversion to the ADP form of mtHsp70 results in the closure of the peptide-binding domain and high-affinity binding of the polypeptide chain [11, 116]. This efficient trapping at the outlet of the channel inhibits backsliding of the polypeptide chain and drives its translocation across the mitochondrial membranes. Tim14 is a membrane-bound DnaJ homologue that is associated with the TIM23 translocase [24, 85, 138]. It interacts with Tim44 and mtHsp70 in an ATP-dependent manner and stimulates the ATPase activity of mtHsp70, thereby promoting the efficient binding of the polypeptide. Then the mtHsp70 bound to the polypeptide chain dissociates from Tim44 [123, 124, 139, 140]. Tim44 is present in the TIM23 translocase as a dimer, and therefore a second mtHsp70 is already present at the translocation pore [87]. Dissociation of the first mtHsp70 and further translocation allow the second one to bind to the next incoming segment of the polypeptide chain. Such "hand-over-hand" action results in efficient translocation. Tim44 can be viewed as an organizer that integrates the import motor with the components of the translocation channel. It recruits mtHsp70, Tim14, and Tim16 to the Tim17-Tim23 subcomplex [85] and interacts with the incoming precursor protein and transfers it to mtHsp70 [123]. Tim16 is a J-domain-related protein that has similarity to Tim14 and forms a stable subcomplex with Tim14 [32, 59]. It is structurally important, as it mediates the interaction of Tim14 with the TIM23 complex. The nucleotide-exchange factor Mge1 mediates the release of ADP from mtHsp70 [8, 65, 89, 151], which allows binding of an ATP and thereby release of mtHsp70 from the polypeptide chain. Thus, the mtHsp70 can be reused for a next cycle of binding and release. Mge1 seems to regulate the binding of mtHsp70 to Tim44 [69, 124]. MtHsp70 also binds in the ADP form to Tim44. Mge1 dissociates such an unproductive interaction, as ADP-bound mtHsp70 cannot bind incoming polypeptides. Indeed, cross-linking experiments showed that mtHsp70 binds to Tim44 in the presence of ATP in intact mitochondria [85].

Following translocation of the precursor protein into the mitochondrial matrix, the targeting signal is cleaved by the matrix-processing peptidase and, in some cases, the mitochondrial intermediate peptidase [90]. Then the extended polypeptide chain folds into its native form, which in many cases is supported by the mitochondrial chaperones mtHsp70, Hsp60, and peptidyl-prolyl *cis/trans* isomerase [14, 55, 95, 110].

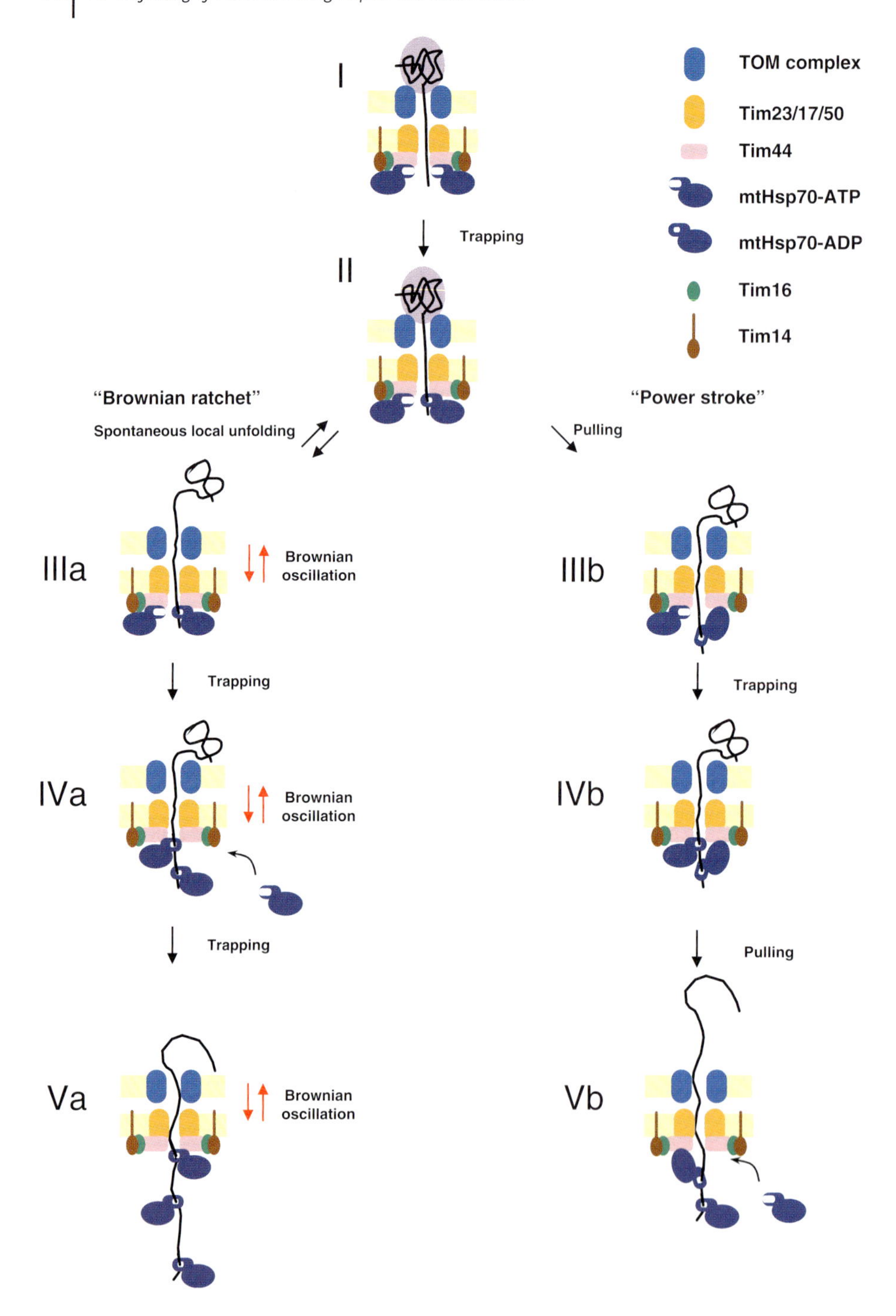

I
TOM complex
Tim23/17/50
Tim44
mtHsp70-ATP
mtHsp70-ADP
Tim16
Tim14
Trapping
II
"Brownian ratchet"
Spontaneous local unfolding
"Power stroke"
Pulling
IIIa
Brownian oscillation
IIIb
Trapping
Trapping
IVa
Brownian oscillation
IVb
Trapping
Pulling
Va
Brownian oscillation
Vb

30.3 Import into Mitochondria Requires Protein Unfolding

The structures of the isolated native TOM complex or the TOM core complex of the mitochondrial outer membrane and the TIM22 complex were analyzed by single-particle electron microscopy. The observed structures suggested pores with an internal diameter of about 20–25 Å for the TOM complex and stain-filled pits with a diameter of 16 Å for the TIM22 complex [2, 63, 83, 112]. The integral membrane core component Tom40 appears to be the subunit forming the pore of the TOM complex, with a diameter of about 22–25 Å [1, 45].

Such a pore diameter is consistent with observations that precursors containing bulky elements can be transported into mitochondria. Examples are precursors containing branched polypeptides or oligonucleotides linked to the C-terminus of a precursor or precursors containing neighboring β-strands that were covalently altered by chemical cross-linking prior to import [125, 142, 144]. It was also reported that the AAC precursor protein and Tim23 cross the TOM complex in a loop con-

Fig. 30.2. The "Brownian ratchet" and the "power-stroke" models for unfolding and translocation of precursor proteins into mitochondria. Stage I: Membrane potential–dependent translocation of the N-terminal targeting signal of a precursor protein into the matrix and its cleavage by the mitochondrial processing peptidase generate a preprotein spanning the mitochondrial outer and inner membranes. The folded domain of this early translocation intermediate is on the surface of mitochondria. Stage II: When the preprotein emerges from the translocation channel, it is bound by the mitochondrial heat shock protein 70 (mtHsp70), which is recruited by Tim44 to the translocation channel. ATP hydrolysis by mtHsp70 is stimulated by Tim14. The conversion of the ATP-bound form of mtHsp70 to the ADP-bound form leads to a conformational change of the peptide-binding domain and the tight binding of the preprotein. "Brownian ratchet" model: Stage IIIa: Thermal fluctuations lead to fast local unfolding and refolding of the preprotein, termed "thermal breathing." Stage IVa: The mtHsp70 bound to the preprotein dissociates from Tim44. Unfolded segments of the preprotein can slide back and forth in the translocation channel due to random thermal fluctuations (Brownian oscillations). This exposes a new segment of the incoming polypeptide chain at the exit of the import channel, which is trapped by a second mtHsp70 bound to the second Tim44 in the TIM23 complex. Tim14 stimulates ATP hydrolysis of this mtHsp70 and allows efficient trapping. Thus, backsliding of the preprotein is prevented, and therefore the preprotein cannot refold outside of mitochondria. Stage Va: Another mtHsp70 is recruited by Tim44 and traps the next segment of the oscillating preprotein at the exit of the channel. "Power-stroke" model: Stage IIIb: MtHsp70 exerts a pulling force perpendicular to the inner membrane, which leads to unfolding of the preprotein on the surface of mitochondria and transport of a segment of the preprotein into the matrix space. The force is generated by a conformational change of mtHsp70. Stage IVb: Tim14 stimulates ATP hydrolysis of the second recruited mtHsp70. The ADP-bound form of mtHsp70 binds tightly to a newly exposed segment of the preprotein. Stage Vb: The first mtHsp70 dissociates from Tim44 and the second mtHsp70 undergoes a conformational change, generating a pulling force.

formation [15, 30, 153]. In other studies experimental evidence suggests that the BCS1 precursor protein is imported across the outer and inner membrane as a loop structure, formed by hydrophobic interactions between its transmembrane domain and the subsequent amphipathic α-helical targeting signal [31, 133]. However, it seems impossible that translocation pores with the observed diameters import fully folded proteins. Indeed, experiments have clearly demonstrated that tightly folded domains cannot be imported. The first evidence was obtained using a fusion protein consisting of the first 22 amino acid residues of the cytochrome oxidase subunit IV presequence fused to mouse dihydrofolate reductase (DHFR). This protein was imported into isolated mitochondria like an authentic mitochondrial precursor protein [50]. Binding of methotrexate, a specific ligand of DHFR, leading to a stably folded DHFR domain inhibited translocation of the precursor protein into mitochondria [26]. Further studies demonstrated that mitochondrial precursor proteins containing the tightly folded heme-binding domain of (HBD) of cytochrome b_2 or the bacterial ribonuclease barnase as folded passenger proteins are transported into mitochondria only in an unfolded state [35, 39, 49, 77]. In an elegant study by Schwartz et al. [125], rigid gold clusters of specific dimensions were cross-linked to the C-terminus of a precursor protein and their import was studied. Above a certain size, these clusters were not imported, confirming that the translocases cannot transport large structures. Clusters of specific dimensions allowed estimating the internal diameter of the TOM complex to be between 20 Å and 26 Å. Comparison with the diameter observed in the electron microscopy studies suggests that the pore diameter of the TOM complex does not alter significantly during translocation. The TIM23 translocase in the inner membrane might behave differently. Calculations based on channel conductance and size exclusion suggest that renatured recombinant Tim23 forms a channel with an even smaller inner diameter of 13 Å [137]. The pore of the TIM23 complex is narrower than the TOM complex. However, the TIM23 complex appears not to have a well-defined rigid pore, as preproteins cross-linked to 20-Å gold clusters were still imported by the TIM23 complex, albeit with strongly reduced efficiency [126].

Most precursor proteins are imported posttranslationally into mitochondria. Therefore, they may fold in the cytoplasm prior to import. The requirement of unfolding of preproteins prior to their import has been observed not only in vitro but also in vivo. When a hybrid protein of the amino-terminal portion of the mitochondrial precursor protein cytochrome b_2 and of DHFR was expressed in intact yeast cells and the specific ligand methotrexate was added to the cells, the precursor protein was arrested in mitochondrial contact sites, demonstrating the importance of protein unfolding for translocation in vivo [154]. It has also been shown that the heme-binding domain, an authentic mitochondrial domain, folds in the cytosol [10]. Further research is needed to determine which endogenous mitochondrial proteins are held in a translocation-competent unfolded state by cytosolic chaperones and which ones fold and have to be unfolded prior to their translocation into mitochondria. However, it is obvious that some precursor proteins have to be unfolded by the mitochondrial import machinery.

30.4
Mechanisms of Unfolding by the Mitochondrial Import Motor

The mitochondrial import motor drives the unfolding of protein domains and the translocation of polypeptides across the mitochondrial membranes. Two models have been proposed to explain how the mitochondrial import motor exerts its function (see Figure 30.2). In the "targeted Brownian ratchet model," the import motor acts as a molecular ratchet that traps the incoming polypeptide chain directly at the outlet of the protein-conducting channel of the inner membrane [92, 123, 127]. In the "power-stroke model" or "pulling model," the mitochondrial import motor exerts a power stroke perpendicular to the membrane and generates a pulling force on the polypeptide chain [38, 39].

30.4.1
Targeted Brownian Ratchet

Proteins partially unfold and fold in the range of milliseconds, a process also termed "thermal breathing". If an unfolded segment is exposed due to "thermal breathing" during protein import, the unfolded segment of the polypeptide chain can move into the translocation channel due to Brownian motion [140]. Within the translocation channel it can oscillate in both directions [140]. MtHsp70 of the import motor binds to segments of the polypeptide chain as they are emerging from the translocation channel at the matrix side. According to this model, the import motor acts as a ratchet that permits forward movement but not backsliding of the polypeptide chain [81, 86, 87, 94, 123, 124]. Thereby it transduces spontaneous Brownian movement into vectorial transport. The polypeptide chain is trapped and refolding of the polypeptide on the outside of mitochondria is impossible. Such a mechanism requires a very efficient trapping process. The attachment of mtHsp70 to the TIM23 complex by Tim44 targets the trapping device to the site of the emerging polypeptide and assures immediate binding and trapping. In summary, the targeted molecular ratchet unfolds proteins, because it inhibits the refolding of spontaneously unfolded segments of the protein at the outside of mitochondria [35, 54].

30.4.2
Power-stroke Model

In the power-stroke model, the action of the mitochondrial import motor generates a force on the polypeptide chain spanning the mitochondrial membranes. This force is transmitted by the polypeptide chain and is required for unfolding of protein domains on the outside of mitochondria [38, 53, 78, 103, 148]. An ATP-induced conformational change of mtHsp70 exerts the pulling force, which is transferred by a lever arm to the polypeptide chain. Following the power stroke, mtHsp70 has to be released from Tim44 so that the next mtHsp70 can bind and exert another power stroke. Backsliding of the polypeptide chain before binding of the next mtHsp70 must be prevented. It was proposed that the protein-conducting channel fulfills such a function [13, 19]. Multiple cycles of a power stroke would result in a regular stepwise translocation.

30.5
Studies to Discriminate between the Models

Studies to discriminate between the two models have addressed the mode of unfolding of preproteins on the outside of mitochondria during the protein translocation as well as mechanistic aspects of the mitochondrial import motor.

30.5.1
Studies on the Unfolding of Preproteins

Before discussing protein unfolding in the context of mitochondrial protein import, two modes of unfolding will be distinguished: global and local unfolding of proteins. Global unfolding results in a completely unfolded polypeptide chain without any structured elements. The equilibrium between the folded and unfolded states is described by the thermodynamic stability of a protein. The thermodynamic stability of folded proteins seems to correlate with their resistance against protease digestion, which requires extensive unfolding of domains [35].

In contrast, local unfolding, also termed "breathing," is characterized by perpetual thermal fluctuations within a protein [159]. These fluctuations occur in the time range of nanoseconds to seconds and result in local unfolding events. Binding of denaturants such as urea [159], chemical modification of side chains of amino acid residues [35], or interaction with other proteins such as chaperones leads to the trapping of such local unfolded segments with varying efficiencies [49, 68]. These segments cannot refold once they are trapped. Further unfolding occurs and eventually a completely unfolded state results within milliseconds to minutes.

30.5.1.1 Comparison of the Import of Folded and Unfolded Proteins

In order to understand the unfolding of precursor proteins during import, proteins containing folded domains were analyzed in comparison with the same proteins harboring the domains in an unfolded state. In most cases mouse cytosolic dihydrofolate reductase (DHFR) was used as passenger protein and was fused to a mitochondrial presequence [50]. The unfolded state was acquired by addition of urea prior to the import reaction [25, 96] or by using a mutant form, which does not stably fold [143]. The import process was started by allowing the presequence to translocate across the TIM23 complex with the help of $\Delta\tilde{\Psi}$. In this way the precursor chain was threaded into the translocation pores of the TOM and TIM23 complexes.

Import of the unfolded form of the protein was always faster than import of the folded counterpart [33, 35, 68, 81, 143, 145, 149]. According to the Brownian ratchet model, movement of the polypeptide chain in the import channels is stopped when a folded domain reaches the TOM complex. Only when local spontaneous unfolding events occur will unfolded segments be exposed and then be able to enter the translocation channels. The kinetics of this breathing determines the average time period for which the forward movement of a folded domain is halted. Experimental observations support this. Folded, but not unfolded, DHFR-containing precursor proteins do pause upon import [35]. The import kinetics of preproteins with folded DHFR show a notable lag phase. After unfolding by urea, no lag phase in the import was observed [68]. Furthermore, passenger proteins such as HBD and titin (see below and Section 30.5.1.3) showed lag phases with different time periods [35, 94]. It should be noted that most significant lag phases were observed for proteins with short amino-terminal presequences in front of the folded domain, such as cyt b_2(47)-DHFR [68]. In these cases the presequences in front of the folded domain are not long enough to interact with the import motor, and therefore local spontaneous unfolding events of the folded domain cannot be trapped or promoted by the import motor in the initial stage of the translocation process. Preproteins with longer presequences interact with the import motor, spontaneous local unfolding events are trapped, and back-sliding of the polypeptide chain is inhibited. In the case of pSu9(1-69)DHFR, a fusion protein consisting of the presequence of the subunit 9 of *Neurospora* F_1F_0-ATPase and mouse DHFR, the rate of import was determined for different forms of the DHFR passenger. When wild-type DHFR was present, import was much slower than when an unfolded mutant form of DHFR was present [35]. A sequence of 69 amino acid residues was shown to be sufficient to reach the mitochondrial import motor; about 50 residues were reported to be required to span both the outer and inner membranes [108, 140]. Therefore, the faster import of the mutant form must have been due to the unfolded state of the passenger DHFR.

One would expect that the kinetics of import of folded proteins after the lag phase is similar to those of urea-unfolded proteins. Indeed, this was observed, demonstrating that a preprotein following the initial unfolding events reflects the behavior of an unfolded protein [68].

According to the power-stroke model, a folded domain will be unfolded by a power stroke of the import motor. One power stroke should be sufficient to cause a nearly complete collapse of the folded structure. The power stroke is supposed to be exerted by a fast conformational change. Assuming one fast power stroke, it is difficult to explain that the lag phases of import are in the range of minutes.

In addition, unfolding and import of preproteins were compared with regard to their temperature dependence [35]. Import of folded and unfolded DHFR at 30 °C occurred with similar rates. The permanently unfolded mutant variant of DHFR has a twofold lower import rate into mitochondria at 10 °C than at 30 °C. This result indicates that the import motor has a moderate temperature dependence. On the other hand, the import rates of folded DHFR were 30-fold faster at 30 °C than at 10 °C, and there was very little import at 4 °C. In the case of a power-stroke mechanism, the temperature dependence of the import would be determined by that of the import motor and is expected to be similar for the import of folded and unfolded proteins. This was not observed.

According to the Brownian ratchet model, import of folded DHFR depends on its spontaneous unfolding at the N-terminus. The kinetics of N-terminal breathing was analyzed by measuring the modification by N-ethylmaleimide (NEM) of the cysteine residue in position 7 of DHFR [35]. This cysteine is buried and therefore inaccessible to NEM in the folded protein. Upon breathing of the DHFR, the N-terminal segment moves out and can be modified by NEM. The modified protein cannot refold and is sensitive to added protease, in contrast to the folded protein. The measured rates of amino-terminal unfolding confirmed, for all temperatures analyzed, that unfolding is faster than import. Furthermore, the amino-terminal breathing is strongly reduced at low temperature. At 30 °C the unfolding at the N-terminus was observed to be ca. 14-fold faster than at 10 °C. The temperature dependence of the import rates can be explained by the temperature dependence of spontaneous unfolding, but not by the temperature dependence of the import motor.

The heme-binding domain of cytochrome b_2 (HBD) was another passenger domain studied in this respect. The domain was fused to mitochondrial presequences and its import characteristics were investigated [35]. The behavior of the HBD differed from that of DHFR. The import rates of HBD passenger proteins were only twofold higher at 30 °C than at 10 °C, and import was still quite efficient at 0–4 °C. Thus, import of the HBD displayed a very weak temperature dependence. This correlates with a faster local unfolding of HBD than of DHFR [35]. The rate of local unfolding of HBD was measured by chemical modification of residues buried in the folded domain. Interestingly, local unfolding of HBD was less temperature-dependent than that of DHFR. This is consistent with the finding that pausing of preproteins containing HBD, as described for DHFR-containing constructs, was not observed. When the global thermodynamic stability of the proteins was assessed by their protease resistance, HBD was found to be much more stable than DHFR [35]. In conclusion, the kinetics of spontaneous local unfolding and not the thermodynamic stability of a folded domain is the crucial parameter for the efficiency of import.

The observed results with DHFR and HBD are in agreement with those from experiments using bacterial RNase barnase as a passenger protein [49]. When the import of barnase fused to mitochondrial presequences was analyzed, it was observed that its import kinetics were faster than that of fusion proteins consisting of DHFR and presequences, although the thermodynamic stability of barnase is higher than that of DHFR. In addition, it was observed that import of barnase into mitochondria results in an unfolding pathway that differs from the one of global unfolding in solution [49]. During import the barnase was unfolded from its N-terminus. The amino-terminal structure of barnase consists of an

α-helix on the surface of the folded barnase structure and can therefore unfold with a relatively high probability [78]. Such an arrangement is also found in the HBD. In contrast, the amino-terminal β-strand of DHFR is buried in a central β-stranded core and therefore does not unfold easily. The amino-terminal structures of DHFR on the one hand and HBD and barnase on the other hand appear to be the reason for their different import and unfolding behavior.

In conclusion, spontaneous unfolding together with a targeted ratchet mechanism is sufficient to explain the temperature dependence of import, the pausing of import at folded domains, and the amino-terminal unfolding behavior of the protein domains analyzed.

30.5.1.2 **Import of Preproteins With Different Presequence Lengths**

Precursor proteins that have segments of different length in front of a folded DHFR domain were imported with very different rates [35, 77]. Rather slow import was observed for precursors with 60 residues in front of DHFR, whereas precursors with segments of 90 or more residues had 10- to 40-fold higher import rates. The import rate increased dramatically at lengths between 60–70 residues due to accessibility of the presequence to the molecular import motor. To span both mitochondrial membranes and to reach the matrix space, a preprotein needs a minimal length of ca. 50–60 residues in front of folded DHFR. This allows binding of mtHsp70, which, together with the other motor proteins, promotes forward movement and unfolding of the DHFR domain. In contrast, precursors containing less than 50–60 residues in front of DHFR (referred to as shorter precursors) are not able to reach the matrix without unfolding of a sufficiently long segment [35, 77]. Without unfolding, mtHsp70 cannot bind and support import.

In the Brownian ratchet model, spontaneous unfolding and forward movement by Brownian motion occur with longer as well as with shorter precursors. However, with a longer precursor mtHsp70 can trap a segment of the precursor at the outlet of the import channel. Therefore, backward sliding is inhibited and refolding of the DHFR is strongly impaired. In contrast, shorter precursors can slide back and refold because mtHsp70 does not reach the precursor. Backsliding and refolding compete with forward movement and import, so that overall import is much slower [35].

According to the power-stroke model, longer precursors have faster import kinetics because an amino-terminal segment of the polypeptide chain reaches into the mitochondrial matrix and mtHsp70 can exert a pulling force and cause a collapse of the folded domain on the mitochondrial surface [77, 78]. Thus, both models explain the higher import rates of precursors having longer presequences in front of a folded domain compared to the rates of those with short presequences. The Brownian ratchet model, in contrast to the power-stroke model, predicts that the dependence of import rates on the length of the N-terminal stretch varies among folded domains, which have a different rate of spontaneous unfolding. This was indeed observed with DHFR and HBD as passenger proteins [35].

30.5.1.3 Import of Titin Domains

Strong evidence in favor of the targeted Brownian ratchet mechanism was obtained using the immunoglobulin (Ig)-like domains of muscle titin as a passenger protein. The Ig-like domains are tightly folded. Experiments with atomic force microscopy indicated that mechanical pulling with forces of 200–300 pN is needed to unfold these domains [12, 67, 72, 128]. However, import of these domains into isolated mitochondria occurs efficiently when they are combined with a mitochondrial targeting sequence [94]. This raises the question of what kind of force a hypothetical power stroke of mtHsp70 could generate. The well-known motor proteins myosin or kinesin generate forces of about 5 pN [129, 130]. Thus, a reasonable assumption would estimate similar forces for mtHsp70. Furthermore, one single stroke and therefore hydrolysis of one ATP should lead to the complete collapse and hence unfolding of a folded domain. Estimating a stroke length of 3.5 nm, the equivalent of ca. 10 extended amino acid residues, the energy in one ATP can exert a force of about 14 pN [94]. Although the force measurements were performed under in vitro conditions and the values may be slightly different in the biological system, it appears impossible that the required force in the range of 200–300 pN could be generated by a power stroke of mtHsp70. It also has to be kept in mind that rapid cycles of mtHsp70 binding and release of the incoming polypeptide chain take place at the import site, when a folded domain on the outside of the mitochondria blocks forward movement [35, 94]. This excludes titin unfolding with low force by a slow power stroke in the minute range.

30.5.1.4 Unfolding by the Mitochondrial Membrane Potential $\Delta\Psi$

Some precursor proteins have a targeting sequence of such length in front of a folded domain that they do not reach into the matrix to interact with mtHsp70. However, if they are of sufficient length that their positively charged presequences are exposed in the import channel of the inner membrane, then the membrane potential across the inner membrane can act on them. Since the electrical potential is negative at the inner surface of the membrane, the positively charged presequence is driven towards the matrix. It has been reported that such an action of $\Delta\Psi$ on the presequence can result in unfolding of a protein on the surface of mitochondria [48]. According to this view, spontaneous local unfolding events at the N-terminus of the folded domain can be trapped by the membrane potential. The activation barrier for spontaneous local unfolding events at the N-terminus of the folded domain may be lowered by the force generated by the action of the membrane potential on the presequence [48, 76]. However, this intriguing possibility has yet to be verified.

30.5.2 Mechanistic Studies of the Import Motor

30.5.2.1 Brownian Movement of the Polypeptide Within the Import Channel

The polypeptide chain can move within the translocation channels of the TOM and TIM23 complexes in forward and reverse direction. Studies using intact mitochondria, isolated outer-membrane vesicles, and purified TOM complex demonstrate that the presequence of a precursor protein can translocate across the TOM complex and bind to the *trans* side of the complex independent of an energy source [79, 132]. In the absence of a membrane potential, the preprotein interacts with a component of the inner-membrane translocase, Tim50, as demonstrated by a cross-linking approach [84, 157]. However, the membrane potential is needed for further translocation across the inner membrane because it appears to be involved in the gating of the TIM23 complex [4, 137]. Subsequently, the polypeptide chain traverses the outer and inner membranes at the same time [26, 107]. Intermediates that span both membranes were used to demonstrate retrograde movement of the polypeptide chain within the translocation channels [35, 124, 139, 140]. When mitochondria were depleted of ATP, mtHsp70 was unable to arrest the intermediates in the matrix, and these intermediates slid back and fell out of the import channels. Forward movement of the polypeptide chain without the function of mtHsp70 was supported by experiments with specific precursor proteins. A certain group of precursor proteins is sorted into the inner membrane by arrest of their single-transmembrane segment in the translocase and subsequent lateral release into the lipid bilayer, a process termed stop-transfer pathway [82, 115]. In these precursors the transmembrane segment follows the amino-terminal presequence in a short distance of about 20 amino acid residues. The enzyme D-lactate dehydrogenase (D-LD) [115] and a mutant form of subunit 5A of cytochrome oxidase [34] are such preproteins. Neither matrix ATP nor a functional mtHsp70 is needed for import of these preproteins. It appears that the polypeptide chain moves in the forward direction until the hydrophobic transmembrane segment is trapped in the translocation channel of the TIM23 complex. These proteins apparently do not require the mtHsp70 motor.

Another group of precursor proteins was genetically engineered. A presequence was fused to a segment of 25 or 50 residues of glutamate or glycine residues followed by the coding sequence of mouse DHFR [94]. Stretches of these amino acid residues could not be bound by mtHsp70, which is consistent with the substrate specificity of the binding pocket of its bacterial homologue DnaK [40, 116]. Although mtHsp70 was not able to bind and therefore to "pull" on the preprotein, import of these fusion proteins into mitochondria occurred efficiently.

These experiments support the view that movement of polypeptide chains within the import channels in a forward and reverse direction can occur without the mtHsp70 and a power stroke being involved.

30.5.2.2 Recruitment of mtHsp70 by Tim44

Tim44 anchors the mtHsp70 chaperone to the TIM23 complex at the matrix side of the inner membrane [62, 109, 123]. An analysis of the interaction of Tim44 with mtHsp70 may help to discriminate between the two models. MtHsp70 consists of an ATPase domain, a peptide-binding domain (PBD), and a carboxy-terminal domain or "lid" [11, 42]. According to the Brownian ratchet model, any of these domains could interact with Tim44 to allow mtHsp70 to act as a targeted trapping device. In the power-stroke model, Tim44 is expected to bind the ATPase domain of mtHsp70, so that an ATP hydrolysis–dependent conformational change could generate a vectorial movement. In this case the PBD could move relative to the membrane anchor and function as a lever arm. Such a pulling of the polypeptide chain would be impossible if the PBD were bound to Tim44.

In two studies, the interaction between Tim44 and mtHsp70 was analyzed by co-immunoprecipitation experiments following the import of domains of mtHsp70 and hybrids between mtHsp70 and mitochondrial homologues of mtHsp70 into isolated mitochondria [61, 134]. The PBD of mtHsp70 in combination with the various ATPase domains were co-precipitated with Tim44 with an efficiency of about 50%, whereas the ATPase domains without the PBD were not precipitated or only minor quantities of about 10% precipitated. When a construct consisting of the peptide-binding and lid domain was imported, no interaction was observed; the functionality of the construct, however, was not shown [61, 134].

Another study was performed with hybrids between mtHsp70 and the closely related DnaK from *E. coli*, which were expressed in mitochondria [86]. In addition, truncated versions of mtHsp70 were imported into mitochondria. Co-immunoprecipitation analysis revealed that the PBD alone binds to Tim44. Together with the lid domain, the PBD was essential to transmit information on the nucleotide state of the ATPase domain to the rest of the chaperone. In addition, a mutant form of mtHsp70, the *ssc1-2* mutant, was isolated [55]. The *ssc1-2* mutation is present within the PBD. The interaction of mtHsp70 with Tim44, and therefore mitochondrial import, is compromised in the *ssc1-2* mutant [33, 55, 123, 147]. Intragenic suppressors were isolated that restore binding of the mutant form of mtHsp70 to Tim44 [145]. The suppressor mutations also reside within the PBD. The position of the mutations in the *ssc1-2* mutant is consistent with an interaction between the PBD and Tim44. In summary, it appears that the PBD of mtHsp70 cannot function as a lever arm, as it seems to bind to Tim44.

30.5.2.3 **Import Without Recruitment of mtHsp70 by Tim44**

According to the targeted Brownian ratchet model, the polypeptide chain will be trapped directly at the outlet of the import channel. Therefore, mtHsp70 has to be recruited to this site.

It was suggested that mtHsp70 can act as a ratchet in the mitochondrial matrix and import unfolded, but not folded, proteins without anchoring of mtHsp70 to Tim44 [68, 80, 145]. A trapping mechanism was proposed to be sufficient for import of unfolded proteins, but import of folded domains would require a Tim44-mtHsp70 interaction supporting the power-stroke model [78, 111, 145]. However, it has to be kept in mind that a Tim44-mtHsp70 interaction is equally essential for a targeted ratchet. The argumentation was based on experiments with two conditional mutant strains. One strain carries a mutant form of Tim44 and the other strain, *ssc1-2*, carries a mutation in the gene encoding mtHsp70. At non-permissive temperature these strains are unable to grow. When isolated mitochondria of the Tim44 mutant were incubated at non-permissive temperature, unfolded preproteins were imported, but import of folded preproteins was strongly reduced [9].

The same was found with *ssc1-2* mutant mitochondria after incubation at non-permissive temperature. At non-permissive temperature a prolonged binding of mtHsp70 to unfolded polypeptide chains was detected, but mtHsp70 did not interact with Tim44 anymore [145]. However, the prolonged binding does not reflect a stronger trapping capacity of the mutant Hsp70, Ssc1-2p, because its association and dissociation rates towards the polypeptide are not strongly altered [70]. Rather, folding of the polypeptide is affected, and therefore more unfolded precursor is present, which then interacts with mtHsp70. Binding to Tim44 and import of folded precursors could be restored by intragenic suppressors of *ssc1-2* [145].

Two explanations for this behavior were discussed. First, binding to the polypeptide chain by mtHsp70 without recruitment to the import channel could be sufficient for import of unfolded, but not folded, proteins. Second, there might be residual activities in the mutant forms that are sufficient for import of unfolded precursors.

The *ssc1-2* strain and a strain with reduced levels of functional Tim44 were used in another study that demonstrated that the activity of import of high amounts of recombinant unfolded precursor (picomoles) was drastically reduced in these mitochondria [81]. The level of reduction in import activity in mitochondria of the *ssc1-2* strain might depend on the precursor protein studied and on subtle experimental differences, as a modest reduction was observed in another report [68]. Still, these mitochondria imported radio labeled, unfolded precursors that were produced in a cell-free translation system (femtomoles) [81]. Residual activities of the conditional mutants were able to promote import of unfolded, but not folded, preproteins in small quantities. These experiments indicate that Tim44 and mtHsp70 are essential for efficient import [81].

In conclusion, both proposed models, the targeted molecular ratchet and the power-stroke mechanism, require the Tim44-mtHsp70 interaction to mediate efficient import of unfolded proteins. Therefore, these studies [68, 80, 81, 145] are not suited to discriminate between these models.

30.5.2.4 MtHsp70 Function in the Import Motor

MtHsp70 in the ATP-bound form is recruited by Tim44 to the import sites and efficiently binds to the incoming polypeptide chains. Tim14 then stimulates ATP hydrolysis, which leads to a conformational change in mtHsp70 so that the PBD of the ADP-bound form closes. Repeated cycles lead to inward movement of the polypeptide chain. Two classes of interactions of the unfolded polypeptide chain with mtHsp70s can be distinguished. The first class is mediated by Tim44 together with Tim14 at the outlet of the import channel. Even low-affinity complexes, which lead to short-lived interactions of substrate with mtHsp70 and which do not contribute to efficient trapping of the chain, may be formed. The on-rate of mtHsp70 to the substrate is high due to the presence of the J-domain protein Tim14 in the import motor, but the off-rate is also high due to the low binding affinity. Thus, there is a high rate of binding and release. Only a few binding sites of very high affinity are present in proteins [117, 122]. High-affinity binding can contribute to efficient trapping of segments exposed in the matrix space. This explains why high ATP levels in the matrix are necessary to find a pool of complexes of mtHsp70 in the import motor associated with the translocating precursor [84, 123, 139]. High ADP levels lead to a loss of this pool, as the ADP form of this mtHsp70 is rapidly released and Tim44-mediated cycling of the ATP form cannot occur.

The second class comprises high-affinity interactions. MtHsp70 entering the precursor at the import motor will remain tightly bound to a few segments of an unfolded polypeptide chain. These longer-lived complexes remain stable even after movement into the mitochondrial matrix. Moreover, dissociation of these complexes is dependent on the action of the nucleotide-exchange factor Mge1 and subsequent ATP uptake. Once mtHsp70 is released, but the protein still does not reach a folded state, mtHsp70 will rebind to the high-affinity binding sites with the help of the mitochondrial matrix co-chaperone Mdj1. High ATP in the matrix favors the open ATP-bound form of mtHsp70, and therefore steady-state levels of high-affinity complexes mediated by Mdj1 are low.

In conclusion, the mode of interaction of mtHsp70 with preproteins apparently depends on its location and its partner proteins. When ATP levels are high, Tim44-mediated interactions with preproteins at the import sites are observed, whereas at low ATP levels, interactions with preproteins in the matrix mediated by Mdj1 are observed [139, 140]. This demonstrates the specialized nature of the mitochondrial import motor, which allows efficient trapping of incoming precursor proteins. The recruitment of mtHsp70 and the J-protein Tim14 by Tim44 directly to the translocation channel allows efficient trapping, and even precursor segments with lower affinity and short-lived interactions can contribute. On the other hand, Mdj1-supported interactions of mtHsp70 with a few high-affinity sites of the polypeptide generate long-lived complexes in the matrix. These latter complexes may be able to prevent large-scale retrograde movements of the preprotein, but not the small-scale oscillations in the import channel that have to be ratcheted, in particular to promote unfolding at the mitochondrial surface.

30.6
Discussion and Perspectives

Mitochondrial precursor proteins are imported in a largely unfolded state. The translocases of the mitochondrial membranes are not able to transport proteins in a folded state across the membranes. There are at least three mechanisms that result in an unfolded polypeptide chain. First, co-translational import makes protein folding prior to translocation impossible. Second, molecular chaperones can interact with newly synthesized proteins. They protect the proteins against aggregation and keep them in an unfolded, translocation-competent state. Third, unfolding of proteins is mediated by the molecular import motor. It remains to be investigated which substrates have to be unfolded by the mitochondrial import motor in vivo and which substrates are kept in a translocation-competent state by cytosolic chaperones. It will be interesting to see how it is determined whether a protein folds or interacts with chaperones prior to the import into mitochondria.

The Brownian ratchet mechanism is widely accepted as a mechanism to drive protein translocation across membranes. This is the case not only for the mitochondrial system but also for translocation into the endoplasmic reticulum [53, 74, 145]. In mitochondria, protein translocation and protein unfolding on the surface of mitochondria are driven by the mitochondrial import motor. The targeted Brownian ratchet mechanism is sufficient to explain all experimental observations on the function and structure of the mitochondrial protein import motor. One of these functions is the unfolding of proteins during mitochondrial import. Folded proteins do not have requirements for their import that are different from or additional to those of folded proteins. It is still debated whether the import motor can exert a power stroke or whether it can switch to a power-stroke mode, in particular for the import of folded proteins. So far direct evidence for a power stroke has not been presented. It cannot be excluded that there is a weak pulling force generated by the import motor that assists in protein translocation. However, such a form does not seem to be either helpful or necessary. In the case of a power stroke, it should be possible to measure directly the force generated by the import motor. In order to support the power-stroke model, experimental proof for a lever-arm function of mtHsp70 would be required. In addition, experimental proof for an import in regular steps that correlates with the lever-arm dimensions would be required. As it was suggested that the import motor could switch from the trapping to the pulling mode to achieve unfolding by one power stroke, such a switching mechanism must be supported by experimental evidence as well.

A promising approach to address the mechanism of the import motor will be to reconstitute the import motor with purified components. Initial experiments have already been performed [69]. The observations made in this study are consistent with a Brownian ratchet, but not with a power-stroke mode. However, this study did not include the complete set of import motor components. The discovery of two new essential components of the import motor indicates that the motor is more complex than previously thought [24, 32, 59, 85, 138]. The search for further possible components of the motor will be a major task in the future, as it might provide us with new insights into the action of the import motor during the translocation process. For example, the membrane integration of the J-protein Tim14, one of the new components, clearly demonstrates that the ATP hydrolysis occurs at the membrane and not in the matrix space. After having identified the complete set of components, it will be necessary to reconstitute the mitochondrial import motor. This will allow us to analyze the system under experimentally defined conditions. One characteristic feature of the import motor is its association with the membrane-integrated translocation channel. Therefore, in the long run the import motor will have to be reconstituted into proteoliposomes together with the components of the translocation channel.

30.7
Experimental Protocols

30.7.1
Protein Import Into Mitochondria in Vitro

Import into isolated mitochondria of *Saccharomyces cerevisiae* is an important tool for analyzing whether a protein has a mitochondrial location. In addition, it is a powerful assay to study the mechanisms of the mitochondrial import machinery and its energetic and structural requirements.

In an in vitro import assay, precursor proteins are incubated together with isolated mitochondria and the uptake of precursor proteins by the mitochondria is analyzed. Precursor proteins are synthesized either in vitro in radiochemical amounts or in vivo in chemical amounts. For in vitro synthesis of radio labeled precursor proteins, purified plasmid DNA encoding the gene of interest under control of the SP6 promoter is first transcribed with SP6 RNA polymerase. The mRNA obtained is used in an in vitro translation reaction in the presence of [^{35}S]methionine (specific activity 1000 Ci/mM) using rabbit reticulocyte lysate (Promega) [16, 100]. Chemical amounts of precursor proteins are produced by expression of the precursor protein in *E. coli* and subsequent purification of the protein from the bacterial lysate (e.g., [19, 81]). Mitochondria are isolated from yeast cells according to published procedures [17, 43]. They are resuspended to a protein concentration of 10 mg mL^{-1} in SEM buffer (250 mM sucrose, 1 mM EDTA, 10 mM MOPS, pH 7.2), aliquoted, quick-frozen in liquid nitrogen, and stored at −80 °C. Mitochondria are thawed directly before use. For import into mitochondria from *N. crassa* and mammalian cells, mitochondria have to be prepared directly before use in the import assay [97, 150, 155]. Freezing them will drastically decrease their import efficiency.

In the in vitro import assay, mitochondria are added to import buffer (0.5 M sorbitol, 80 mM KCl, 10 mM magnesium acetate, 2 mM KH_2PO_4, 2.5 mM EDTA, 5 mM $MnCl_2$, 2 mM ATP, 2 mM NADH, 0–3% [w/v] bovine serum albumin [BSA, fatty acid-free], 50 mM HEPES, adjusted to pH 7.2 with KOH) in a 1.5-mL microcentrifuge tube. In general, 50–200 μg mitochondrial protein is used per 200 μL of buffer. The import efficiency of some precursor proteins is increased by the addition of BSA. The BSA has to be free of fatty acids, as these damage the mitochondrial membranes. Normally, ATP and NADH are added to energize the mitochondria and to maintain a high mitochondrial membrane potential.

The samples are mixed and then adjusted to the temperature of the import reaction for 3 min. The reaction is started by the addition of reticulocyte lysate containing the precursor protein (1–4% [v/v]) or the purified precursor protein (about 800 pmol mg^{-1} mitochondria). Following the import reaction, samples are placed on ice and 1 μM valinomycin, which destroys the membrane potential across the inner membrane, is added (stock: 100 μM in EtOH, store at −20 °C) to stop the reaction. The sample is divided into two aliquots and diluted with 400 μL ice-cold SHKCl (0.6 M sorbitol, 80 mM KCl 20 mM HEPES, adjusted to pH 7.2 with

KOH). One aliquot is treated with 50–100 μg mL^{-1}proteinase K (PK, stock solution: 10 mg mL^{-1}, stored in aliquots at −20 °C) for 30 min on ice. PK action is stopped by addition of 1 mM phenylmethylsulfonyl fluoride (PMSF 200 mM stock in EtOH, prepare fresh) and incubated for 5 min on ice. Then mitochondria are re-isolated by centrifugation at 15 000 g for 15 min at 4 °C. The pellet is washed, but not resuspended, with 300 μL SHKCl containing 0.3 mM PMSF and centrifuged as before. Pelleted mitochondria are resuspended in SDS-polyacrylamide gel electrophoresis (SDS-PAGE) buffer and heated immediately to 95 °C for 5 min to inhibit the protease completely. Samples are analyzed by SDS-PAGE followed by autoradiography and quantification using a phosphoimaging system or by immunodecoration and quantification by densitometry. One lane of the SDS-PAGE is loaded with 5–20% of the solution containing the precursor protein used per import sample to provide a standard that allows determination of the amount of imported radio labeled precursor protein. In order to determine the amount of imported recombinant precursor protein, several amounts (1–20%) of recombinant protein are loaded onto the SDS-PAGE gel and the values obtained by immunodecoration and quantification are used as standards. However, analysis by immunodecoration is not possible if a mitochondrial protein is imported, as the signal of the endogenous protein would interfere with the imported protein in the immunodecoration. In this case the protein has to be radio labeled.

Each precursor protein has a linear time range of import. To analyze and compare different import parameters, the import of the precursor protein should be performed in this linear range of import. To obtain import kinetics for a precursor protein, one larger sample is prepared in a 1.5-mL microcentrifuge tube as above and started by the addition of precursor protein. At certain time points (normally about five points in the range of the first 30 min), aliquots are removed and then processed and analyzed as described above.

In the case of precursor proteins with a cleavable presequence, import can be monitored by their processing. The precursor protein (p-form) is converted and catalyzed by the matrix-processing peptidase (MPP) to a lower-molecular-weight mature protein (m-form). Protection against added protease indicates that the mature form or a non-processed precursor protein with internal targeting signal is translocated into mitochondria. As a control for specific translocation of the precursor protein, the import reaction should be performed in the absence of a membrane potential, $\Delta\Psi$ across the inner membrane. Without $\Delta\Psi$ import of presequence-containing proteins and inner-membrane proteins into mitochondria does not take place. $\Delta\Psi$ can be depleted by the addition of uncouplers, such as 1 μM valinomycin.

In order to analyze whether a protein is imported into the mitochondrial matrix or is located completely or partly in the intermembrane space, the outer membrane is selectively opened by hypotonic treatment of mitochondria in the presence of protease. To this end one aliquot of the import reaction is diluted with nine volumes of 20-mM HEPES, pH 7.2, and treated with 50–100 μg mL^{-1} PK (see above). The resulting mitoplasts are isolated and analyzed as described for the mitochondria. If the protein or parts of it are located in the IMS, the protein will be digested completely or to smaller fragments. For control, mitochondria are lysed with 0.2% (w/v) of Triton X-100 and treated with the protease. Then total protein is precipitated with trichloroacetic acid. This confirms that no protease-resistant fragment is generated due to a stably folded structure but rather due to protection by a mitochondrial membrane.

30.7.2
Stabilization of the DHFR Domain by Methotrexate

The enzyme DHFR catalyzes the reduction of dihydrofolate to tetrahydrofolate using NADPH as co-substrate [6]. Methotrexate (MTX) is a substrate analogue of dihydrofolate that binds to DHFR, stabilizes its folded structure, and inhibits the enzyme activity [26, 135]. DHFR is not a mitochondrial protein. However, a fusion protein of DHFR and a mitochondrial targeting signal can be efficiently translocated into mitochondria. The fused targeting signal usually does not interfere with the folding of the DHFR [26]. MTX blocks the unfolding of the DHFR and thus inhibits translocation across the mitochondrial membranes [26]. The DHFR-containing precursor protein is incubated in the presence of 1 mM NADPH (stock solution: 100 mM in H_2O, make fresh each time) and 1 μM MTX (10 mM in SEM buffer stock solution; store at −20 °C and dilute further before use) in import buffer for 10 min on ice. Then isolated mitochondria are added and the samples are further treated as in a normal in vitro import assay.

If the N-terminal segment of a DHFR-containing precursor protein starting with the targeting signal is long enough to reach the mitochondrial matrix, a translocation intermediate spanning both mitochondrial membranes can be generated in the presence of MTX. Such translocation intermediates have proved to be a very useful tool for analyzing the characteristics of mitochondrial import. It has been demonstrated that precursor proteins with N-terminal targeting sequences traverse the mitochondrial membrane as an extended chain in an unfolded conformation and that ca. 50 amino acid residues are needed to span both membranes [26, 108]. In addition to the conformational state of the translocation intermediate, the energetic requirements, the sequential steps, and the mechanistic aspects of the translocation process also have been defined with the help of translocation intermediates [139, 140]. To this end, MTX-arrested translocation intermediates were generated in a first reaction [16]. They can be completely imported into mitochondria in a second reaction, if the MTX is removed from the reaction. If it is planned to remove MTX in the second reaction, the import of the DHFR-containing precursor protein is blocked with lower concentrations of MTX (5 nM MTX) in the first reaction. Then mitochondria are re-isolated, washed to remove the MTX, and incubated for the second import reaction [16].

Translocation intermediates were also used to define the import pathway of preproteins and to identify components of the translocation machinery by chemical cross-linking to and co-immunoprecipitation of the arrested intermediate [64, 84, 141, 157].

Studies with MTX-arrested translocation intermediates are not restricted to proteins with N-terminal presequences. Experiments performed with MTX-arrested

DHFR fusion proteins containing internal targeting signals have demonstrated that they enter mitochondria in a loop conformation [15, 30, 133, 153].

30.7.3
Import of Precursor Proteins Unfolded With Urea

Unfolding of precursor proteins with urea allows comparing the import characteristics of proteins in the unfolded state and the folded state. In order to unfold precursor proteins, they are first precipitated by the addition of a saturated ammonium sulfate solution [95]. Two volumes of saturated ammonium sulfate solution are added to the reticulocyte lysate containing the precursor protein in three steps with vigorous mixing following each step. The sample is incubated for 30 min on ice and centrifuged at 20 000 g for 15 min at 4 °C, and the pellet is resuspended in the original volume of urea-containing buffer (8 M urea 20 mM HEPES/KOH, pH 7.4, and freshly added 100 mM DTT). The precursor protein–containing sample is diluted into the import reaction so that the final urea concentration is below 300 mM. Then the import reaction is performed as described above.

30.7.4
Kinetic Analysis of the Unfolding Reaction by Trapping of Intermediates

Thermal breathing results in transient partial unfolding events of a protein. Partial unfolding leads to the exposure of amino acid side chains to the solvent, which are normally buried and inaccessible in the folded state of the protein. In the unfolded state chemical reagents have access to these amino acid side chains and can chemically modify them [35]. Determination of the kinetics of the chemical modification allows estimating the rates of the transient unfolding events. The minimal rate of transient unfolding is obtained, as reversible unfolding events that are faster than the chemical modification reaction cannot be trapped. The chemical modification prevents refolding of the protein domain, and therefore the partial unfolded intermediate is trapped. In the unfolded state, in contrast to the folded state, the protein is accessible to added protease, and the ratio of protease-sensitive to protease-resistant protein reflects the percentage of chemically modified protein and hence the amount of the transiently unfolded protein domain. The modification is performed for increasing time periods, and the data obtained will give the kinetics of the transient unfolding process.

Transient local unfolding events might occur throughout the protein. However, only the local unfolding of the protein domain in which the modified amino acid residue is located will be analyzed. Preproteins carrying an N-terminal presequence are imported in an N-to-C-terminal direction. Therefore, the unfolding of the N-terminal segment of the protein domain is the limiting step for the import. Unfolding kinetics of this segment has to be compared with the kinetics of import.

Transient Unfolding of DHFR

The sequence of mouse DHFR has one unique cysteine residue close to the N-terminus at position 7 (Cys7). Normally, Cys7 is buried inside the folded protein

[93, 131]. We analyzed the accessibility of this cysteine to the water-soluble, alkylating reagent *N*-ethylmaleimide (NEM). In the folded state of DHFR, Cys7 is not modified by NEM, whereas it is modified by NEM in the unfolded state [35]. Therefore, NEM modification of Cys7 can be used to monitor the thermal unfolding of DHFR. This could be shown by the concentration-dependent aggregation of DHFR after NEM treatment following unfolding at higher temperature [35]. Alkylation of Cys7 interferes with refolding of the DHFR and destabilizes its structure, resulting in increased protease sensitivity of the modified DHFR.

To estimate the rates of transient unfolding, reticulocyte lysate containing the in vitro–synthesized radio labeled precursor protein pSu9(1-69)-DHFR [35] is diluted 10-fold with import buffer containing 0.1% (w/v) BSA and preincubated for 4 min at the reaction temperature. Then 3 mM NEM is added to the sample, and after 0, 1, 3, 9, and 30 min aliquots are taken, put on ice, and immediately treated with 45 mM DTT to quench the NEM. Samples are incubated with 50 μg mL^{-1} PK for 25 min on ice. The protease treatment is stopped by the addition of 2 mM PMSF and samples are analyzed by SDS-PAGE, autoradiography, and quantification using a phosphoimaging system. Protease treatment results in generation of a fragment if the DHFR domain was folded and not modified by NEM. The amount of this PK-resistant fragment was quantified for all time points, corrected for the loss of radio labeled methionine, and compared to the signal of the DHFR in the untreated sample. The percentage of protease-resistant fragment was plotted versus the time, and the unfolding rate (percentage of degraded protein per minute) was determined by fitting the data to a first-order equation.

Transient Unfolding of the Heme-binding Domain (HBD)

Yeast cytochrome b_2 contains a tightly folded domain of about 99 amino acid residues, i.e., the heme-binding domain (HBD). The HBD containing bound heme is degraded by PK to a core fragment (amino acid residues 9 to approximately 90) under native conditions, whereas it is totally degraded in the heme-free form [35]. This indicates that the HBD is not tightly folded in the absence of its ligand. The HBD contains three histidine residues at positions 19, 43, and 66. His43 and His66 are located in the heme-binding pocket and are involved in ligand binding [156]. Modification of the imidazole group of these histidine residues by treatment with diethylpyrocarbonate (DEPC) results in protease sensitivity of the HBD. This indicates that the modified HBD is not folded [35]. The rate of the DEPC modification of the HBD monitors the amount of transient unfolding of the HBD. In order to estimate the rate of transient unfolding of the HBD, the kinetics of the DEPC modification of the histidine residues in the HBD is analyzed by protease resistance. The analysis is performed with a construct, pSu9(1-69)HBD, containing a mitochondrial targeting signal fused to the full-sized HBD (residues 1–99 of mature cytochrome b_2) [35]. Reticulocyte lysate containing this radio labeled precursor protein (10 μL) is diluted 50-fold in 100 mM KH_2PO_4, pH 7.4. Samples are pretreated for 5 min and then incubated with 6.9 mM DEPC at the temperature chosen. After 0, 1, 3, 9, and 30 min, an aliquot (50 μL) is withdrawn and the DEPC is quenched with an excess of imidazole (final concentration 200 mM). The samples (40% of the total volume) are treated with PK (1 mg mL^{-1}) for 25 min at 37 °C and 2 mM PMSF is added to stop PK. Local protein unfolding rates are determined by quantification of the protease-resistant fragment as described above for the kinetic determination of the NEM modification of DHFR.

DEPC is a clear, colorless liquid with a molarity of 6.9 M. It is very sensitive to moisture. Once opened, the DEPC should be stored under a nitrogen layer and the closed bottle should be stored at 0–4 °C to avoid hydrolysis of the DEPC. DEPC should be used carefully, as it is suspected to be carcinogenic.

References

1 Ahting, U., Thieffry, M., Engelhardt, H., Hegerl, R., Neupert, W. and Nussberger, S. (2001) Tom40, the pore-forming component of the protein-conducting TOM channel in the outer membrane of mitochondria. *J. Cell Biol.*, **153**, 1151–1160.

2 Ahting, U., Thun, C., Hegerl, R., Typke, D., Nargang, F. E., Neupert, W. and Nussberger, S. (1999) The TOM core complex: the general protein import pore of the outer membrane of mitochondria. *J. Cell Biol.*, **147**, 959–968.

3 Bauer, M. F., Hofmann, S., Neupert, W. and Brunner, M. (2000) Protein translocation into mitochondria: the role of TIM complexes. *Trends Cell Biol.*, **10**, 25–31.

4 Bauer, M. F., Sirrenberg, C., Neupert, W. and Brunner, M. (1996) Role of Tim23 as voltage sensor and presequence receptor in protein import into mitochondria. *Cell*, **87**, 33–41.

5 Berthold, J., Bauer, M. F., Schneider, H. C., Klaus, C., Dietmeier, K., Neupert, W. and Brunner, M. (1995) The MIM complex mediates preprotein translocation across the mitochondrial inner membrane and couples it to the mt-Hsp70/ATP driving system. *Cell*, **81**, 1085–1093.

6 Blakley, R. L. (1995) Eukaryotic dihydrofolate reductase. *Adv. Enzymol. Relat. Areas Mol. Biol.*, **70**, 23–102.

7 Bolliger, L., Deloche, O., Glick, B. S., Georgopoulos, C., Jeno, P., Kronidou, N., Horst, M., Morishima, N. and Schatz, G. (1994) A mitochondrial homolog of bacterial GrpE interacts with mitochondrial hsp70 and is essential for viability. *EMBO J.*, **13**, 1998–2006.

8 Bolliger, L., Junne, T., Schatz, G. and Lithgow, T. (1995) Acidic receptor domains on both sides of the outer membrane mediate translocation of precursor proteins into yeast mitochondria. *EMBO J.*, **14**, 6318–6326.

9 Bomer, U., Maarse, A. C., Martin, F., Geissler, A., Merlin, A., Schonfisch, B., Meijer, M., Pfanner, N. and Rassow, J. (1998) Separation of structural and dynamic functions of the mitochondrial translocase: Tim44 is crucial for the inner membrane import sites in translocation of tightly folded domains, but not of loosely folded preproteins. *EMBO J.*, **17**, 4226–4237.

10 Bomer, U., Meijer, M., Guiard, B., Dietmeier, K., Pfanner, N. and Rassow, J. (1997) The sorting route of cytochrome b2 branches from the general mitochondrial import pathway at the preprotein translocase of the inner membrane. *J. Biol. Chem.*, **272**, 30439–30446.

11 Bukau, B. and Horwich, A. L. (1998) The Hsp70 and Hsp60 chaperone machines. *Cell*, **92**, 351–366.

12 Carrion-Vazquez, M., Oberhauser, A. F., Fowler, S. B., Marszalek, P. E., Broedel, S. E., Clarke, J. and Fernandez, J. M. (1999) Mechanical

and chemical unfolding of a single protein: a comparison. *Proc. Natl. Acad. Sci. USA*, **96**, 3694–3699.

13 Chauwin, J. F., Oster, G. and Glick, B. S. (1998) Strong precursor-pore interactions constrain models for mitochondrial protein import. *Biophys. J*, **74**, 1732–1743.

14 Cheng, M. Y., Hartl, F. U., Martin, J., Pollock, R. A., Kalousek, F., Neupert, W., Hallberg, E. M., Hallberg, R. L. and Horwich, A. L. (1989) Mitochondrial heat-shock protein hsp60 is essential for assembly of proteins imported into yeast mitochondria. *Nature*, **337**, 620–625.

15 Curran, S. P., Leuenberger, D., Schmidt, E. and Koehler, C. M. (2002) The role of the Tim8p-Tim13p complex in a conserved import pathway for mitochondrial polytopic inner membrane proteins. *J. Cell Biol.*, **158**, 1017–1027.

16 Cyr, D. M., Ungermann, C. and Neupert, W. (1995) Analysis of mitochondrial protein import pathway in Saccharomyces cerevisiae with translocation intermediates. *Methods Enzymol.*, **260**, 241–252.

17 Daum, G., Bohni, P. C. and Schatz, G. (1982) Import of proteins into mitochondria. Cytochrome b2 and cytochrome c peroxidase are located in the intermembrane space of yeast mitochondria. *J. Biol. Chem.*, **257**, 13028–13033.

18 Dekker, P. J., Keil, P., Rassow, J., Maarse, A. C., Pfanner, N. and Meijer, M. (1993) Identification of MIM23, a putative component of the protein import machinery of the mitochondrial inner membrane. *FEBS Lett.*, **330**, 66–70.

19 Dekker, P. J., Martin, F., Maarse, A. C., Bomer, U., Muller, H., Guiard, B., Meijer, M., Rassow, J. and Pfanner, N. (1997) The Tim core complex defines the number of mitochondrial translocation contact sites and can hold arrested preproteins in the absence of matrix Hsp70-Tim44. *EMBO J.*, **16**, 5408–5419.

20 Dekker, P. J., Ryan, M. T., Brix, J., Muller, H., Honlinger, A. and Pfanner, N. (1998) Preprotein translocase of the outer mitochondrial membrane: molecular dissection and assembly of the general import pore complex. *Mol. Cell. Biol.*, **18**, 6515–6524.

21 Deshaies, R. J., Koch, B. D., Werner-Washburne, M., Craig, E. A. and Schekman, R. (1988) A subfamily of stress proteins facilitates translocation of secretory and mitochondrial precursor polypeptides. *Nature*, **332**, 800–805.

22 Donzeau, M., Kaldi, K., Adam, A., Paschen, S., Wanner, G., Guiard, B., Bauer, M. F., Neupert, W. and Brunner, M. (2000) Tim23 links the inner and outer mitochondrial membranes. *Cell*, **101**, 401–412.

23 Driessen, A. J., Manting, E. H. and van der Does, C. (2001) The structural basis of protein targeting and translocation in bacteria. *Nat. Struct. Biol.*, **8**, 492–498.

24 D'Silva, P. D., Schilke, B., Walter, W., Andrew, A. and Craig, E. A. (2003) J protein cochaperone of the mitochondrial inner membrane required for protein import into the mitochondrial matrix. *Proc. Natl. Acad. Sci. USA*, **100**, 13839–13844.

25 Eilers, M., Hwang, S. and Schatz, G. (1988) Unfolding and refolding of a purified precursor protein during import into isolated mitochondria. *EMBO J.*, **7**, 1139–1145.

26 Eilers, M. and Schatz, G. (1986) Binding of a specific ligand inhibits import of a purified precursor protein into mitochondria. *Nature*, **322**, 228–232.

27 Emtage, J. L. and Jensen, R. E. (1993) MAS6 encodes an essential inner membrane component of the yeast mitochondrial protein import pathway. *J. Cell Biol.*, **122**, 1003–1012.

28 Endo, T. and Kohda, D. (2002) Functions of outer membrane receptors in mitochondrial protein import. *Biochim. Biophys. Acta*, **1592**, 3.

29 Endo, T., Yamamoto, H. and Esaki, M. (2003) Functional cooperation and separation of translocators in protein

import into mitochondria, the double-membrane bounded organelles. *J. Cell Sci.*, **116**, 3259–3267.

30 Endres, M., Neupert, W. and Brunner, M. (1999) Transport of the ADP/ATP carrier of mitochondria from the TOM complex to the TIM22.54 complex. *EMBO J.*, **18**, 3214–3221.

31 Foelsch, H., Guiard, B., Neupert, W. and Stuart, R. A. (1996) Internal targeting signal of the BCS1 protein: a novel mechanism of import into mitochondria. *EMBO J.*, **15**, 479–487.

32 Frazier, A., Dudek, J., Guiard, B., Voos, W., Li, Y., Lind, M., Meisinger, C., Geissler, A., Sickmann, A., Meyer, H. E., Bilanchone, V., Cumsky, M. G., Truscott, K. N., Pfanner, N. and Rehling, P. (2004) Pam16 plays an essential role in the mitochondrial protein import motor. *Nature Struct. Mol. Biol.*, **11**, 226–233.

33 Gambill, B. D., Voos, W., Kang, P. J., Miao, B., Langer, T., Craig, E. A. and Pfanner, N. (1993) A dual role for mitochondrial heat shock protein 70 in membrane translocation of preproteins. *J. Cell Biol.*, **123**, 109–117.

34 Gärtner, F., Voos, W., Querol, A., Miller, B. R., Craig, E. A., Cumsky, M. G. and Pfanner, N. (1995) Mitochondrial import of subunit Va of cytochrome c oxidase characterized with yeast mutants – Independence from receptors, but requirement for matrix hsp70 translocase function. *J. Biol. Chem.*, **270**, 3788–3795.

35 Gaume, B., Klaus, C., Ungermann, C., Guiard, B., Neupert, W. and Brunner, M. (1998) Unfolding of preproteins upon import into mitochondria. *EMBO J.*, **17**, 6497–6507.

36 Gentle, I., Gabriel, K., Beech, P., Waller, R. and Lithgow, T. (2004) The Omp85 family of proteins is essential for outer membrane biogenesis in mitochondria and bacteria. *J. Cell Biol.*, **164**, 19–24.

37 Geissler, A., Chacinska, A., Truscott, K. N., Wiedemann, N., Brandner, K., Sickmann, A., Meyer, H. E., Meisinger, C., Pfanner, N. and Rehling, P. (2002) The mitochondrial presequence translocase: an essential role of Tim50 in directing preproteins to the import channel. *Cell*, **111**, 507–518.

38 Glick, B. S. (1995) Can Hsp70 proteins act as force-generating motors? *Cell*, **80**, 11–14.

39 Glick, B. S., Wachter, C., Reid, G. A. and Schatz, G. (1993) Import of cytochrome b2 to the mitochondrial intermembrane space: the tightly folded heme-binding domain makes import dependent upon matrix ATP. *Protein Sci.*, **2**, 1901–1917.

40 Gragerov, A., Zeng, L., Zhao, X., Burkholder, W. and Gottesman, M. E. (1994) Specificity of DnaK-peptide binding. *J. Mol. Biol.*, **235**, 848–854.

41 Hachiya, N., Komiya, T., Alam, R., Iwahashi, J., Sakaguchi, M., Omura, T. and Mihara, K. (1994) MSF, a novel cytoplasmic chaperone which functions in precursor targeting to mitochondria. *EMBO J.*, **13**, 5146–5154.

42 Hartl, F. U. and Hayer-Hartl, M. (2002) Molecular chaperones in the cytosol: from nascent chain to folded protein. *Science*, **295**, 1852–1858.

43 Herrmann, J. M., Foelsch, H., Neupert, W. and Stuart, R. A. (1994) Isolation of yeast mitochondria and study of mitochondrial protein translation. In Celis, J. E. (ed.), *Cell Biology: A Laboratory Handbook.* Academic Press, San Diego, Vol. 1, pp. 538–544.

44 Herrmann, J. M. and Neupert, W. (2000) What fuels polypeptide translocation? An energetical view on mitochondrial protein sorting. *Biochim. Biophys. Acta*, **1459**, 331–338.

45 Hill, K., Model, K., Ryan, M. T., Dietmeier, K., Martin, F., Wagner, R. and Pfanner, N. (1998) Tom40 forms the hydrophilic channel of the mitochondrial import pore for preproteins. *Nature*, **395**, 516–521.

46 Holroyd, C. and Erdmann, R. (2001) Protein translocation machineries of peroxisomes. *FEBS Lett.*, **501**, 6–10.

47 Hönlinger, A., Kübrich, M., Moczko, M., Gärtner, F., Mallet, L., Bussereau, F., Eckerskorn, C., Lottspeich, F., Dietmeier, K., Jacquet, M. and Pfanner, N. (1995) The mitochondrial receptor complex: Mom22 is essential for cell viability and directly interacts with preproteins. *Mol. Cell. Biol.*, **15**, 3382–3389.
48 Huang, S., Ratliff, K. S. and Matouschek, A. (2002) Protein unfolding by the mitochondrial membrane potential. *Nat. Struct. Biol.*, **9**, 301–307.
49 Huang, S., Ratliff, K. S., Schwartz, M. P., Spenner, J. M. and Matouschek, A. (1999) Mitochondria unfold precursor proteins by unraveling them from their N-termini. *Nat. Struct. Biol.*, **6**, 1132–1138.
50 Hurt, E. C., Pesold-Hurt, B. and Schatz, G. (1984) The cleavable prepiece of an imported mitochondrial protein is sufficient to direct cytosolic dihydrofolate reductase into the mitochondrial matrix. *FEBS Lett.*, **178**, 306–310.
51 Jarvis, P. and Soll, J. (2002) Toc, tic, and chloroplast protein import. *Biochim. Biophys. Acta*, **1590**, 177–189.
52 Jensen, R. and Dunn, C. (2002) Protein import into and across the mitochondrial inner membrane: role of the TIM23 and TIM22 translocons. *Biochim. Biophys. Acta*, **1592**, 25.
53 Jensen, R. E. and Johnson, A. E. (1999) Protein translocation: is Hsp70 pulling my chain? *Curr. Biol.*, **9**, R779–782.
54 Johnson, A. E. and Haigh, N. G. (2000) The ER translocon and retrotranslocation: is the shift into reverse manual or automatic? *Cell*, **102**, 709–712.
55 Kang, P. J., Ostermann, J., Shilling, J., Neupert, W., Craig, E. A. and Pfanner, N. (1990) Requirement for hsp70 in the mitochondrial matrix for translocation and folding of precursor proteins. *Nature*, **348**, 137–143.
56 Keegstra, K. and Froehlich, J. E. (1999) Protein import into chloroplasts. *Curr. Opin. Plant Biol.*, **2**, 471–476.
57 Kiebler, M., Keil, P., Schneider, H., van der Klei, I. J., Pfanner, N. and Neupert, W. (1993) The mitochondrial receptor complex: a central role of MOM22 in mediating preprotein transfer from receptors to the general insertion pore. *Cell*, **74**, 483–492.
58 Koehler, C. M. (2000) Protein translocation pathways of the mitochondrion. *FEBS Lett.*, **476**, 27–31.
59 Kozany, C., Mokranjac, D., Sichting, M., Neupert, W., and Hell, K. (2004) The J-domain related co-chaperone Tim16 is a constituent of the mitochondrial TIM23 preprotein translocase. *Nature Struct. Mol. Biol.*, **11**, 234–241.
60 Kozjak, V., Wiedemann, N., Milenkovic, D., Lohaus, C., Meyer, H. E., Guiard, B., Meisinger, C. and Pfanner, N. (2003) An essential role of Sam50 in the protein sorting and assembly machinery of the mitochondrial outer membrane. *J. Biol. Chem.*, **278**, 48520–48523.
61 Krimmer, T., Rassow, J., Kunau, W. H., Voos, W. and Pfanner, N. (2000) Mitochondrial protein import motor: the ATPase domain of matrix Hsp70 is crucial for binding to Tim44, while the peptide binding domain and the carboxy-terminal segment play a stimulatory role. *Mol. Cell. Biol.* **20**, 5879–5887.
62 Kronidou, N. G., Oppliger, W., Bolliger, L., Hannavy, K., Glick, B. S., Schatz, G. and Horst, M. (1994) Dynamic interaction between Isp45 and mitochondrial hsp70 in the protein import system of the yeast mitochondrial inner membrane. *Proc. Natl. Acad. Sci. U.S.A.*, **91**, 12818–12822.
63 Kunkele, K. P., Heins, S., Dembowski, M., Nargang, F. E., Benz, R., Thieffry, M., Walz, J., Lill, R., Nussberger, S. and Neupert, W. (1998) The preprotein translocation channel of the outer membrane of mitochondria. *Cell*, **93**, 1009–1019.
64 Kurz, M., Martin, H., Rassow, J.,

Pfanner, N. and Ryan, M. T. (1999) Biogenesis of Tim proteins of the mitochondrial carrier import pathway: differential targeting mechanisms and crossing over with the main import pathway. *Mol. Biol. Cell*, **10**, 2461–2474.

65 Laloraya, S., Gambill, B. D. and Craig, E. A. (1994) A role for a eukaryotic GrpE-related protein, Mge1p, in protein translocation. *Proc. Natl. Acad. Sci. U.S.A.*, **91**, 6481–6485.

66 Lee, C. M., Neupert, W. and Stuart, R. A. (2000) Mitochondrial targeting signals. In J. A. F. op den Kamp, e. (ed.), *Protein, Lipid and Membrane Traffic: Pathways and Targeting*. IOS Press, Amsterdam, pp. 151–159.

67 Li, H., Oberhauser, A. F., Fowler, S. B., Clarke, J. and Fernandez, J. M. (2000) Atomic force microscopy reveals the mechanical design of a modular protein. *Proc. Natl. Acad. Sci. USA*, **97**, 6527–6531.

68 Lim, J. H., Martin, F., Guiard, B., Pfanner, N. and Voos, W. (2001) The mitochondrial Hsp70-dependent import system actively unfolds preproteins and shortens the lag phase of translocation. *EMBO J.*, **20**, 941–950.

69 Liu, Q., D'Silva, P., Walter, W., Marszalek, J. and Craig, E. A. (2003) Regulated cycling of mitochondrial Hsp70 at the protein import channel. *Science*, **300**, 139–141.

70 Liu, Q., Krzewska, J., Liberek, K. and Craig, E. A. (2001) Mitochondrial Hsp70 Ssc1: role in protein folding. *J. Biol. Chem.*, **276**, 6112–6118.

71 Maarse, A. C., Blom, J., Keil, P., Pfanner, N. and Meijer, M. (1994) Identification of the essential yeast protein MIM17, an integral mitochondrial inner membrane protein involved in protein import. *FEBS Lett.*, **349**, 215–221.

72 Marszalek, P. E., Lu, H., Li, H., Carrion-Vazquez, M., Oberhauser, A. F., Schulten, K. and Fernandez, J. M. (1999) Mechanical unfolding intermediates in titin modules. *Nature*, **402**, 100–103.

73 Martin, J., Mahlke, K. and Pfanner, N. (1991) Role of an energized inner membrane in mitochondrial protein import. Delta psi drives the movement of presequences. *J. Biol. Chem.*, **266**, 18051–18057.

74 Matlack, K. E., Misselwitz, B., Plath, K. and Rapoport, T. A. (1999) BiP acts as a molecular ratchet during posttranslational transport of prepro-alpha factor across the ER membrane. *Cell*, **97**, 553–564.

75 Matlack, K. E., Mothes, W. and Rapoport, T. A. (1998) Protein translocation: tunnel vision. *Cell*, **92**, 381–390.

76 Matouschek, A. (2003) Protein unfolding – an important process in vivo? *Curr. Opin. Struct. Biol.*, **13**, 98–109.

77 Matouschek, A., Azem, A., Ratliff, K., Glick, B. S., Schmid, K. and Schatz, G. (1997) Active unfolding of precursor proteins during mitochondrial protein import. *EMBO J.*, **16**, 6727–6736.

78 Matouschek, A., Pfanner, N. and Voos, W. (2000) Protein unfolding by mitochondria. The Hsp70 import motor. *EMBO Rep.*, **1**, 404–410.

79 Mayer, A., Neupert, W. and Lill, R. (1995) Mitochondrial protein import: Reversible binding of the presequence at the trans side of the outer membrane drives partial translocation and unfolding. *Cell*, **80**, 127–137.

80 Merlin, A., Voos, W., Maarse, A. C., Meijer, M., Pfanner, N. and Rassow, J. (1999) The J-related segment of tim44 is essential for cell viability: a mutant Tim44 remains in the mitochondrial import site, but inefficiently recruits mtHsp70 and impairs protein translocation. *J. Cell Biol.*, **145**, 961–972.

81 Milisav, I., Moro, F., Neupert, W. and Brunner, M. (2001) Modular structure of the TIM23 preprotein translocase of mitochondria. *J. Biol. Chem.*, **276**, 25856–25861.

82 Miller, B. R. and Cumsky, M. G. (1993) Intramitochondrial sorting of the precursor to yeast cytochrome c oxidase subunit Va. *J. Cell Biol.*, **121**, 1021–1029.

83 Model, K., Prinz, T., Ruiz, T.,

Radermacher, M., Krimmer, T., Kuhlbrandt, W., Pfanner, N. and Meisinger, C. (2002) Protein Translocase of the Outer Mitochondrial Membrane: Role of Import Receptors in the Structural Organization of the TOM Complex. *J. Mol. Biol.*, **316**, 657–666.

84 Mokranjac, D., Paschen, S. A., Kozany, C., Prokisch, H., Hoppins, S. C., Nargang, F. E., Neupert, W. and Hell, K. (2003a) Tim50, a novel component of the TIM23 preprotein translocase of mitochondria. *EMBO J.*, **22**, 816–825.

85 Mokranjac, D., Sichting, M., Neupert, W. and Hell, K. (2003b) Tim14, a novel key component of the import motor of the TIM23 protein translocase of mitochondria. *EMBO J.*, **22**, 4945–4956.

86 Moro, F., Okamoto, K., Donzeau, M., Neupert, W. and Brunner, M. (2002) Mitochondrial protein import: molecular basis of the ATP-dependent interaction of MtHsp70 with Tim44. *J. Biol. Chem.*, **277**, 6874–6880.

87 Moro, F., Sirrenberg, C., Schneider, H. C., Neupert, W. and Brunner, M. (1999) The TIM17.23 preprotein translocase of mitochondria: composition and function in protein transport into the matrix. *EMBO J.*, **18**, 3667–3675.

88 Murakami, H., Pain, D. and Blobel, G. (1988) 70-kD heat shock-related protein is one of at least two distinct cytosolic factors stimulating protein import into mitochondria. *J. Cell. Biol.*, **107**, 2051–2057.

89 Nakai, M., Kato, Y., Ikeda, E., Toh-e, A. and Endo, T. (1994) Yge1p, a eukaryotic Grp-E homolog, is localized in the mitochondrial matrix and interacts with mitochondrial Hsp70. *Biochem. Biophys. Res. Commun.* **200**, 435–442.

90 Neupert, W. (1997) Protein import into mitochondria. *Annu. Rev. Biochem.*, **66**, 863–917.

91 Neupert, W. and Brunner, M. (2002) The protein import motor of mitochondria. *Nat. Rev. Mol. Cell Biol.*, **3**, 555–565.

92 Neupert, W., Hartl, F. U., Craig, E. A. and Pfanner, N. (1990) How do polypeptides cross the mitochondrial membranes? *Cell*, **63**, 447–450.

93 Oefner, C., D'Arcy, A. and Winkler, F. K. (1988) Crystal structure of human dihydrofolate reductase complexed with folate. *Eur. J. Biochem.*, **174**, 377–385.

94 Okamoto, K., Brinker, A., Paschen, S. A., Moarefi, I., Hayer-Hartl, M., Neupert, W. and Brunner, M. (2002) The protein import motor of mitochondria: a targeted molecular ratchet driving unfolding and translocation. *EMBO J.*, **21**, 3659–3671.

95 Ostermann, J., Horwich, A. L., Neupert, W. and Hartl, F. U. (1989) Protein folding in mitochondria requires complex formation with hsp60 and ATP hydrolysis. *Nature*, **341**, 125–130.

96 Ostermann, J., Voos, W., Kang, P. J., Craig, E. A., Neupert, W. and Pfanner, N. (1990) Precursor proteins in transit through mitochondrial contact sites interact with hsp70 in the matrix. *FEBS Lett.*, **277**, 281–284.

97 Pallotti, F. and Lenaz, G. (2001) Isolation and subfractionation of mitochondria from animal cells and tissue culture lines. *Methods Cell Biol.*, **65**, 1–35.

98 Paschen, S. A. and Neupert, W. (2001) Protein import into mitochondria. *IUBMB Life*, **52**, 101–112.

99 Paschen, S. A., Waizenegger, T., Stan, T., Preuss, M., Cyrklaff, M., Hell, K., Rapaport, D. and Neupert, W. (2003) Evolutionary conservation of biogenesis of beta-barrel membrane proteins. *Nature*, **426**, 862–866.

100 Pelham, H. R. B. and Jackson, R. J. (1976) An efficient mRNA-dependent translation system from reticulocyte lysates. *Eur. J. Biochem.*, **67**, 247–256.

101 Pfanner, N. and Chacinska, A. (2002) The mitochondrial import machinery: preprotein-conducting channels with binding sites for presequences. *Biochim. Biophys. Acta*, **1592**, 15.

102 PFANNER, N. and GEISSLER, A. (2001) Versatility of the mitochondrial protein import machinery. *Nat. Rev. Mol. Cell. Biol.*, **2**, 339–349.

103 PFANNER, N. and MEIJER, M. (1995) Protein sorting. Pulling in the proteins. *Curr. Biol.*, **5**, 132–135.

104 RAPAPORT, D. (2002) Biogenesis of the mitochondrial TOM complex. *Trends Biochem. Sci.*, **27**, 191–197.

105 RAPAPORT, D., KUNKELE, K. P., DEMBOWSKI, M., AHTING, U., NARGANG, F. E., NEUPERT, W. and LILL, R. (1998) Dynamics of the TOM complex of mitochondria during binding and translocation of preproteins. *Mol. Cell. Biol.*, **18**, 5256–5262.

106 RAPAPORT, D., NEUPERT, W. and LILL, R. (1997) Mitochondrial protein import. Tom40 plays a major role in targeting and translocation of preproteins by forming a specific binding site for the presequence. *J. Biol. Chem.*, **272**, 18725–18731.

107 RASSOW, J., GUIARD, B., WIENHUES, U., HERZOG, V., HARTL, F. U. and NEUPERT, W. (1989) Translocation arrest by reversible folding of a precursor protein imported into mitochondria. A means to quantitate translocation contact sites. *J. Cell Biol.*, **109**, 1421–1428.

108 RASSOW, J., HARTL, F. U., GUIARD, B., PFANNER, N. and NEUPERT, W. (1990) Polypeptides traverse the mitochondrial envelope in an extended state. *FEBS Lett.*, **275**, 190–194.

109 RASSOW, J., MAARSE, A. C., KRAINER, E., KUBRICH, M., MULLER, H., MEIJER, M., CRAIG, E. A. and PFANNER, N. (1994) Mitochondrial protein import: biochemical and genetic evidence for interaction of matrix hsp70 and the inner membrane protein MIM44. *J. Cell Biol.*, **127**, 1547–1556.

110 RASSOW, J., MOHRS, K., KOIDL, S., BARTHELMESS, I. B., PFANNER, N. and TROPSCHUG, M. (1995) Cyclophilin 20 is involved in mitochondrial protein folding in cooperation with molecular chaperones Hsp70 and Hsp60. *Mol. Cell. Biol.*, **15**, 2654–2662.

111 RASSOW, J. and PFANNER, N. (2000) The protein import machinery of the mitochondrial membranes. *Traffic*, **1**, 457–464.

112 REHLING, P., MODEL, K., BRANDNER, K., KOVERMANN, P., SICKMANN, A., MEYER, H. E., KUHLBRANDT, W., WAGNER, R., TRUSCOTT, K. N. and PFANNER, N. (2003a) Protein insertion into the mitochondrial inner membrane by a twin-pore translocase. *Science*, **299**, 1747–1751.

113 REHLING, P., PFANNER, N. and MEISINGER, C. (2003b) Insertion of hydrophobic membrane proteins into the inner mitochondrial membrane – a guided tour. *J. Mol. Biol.*, **326**, 639–657.

114 ROBINSON, C. and BOLHUIS, A. (2001) Protein targeting by the twin-arginine translocation pathway. *Nat. Rev. Mol. Cell Biol.*, **2**, 350–356.

115 ROJO, E. E., GUIARD, B., NEUPERT, W. and STUART, R. A. (1998) Sorting of D-lactate dehydrogenase to the inner membrane of mitochondria. Analysis of topogenic signal and energetic requirements. *J. Biol. Chem.*, **273**, 8040–8047.

116 RUDIGER, S., BUCHBERGER, A. and BUKAU, B. (1997a) Interaction of Hsp70 chaperones with substrates. *Nat. Struct. Biol.*, **4**, 342–349.

117 RUDIGER, S., GERMEROTH, L., SCHNEIDER-MERGENER, J. and BUKAU, B. (1997b) Substrate specificity of the DnaK chaperone determined by screening cellulose-bound peptide libraries. *EMBO J.*, **16**, 1501–1507.

118 RYAN, K. R. and JENSEN, R. E. (1993) Mas6p can be cross-linked to an arrested precursor and interacts with other proteins during mitochondrial protein import. *J. Biol. Chem.*, **268**, 23743–23746.

119 RYAN, K. R., MENOLD, M. M., GARRETT, S. and JENSEN, R. E. (1994) SMS1, a high-copy suppressor of the yeast mas6 mutant, encodes an essential inner membrane protein required for mitochondrial protein import. *Mol. Biol. Cell*, **5**, 529–538.

120 SCHATZ, G. (1997) Just follow the acid chain. *Nature*, **388**, 121–122.

121 SCHATZ, G. and DOBBERSTEIN, B. (1996) Common principles of protein

translocation across membranes. *Science*, **271**, 1519–1526.

122 Schmid, D., Baici, A., Gehring, H. and Christen, P. (1994) Kinetics of molecular chaperone action. *Science*, **263**, 971–973.

123 Schneider, H. C., Berthold, J., Bauer, M. F., Dietmeier, K., Guiard, B., Brunner, M. and Neupert, W. (1994) Mitochondrial Hsp70/MIM44 complex facilitates protein import. *Nature*, **371**, 768–774.

124 Schneider, H. C., Westermann, B., Neupert, W. and Brunner, M. (1996) The nucleotide exchange factor MGE exerts a key function in the ATP-dependent cycle of mt-Hsp70-Tim44 interaction driving mitochondrial protein import. *EMBO J.*, **15**, 5796–5803.

125 Schwartz, M. P., Huang, S. and Matouschek, A. (1999) The structure of precursor proteins during import into mitochondria. *J. Biol. Chem.*, **274**, 12759–12764.

126 Schwartz, M. P. and Matouschek, A. (1999) The dimensions of the protein import channels in the outer and inner mitochondrial membranes. *Proc. Natl. Acad. Sci. U.S.A.*, **96**, 13086–13090.

127 Simon, S. M., Peskin, C. S. and Oster, G. F. (1992) What drives the translocation of proteins? *Proc. Natl. Acad. Sci. USA*, **89**, 3770–3774.

128 Smith, D. A. and Radford, S. E. (2000) Protein folding: pulling back the frontiers. *Curr. Biol.*, **10**, R662–664.

129 Spudich, J. A. (1994) How molecular motors work. *Nature*, **372**, 515–518.

130 Spudich, J. A. (2001) The myosin swinging cross-bridge model. *Nat. Rev. Mol. Cell Biol.*, **2**, 387–392.

131 Stammers, D. K., Champness, J. N., Beddell, C. R., Dann, J. G., Eliopoulos, E., Geddes, A. J., Ogg, D. and North, A. C. (1987) The structure of mouse L1210 dihydrofolate reductase-drug complexes and the construction of a model of human enzyme. *FEBS Lett.*, **218**, 178–184.

132 Stan, T., Ahting, U., Dembowski, M., Kunkele, K. P., Nussberger, S., Neupert, W. and Rapaport, D. (2000) Recognition of preproteins by the isolated TOM complex of mitochondria. *EMBO J.*, **19**, 4895–4902.

133 Stan, T., Brix, J., Schneider-Mergener, J., Pfanner, N., Neupert, W. and Rapaport, D. (2003) Mitochondrial protein import: recognition of internal import signals of BCS1 by the TOM complex. *Mol. Cell. Biol.*, **23**, 2239–2250.

134 Strub, A., Rottgers, K. and Voos, W. (2002) The Hsp70 peptide-binding domain determines the interaction of the ATPase domain with Tim44 in mitochondria. *EMBO J.*, **21**, 2626–2635.

135 Touchette, N. A., Perry, K. M. and Matthews, C. R. (1986) Folding of dihydrofolate reductase from Escherichia coli. *Biochemistry*, **25**, 5445–5452.

136 Truscott, K. N., Brandner, K. and Pfanner, N. (2003a) Mechanisms of protein import into mitochondria. *Curr. Biol.*, **13**, R326–337.

137 Truscott, K. N., Kovermann, P., Geissler, A., Merlin, A., Meijer, M., Driessen, A. J., Rassow, J., Pfanner, N. and Wagner, R. (2001) A presequence- and voltage-sensitive channel of the mitochondrial preprotein translocase formed by Tim23. *Nat. Struct. Biol.*, **8**, 1074–1082.

138 Truscott, K. N., Voos, W., Frazier, A. E., Lind, M., Li, Y., Geissler, A., Dudek, J., Muller, H., Sickmann, A., Meyer, H. E., Meisinger, C., Guiard, B., Rehling, P. and Pfanner, N. (2003b) A J-protein is an essential subunit of the presequence translocase-associated protein import motor of mitochondria. *J. Cell Biol.*, **163**, 707–713.

139 Ungermann, C., Guiard, B., Neupert, W. and Cyr, D. M. (1996) The delta psi- and Hsp70/MIM44-dependent reaction cycle driving early steps of protein import into mitochondria. *EMBO J.*, **15**, 735–744.

140 Ungermann, C., Neupert, W. and Cyr, D. M. (1994) The role of Hsp70 in conferring unidirectionality on protein translocation into mitochondria. *Science*, **266**, 1250–1253.

141 Vestweber, D., Brunner, J., Baker, A. and Schatz, G. (1989) A 42K outer-membrane protein is a component of the yeast mitochondrial protein import site. *Nature*, **341**, 205–209.

142 Vestweber, D. and Schatz, G. (1988a) Mitochondria can import artificial precursor proteins containing a branched polypeptide chain or a carboxy-terminal stilbene disulfonate. *J. Cell Biol.*, **107**, 2045–2049.

143 Vestweber, D. and Schatz, G. (1988b) Point mutations destabilizing a precursor protein enhance its post-translational import into mitochondria. *EMBO J.*, **7**, 1147–1151.

144 Vestweber, D. and Schatz, G. (1989) DNA-protein conjugates can enter mitochondria via the protein import pathway. *Nature*, **338**, 170–172.

145 Voisine, C., Craig, E. A., Zufall, N., von Ahsen, O., Pfanner, N. and Voos, W. (1999) The protein import motor of mitochondria: Unfolding and trapping of preproteins are distinct and separable functions of matrix Hsp70. *Cell*, **97**, 565–574.

146 Von Heijne, G. (1986) Mitochondrial targeting sequences may form amphiphilic helices. *EMBO J.*, **5**, 1335–1342.

147 Voos, W., Gambill, B. D., Guiard, B., Pfanner, N. and Craig, E. A. (1993) Presequence and mature part of preproteins strongly influence the dependence of mitochondrial protein import on heat shock protein 70 in the matrix. *J. Cell Biol.*, **123**, 119–126.

148 Voos, W. and Rottgers, K. (2002) Molecular chaperones as essential mediators of mitochondrial biogenesis. *Biochim. Biophys. Acta*, **1592**, 51.

149 Voos, W., von Ahsen, O., Muller, H., Guiard, B., Rassow, J. and Pfanner, N. (1996) Differential requirement for the mitochondrial Hsp70-Tim44 complex in unfolding and translocation of preproteins. *EMBO J.*, **15**, 2668–2677.

150 Werner, S. and Neupert, W. (1972) Functional and biogenetical heterogeneity of the inner membrane of rat-liver mitochondria. *Eur. J. Biochem.*, **25**, 379–396.

151 Westermann, B., Prip-Buus, C., Neupert, W. and Schwarz, E. (1995) The role of the GrpE homologue, Mge1p, in mediating protein import and protein folding in mitochondria. *EMBO J.*, **14**, 3452–3460.

152 Wiedemann, N., Kozjak, V., Chacinska, A., Schonfisch, B., Rospert, S., Ryan, M. T., Pfanner, N. and Meisinger, C. (2003) Machinery for protein sorting and assembly in the mitochondrial outer membrane. *Nature*, **424**, 565–571.

153 Wiedemann, N., Pfanner, N. and Ryan, M. T. (2001) The three modules of ADP/ATP carrier cooperate in receptor recruitment and translocation into mitochondria. *EMBO J.*, **20**, 951–960.

154 Wienhues, U., Becker, K., Schleyer, M., Guiard, B., Tropschug, M., Horwich, A. L., Pfanner, N. and Neupert, W. (1991) Protein folding causes an arrest of preprotein translocation into mitochondria in vivo. *J. Cell Biol.*, **115**, 1601–1609.

155 Wienhues, U., Koll, H., Becker, K., Guiard, B. and Hartl, F.-U. (1992) Protein targeting to mitochondria. In *A Practical Approach to Protein Targeting*. IRL (Oxford University Press), pp. 135–159.

156 Xia, Z. X. and Mathews, F. S. (1990) Molecular structure of flavocytochrome b2 at 2.4 A resolution. *J. Mol. Biol.*, **212**, 837–863.

157 Yamamoto, H., Esaki, M., Kanamori, T., Tamura, Y., Nishikawa, S. and Endo, T. (2002) Tim50 is a subunit of the TIM23 complex that links protein translocation across the outer and inner mitochondrial membranes. *Cell*, **111**, 519–528.

158 Young, J. C., Hoogenraad, N. J. and Hartl, F. U. (2003) Molecular chaperones Hsp90 and Hsp70 deliver preproteins to the mitochondrial import receptor Tom70. *Cell*, **112**, 41–50.

159 Zhuang, X., Ha, T., Kim, H. D., Centner, T., Labeit, S. and Chu, S. (2000) Fluorescence quenching: A tool for single-molecule protein-folding study. *Proc. Natl. Acad. Sci. USA*, **97**, 14241–14244.

31
The Chaperone System of Mitochondria

Wolfgang Voos and Nikolaus Pfanner

31.1
Introduction

Mitochondrial protein biogenesis comprises a complex set of reactions starting with the import of cytosolically synthesized preproteins, followed by folding and assembly processes to acquire a functional conformation, and ending with proteolysis to remove unwanted proteins and to reutilize their amino acids. All steps require specific proteinaceous machineries that assist the progress of the respective substrate proteins along this pathway and prevent irregular side reactions. Based on the endosymbiotic origin of the organelle, mitochondria utilize in part mechanisms that are similar to bacterial organisms. However, their intracellular localization requires the evolution of organelle-specific processes that are distinct from their bacterial ancestors and also different from the reaction pathways in the eukaryotic cytosol. Due to the limited capacity to produce endogenous proteins, mitochondria face specific problems in importing proteins that were synthesized in the cytosol and in the coordination of mitochondrial and cytosolic protein expression. For these reasons, mitochondria contain a complex system of chaperone proteins that are involved in practically all steps of protein biogenesis. During many processes a closely coordinated function of chaperone proteins from different subclasses and with different functional specificities has been observed, resulting in the description of the chaperone system as an essential functional network [1]. Most chaperone proteins also require interaction with specific partner proteins that have no genuine chaperone function on their own but are required for the regulation of activity and functional specificity of their corresponding chaperone partners (Table 31.1). The specific roles of the mitochondrial chaperone proteins and their co-chaperones are discussed in this chapter.

31.2
Membrane Translocation and the Hsp70 Import Motor

Most mitochondrial polypeptides are synthesized by cytosolic ribosomes as precursor proteins and transported posttranslationally to their final location. Correct pro-

Protein Folding Handbook. Part II. Edited by J. Buchner and T. Kiefhaber

ISBN: 3-527-30784-2

Tab. 31.1. Mitochondrial chaperones and partner proteins in *S. cerevisiae*.

Chaperone	*Protein*	*Function*	*Partner proteins*
Hsp70	mtHsp70/Ssc1	Preprotein import, protein folding, stress protection	Tim44, Mdj1, Pam18, Mge1
	Ssq1	Protein complex assembly, Fe/S cluster biogenesis	Jac1, Mge1
	Ssc3 (Ecm10)	Unknown	Mge1
Co-chaperone (DnaJ type)	Mdj1	Protein folding, regulation of mtHsp70 activity, mitochondrial DNA replication	Ssc1
	Mdj2	Unknown	
	Jac1	Protein assembly, Fe/S cluster biogenesis	Ssq1
	Pam16	Preprotein import, Ssc1-Tim44 interaction	Ssc1
	Pam18	Preprotein import, stimulation of mtHsp70 activity	Ssc1
Co-chaperone	Tim44	Preprotein import, membrane anchor for mtHsp70	Ssc1, Tim23, Tim17
Co-chaperone (GrpE type)	Mge1	Nucleotide-exchange factor for mtHsp70	Ssc1, Ssq1, Ssc3
Hsp60	Hsp60	Protein folding	Hsp10
	Hsp10	Regulation of Hsp60 activity	Hsp60
Cyclophilin	Cpr3	Protein folding, peptidyl-prolyl isomerization	
Hsp100/Clp	Hsp78	Protection against aggregation, protein degradation	Ssc1, Pim1
	Mcx1	Unknown	
Proteases with putative chaperone activity	Pim1	ATP-dependent proteolysis of soluble proteins in the matrix	Hsp78
	Yme1	Proteolysis of inner-membrane proteins, assembly of respiratory chain complexes	
	Yta12/Yta10	Proteolysis of inner-membrane proteins	

tein localization is ensured by specific targeting sequences contained within the proteins that are recognized by dedicated receptor proteins in the outer mitochondrial membrane. Two general types of targeting sequences have been identified. Proteins destined for the mitochondrial matrix as well as some inner-membrane proteins contain amino-terminal extensions (presequences) that are removed after import by specific processing peptidases [2]. Polytopic membrane proteins of the inner membrane contain non-cleavable internal targeting sequences that enable an insertion into the inner membrane. The transport of the polypeptide chain is in both cases initiated by the insertion of the precursor protein into the outer and inner membranes. These steps are performed with the help of specialized membrane-integrated translocase complexes, which form pore structures in the membrane [3, 4]. In addition, many accessory proteins of the translocase complexes have significant roles in enhancing translocation activity. Interestingly, the dimensions of the pore structures with a diameter of 1.2–2 nm limit the possible conformations a precursor protein can adopt during the translocation process, as only unfolded or at most α-helical structures can be accommodated [5, 6]. Derived from these structural prerequisites, two aspects of the import process require the input of exogenous energy for efficient polypeptide transport. First, energy is required for the vectorial movement of the polypeptide chain through the translocation channels. Second, the unfolding of the preprotein polypeptide required for the passage through the pore structures in the membranes has to be coupled to energy consumption. The strong correlation between precursor folding state and energy consumption in preprotein translocation suggests that chaperone proteins are involved in several steps of the import reaction.

While chaperone proteins may interact with newly synthesized preproteins early in the targeting process, the initial insertion of preproteins into the outer membrane does not require an involvement of mitochondrial chaperone proteins. The participation of cytosolic chaperones in the translocation process is described in chapter 30. The insertion of the amino-terminal presequences into the inner membrane is driven via an electrophoretic mechanism by the membrane potential ($\Delta\psi$) generated by the respiratory chain, as amino-terminal targeting signals have an overall net positive charge [7]. Full translocation of the polypeptide chain into the mitochondrial matrix, however, does require an additional energy source. The mitochondrial Hsp70 (mtHsp70), in *Saccharomyces cerevisiae* encoded by the gene *SSC1*, is localized in the matrix compartment and utilizes the hydrolysis of ATP to drive completion of preprotein transport [8]. As mtHsp70 is the only ATPase implicated in the protein import process to date, it is a key component of the import pathway. The energy generated by ATP hydrolysis is utilized in two separate processes: the vectorial movement of the polypeptide chain through the translocation pores and unfolding of carboxy-terminal domains of preproteins on the outer side of the mitochondrial membranes [9–11]. After mtHsp70-driven translocation, preproteins are processed to their mature forms by the matrix-processing peptidase and are folded and assembled. All matrix proteins analyzed so far are dependent on mtHsp70 for translocation through the membranes. Membrane proteins can usually insert into the inner membrane independent of matrix ATP and mtHsp70 [12], but some inner-membrane proteins, primarily those that expose bulky domains on the matrix face of the inner membrane, do require involvement of mtHsp70 [13].

During the import process, mtHsp70 cooperates with two additional proteins, Mge1 and Tim44 (Figure 31.1). These proteins have been shown to form a stable protein complex that is sensitive to the presence of ATP [14]. Since the complex is

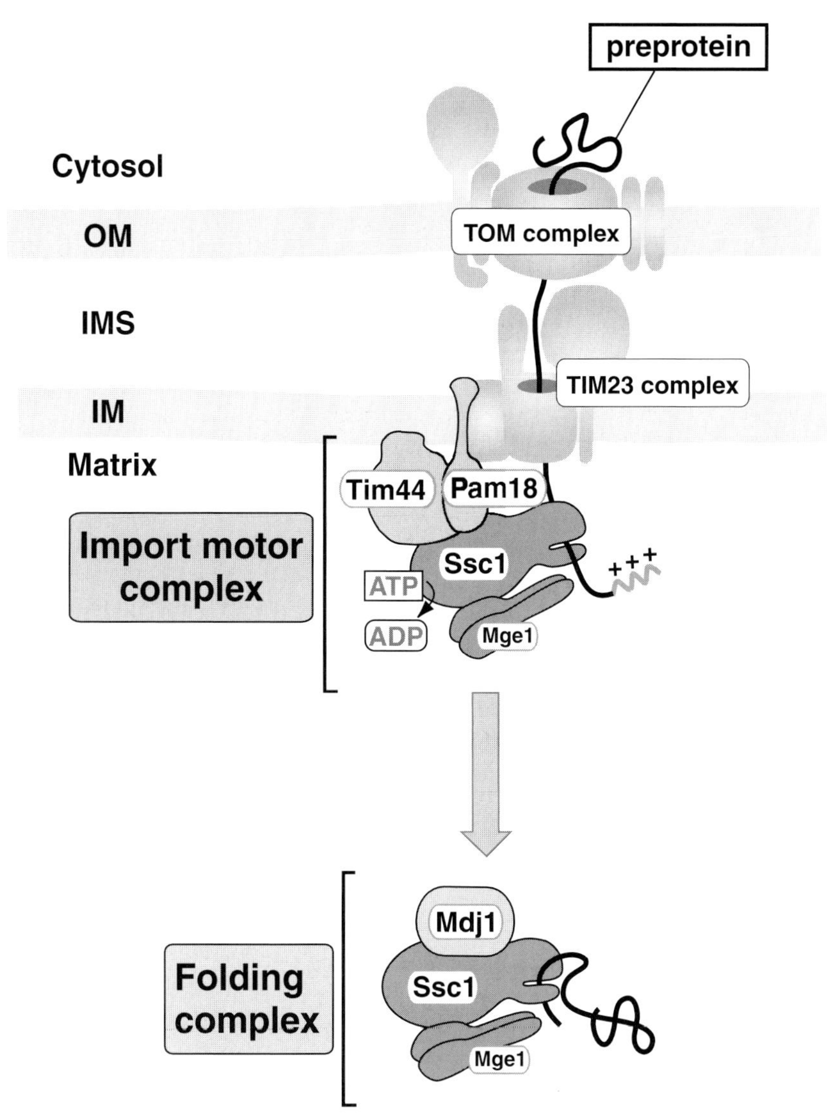

Fig. 31.1. The role of mtHsp70 (Ssc1) and its specific partner proteins during protein import and folding in the mitochondrial matrix. ATP-dependent translocation of the incoming polypeptide chain through the mitochondrial membranes (OM, outer membrane; IMS, intermembrane space; IM, inner membrane) into the matrix is driven by the import motor complex composed of mtHsp70 (Ssc1) as the core component, the nucleotide-exchange factor Mge1, the membrane anchor Tim44, and the DnaJ-type protein Pam18. The pore structures for preprotein import are provided by the TOM (translocase of the outer membrane) and TIM23 (presequence translocase of the inner membrane) complexes. After completion of polypeptide transport, the amino-terminal presequence is cleaved off and folding reactions are carried out with the assistance of the folding complex composed of mtHsp70 (Ssc1), Mge1, and the mitochondrial DnaJ homologue Mdj1.

in a close local and functional proximity to the inner-membrane translocase complex, the functional unit of mtHsp70, Mge1, and Tim44 has been dubbed import motor complex [15]. MtHsp70, Mge1, and Tim44 are essential proteins in yeast, with null mutations leading to lethal phenotypes. This observation enhances the significance of the import motor components for the overall import process, as the only other proteins of the matrix translocation pathway encoded by essential genes are the core components of the membrane translocase complexes, Tom40 and Tim23/17. The protein Mge1, a member of the bacterial GrpE protein family, acts as a nucleotide-exchange factor for mitochondrial Hsp70s [16, 17]. Mge1 favors the release of hydrolyzed ADP at the end of the Hsp70 reaction cycle and facilitates binding of a new ATP molecule, enhancing the ATPase activity of mtHsp70 significantly. Thus, Mge1 is a key regulator of the ATP-dependent import reaction cycle of mtHsp70. It not only enhances the activity of mtHsp70 but also regulates the interaction with protein substrates and the membrane anchor Tim44 [18]. To further illustrate the importance of Mge1, it has been shown that the protein, via its interaction with mtHsp70, associates with preproteins in transit but not with preproteins that have been completely translocated [19]. The essential role of Mge1 is certainly represented by its assistance in preprotein translocation; however, Mge1 is also involved in protein-folding reactions that occur subsequent to the transport into the matrix [20].

For efficient import of matrix-targeted proteins, the soluble chaperone machinery has to coordinate its activity closely with the inner-membrane translocase. MtHsp70 and the membrane protein Tim44 tightly interact in a nucleotide-regulated manner, essentially tethering a portion of the soluble mtHsp70 pool to the inner membrane [21–23]. Tim44 is also peripherally associated with the inner-membrane translocase, and via the interaction with Tim44, mtHsp70 is brought into the direct vicinity of the import pore. Several studies have established the importance of the Tim44/mtHsp70 interaction for the overall translocation process. It has been shown that mtHsp70 is found bound to imported polypeptides in the absence of functional Tim44 but that efficient translocation requires the reversible binding of mtHsp70 to Tim44 [11, 24]. MtHsp70 must interact with Tim44 for efficient preprotein translocation since a deletion of a potential mtHsp70 interaction site in Tim44 resulted in a strong translocation defect in vitro and a lethal phenotype in vivo [25].

Several models have been proposed to explain the function of mtHsp70 during the preprotein import. One model is based on the tight interaction of mtHsp70 with translocating unfolded preprotein polypeptide chains [26]. The interaction of mtHsp70 with the preprotein would trap the membrane-spanning preprotein in the translocation pore and prevent backward movement of the polypeptide chain. Any forward movement, generated by random Brownian motion, would expose further binding sites for additional mtHsp70s. Several repetitions of this binding cycle would result in the complete translocation of the preprotein. By influencing the affinity of mtHsp70 to the preprotein, the hydrolysis of ATP would lock the preprotein bound to the peptide-binding domain of mtHsp70, while the release of ADP and rebinding of ATP would destabilize the substrate interaction, resulting in the dissociation of the bound preprotein. This model describes the action of mtHsp70 essentially as a passive "ratchet" that traps preprotein segments in the matrix [24, 27, 28]. The interaction of mtHsp70 with Tim44 serves in this model essentially to increase the local concentration of mtHsp70 at the import site. A passive mechanism of mtHsp70 as a "Brownian ratchet" for the translocation of preprotein chains has been supported by several studies. However, a significant translocation activity in an mtHsp70 mutant that is restricted to a ratchet mechanism could be observed only with preproteins that either were completely unfolded, e.g., after denaturation by urea, or that had a low overall thermodynamic stability [8, 11]. Another study has shown that mitochondria were able to import preprotein constructs containing Hsp70 binding sites separated by large patches of non-binding residues. Again, the import efficiencies of stably folded C-terminal domains were significantly reduced [29].

Several experimental observations indicate that mtHsp70 may perform a more active role during the translocation of precursor proteins. For preproteins with stably folded domains, binding to mtHsp70 alone was not sufficient to effectively drive preprotein import into the matrix [30, 31]. By detailed analysis of preprotein import kinetics, it was determined that the rates of preprotein translocation are faster than the rate of spontaneous unfolding of preproteins [32, 33]. A prerequisite for active unfolding by the translocation machinery is an efficient interaction with the mtHsp70 motor complex. Especially the import of preproteins with short N-terminal extensions that do not expose sufficient Hsp70 binding sites in the matrix becomes susceptible to mutations in the motor complex components [33, 34]. It was also observed that the import process, proceeding in an N-to-C-terminal direction, can significantly alter the unfolding pathway of a preprotein [35]. These experiments led to the conclusion that mtHsp70 performs an active role in preprotein transport. It was suggested that the unfolding of preproteins was achieved by mtHsp70 by generating an inward-directed translocation force on the polypeptide chain. The force exerted on the preprotein in transit might be small but may be sufficient to lower the activation energy necessary to unfold preprotein domains on the outer face of the outer membrane [36]. Interestingly, by this "pulling" on the polypeptide chain, the energy of ATP hydrolysis in the matrix can be utilized to catalyze the required unfolding of protein domains on the outer face of the mitochondrial membrane system. In this model, termed the "active motor" mechanism, interaction of mtHsp70 with Tim44 would serve as an anchor point to produce the necessary leverage for a force generation [37–39]. In support of this model analysis of mitochondria carrying a conditional mutant of Tim44 showed defects mainly in the import of preproteins containing stably folded domains [40]. Taken together, these results suggest that the ATP hydrolysis cycle induces a conformational change in mtHsp70, resulting in the "pulling" of preproteins into the matrix. As there is evidence in support of both models, it is likely that protein translocation in vivo is driven by a combination of the pulling and trapping mechanisms.

However, direct evidence for a force generation by an mtHsp70 power stroke is still lacking. Recent experiments to reconstitute the import motor with purified components consisting of mtHsp70, Tim44, and Mge1 in vitro failed to reproduce a crucial prerequisite for the "active motor" mechanism, the formation of a ternary complex among mtHsp70, substrate proteins, and Tim44 [41]. However, it is possible that the mitochondrial import motor might be more complex than previously noted, as recently two new essential components of the translocation machinery have been identified. The small, membrane-embedded protein Pam18 (presequence translocase associated motor subunit of 18 kDa) was found in association with the inner-membrane translocase TIM23 [42, 43]. Pam18 (also named Tim14) contains a so-called J-domain that is found in Hsp70 co-chaperones of the DnaJ-family. The J-domain is exposed to the matrix, indicating a functional relationship with mtHsp70 during the import reaction. Indeed, Pam18 was shown to activate the ATPase activity of mtHsp70 substantially and is required for the mtHsp70-dependent import of preproteins into the matrix [42, 44]. Interestingly, Pam18 forms a complex with the second new essential translocation component, Pam16 [45, 46]. Similar to Pam18, Pam16 (or Tim16) is a small, membrane-bound protein with some similarity to J-domain-containing proteins. Pam16 is required for the preprotein translocation into the matrix, although it cannot activate the ATPase activity of mtHsp70. It becomes evident that both proteins potentially form a functional interface between the polypeptide pore in the inner membrane and the mtHsp70-dependent translocation motor complex. Despite the progress made in the biochemical characterization of the mitochondrial translocation reaction, the molecular mechanism of mtHsp70 during import is still only partially understood. Future studies need to integrate the novel components into the mtHsp70 translocation-specific reaction cycle.

31.3
Folding of Newly Imported Proteins Catalyzed by the Hsp70 and Hsp60 Systems

Mitochondrial preproteins must cross the membranes as an unfolded polypeptide chain. As a result, the acquisition of the native protein conformation must occur in the matrix of mitochondria following completion of the translocation reaction. Since the matrix chaperone mtHsp70 is crucial to the preprotein translocation reaction, the interaction of preproteins with the mtHsp70 system represents the first step of the protein-folding pathway in the matrix [47]. Folding of imported proteins therefore follows the "classical" order of chaperone-catalyzed folding reactions [48]. First, interaction of preproteins with the Hsp70 system stabilizes and protects unfolded substrate proteins, while the acquisition of the native conformation is mainly assisted by the Hsp60 chaperonin system (Figure 31.2). In that respect, the folding process in the mitochondrial matrix shares similar features with other organellar systems [49], including folding of newly synthesized proteins at the ribosome [50] and protein folding in the bacterial cytosol [51].

Interestingly, the dual function of mtHsp70 in protein translocation and folding in mitochondria is reflected in its differential interaction with specific partner proteins (Figure 31.1). As stated above, interaction of mtHsp70 with the inner-

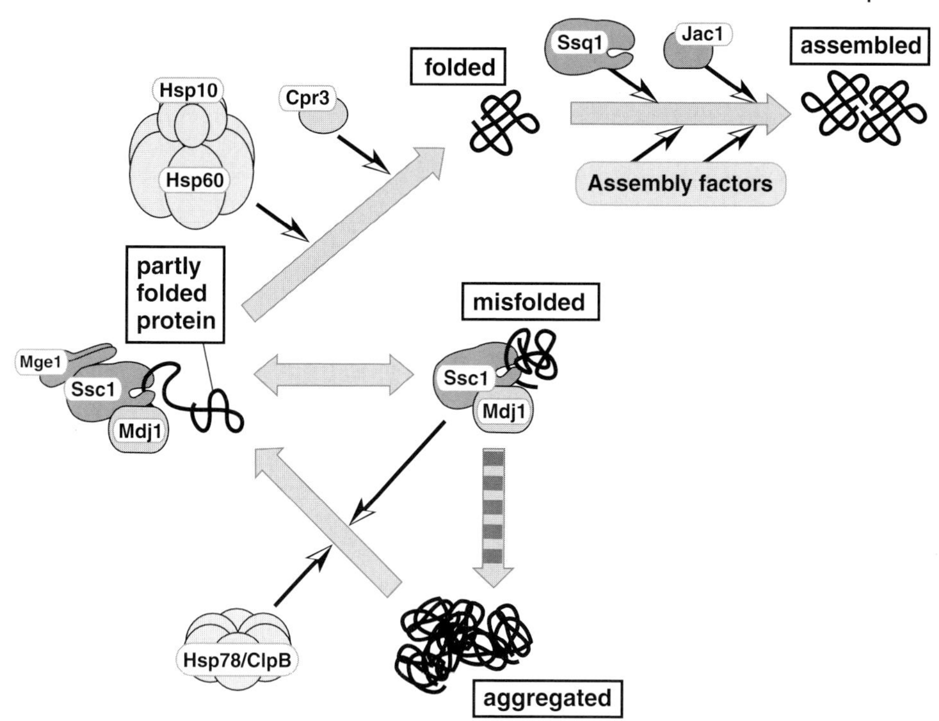

Fig. 31.2. The mitochondrial chaperone network employed for protein folding and complex assembly. Protein folding in the mitochondrial matrix is assisted by the mtHsp70 system and the Hsp60/Hsp10 chaperonin complex in the matrix. Cpr3 is a peptidyl-prolyl isomerase that accelerates folding by the *cis/trans* isomerization of prolyl bonds. Further maturation reactions and the assembly of macromolecular protein complexes are catalyzed by the Hsp70 family member Ssq1 with its DnaJ-type co-chaperone Jac1 and other numerous assembly factors (see text). Denatured and/or misfolded proteins are stabilized mainly by the mtHsp70 system. Aggregated proteins can be resolubilized and renatured by a concerted action of the mitochondrial ClpB homologue Hsp78 and the mtHsp70 system.

membrane protein Tim44 is required for the translocation function, but during protein folding mtHsp70 forms a complex with the co-chaperone Mdj1 [47]. Mdj1 is the mitochondrial homologue of the bacterial DnaJ protein [52]. Members of this diverse protein family are generally involved in protein-folding reactions by assisting substrate binding by the corresponding Hsp70s and activation of their ATPase activity [53]. Mutations in Mdj1 do not result in defects in protein translocation but reduce or abolish folding of newly imported proteins [54]. The role of Mdj1 and mtHsp70 in protein folding reactions could be demonstrated directly in in vitro experiments [55]. It is assumed that the mechanism of Mdj1 during protein-folding reactions closely resembles the action of its bacterial relative DnaJ [56]. A dysfunction of Mdj1 has direct or indirect deleterious consequences on many mitochondrial functions that result in a respiratory defect of the respective mutations [57].

In recent years, other mitochondrial members of the mitochondrial DnaJ protein family have been identified, although information about their functional role is still scarce. Of all mitochondrial proteins containing homology to the bacterial DnaJ, only the above-mentioned Pam18 is essential for cellular viability. Its essential role is most probably related to the translocation function in the import motor during preprotein import. Another integral membrane protein of the inner membrane, termed Mdj2, exposes a J-homology domain to the matrix compartment. However, a deletion mutant in yeast did not show any phenotype. As a result, no specific function could be attributed to Mdj2 [58]. A small soluble protein, Jac1, also contains a J-homology domain. A mutation of *JAC1* has strong negative effects on growth on both fermentable and non-fermentable carbon sources, indicating an involvement in a prominent metabolic function of mitochondria. Recent data showed that Jac1 is involved in the formation of Fe/S clusters, required for the function of several mitochondrial and cytosolic enzymes [59–61]. In this role, Jac1 cooperates with a second member of the mitochondrial Hsp70 family, termed Ssq1 in yeast [62, 63]. Under normal conditions, Ssq1 is far less abundant than the major mitochondrial Hsp70 Ssc1. In contrast to Ssc1 null mutations, a deletion mutant of *SSQ1* is viable but shows a cold-sensitive phenotype. Ssq1 seems to be able to perform standard chaperone functions and is also regulated in its activity by the nucleotide-exchange factor Mge1 [64]. Despite many similarities between the two mitochondrial Hsp70 systems, Ssq1 with its co-chaperone Jac1 cannot substitute for the functions of Ssc1 in protein translocation and folding reactions even upon overexpression of the protein. The main reason for this inability seems to be the lack of interaction of Ssq1 with the typical co-chaperones of Ssc1: Tim44 for preprotein translocation reaction or Mdj1 for protein-folding events in the matrix. Ssq1 seems to play a more specialized role during the assembly reaction of Fe/S clusters, but its mechanistic role in that process remains to be clarified.

While the mtHsp70 system is the first to interact with imported preproteins, the main chaperone involved in protein folding in the matrix is certainly the mitochondrial Hsp60. Preproteins have been shown to follow a well-defined pathway of chaperone interactions in the mitochondrial matrix [65, 66]. Hsp70 is generally regarded as a stabilizing component, preventing irregular interactions of unfolded preproteins while the actual folding is catalyzed by the oligomeric Hsp60 protein complex. The important role of Hsp60 in mitochondrial protein biogenesis was identified before that of Hsp70 by the analysis of mutant Hsp60 yeast strains. Mutants in Hsp60 showed strong defects in the mitochondrial metabolism due to the importance of Hsp60 for protein biogenesis [67, 68]. A deletion of Hsp60 in yeast is lethal, also emphasizing the importance of this component for mitochondrial and, indirectly, cellular metabolism. It has been shown that proteins imported in vivo physically interact with Hsp60 after their import is completed [69]. Mitochondria with defective Hsp60 are able to import preproteins, but imported proteins do not fold to their native conformation and likely end up as insoluble aggregates. The molecular mechanism of the mitochondrial Hsp60 is similar to the well-studied bacterial homologue, GroEL [70]. Hsp60 forms a homo-oligomeric protein complex with a large central cavity. Nonnative proteins up to a certain size threshold are able to enter this cavity and to interact with hydrophobic residues at the inner face of the cavity wall. As in the bacterial system, the cavity can be closed by an interaction with a second essential component, the mitochondrial Hsp10 [71]. Hsp10 is a homologue of the bacterial GroES and forms as a homoheptameric complex, which serves as a lid to block the entrance of the Hsp60 ring complex. Since Hsp60 and Hsp10 form a functional unit, deletion of Hsp10 in yeast is also lethal [72]. Its involvement in protein folding has been primarily analyzed through the use of temperature-sensitive mutants. Hsp10 has been shown to be involved not only in the folding of many soluble imported proteins but also in the sorting of precursor proteins destined for the inner mitochondrial membrane [73]. The Hsp60-Hsp10 complex thereby provides a protected environment where the spontaneous folding processes of the substrate protein can occur without the ability to interact with other proteins. Activity of Hsp60 is strictly regulated by an ATP-dependent reaction cycle. Binding of ATP induces a conformational change in the Hsp60 complex, which results in substrate release from the cavity wall, allowing the substrate to undergo folding reactions. After a certain time, determined by the velocity of ATP hydrolysis, the substrate protein is discharged from the Hsp60 cavity. It has been shown that apart from Hsp10's structural role in the interaction with Hsp60, it also has a regulatory effect on the ATPase activity of the Hsp60 complex [74].

Despite general similarities between the bacterial GroEL system and the mitochondrial Hsp60, there is one major structural difference between the two chaperones. The mitochondrial Hsp60 functions as a single heptameric ring complex, while GroEL forms a double-ring system where both rings are functionally interconnected. The single-ring system of mitochondrial Hsp60 seems to be fully competent in protein folding [75] and can substitute for the function of GroEL in bacteria [76]. Similar to the bacterial homologue, Hsp60 is not absolutely required for the folding of all imported proteins, as the analysis of the folding process of several substrate proteins with different biochemical properties revealed [77]. Dependence on Hsp60 during the folding process is largely determined by the specific folding characteristics of the individual proteins. An elegant assay using the complete set of mitochondrial precursor proteins generated by in vitro translation of isolated cellular mRNA revealed the general substrate specificity range of Hsp60 [78]. It is interesting to note that even substrate proteins that interact with the Hsp60 complex may vary in the requirement for the co-chaperone Hsp10 for folding. This indicates that the interaction of Hsp10 with Hsp60 is not strictly required for catalyzing successful protein folding. Although not every imported preprotein depends on Hsp60 for folding, the lethal phenotype of the *hsp60* null mutation indicates that at least some important mitochondrial proteins are strictly dependent on Hsp60. An alternate hypothesis would be that a deactivation of folding catalysis in general would generate significant damage to the mitochondrial functions to be incompatible with cellular survival.

The prevention of unproductive side reactions through the stabilization of nonnative folding intermediates is the typical function of molecular chaperones such as Hsp70 and Hsp60. In mitochondria, however, protein folding in the matrix is facilitated by a second mechanism. The peptidyl-prolyl isomerase cyclophilin accelerates a rate-limiting step of protein folding and has been shown to increase the folding rate of imported preproteins [79, 80]. In mitochondria lacking cyclophilin, encoded by the gene *CPR3* in yeast, the folding of a DHFR fusion protein was delayed although not completely abolished. Similar observations were made with wild-type mitochondria that were treated with a specific inhibitor of cyclophilin, cyclosporin A. The fact that cyclophilin is not absolutely required for protein folding correlates well with its proposed rate-enhancing role. As a result of the complementary role of cyclophilin in enhancing the rate of protein folding with the stabilizing role of chaperone machineries in the mitochondrial matrix, it is no surprise that a close functional cooperation has been observed between the two machineries. Chaperones can partially substitute for the function of cyclophilin by binding and stabilizing the incompletely folded substrate protein, while cyclophilin enhances the acquisition of a native conformation and as a result decreases the amount of chaperone-bound substrate proteins.

31.4 Mitochondrial Protein Synthesis and the Assembly Problem

Mitochondria synthesize a small subset of about a dozen proteins by their endogenous protein-synthesis machinery. Most of the mitochondrially encoded proteins are very hydrophobic in nature and are integral membrane proteins of the inner membrane. These proteins are subunits either of respiratory chain complexes or of the membrane-integrated F_0-ATPase complex. The hydrophobicity of these components is the likely reason for the evolutionary retention of a mitochondrial protein-synthesis capability. Despite the synthesis in the mitochondrial matrix, mechanisms have to be employed to prevent aggregation of the nascent hydrophobic polypeptides and to ensure their functional assembly into the inner-membrane complexes. As the primary role of chaperone proteins is the prevention of protein aggregation, it is not surprising that the mtHsp70 chaperone system is involved in the stabilization of the mitochondrially encoded proteins [81]. As with the folding catalysis of soluble proteins, mtHsp70 closely cooperates with its partner protein Mdj1 in the assembly reaction of mitochondrially encoded proteins [82]. Similar to its counterpart in the cytosol, it is thought that the mtHsp70 system can interact with nascent polypeptide chains as they emerge from mitochondrial ribosomes. This interaction stabilizes the polypeptide chain during synthesis until they fold to their native conformation or assemble into the respective protein complexes.

The large size and complex protein composition of the respiratory chain enzymes in the inner mitochondrial membrane not only require coordinated protein synthesis of nuclear and mitochondrially encoded subunits but also pose significant challenges for the assembly of these complexes. As a result, it is not surprising that several proteins have been identified that are required for the enzymatic function of the inner-membrane complexes but are not components of the functional oligomers. Based on this observation they have been described as assembly factors, but to date their functional role in the assembly reaction is unclear. For example, the assembly factors Atp11 and Atp12 are required for the function of the F_1F_o-ATPase [83, 84], while the cytochrome bc_1 complex requires the factor Bcs1 for assembly. The cytochrome *c*–oxidase (COX) complex is formed with the help of a different set of specific factors [85–87]. Interestingly, Tcm62, a mitochondrial membrane protein with a small but significant homology to Hsp60, has been identified and has been shown to be required for the functional assembly of the succinate dehydrogenase (SDH) complex [88]. It is uncertain whether these assembly factors can be designated as genuine "molecular chaperones" since their function seems to be restricted to a single function and seems to be specific for the assembly of a defined set of inner membrane components.

Biogenesis of the mitochondrial respiratory chain complexes requires a close coordination of mitochondrial and cytosolic protein synthesis since components encoded in both cellular compartments must be assembled into the same oligomeric structures. The presence of excess subunits from either source would present the danger of irregular aggregation reactions and eventual mitochondrial damage. A set of proteins has been described in mitochondria that combines both chaperone and protease functions in one polypeptide to prevent damage by the accumulation of unassembled subunits. Two of these proteins, Rca1 (Yta12) and Afg3 (Yta10), were found by the analysis of mutants that show defects in mitochondrial respiration [89, 90]. Mutations of Rca1 or Afg3 cause defects in the assembly of functional inner-membrane protein complexes such as the F_1F_o-ATPase [91]. A third protein, Yme1, was initially shown to be required for maintenance of mitochondrial DNA [92] and was later shown to have protease properties [93].

All three proteins have ATPase activity and belong to the widespread family of AAA proteins (ATPases associated with various cellular activities) that are involved in diverse cellular processes such as vesicular transport, organelle biogenesis, microtubule rearrangement, and protein degradation. The AAA proteases have a metal-dependent peptidase activity and mediate the degradation of non-assembled membrane proteins. Rca1 and Afg3 form a hetero-oligomeric protein complex, named the m-AAA protease, that is located in the inner mitochondrial membrane with its proteolytic sites exposed to the matrix [94]. In contrast, the active site of the homo-oligomeric complex formed by Yme1 faces the intermembrane space but is also inserted into the inner membrane. The Yme1 complex has therefore been termed i-AAA protease [95]. It has been shown that Yme1 is able to sense the folding state of protein domains exposed to the intermembrane space and to specifically degrade unfolded membrane proteins [96]. The affinity of Yme1 for unfolded polypeptides closely resembles the activity of molecular chaperones. The purified AAA domain of Yme1 was shown to bind unfolded polypeptides and to suppress their aggregation, which are features typical for chaperone proteins. Yme1 displays a modular structure with a separate proteolytic domain and a chaperone-like ATPase domain. Although the molecular mechanism of Yme1 function has not been determined in detail, it is hypothesized that the ATPase domain

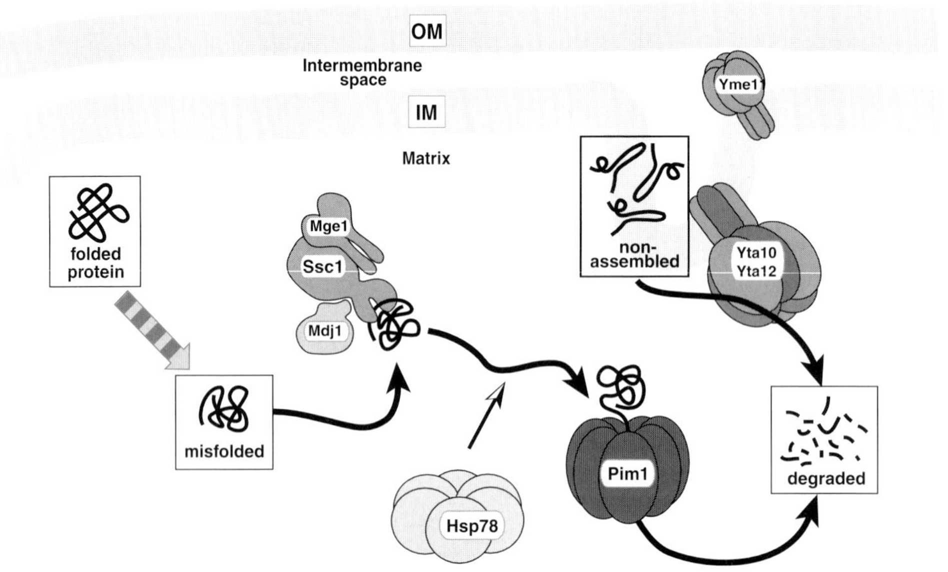

Fig. 31.3. Through the cooperation of chaperones and proteases, a mitochondrial protein quality-control system is formed. Proteins of the mitochondrial matrix that are damaged and/or misfolded are recognized by the mtHsp70 (Ssc1) system and are targeted for degradation by the soluble protease Pim1/LON. The degradation efficiency is enhanced by the mitochondrial ClpB homologue Hsp78. Membrane proteins of the inner membrane (IM) that are misfolded or fail to assemble are degraded by two protease complexes located in the inner membrane. The i-AAA protease (Yme1) degrades proteins exposed to the intermembrane space, while the m-AAA protease (Yta10/Yta12) acts on proteins with domains facing the matrix. Both complexes are also involved in the biogenesis of inner-membrane protein complexes.

of the protein might extract substrate proteins from the membrane and stabilize the polypeptide chain until proteolysis is complete.

A different proteolytic system exists in the mitochondrial matrix (Figure 31.3). A high-molecular-weight complex of the Pim1 protein forms an ATP-dependent protease that is highly homologous to the bacterial LON protease and also belongs to the AAA protein family [97, 98]. Pim1 is primarily involved in the degradation of soluble unfolded proteins in the matrix and shows a close functional coordination with the mtHsp70 system, especially under stress conditions [99]. It is thought that the chaperone proteins mtHsp70 and Mdj1 stabilize substrate proteins against aggregation, but if the attempts to refold the proteins fail, the damaged proteins are removed by the Pim1 protease. Similar to membrane-bound AAA proteases, Pim1 may exhibit chaperone properties, as it was found that a proteolytically inactive Pim1 could partially substitute for Rca1 and Afg3 in assembly of the inner-membrane respiratory complexes [100]. However, the mechanistic details of the possible chaperone function still have to be clarified. In their unique combination of protease and chaperone functions, the AAA proteases serve as a dedicated protein quality-control system of mitochondria, responsible for both the coordinated assembly of protein complexes and the removal of surplus polypeptides.

31.5
Aggregation versus Degradation: Chaperone Functions Under Stress Conditions

The functional network of chaperone proteins and proteases in mitochondria plays an even more important role under stress conditions such as elevated temperature and high concentrations of reactive oxygen species or other toxic chemicals. These conditions lead to damage and/or denaturation of endogenous proteins. Apart from the loss of the enzymatic activities, high amounts of these damaged proteins favor aggregation reactions that have indirect but significant deleterious effects on mitochondrial function. In a process similar to other cellular systems, mitochondria employ a two-pronged approach to tackle the problem of protein aggregation. First, interaction of damaged proteins with chaperones will keep them in solution and increase the probability of refolding them to the native conformation under more benign conditions (Figure 31.2). Chaperone systems are even capable of resolubilizing aggregated protein material and reactivating them for refolding. Second, terminally damaged proteins will have to be removed from the mitochondrial protein pool by proteolytic reactions coordinating chaperone and protease functions (Figure 31.3). The ultimate fate of the substrate proteins is likely determined by kinetic partitioning. A short interaction of folding-competent substrate proteins leads to refolding, while an extended binding of misfolded proteins, which fail to refold, favors degradation.

The main chaperone family involved in protection from protein aggregation is the Hsp100/Clp family. Initial characterization in bacteria identified these chaperones as components of a proteolytic system, the Clp protease (caseinolytic protease). In this system the protease subunit, ClpP, forms a ring-shaped oligomeric complex that interacts with a chaperone component, either the ClpA or ClpX proteins. The chaperone component recognizes and unfolds potential substrate proteins in an ATP-dependent reaction. The unfolded polypeptide chain is then translocated to the interior of the ClpP protease subunit, where it is degraded [101, 102]. In bacteria, ClpA and ClpX have been shown to act as chaperones independently of an association with the ClpP complex. In mitochondria the only homologue of this system identified to date is Mcx1, a homologue of the bacterial ClpX. A deletion of this gene in yeast did not result in a discernible phenotype; therefore, its cellular function is unclear.

In bacteria an additional member of the Hsp100/Clp family has been identified that has no direct role in proteolysis. This protein, ClpB, forms a ring-shaped oligomeric complex and seems to be involved in protective reactions under stress conditions. In eukaryotes, both a cytosolic homologue, Hsp104, and a mitochondrial homologue, Hsp78, have been identified [103, 104]. Both ClpB and Hsp104 have been shown to reduce protein aggregation in vitro and in vivo [105, 106]. ClpB's main activity seems to be the resolubilization of aggregated proteins, not the prevention of aggregation per se. Although ClpB and homologues can bring aggregated proteins back into solution, refolding of proteins to their native conformation requires the Hsp70 chaperone system [107, 108]. ClpB may act to resolubilize proteins by actively exposing new Hsp70 binding sites in the aggregated proteins in an ATP-driven process [109]. However, the details of the molecular mechanism of the disaggregation reaction remain to be determined.

In yeast mitochondria, an Hsp78 deletion mutant shows no phenotype at normal growth conditions, but Hsp78 plays a role in thermotolerance and in the protection of mitochondrial functions under stress conditions [110, 111]. Again, a functional cooperation of Hsp78 and the mitochondrial Hsp70 system in the maintenance of respiratory competence and genome integrity was observed. The bi-chaperone system of Hsp78 and mtHsp70 is able to efficiently refold heat-inactivated substrate proteins in vitro [112]. It is interesting to note that mtHsp70 could not be replaced by the bacterial homologue DnaK in these experiments, indicating a specific functional cooperation between mtHsp70 and Hsp78. The role of Hsp78 in genome maintenance is probably indirect and is a result of its function in protein stabilization, as it has a direct effect on the restoration of DNA polymerase activity after heat shock [113]. In addition, Hsp78 is also involved in the degradation of proteins in the matrix by the protease Pim1 [114]. A deletion mutant of Hsp78 showed a significant reduction in Pim1-dependent proteolysis of imported reporter proteins. This effect was already obvious at normal growth conditions and was not correlated with the function of Hsp78 in the protection from protein aggregation. The involvement of Hsp78 in proteolytic reactions represents a novel level of cooperation among different protein families, expanding the concept of a closely coordinated functional network composed of chaperone proteins and proteases in mitochondria.

31.6
Experimental Protocols

31.6.1
Chaperone Functions Characterized With Yeast Mutants

Selection of Temperature-sensitive Mutants

The major mitochondrial chaperones Hsp60 and Hsp70 are encoded by essential genes. Since null mutations of these genes result in a lethal phenotype, a functional analysis of these proteins requires the use of conditional mutants. Usually temperature-sensitive alleles are generated for the biochemical analysis. The procedure follows standard yeast genetic procedures and will be described only briefly. Temperature-sensitive mutations in yeast are generated by a random mutagenesis (usually error-prone PCR) of the respective gene cloned into a single-copy yeast vector. The mutated plasmids are transformed into a strain that has a deletion of the original gene on the chromosome but carries the wild-type gene on another plasmid. By a plasmid-shuffling procedure, the plasmid with the wild-type gene is removed and the appropriate mutant is selected [115].

Purification of Mitochondria and Preproteins

The effects of chaperone mutations on the import and folding of mitochondrial proteins are best analyzed by an in vitro assay system utilizing isolated intact mitochondria. This approach allows the use of specifically designed preproteins that are imported in the in vitro system and act as reporter molecules for the translocation and folding reaction. In addition, the in vitro assay permits the manipulation of experimental conditions and the import activity of mitochondria to a high extent while leaving the biochemical environment inside the mitochondria largely intact.

Purification of yeast mitochondria and preprotein import reactions is performed by standard procedures [116]. The precursor proteins can be used in radio-labeled form after in vitro transcription-translation or in purified form as recombinant proteins. In most cases, fusion proteins with an amino-terminal part derived from an authentic mitochondrial protein and a carboxy-terminal dihydrofolate reductase (DHFR) domain are used. While the targeting information is supplied by the mitochondrial protein, the import and folding of the heterologous DHFR domain can be followed directly.

Protein Import Reaction, Non-Permissive Conditions

Import reactions are performed in a volume of 100 μL import buffer (3% bovine serum albumin, 250 mM sucrose, 80 mM KCl, 5 mM $MgCl_2$, 2 mM KH_2PO_4, 5 mM methionine, 10 mM MOPS-KOH, pH 7.2) containing mitochondria (25 μg mitochondrial protein) that were isolated under permissive growth conditions. To induce the temperature-sensitive phenotype, the reaction mix is heated to 37 °C for 15 min. After cooling to 25 °C, 2 mM ATP and 2 mM NADH are added. The import reaction is started by the addition of up to 10 μL of reticulocyte lysate from the in vitro translation reaction containing the radio-labeled preprotein. Import reactions are typically performed at 25 °C for 2–30 min, depending on the translocation efficiency of the preprotein used.

Precursor proteins imported into the matrix are processed by the matrix-processing peptidase (MPP) to their mature form, indicated by a characteristic increase in electrophoretic mobility. In addition, complete import of the polypeptide chain is indicated by the resistance of the imported protein against external proteases. Non-imported proteins are digested by a treatment of the mitochondria with 20–150 μg mL^{-1} proteinase K for 10 min at 0 °C. The protease is inactivated by 2 mM phenylmethylsulfonyl fluoride (PMSF) for 5 min at 0 °C. Mitochondria are then pelleted by centrifugation C at 13 000 g for 10 min at 2°. The supernatant is discarded and the mitochondria are solubilized by electrophoresis sample buffer (2% [w/v] SDS, 5% [w/v] 2-mercaptoethanol, 10% [v/v] glycerol, 60 mM TRIS/HCl, pH 6.8, 0.02% [w/v] bromophenol blue). Imported proteins are analyzed by standard SDS-PAGE. A useful control reaction is to perform the import reaction in the presence of a mixture of inhibitors of the mitochondrial respiratory chain. In the absence of an inner-membrane potential, insertion of the preprotein into the inner membrane is blocked and no processing or protease resistance is observed. The following concentrations are used: 8 μM antimycin A, 1 mM valinomycin, and 20 μM oligomycin. The three inhibitors can be mixed together in a 100× stock solution in ethanol and stored at −20 °C.

Temperature-sensitive mutants of the mtHsp70, Ssc1, show two types of translocation defects [9]. A complete inactivation of Ssc1, e.g., using a mutation in the ATPase domain (mutant strain *ssc1-3*), results in a block of translocation of matrix proteins. Under these conditions, preproteins remain bound to the surface of the outer membrane. Although insertion of preproteins into the inner membrane driven by the membrane potential is possible, typically no processing or protease resistance is observed for matrix proteins due to the lack of import motor activity in the matrix. A second type of mutant is affected in its binding to protein substrates. An example is the mutant *ssc1-2*, which contains an alteration in the peptide-binding domain. This mutation locks the mtHsp70 in its high-affinity state. In principle, mutant *ssc1-2* mitochondria are import-competent as a result of a high binding efficiency to imported proteins, but significant translocation efficiency was observed only with preproteins unfolded prior to import by treatment with 7 M urea.

31.6.2 Interaction of Imported Proteins With Matrix Chaperones

Co-immunoprecipitation

Physical interactions of imported preproteins with mitochondrial chaperones can be detected by co-immunoprecipitation with specific antibodies. The binding conditions described here have been optimized for the mtHsp70. The binding conditions with other chaperones or other substrate proteins might be significantly different; therefore, the binding conditions need to be closely monitored and changed if appropriate.

First, the respective antibodies need to be pre-bound to protein A Sepharose (PAS). Per import reaction, 10 μL wet volume of PAS is incubated with 5–20 μL of antiserum for 1 h in a volume of 500 μL TBS (0.9% NaCl, 10 mM TRIS/HCl, pH 7.4). The amount of antiserum needed depends on the avidity of the individual antibody and has to be established by test experiments. Two washing steps with TBS result in the removal of unbound material, and the wet PAS pellet can be stored for several hours at 4 °C. Second, an import reaction with radio-labeled precursor proteins is performed as described above. The import reaction is stopped by the addition of 1 μM valinomycin after a short incubation time of 1–5 min. In order to analyze the kinetics of the preprotein-chaperone association, a further incubation at 25 °C can be included. After completion of the import reaction, the mitochondrial pellet is solubilized in 200 μL IP buffer (0.1% Triton X-100, 10 mM TRIS/HCl, pH 7.4, 100 mM NaCl, 5 mM EDTA, proteinase inhibitors) under native conditions. After 5 min vigorous shaking at 4 °C, the mitochondrial lysate is centrifuged at 13 000 g for 5 min at 2 °C to remove insoluble material. The supernatants are incubated with the PAS material containing the pre-bound antibodies for 30 min at 4 °C with gentle agitation. The PAS is washed three times with lysis buffer and once with 10 mM TRIS/HCl (pH 7.4). Bound proteins are eluted by electrophoresis sample buffer and separated by SDS-PAGE.

Gel Filtration

Physical interactions of substrate proteins with the mitochondrial Hsp60 system can also be analyzed by gel filtration [69]. Due to the large size of the oligomeric Hsp60 complex, bound and unbound substrate proteins can be easily separated by a simple one-step gel filtration procedure. In principle, the method can be employed for any substrate or chaperone provided that the molecular weight difference between unbound and bound substrate is large enough. For each reaction, a small column of about 10 cm in length and 0.5 cm in diameter is filled with S-200 Sepharose (Amersham Biosciences) and buffered in column buffer (10 mM TRIS/HCl, pH 7.4, 100 mM NaCl, 5 mM EDTA). Radio-labeled substrate proteins are imported into isolated mitochondria as described above, and after solubilization under native conditions and removal of insoluble material, the supernatant is passed over the gel-filtration column. The column is run with column buffer, and fractions of about 1 mL are collected. The fractions are then analyzed for the presence of Hsp60 and substrate proteins by SDS-PAGE and Western blot.

31.6.3
Folding of Imported Model Proteins

Folding reactions of imported preproteins can be followed if a specific assay for the conformational state is available. In most experiments to date, the folding state of an imported DHFR domain has been monitored [30, 69, 80]. In its native conformation, DHFR is resistant to proteolysis by proteinase K and can be easily distinguished as a protein with a molecular weight of about 20 kDa on SDS-PAGE. In principle any other protein domain with a behavior similar to DHFR can be used in the assay.

Since DHFR lacks the proper targeting information for mitochondria, an in vitro import reaction is performed with DHFR fusion proteins containing an amino-terminal segment containing a mitochondrial presequence that directs the resulting preprotein into the matrix. After a standard import reaction, the mitochondrial pellet is lysed in 200 μL lysis buffer (0.3% Triton X-100, 30 mM TRIS/HCl, pH 7.4, 80 mM KCl, 5% glycerol) in the presence of 50 μg mL^{-1} proteinase K and incubated for 2–10 min at 0 °C. After inactivation of the protease with PMSF, insoluble material is removed, and the supernatant is treated with trichloroacetic (TCA) acid to precipitate the proteins and then is analyzed by SDS-PAGE and autoradiography.

In the case of the DHFR-based assay, it is absolutely essential to perform a control import reaction where the membrane potential of the inner membrane has been dissipated, because excess preprotein in the assay reaction results in a background of folded DHFR in the absence of an import reaction. The folding assays are usually performed at low temperatures (4–15 °C), as the folding reactions in wild-type mitochondria proceed at a high rate.

31.6.4
Assaying Mitochondrial Degradation of Imported Proteins

Similar to the assays described above, mitochondrial degradation reactions can be analyzed by reduction of the amount of imported radio-labeled preprotein over time. The mitochondria are preloaded with suitable substrate proteins by an import reaction in vitro. In this assay, fusion proteins between the mitochondrial cytochrome b_2 and the DHFR domain from mouse have been used extensively. For this precursor, translocation into the matrix is ensured by a deletion of 19 amino acids from the intermembrane space sorting signal of cytochrome b_2 [117]. The fusion proteins are recognized as foreign proteins and are degraded by the Pim1 protease in the matrix. The influence of certain protein components such as chaperones can be assessed by using mutant mitochondria [99, 114]. Authentic mitochondrial proteins usually show a very slow rate of degradation at normal growth conditions, which may not be detected over the time this assay is performed. However, damage inflicted on mitochondrial proteins, which leads to protein denaturation or unfolding, is recognized by the proteolytic system by an unknown mechanism, and these proteins are degraded [118, 119]. Per time point, 25 μg mitochondria (protein amount) is used for the import reaction. Following an incubation of 15 min at 25 °C, the translocation is stopped and unimported preprotein is removed by a short treatment with proteinase K. The mitochondria are then pelleted and resuspended in 100 μL import buffer per sample. To maintain a maximum level of activity, an ATP-regenerating system (3 mM ATP, 20 mM creatine phosphate, 200 μg mL^{-1} creatine kinase) should be included in the degradation reaction. Proteolysis of the imported protein is assessed by removing aliquots of mitochondria (25 μg of mitochondrial protein each) at several time points over the course of about 2 h. The collected aliquots are centrifuged at 13 000 g for 10 min at 2 °C, and the mitochondria are lysed in electrophoresis sample buffer and analyzed by SDS-PAGE and autoradiography.

31.6.5
Aggregation of Proteins in the Mitochondrial Matrix

If the proteolytic system is overwhelmed by high amounts of destabilized or unfolded proteins, aggregation of substrate proteins in the matrix compartment can occur. The aggregation of imported reporter substrates can be tested based on the in vitro import assay. Depending on the particular properties of the reporter protein used, only a low level of aggregation is observed. As a result, proteins containing mutations that affect conformational instability should be used in this assay. This can be achieved by the import of either mutant preproteins or aggregation-prone reporter domains that are targeted to mitochondria. For example, a destabilized mutant of DHFR, containing the mutations Cys7Ser, Ser42Cys, and Asn49Cys [120], can be used for the assay because it shows significant aggregation at elevated temperature (37 °C or higher). After the import reaction, mitochondria are re-isolated and subjected to different experimental conditions. Protein aggregation is tested by differential centrifugation. Per sample, 25 μg mitochondria (protein amount) is lysed in 200 μL lysis buffer (see above) containing a proteinase inhibitor cocktail and subjected to a high-spin centrifugation (100 000 g for 15 min). Soluble proteins stay in the supernatant, while protein aggregates are sedimented and can be recovered in the pellet. The pellet is extracted once with fresh lysis buffer to remove contaminating membrane proteins, and the aggregated material is resolubilized in electrophoresis sample buffer and analyzed by SDS-PAGE and autoradiography.

References

1 Voos, W. & Röttgers, K. (2002). Molecular chaperones as essential mediators of mitochondrial biogenesis. *Biochim. Biophys. Acta* **1592**, 51–62.

2 Gakh, O., Cavadini, P. & Isaya, G. (2002). Mitochondrial processing peptidases. *Biochim. Biophys. Acta* **1592**, 63–77.

3 Pfanner, N. & Wiedemann, N. (2002). Mitochondrial protein import: two membranes, three translocases. *Curr. Opin. Cell Biol.* **14**, 400–411.

4 Jensen, R. & Dunn, C. (2002). Protein import into and across the mitochondrial inner membrane: role of the TIM23 and TIM22 translocons. *Biochim. Biophys. Acta* **1592**, 25–34.

5 Hill, K., Model, K., Ryan, M. T., Dietmeier, K., Martin, F., Wagner, R. & Pfanner, N. (1998). Tom40 forms the hydrophilic channel of the mitochondrial import pore for preproteins. *Nature* **395**, 516–521.

6 Truscott, K. N., Kovermann, P., Geissler, A., Merlin, A., Meijer, M., Driessen, A. J. M., Rassow, J., Pfanner, N. & Wagner, R. (2001). A presequence- and voltage-sensitive channel of the mitochondrial preprotein translocase formed by Tim23. *Nat. Struct. Biol.* **8**, 1074–1082.

7 Voos, W., Martin, H., Krimmer, T. & Pfanner, N. (1999). Mechanisms of protein translocation into mitochondria. *Biochim. Biophys. Acta* **1422**, 235–254.

8 Kang, P. J., Ostermann, J., Shilling, J., Neupert, W., Craig, E. A. & Pfanner, N. (1990). Requirement for hsp70 in the mitochondrial matrix for translocation and folding of precursor proteins. *Nature* **348**, 137–143.

9 Gambill, B. D., Voos, W., Kang, P. J., Miao, B., Langer, T., Craig, E. A. & Pfanner, N. (1993). A dual role for mitochondrial heat shock protein 70 in membrane translocation of preproteins. *J. Cell Biol.* **123**, 109–117.

10 Ungermann, C., Neupert, W. & Cyr, D. M. (1994). The role of hsp70 in conferring unidirectionality on protein translocation. *Science* **266**, 1250–1253.

11 Voos, W., von Ahsen, O., Müller, H., Guiard, B., Rassow, J. & Pfanner, N. (1996). Differential requirement for the mitochondrial Hsp70-Tim44 complex in unfolding and translocation of preproteins. *EMBO J.* **15**, 2668–2677.

12 Wachter, C., Schatz, G. & Glick, B. S. (1992). Role of ATP in the intramitochondrial sorting of cytochrome c_1 and the adenine nucleotide transporter. *EMBO J.* **11**, 4787–4794.

13 Herrmann, J. M., Neupert, W. & Stuart, R. A. (1997). Insertion into

the mitochondrial inner membrane of a polytopic protein, the nuclear-encoded Oxa1p. *EMBO J.* **16**, 2217–2226.

14 von Ahsen, O., Voos, W., Henninger, H. & Pfanner, N. (1995). The mitochondrial protein import machinery. Role of ATP in dissociation of the Hsp70.Mim44 complex. *J. Biol. Chem.* **270**, 29848–29853.

15 Strub, A., Lim, J. H., Pfanner, N. & Voos, W. (2000). The mitochondrial protein import motor. *Biol. Chem.* **381**, 943–949.

16 Miao, B., Davis, J. E. & Craig, E. A. (1997). Mge1 functions as a nucleotide release factor for Ssc1, a mitochondrial Hsp70 of *Saccharomyces cerevisiae*. *J. Mol. Biol.* **265**, 541–552.

17 Dekker, P. J. T. & Pfanner, N. (1997). Role of mitochondrial GrpE and phosphate in the ATPase cycle of matrix Hsp70. *J. Mol. Biol.* **270**, 321–327.

18 Schneider, H. C., Westermann, B., Neupert, W. & Brunner, M. (1996). The nucleotide exchange factor MGE exerts a key function in the ATP-dependent cycle of mt-Hsp70-Tim44 interaction driving mitochondrial protein import. *EMBO J.* **15**, 5796–5803.

19 Voos, W., Gambill, B. D., Laloraya, S., Ang, D., Craig, E. A. & Pfanner, N. (1994). Mitochondrial GrpE is present in a complex with hsp70 and preproteins in transit across membranes. *Mol. Cell. Biol.* **14**, 6627–6634.

20 Westermann, B., Prip-Buus, C., Neupert, W. & Schwarz, E. (1995). The role of the GrpE homologue, Mge1p, in mediating protein import and protein folding in mitochondria. *EMBO J.* **14**, 3452–3460.

21 Kronidou, N. G., Oppliger, W., Bolliger, L., Hannavy, K., Glick, B. S., Schatz, G. & Horst, M. (1994). Dynamic interaction between Isp45 and mitochondrial hsp70 in the protein import system of the yeast mitochondrial inner membrane. *Proc. Natl. Acad. Sci. USA* **91**, 12818–12822.

22 Schneider, H.-C., Berthold, J., Bauer, M. F., Dietmeier, K., Guiard, B., Brunner, M. & Neupert, W. (1994). Mitochondrial Hsp70/MIM44 complex facilitates protein import. *Nature* **371**, 768–774.

23 Rassow, J., Maarse, A. C., Krainer, E., Kübrich, M., Müller, H., Meijer, M., Craig, E. A. & Pfanner, N. (1994). Mitochondrial protein import: biochemical and genetic evidence for interaction of matrix hsp70 and the inner membrane protein MIM44. *J. Cell Biol.* **127**, 1547–1556.

24 Ungermann, C., Guiard, B., Neupert, W. & Cyr, D. M. (1996). The delta psi- and Hsp70/MIM44-dependent reaction cycle driving early steps of protein import into mitochondria. *EMBO J.* **15**, 735–744.

25 Merlin, A., Voos, W., Maarse, A. C., Meijer, M., Pfanner, N. & Rassow, J. (1999). The J-related segment of Tim44 is essential for cell viability: a mutant Tim44 remains in the mitochondrial import site, but inefficiently recruits mtHsp70 and impairs protein translocation. *J. Cell Biol.* **145**, 961–972.

26 Simon, M. S., Peskin, C. S. & Oster, G. F. (1992). What drives the translocation of proteins? *Proc. Natl. Acad. Sci. USA* **89**, 3770–3774.

27 Bauer, M. F., Hofmann, S., Neupert, W. & Brunner, M. (2000). Protein translocation into mitochondria: the role of TIM complexes. *Trends Cell Biol.* **10**, 25–31.

28 Neupert, W. & Brunner, M. (2002). The protein import motor of mitochondria. *Nat. Rev. Mol. Cell Biol.* **3**, 555–565.

29 Okamoto, K., Brinker, A., Paschen, S. A., Moarefi, I., Hayer-Hartl, M., Neupert, W. & Brunner, M. (2002). The protein import motor of mitochondria: a targeted molecular ratchet driving unfolding and translocation. *EMBO J.* **21**, 3659–3671.

30 Voisine, C., Craig, E. A., Zufall, N., von Ahsen, O., Pfanner, N. & Voos, W. (1999). The protein import motor of mitochondria: unfolding and

trapping of preproteins are distinct and separable functions of matrix Hsp70. *Cell* **97**, 565–574.

31 Geissler, A., Rassow, J., Pfanner, N. & Voos, W. (2001). Mitochondrial import driving forces: enhanced trapping by matrix hsp70 stimulates translocation and reduces the membrane potential dependence of loosely folded preproteins. *Mol. Cell. Biol.* **21**, 7097–7104.

32 Matouschek, A., Azem, A., Ratliff, K., Glick, B. S., Schmid, K. & Schatz, G. (1997). Active unfolding of precursor proteins during mitochondrial protein import. *EMBO J.* **16**, 6727–6736.

33 Lim, J. H., Martin, F., Guiard, B., Pfanner, N. & Voos, W. (2001). The mitochondrial Hsp70-dependent import system actively unfolds preproteins and shortens the lag phase of translocation. *EMBO J.* **20**, 941–950.

34 Milisav, I., Moro, F., Neupert, W. & Brunner, M. (2001). Modular structure of the TIM23 preprotein translocase of mitochondria. *J. Biol. Chem.* **276**, 25856–25861.

35 Huang, S., Ratliff, K. S., Schwartz, M. P., Spenner, J. M. & Matouschek, A. (1999). Mitochondria unfold precursor proteins by unraveling them from their N-termini. *Nat. Struct. Biol.* **6**, 1132–1138.

36 Matouschek, A., Pfanner, N. & Voos, W. (2000). Protein unfolding by mitochondria. The Hsp70 import motor. *EMBO Rep.* **1**, 404–410.

37 Pfanner, N. & Meijer, M. (1995). Protein sorting: Pulling in the proteins. *Curr. Biol.* **5**, 132–135.

38 Glick, B. S. (1995). Can Hsp70 proteins act as force-generating motors? *Cell* **80**, 11–14.

39 Jensen, R. E. & Johnson, A. E. (1999). Protein translocation: Is hsp70 pulling my chain? *Curr. Biol.* **9**, R779–782.

40 Bömer, U., Maarse, A. C., Martin, F., Geissler, A., Merlin, A., Schönfisch, B., Meijer, M., Pfanner, N. & Rassow, J. (1998). Separation of structural and dynamic functions of the mitochondrial translocase: Tim44 is crucial for the inner membrane import sites in translocation of tightly folded domains, but not of loosely folded preproteins. *EMBO J.* **17**, 4226–4237.

41 Liu, Q., D'Silva, P., Walter, W., Marszalek, J. & Craig, E. A. (2003). Regulated cycling of mitochondrial Hsp70 at the protein import channel. *Science* **300**, 139–141.

42 Truscott, K. N., Voos, W., Frazier, A. E., Lind, M., Li, Y., Geissler, A., Dudek, J., Müller, H., Sickmann, A., Meyer, H. E., Meisinger, C., Guiard, B., Rehling, P. & Pfanner, N. (2003). A J-protein is an essential subunit of the presequence translocase associated protein import motor of mitochondria. *J. Cell Biol.* **163**, 707–713.

43 Mokranjac, D., Sichting, M., Neupert, W. & Hell, K. (2003). Tim14, a novel key component of the import motor of the TIM23 protein translocase of mitochondria. *EMBO J.* **22**, 4945–4956.

44 D'Silva, P., Schilke, B., Walter, W., Andrew, A. & Craig, E. (2003). J-protein co-chaperone of the mitochodrial inner membrane required for protein import into the mitochondrial matrix. *Proc. Natl. Acad. Sci. USA* **100**, 13839–13844.

45 Frazier, A. E., Dudek, J., Guiard, B., Voos, W., Li, Y., Lind, M., Meisinger, C., Geissler, A., Sickmann, A., Meyer, H. E., Bilanchone, V., Cumsky, M. G., Truscott, K. N., Pfanner, N. & Rehling, P. (2004). Pam16 has an essential role in the mitochondrial protein import motor. *Nat. Struct. Mol. Biol.* **11**, 226–233.

46 Kozany, C., Mokranjac, D., Sichting, M., Neupert, W. & Hell, K. (2004). The J domain-related cochaperone Tim16 is a constituent of the mitochondrial TIM23 preprotein translocase. *Nat. Struct. Mol. Biol.* **11**, 234–241.

47 Horst, M., Oppliger, W., Rospert, S., Schönfeld, H. J., Schatz, G. & Azem, A. (1997). Sequential action of

two hsp70 complexes during protein import into mitochondria. *EMBO J.* **16**, 1842–1849.

48 Langer, T., Lu, C., Echols, H., Flanagan, J., Hayer-Hartl, M. K. & Hartl, F.-U. (1992). Successive action of DnaK (Hsp70), DnaJ, and GroEL (Hsp60) along the pathway of chaperone-assisted protein folding. *Nature* **356**, 683–689.

49 Rapoport, T. A., Jungnickel, B. & Kutay, U. (1996). Protein transport across the eukaryotic endoplasmic reticulum and bacterial inner membranes. *Annu. Rev. Biochem.* **65**, 271–303.

50 Hartl, F. U. & Hayer-Hartl, M. (2002). Molecular chaperones in the cytosol: from nascent chain to folded protein. *Science* **295**, 1852–1858.

51 Bukau, B. & Horwich, A. L. (1998). The Hsp70 and Hsp60 chaperone machines. *Cell* **92**, 351–366.

52 Deloche, O., Kelley, W. L. & Georgopoulos, C. (1997). Structure-function analyses of the Ssc1p, Mdj1p, and Mge1p *Saccharomyces cerevisiae* mitochondrial proteins in *Escherichia coli*. *J. Bacteriol.* **179**, 6066–6075.

53 Cyr, D. M., Langer, T. & Douglas, M. G. (1994). DnaJ-like proteins: molecular chaperones and specific regulators of Hsp70. *Trends Biochem. Sci.* **19**, 176–181.

54 Rowley, N., Prip-Buus, C., Westermann, B., Brown, C., Schwarz, E., Barrell, B. & Neupert, W. (1994). Mdj1p, a novel chaperone of the DnaJ family, is involved in mitochondrial biogenesis and protein folding. *Cell* **77**, 249–259.

55 Kubo, Y., Tsunehiro, T., Nishikawa, S., Nakai, M., Ikeda, E., Toh-e, A., Morishima, N., Shibata, T. & Endo, T. (1999). Two distinct mechanisms operate in the reactivation of heat-denatured proteins by the mitochondrial Hsp70/Mdj1p/Yge1p chaperone system. *J. Mol. Biol.* **286**, 447–464.

56 Deloche, O., Liberek, K., Zylicz, M. & Georgopoulos, C. (1997). Purification and biochemical properties of *Saccharomyces cerevisiae* Mdj1p, the mitochondrial DnaJ homologue. *J. Biol. Chem.* **272**, 28539–28544.

57 Duchniewicz, M., Germaniuk, A., Westermann, B., Neupert, W., Schwarz, E. & Marszalek, J. (1999). Dual Role of the Mitochondrial Chaperone Mdj1p in Inheritance of Mitochondrial DNA in Yeast. *Mol. Cell. Biol.* **19**, 8201–8210.

58 Westermann, B. & Neupert, W. (1997). Mdj2p, a novel DnaJ homolog in the mitochondrial inner membrane of the yeast *Saccharomyces cerevisiae*. *J. Mol. Biol.* **272**, 477–483.

59 Voisine, C., Cheng, Y. C., Ohlson, M., Schilke, B., Hoff, K., Beinert, H., Marszalek, J. & Craig, E. A. (2001). Jac1, a mitochondrial J-type chaperone, is involved in the biogenesis of Fe/S clusters in *Saccharomyces cerevisiae*. *Proc. Natl. Acad. Sci. USA* **98**, 1483–1488.

60 Lutz, T., Westermann, B., Neupert, W. & Herrmann, J. M. (2001). The mitochondrial proteins Ssq1 and Jac1 are required for the assembly of iron sulfur clusters in mitochondria. *J. Mol. Biol.* **307**, 815–825.

61 Kim, R., Saxena, S., Gordon, D. M., Pain, D. & Dancis, A. (2001). J-domain protein, Jac1p, of yeast mitochondria required for iron homeostasis and activity of Fe-S cluster proteins. *J. Biol. Chem.* **276**, 17524–17532.

62 Schilke, B., Forster, J., Davis, J., James, P., Walter, W., Laloraya, S., Johnson, J., Miao, B. & Craig, E. A. (1996). The cold sensitivity of a mutant of *Saccharomyces cerevisiae* lacking a mitochondrial heat shock protein 70 is suppressed by loss of mitochondrial DNA. *J. Cell Biol.* **134**, 603–613.

63 Knight, S. A. B., Sepuri, N. B. V., Pain, D. & Dancis, A. (1998). Mt-Hsp70 homolog, Ssc2p, required for maturation of yeast frataxin and mitochondrial iron homeostasis. *J. Biol. Chem.* **273**, 18389–18393.

64 Schmidt, S., Strub, A., Röttgers, K., Zufall, N. & Voos, W. (2001). The two mitochondrial heat shock proteins

70, Ssc1 and Ssq1, compete for the cochaperone Mge1. *J. Mol. Biol.* **313**, 13–26.

65 Manning-Krieg, U. C., Scherer, P. E. & Schatz, G. (1991). Sequential action of mitochondrial chaperones in protein import into mitochondria. *EMBO J.* **10**, 3273–3280.

66 Heyrovska, N., Frydman, J., Höhfeld, J. & Hartl, F. U. (1998). Directionality of polypeptide transfer in the mitochondrial pathway of chaperone-mediated protein folding. *Biol. Chem.* **379**, 301–309.

67 Cheng, M. Y., Hartl, F. U., Martin, J., Pollock, R. A., Kalusek, F., Neupert, W., Hallberg, E. M., Hallberg, R. L. & Horwich, A. L. (1989). Mitochondrial heat-shock protein hsp60 is essential for assembly of proteins imported into yeast mitochondria. *Nature* **337**, 620–625.

68 Reading, D. S., Hallberg, R. L. & Myers, A. M. (1989). Characterization of the yeast HSP60 gene coding for a mitochondrial assembly factor. *Nature* **337**, 655–659.

69 Ostermann, J., Horwich, A. L., Neupert, W. & Hartl, F. U. (1989). Protein folding in mitochondria requires complex formation with hsp60 and ATP hydrolysis. *Nature* **341**, 125–130.

70 Walter, S. (2002). Structure and function of the GroE chaperone. *Cell. Mol. Life Sci.* **59**, 1589–1597.

71 Lubben, T., Gatenby, A., Donaldson, G., Lorimer, G. & Viitanen, P. (1990). Identification of a groES-like chaperonin in mitochondria that facilitates protein folding. *Proc. Natl. Acad. Sci. USA* **87**, 7683–7687.

72 Rospert, S., Junne, T., Glick, B. S. & Schatz, G. (1993). Cloning and disruption of the gene encoding yeast mitochondrial chaperonin 10, the homolog of *E. coli* groES. *FEBS Lett.* **335**, 358–360.

73 Höhfeld, J. & Hartl, F. U. (1994). Role of the chaperonin cofactor Hsp10 in protein folding and sorting in yeast mitochondria. *J. Cell Biol.* **126**, 305–315.

74 Dubaquie, Y., Looser, R. & Rospert, S. (1997). Significance of chaperonin 10-mediated inhibition of ATP hydrolysis by chaperonin 60. *Proc. Natl. Acad. Sci. USA* **94**, 9011–9016.

75 Nielsen, K. L. & Cowan, N. J. (1998). A single ring is sufficient for productive chaperonin-mediated folding *in vivo*. *Mol. Cell* **2**, 93–99.

76 Nielsen, K. L., McLennan, N., Masters, M. & Cowan, N. J. (1999). A single-ring mitochondrial chaperonin (Hsp60-Hsp10) can substitute for GroEL-GroES *in vivo*. *J. Bacteriol.* **181**, 5871–5875.

77 Rospert, S., Looser, R., Dubaquié, Y., Matouschek, A., Glick, B. S. & Schatz, G. (1996). Hsp60-independent protein folding in the matrix of yeast mitochondria. *EMBO J.* **15**, 764–774.

78 Dubaquie, Y., Looser, R., Fünfschilling, U., Jenö, P. & Rospert, S. (1998). Identification of *in vivo* substrates of the yeast mitochondrial chaperonins reveals overlapping but non-identical requirement for hsp60 and hsp10. *EMBO J.* **17**, 5868–5876.

79 Matouschek, A., Rospert, S., Schmid, K., Glick, B. S. & Schatz, G. (1995). Cyclophilin catalyzes protein folding in yeast mitochondria. *Prc. Natl. Acad. Sci. USA* **92**, 6319–6323.

80 Rassow, J., Mohrs, K., Koidl, S., Barthelmess, I. B., Pfanner, N. & Tropschug, M. (1995). Cyclophilin 20, is involved in mitochondrial protein folding in cooperation with molecular chaperones Hsp70 and Hsp60. *Mol. Cell. Biol.* **15**, 2654–2662.

81 Herrmann, J. M., Stuart, R. A., Craig, E. A. & Neupert, W. (1994). Mitochondrial heat shock protein 70, a molecular chaperone for proteins encoded by mitochondrial DNA. *J. Cell Biol.* **127**, 893–902.

82 Westermann, B., Gaume, B., Herrmann, J. M., Neupert, W. & Schwarz, E. (1996). Role of the mitochondrial DnaJ homolog Mdj1p as a chaperone for mitochondrially synthesized and imported proteins. *Mol. Cell. Biol.* **16**, 7063–7071.

83 Bowman, S., Ackerman, S. H., Griffiths, D. E. & Tzagoloff, A. (1991). Characterization of *ATP12*, a yeast nuclear gene required for the assembly of the mitochondrial F_1-ATPase. *J. Biol. Chem.* **266**, 7517–7523.

84 Ackerman, S. H., Martin, J. & Tzagoloff, A. (1992). Characterization of *ATP11* and detection of the encoded protein in mitochondria of *Saccharomyces cerevisiae*. *J. Biol. Chem.* **267**, 7386–7394.

85 Glerum, D. M., Koerner, T. J. & Tzagoloff, A. (1995). Cloning and characterization of *COX14*, whose product is required for assembly of yeast cytochrome oxidase. *J. Biol. Chem.* **270**, 15585–15590.

86 Glerum, D. M., Muroff, I., Jin, C. & Tzagoloff, A. (1997). *COX15* codes for a mitochondrial protein essential for the assembly of yeast cytochrome oxidase. *J. Biol. Chem.* **272**, 19088–19094.

87 Souza, R. L., Green-Willms, N. S., Fox, T. D., Tzagoloff, A. & Nobrega, F. G. (2000). Cloning and characterization of *COX18*, a *Saccharomyces cerevisiae* PET gene required for the assembly of cytochrome oxidase. *J. Biol. Chem.* **275**, 14898–14902.

88 Dibrov, E., Fu, S. & Lemire, B. D. (1998). The Saccharomyces cerevisiae *TCM62* gene encodes a chaperone necessary for the assembly of the mitochondrial succinate dehydrogenase (complex II). *J. Biol. Chem.* **273**, 32042–32048.

89 Guélin, E., Rep, M. & Grivell, L. A. (1994). Sequence of the AFG3 gene encoding a new member of the FtsH/Yme1/Tma subfamily of the AAA-protein family. *Yeast* **10**, 1389–1394.

90 Tzagoloff, A., Yue, J., Jang, J. & Paul, M. F. (1994). A new member of a family of ATPases is essential for assembly of mitochondrial respiratory chain and ATP synthetase complexes in Saccharomyces cerevisiae. *J. Biol. Chem.* **269**, 26144–26151.

91 Paul, M. F. & Tzagoloff, A. (1995). Mutations in RCA1 and AFG3 inhibit F1-ATPase assembly in Saccharomyces cerevisiae. *FEBS Lett.* **373**, 66–70.

92 Thorsness, P. E., White, K. H. & Fox, T. D. (1993). Inactivation of YME1, a member of the ftsH-SEC18-PAS1-CDC48 family of putative ATPase-encoding genes, causes increased escape of DNA from mitochondria in Saccharomyces cerevisiae. *Mol. Cell Biol.* **13**, 5418–5426.

93 Weber, E. R., Hanekamp, T. & Thorsness, P. E. (1996). Biochemical and functional analysis of the YME1 gene product, an ATP and zinc-dependent mitochondrial protease from S. cerevisiae. *Mol. Biol. Cell* **7**, 307–317.

94 Arlt, H., Tauer, R., Feldmann, H., Neupert, W. & Langer, T. (1996). The YTA10-12 complex, an AAA protease with chaperone-like activity in the inner membrane of mitochondria. *Cell* **85**, 875–885.

95 Leonhard, K., Herrmann, J. M., Stuart, R. A., Mannhaupt, G., Neupert, W. & Langer, T. (1996). AAA proteases with catalytic sites on opposite membrane surfaces comprise a proteolytic system for the ATP-dependent degradation of inner membrane proteins in mitochondria. *EMBO J.* **15**, 4218–4229.

96 Leonhard, K., Stiegler, A., Neupert, W. & Langer, T. (1999). Chaperone-like activity of the AAA domain of the yeast Yme1 AAA protease. *Nature* **398**, 348–351.

97 Van Dyck, L., Pearce, D. A. & Sherman, F. (1994). PIM1 encodes a mitochondrial ATP-dependent protease that is required for mitochondrial function in the yeast *Saccharomyces cerevisiae*. *J. Biol. Chem.* **269**, 238–242.

98 Suzuki, C. K., Suda, K., Wang, N. & Schatz, G. (1994). Requirement for the yeast gene *LON* in intramitochondrial proteolysis and maintenance of respiration. *Science* **264**, 273–276.

99 Wagner, I., Arlt, H., van Dyck, L., Langer, T. & Neupert, W. (1994).

Molecular chaperones cooperate with PIM1 protease in the degradation of misfolded proteins in mitochondria. *EMBO J.* **13**, 5135–5145.

100 Rep, M., van Dijl, J. M., Suda, K., Schatz, G., Grivell, L. A. & Suzuki, C. K. (1996). Promotion of mitochondrial membrane complex assembly by a proteolytically inactive yeast Lon. *Science* **274**, 103–106.

101 Wawrzynow, A., Banecki, B. & Zylicz, M. (1996). The Clp ATPases define a novel class of molecular chaperones. *Mol. Microbiol.* **21**, 895–859.

102 Porankiewicz, J., Wang, J. & Clarke, A. K. (1999). New insights into the ATP-dependent Clp protease: *Escherichia coli* and beyond. *Mol. Microbiol.* **32**, 449–458.

103 Parsell, D. A., Sanchez, Y., Stitzel, J. D. & Lindquist, S. (1991). Hsp104 is a highly conserved protein with two essential nucleotide- binding sites. *Nature* **353**, 270–273.

104 Leonhardt, S. A., Fearon, K., Danese, P. N. & Mason, T. L. (1993). *HSP78* encodes a yeast mitochondrial heat shock protein in the Clp family of ATP-dependent proteases. *Mol. Cell. Biol.* **13**, 6304–6313.

105 Mogk, A., Tomoyasu, T., Goloubinoff, P., Rüdiger, S., Röder, D., Langen, H. & Bukau, B. (1999). Identification of thermolabile Escherichia coli proteins: prevention and reversion of aggregation by DnaK and ClpB. *EMBO J.* **18**, 6934–6949.

106 Parsell, D. A., Kowal, A. S., Singer, M. A. & Lindquist, S. (1994). Protein disaggregation mediated by heat-shock protein Hsp104. *Nature* **372**, 475–478.

107 Goloubinoff, P., Mogk, A., Zvi, A. P., Tomoyasu, T. & Bukau, B. (1999). Sequential mechanism of solubilization and refolding of stable protein aggregates by a bichaperone network. *Proc. Natl. Acad. Sci. USA* **96**, 13732–13737.

108 Zolkiewski, M. (1999). ClpB cooperates with DnaK, DnaJ, and GrpE in suppressing protein aggregation. A novel multi-chaperone system from *Escherichia coli*. *J. Biol. Chem.* **274**, 28083–28086.

109 Ben-Zvi, A. P. & Goloubinoff, P. (2001). Mechanisms of disaggregation and refolding of stable protein aggregates by molecular chaperones. *J. Struct. Biol.* **135**, 84–93.

110 Moczko, M., Schönfisch, B., Voos, W., Pfanner, N. & Rassow, J. (1995). The mitochondrial ClpB homolog Hsp78 cooperates with matrix Hsp70 in maintenance of mitochondrial function. *J. Mol. Biol.* **254**, 538–543.

111 Schmitt, M., Neupert, W. & Langer, T. (1996). The molecular chaperone Hsp78 confers compartment-specific thermotolerance to mitochondria. *J. Cell Biol.* **134**, 1375–1386.

112 Krzewska, J., Langer, T. & Liberek, K. (2001). Mitochondrial Hsp78, a member of the Clp/Hsp100 family in *Saccharomyces cerevisiae*, cooperates with Hsp70 in protein refolding. *FEBS Lett.* **489**, 92–96.

113 Germaniuk, A., Liberek, K. & Marszalek, J. (2002). A bi-chaperone (Hsp70-Hsp78) system restores mitochondrial DNA synthesis following thermal inactivation of Mip1p polymerase. *J. Biol. Chem.* **277**, 27801–27808.

114 Röttgers, K., Zufall, N., Guiard, B. & Voos, W. (2002). The ClpB homolog Hsp78 is required for the efficient degradation of proteins in the mitochondrial matrix. *J. Biol. Chem.* **277**, 45829–45837.

115 Strub, A., Röttgers, K. & Voos, W. (2002). The Hsp70 peptide-binding domain determines the interaction of the ATPase domain with Tim44 in mitochondria. *EMBO J.* **21**, 2626–2635.

116 Ryan, M. T., Voos, W. & Pfanner, N. (2001). Assaying protein import into mitochondria. *Methods Cell Biol.* **65**, 189–215.

117 Koll, H., Guiard, B., Rassow, J., Ostermann, J., Horwich, A. L., Neupert, W. & Hartl, F.-U. (1992). Antifolding activity of hsp60 couples protein import into the mitochondrial matrix with export to the intermembrane space. *Cell* **68**, 1163–1175.

118 Langer, T. & Neupert, W. (1996). Regulated protein degradation in mitochondria. *Experientia* **52**, 1069–1076.

119 Rep, M. & Grivell, L. A. (1996). The role of protein degradation in mitochondrial function and biogenesis. *Curr. Genet.* **30**, 367–380.

120 Vestweber, D. & Schatz, G. (1988). Point mutations destabilizing a precursor protein enhance its post-translational import into mitochondria. *EMBO J.* **7**, 1147–1151.

32
Chaperone Systems in Chloroplasts

Thomas Becker, Jürgen Soll, and Enrico Schleiff

32.1
Introduction

More than 25 years ago the terminology of molecular chaperones was invented by Laskey [1] to describe nucleoplasmin required for the assembly of nucleosomes from DNA and histones. Ten years later Pelham [2] offered a refined definition. Molecular chaperones prevent the formation of incorrect structures and disassemble aggregated structures, but they do not form part of the final structure, nor do they possess steric information–specifying assembly. At this time, several proteins of the 70-kDa heat shock protein (Hsp70) class were identified in mammal systems [3, 4], and the first homologue of the chaperonin of the 60-kDa (Cpn60) family was discovered in plants in the form of the Rubisco large subunit-binding protein in chloroplasts [5–7]. Since then, much effort has been made to understand the molecular composition and action of the chaperone machineries. It became clear that the chaperone activity was not limited to the cytoplasm but was also found in organelles, for example, in plastids, which are endosymbiotically derived organelles [8]. Plastids evolved from prokaryotic cyanobacteria, which were captured by a mitochondria-containing host cell. During evolution most but not all of the genes of the endosymbiont were transferred to the nucleus [9]. In turn, some of the gene products are now targeted back toward plastids and others are replaced by gene products of the host cell [10]. Therefore, present-day molecular machines in plastids are often composed of plastid and nuclear gene products of prokaryotic and eukaryotic origin. The molecular chaperones essential for plastid biogenesis in higher plants are generally nuclear-encoded but assemble proteins of both prokaryotic and eukaryotic origin.

Whereas many components of the chaperone family were identified in plastids, not much is known about their specific function in this organelle. In most cases the function was proposed to be similar to the prokaryotic or eukaryotic homologues of these proteins. In plastids these machineries are involved in many essential processes such as protein translocation, intra-organellar transport, protein folding and storage, complex assembly, and stress adaptation (Figure 32.1). In the past the research was focused on the regulation of the chaperone family and their developmental expression profiles. In this chapter we describe the identified proteins

Protein Folding Handbook. Part II. Edited by J. Buchner and T. Kiefhaber

ISBN: 3-527-30784-2

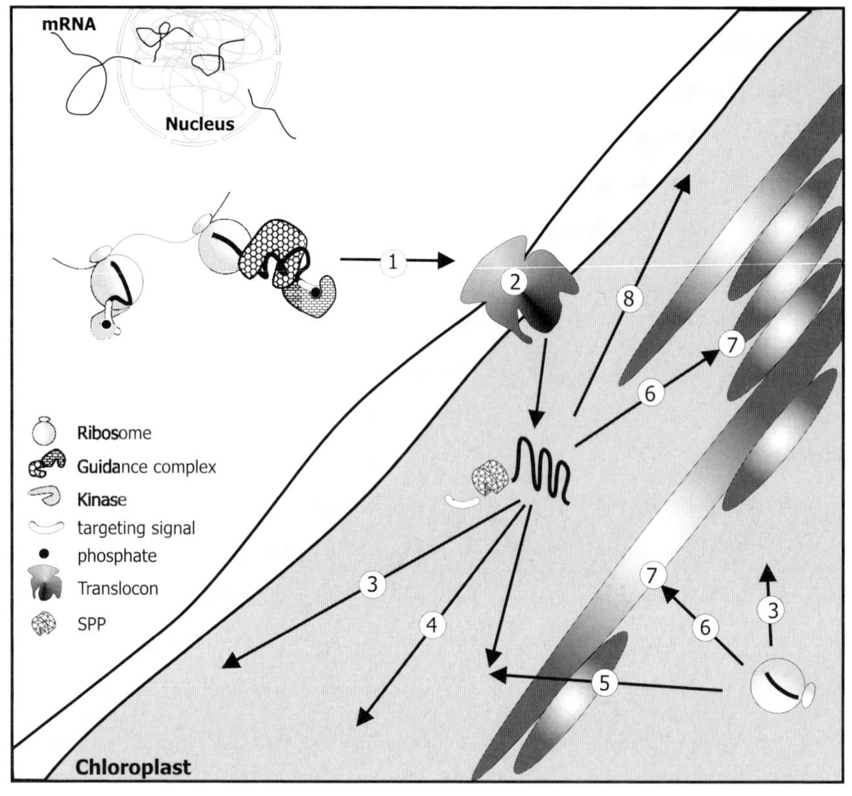

Fig. 32.1. Chaperones are involved in many processes essential for the function of plastids. Chaperones are found to be involved in protein targeting of preproteins toward plastids (1) and in their translocation across the two envelope membranes (2). Molecular chaperones facilitate folding (3) of imported and plastidial-synthesized proteins in the stroma. Furthermore, chaperones are found to be necessary for the storage of proteins (4) and for the assembly of proteinaceous complexes (5). They are also needed for intra-organellar transport of proteins toward the thylakoid membranes (6) and for protein insertion into the thylakoid membrane and lumen (7). Last but not least, molecular chaperones are involved in protein degradation (8) and in stress tolerance.

and their expression regulation and discuss the functional pathways where molecular chaperones are involved.

32.2
Chaperone Systems within Chloroplasts

32.2.1
The Hsp70 System of Chloroplasts

Hsp70 proteins fulfill a set of different functions in prokaryotes and eukaryotes. They are involved in folding of newly synthesized proteins, in protein translocation into cell organelles, in protein degradation, and in protection of cells against aggregation of proteins with nonnative conformations caused, for example, by heat. Hsp70s are characterized by an N-terminal ATP-binding and a C-terminal substrate-binding domain. Thus, the chaperone activity comprises substrate binding and release, which is coupled to the ATPase activity of the Hsp70 protein [11]. Cell organelles such as the endoplasmatic reticulum, mitochondria, and chloroplasts harbor their own Hsp70 isoforms. In higher plants, several Hsp70 homologues are expressed. For example, in *A. thaliana*, 17 Hsp70 proteins and in spinach, 12 Hsp70 proteins are annotated [12, 13]. The number of Hsps in higher plants often exceeds that of other eukaryotes such as *Saccharomyces cerevisiae*, with 14 annotated Hsp70 homologues, or *Homo sapiens*, with 12.

The difference can be explained by the existence of plastids as an additional cell organelle with various biochemical functions. Consequently, several Hsp70 proteins have been identified in chloroplasts of higher plants. The majority of the plastid-localized Hsp70s are nuclear-encoded and posttranslationally imported into the organelle. Only one chloroplast-encoded Hsp70 has so far been identified in the chromophytic algae *Pavlova lutherii* [14]. Two stromal and two outer envelope –localized Hsp70s are present in *Pisum sativum* [15, 16]. By screening the *A. thaliana* genome, three Hsp70s were predicted with chloroplast localization (Table 32.1) [13]. Furthermore, at least one Hsp70 was identified in spinach chloroplasts, and the existence of further homologues remains likely [12]. In the green algae *Chlamydomonas reinhardtii*, one Hsp70 is localized in chloroplasts [17]. Sequence analysis revealed that stromal Hsp70s are more related to the DnaK of *Synechocystis* than to Hsp70 proteins of mitochondria, the cytosol, or the endoplasmatic reticulum, suggesting an endosymbiotic origin of the plastid Hsp70 proteins [18].

32.2.1.1 **The Chloroplast Hsp70s**

One major stromal Hsp70 is identified in pea (Css1), in spinach (Hsp70-9), and in *Chlamydomonas reinhardtii* (Hsp70B) [15, 17, 19]. They are constitutively expressed but can be further induced upon heat stress. After heat treatment, pea Css1 transcripts were increased ninefold [20]. Likewise, the transcript level of Hsp70-9 of spinach was increased upon heat stress [17, 19]. For Hsp70B from *Chlamydomonas reinhardtii*, a 16-fold increase of the transcript level was reported upon heat treatment [21]. Interestingly, a heat-independent induction of Hsp70 expression by light was described for Hsp70B [22–24]. Further studies revealed that an intermediate of chlorophyll biosynthesis is involved in a signal transduction chain for induction of Hsp70B expression [22, 23]. In a *Chlamydomonas reinhardtii* mutant defective in MgPROTO synthesis, Hsp70B expression remained heat- but not light-inducible [25]. Exogenous application of MgPROTO or $MgPROTOMe_2$ to *Chlamydomonas reinhardtii* cells induced expression of Hsp70B, whereas the application of other intermediates of the chlorophyll biosynthesis pathway showed no effect. Furthermore, mutant cells defective in a later stage of chlorophyll biosynthesis, namely, in the formation of chlorophyllide from protochlorohyllide, show light-inducible Hsp70 expression [25]. Therefore, it is concluded that MgPROTO or $MgPROTOMe_2$ is a part of a signaling chain leading to increased Hsp70B expression in *Chlamydomonas reinhardtii* upon illumination.

Tab. 32.1. Chloroplast chaperones in the *A. thaliana* genome. In the following the chloroplast and putative chloroplast chaperones, chaperonins, and chaperone-like proteins are listed. The localization of the proteins was predicted by using the TargetP prediction program, which is available on the Internet (*www.cbs.dtu.dk/services/TargetP/*). Name, AtG code classification, size of the protein in number of amino acids, and the presence of EST clones are listed for each predicted chloroplast chaperone.

Class	*Atg number*	*Name*	*EST clones*	*Length*
FKBP proteins	At5g45680	FKBP-1	+	208
	At4g39710	Unknown protein	+	217
	At1g73655	PPIase-like protein	+	234
	At4g26555	Unknown protein	–	191
Cyclophilins	At3g62030	Roc4	+	260
	At5g13120	Roc8 (TLP 20)	+	259
	At1g74070	Putative cyclophilin	+	317
	At3g01480	AtCyp38	+	437
	At3g15520	AtCyp37	+	406
Hsp70	At4g24280	AtHsp70-7	+	718
	At5g49910	AtHsp70-7	+	718
	At2g32120	AtHsp70-8	+	563
GrpE	At1g36390	Putative Hsp	+	279
	At5g17710	Chloroplast GrpE	+	326
DnaJ	At1g80920	Putative J8	+	163
	At4g36040	AtJ11	+	161
	At5g23240	Unknown protein	+	465
	At2g42750	Unknown protein	+	344
	At2g17880	Putative DnaJ	–	160
	At3g17830	Putative DnaJ	–	493
	At1g77930	DnaJ-like protein	+	271
	At1g80030	Unknown protein	+	500
	At5g05750	DnaJ-like protein	+	294
GroEL	At3g13470	Cpn60β like	+	596
	At1g55490	Cpn60β	+	600
	At5g56500	Cpn60β precursor	+	596
	At2g28000	Cpn60α	+	586
	At1g26230	Putative chaperonin	+	611
	At5g18820	Cpn60α-like protein	–	575
GroES	At5g20720	Cpn20	+	253
	At3g60210	Unknown protein	+	138
	At2g44650	Cpn10	+	139
Hsp90	At2g04030	Hsp-like	+	780
Hsp21	At4g27670	Hsp21	+	227
ClpA	At5g50920	ATP-binding Clp Subunit	+	929
	At3g48870	AtClpC	+	952
	At5g51070	Erd1	+	945
	At5g15450	ClpB-like	+	968

Tab. 32.1. *(continued)*

Class	*Atg number*	*Name*	*EST clones*	*Length*
	At4g12060	Unknown protein	+	267
	At4g25370	Unknown protein	+	238
	At5g49840	ClpX-like	+	608
	At4g30350	Hsp101-like	+	924

Pea Css1 and spinach Hsp70-9 are constitutively expressed in both photosynthetic and non-photosynthetic tissues [12, 20]. The expression level in leaves is higher than in roots, suggesting a role in photosynthesis [12, 20]. Furthermore, a diurnally regulated expression of spinach Hsp70-9 is reported. From this observation, the authors suggested a light-controlled expression of the gene [26]. Although for the pea CSS1 such investigations are still missing, the light induction of the major stromal Hsp70 is probably a common feature among plants.

Using polyclonal Hsp70 antibodies, a second 75-kDa stromal Hsp70 homologue beside the 78-kDa Css1 was detected in the stroma of pea chloroplasts. The expression level of this Hsp70 is not affected by heat treatment [20]. In line with the presence of multiple stromal Hsp70 proteins, three open reading frames of Hsp70 homologues with a predicted chloroplast-targeting signal were found in the *A. thaliana* genome (Table 32.1) [13]. The implications of the presence of multiple stromal Hsp70 remain to be clarified.

In addition to the stromal Hsp70s, two outer-envelope membrane–localized Hsp70 proteins are also described in pea chloroplasts [15, 16]. The one localized on the cytosolic site of the outer membrane of chloroplasts, Com70, was not induced upon heat stress [16]. Com70 is thought to be involved in protein translocation [27], which might explain the missing heat sensitivity. Similarly, the expression of another Hsp70 involved in protein translation across the outer envelope of pea chloroplasts is also not induced upon heat treatment. This chaperone is localized in the outer-envelope membrane and faces the intermembrane space [15].

32.2.1.2 **The Co-chaperones of Chloroplastic Hsp70s**

The ATP-dependent cycle of substrate binding and releasing by Hsp70 proteins is regulated by co-chaperones. The best-known Hsp70 homologue, DnaK of *E. coli*, requires DnaJ, a so-called J-protein, and GrpE as co-chaperones. DnaJ stimulates ATPase activity of Hsp70, which leads to a strong binding of Hsp70 to its substrate. The stimulating activity is mediated by the interaction of the characteristic J-domain of DnaJ proteins with the ATP-binding pocket of Hsp70 homologues. For substrate release, ADP is exchanged for ATP, which is catalyzed by GrpE proteins [11]. Since stromal Hsp70s show high homology to prokaryotic Hsp70 such as the DnaK of *Synechocystis*, co-chaperones from prokaryotic types are expected in the stroma. Indeed, use of polyclonal DnaJ antibody led to detection of a stromal-localized J-protein in isolated pea chloroplasts [28], which was subsequently termed PCJ1. The nuclear-encoded 431-amino-acid PCJ1 consists of an N-terminal

J-domain as well as a glycine- and cysteine-rich region in the C-terminal region [28]. The glycine- and cysteine-rich region is proposed to be involved in substrate binding. The identity of the interacting Hsp70 is not yet known. Another J-protein, called Toc12, was identified in the outer envelope of chloroplasts [29]. The 103 amino acids of Toc12 comprise an N-terminal membrane anchor and a C-terminal J-domain facing the intermembrane space [29]. The J-domain of Toc12 is capable of stimulating DnaK ATPase activity. Furthermore, computer-supported structural analysis predicts the formation of a disulfide bridge in order to provide a functional J-domain [29]. The outer-envelope membrane–localized intermembrane space-facing Hsp70 was identified as the interaction partner of Toc12.

Despite the identification of nine DnaJ proteins in the *A. thaliana* genome with predicted chloroplast localization (Table 32.1), only one stromal J-protein is characterized in chloroplasts [30]. The nuclear-encoded 125-amino-acid AtJ11 comprises only a J-domain. AtJ11 can stimulate the ATPase activity of DnaK twofold. However, AtJ11 does not stimulate the refolding activity of DnaK [30]. The protein is strongly expressed in roots, stems, flowers and siliques, whereas the expression level in leaves is rather low compared to the other tissues [30]. However, the function and interaction partner remain to be investigated. Remarkably, DnaJ co-chaperones of the major stromal Hsp70s in pea, spinach, and *Chlamydomonas reinhardtii* are not identified.

The second class of co-chaperones of Hsp70 proteins is the family of GrpE homologues. Two open reading frames of GrpE homologues with predicted chloroplast localization were identified in the *A. thaliana* genome (Table 32.1). Using antiserum against *E. coli* GrpE, a GrpE homologue of 19 kDa was detected in a stromal fraction of pea [28]. Further analyses are still missing. In *Chlamydomonas reinhardtii* a 26-kDa GrpE homologue named CGE1 was co-immunoprecipitated by Hsp70B antibodies [21]. Hsp70B and CGE1 form complexes of 120 kDa and 230 kDa. It is assumed that one Hsp70B together with two CGE1 build up the 120-kDa complex, since CGE1 was found to form dimers and tetramers. Therefore, it was further suggested that two Hsp70 and four CGE1 proteins form the 230-kDa complex [21]. The nuclear-encoded CGE1 is localized in the stroma, with a minor fraction associated with membranes [21]. Further sequence analysis revealed that this protein exists in two isoforms, CGE1a and CGE1b, which is longer by two amino acids. The isoforms emerged from differential RNA splicing. Interestingly, CGE1a is predominantly transcribed in response to light stress and under non-stress conditions, whereas under heat stress the transcript level of CGE1b is higher than that of CGE1a [21]. The transcript level of both isoforms together is enhanced 16 times during heat stress and twofold upon light stress. The functionality of both CGE1 isoforms was further demonstrated by complementation of a GrpE-deficient mutation in *E. coli* [21].

32.2.2
The Chaperonins

Chaperonins are divided into two classes. The eubacterial, mitochondrial, and chloroplastic chaperonins form the first class. They require GroES homologues for proper function. The second class includes eukaryotic cytosolic and archaebacterial chaperonins, which act independently of GroES proteins [31]. The best-characterized chaperonin is the GroEL/GroES machinery of eubacteria. It is involved in the folding of approximately 10–15% of the newly synthesized proteins after the folding reaction by the Hsp70 system [31]. GroEL forms two heptameric rings that enclose a central cavity. This barrel-like structure can enclose proteins up to 60 kDa to provide an appropriate environment for folding. The ATP-dependent folding reaction is regulated by the transient association of a homoheptameric ring of GroES proteins [31].

Chloroplastic Cpn60 was the first identified chaperonin. It was found binding to newly synthesized large subunits of Rubisco in isolated pea chloroplasts before holo-enzyme assembly of Rubisco occurred [32]. Cpn60 proteins share high homology to cyanobacterial Cpn60s, which further supports the notion of the endosymbiotic origin of chloroplasts [33–35]. Sequence analysis showed that Cpn60s are highly conserved among plants [7]. Cpn60s of higher plants are nuclear-encoded and are translated in the cytosol with an N-terminal transit peptide for posttranslational import [7]. So far only one plastome-encoded Cpn60 was identified, namely, a Cpn60 in the cryptomonad *Pyrenomonas salino* [33].

Cpn60s were found at a concentration of 90 µM in the chloroplast stroma of pea chloroplasts [36]. Surprisingly, there are two Cpn60 isoforms present in a 1:1 ratio [36], which is unique in comparison to non-plastid chaperonin systems. These isoforms, the 61-kDa αCpn60 and the 60-kDa βCpn60, form mixed tetradecameric complexes in vitro and in situ [37, 38]. The amino acid sequences are 50% identical [39]. In the *A. thaliana* genome, six open reading frames with predicted chloroplast localization were found (Table 32.1), indicating the presence of further Cpn60 isoforms. Furthermore, αCpn60 has 46% and βCpn60 has 50% sequence identity to the bacterial homologues [7]. Cpn60 was found to be heat-inducible in barley [40] but not in pea [35]. A crucial function of βCpn60 is further indicated by observed pleiotropic effects in transgenic plants with low βCpn60 expression level [41].

Since plastid Cpn60 requires co-chaperones for proper refolding of Rubisco, size-exclusion fractions of stromal extract from *Rhodospirillum rubrum* were incubated with GroEL, ATP, and bacterial Rubisco. Refolding of the bacterial Rubisco was attributed to a 55-kDa fraction, which contained a 24-kDa protein interacting with GroEL in the presence of ADP, subsequently identified as Cpn20 [42]. Cpn20 is widely distributed among mono- and dicots, mosses and liverworts [43]. The spinach Cpn20 is localized in the stroma and forms tetramers in vitro and in situ. These tetramers stimulate refolding of citrate synthase by GroEL in vitro [44]. This observation suggests that the tetramer is the functional form of Cpn20. In line with this proposal is the observation that from multiple oligomers of recombinantly expressed Cpn20, only one oligomer is competent to form a complex in an ATP-dependent manner with the *E. coli* GroEL [45]. The protein comprises two GroES domains organized in tandem. The N-terminal GroES-like domain shows prokaryotic origin, whereas the C-terminal GroES-like domain is more similar to eukaryotic chaperonins [42]. Each of the domains alone is able to replace GroES in order to support phage assembly, but each failed to stimulate Rubisco refolding by GroEL in vitro [43]. Whether the double-sized GroES homologue Cpn20 refers to a specific requirement of Cpn60 complexes composed of two different subunits remains to be investigated.

In addition, a normal-sized GroES-like protein, Cpn10, was detected in the thylakoid lumen of pea [46]. A later report describes a Cpn10 with putative plastid localization in the EST database of *A. thaliana*, tomato, and soybean [47]. Performing immuno-gold labeling of chloroplasts of leaves from a transgenic *A. thaliana* plant reveals a stromal localization of Cpn10 with a thylakoid-associated population. Whether this protein is homologous to the previously described thylakoid lumen Cpn10 of pea chloroplasts is still not known. In *A. thaliana* two open reading frames encode for Cpn10 proteins with predicted plastid localization (Table 32.1). Sequence analysis reveals that Cpn10 is largely different from other GroES homologues, including Cpn20 [47]. The stromal Cpn10 forms heptameric structures, which can stimulate the refolding activity of *E. coli* GroEL in vitro [45, 47]. Cpn10 seems to be expressed in later stages of plant development than is Cpn20, and its expression is less stimulated by light in comparison to Cpn20. Northern blot analysis revealed that Cpn10 is present in leaves and stems but not in roots [47]. The expression pattern of the chloroplastic GroES-like proteins indicates a differential requirement of the two GroES homologues in plastids.

The requirement of different Cpn10 homologues within chloroplasts is not yet understood. It might be connected to the expression of various Cpn60s in chloroplasts. In the *A. thaliana* genome, two open reading frames coding for αCpn60 and four open reading frames coding for βCpn60 are reported. EST clones are available for only one αCpn60 and three βCpn60s (Table 32.1) [47]. Furthermore, there is also a report about a thylakoid lumen–localized Cpn60 with an isoelectric point different from that of the stromal isoforms [46]. However, apart from the expression of different Cpn60s, different arrangements of the Cpn60 subunits in the tetradecameric complex remain possible, which is implicated by the results of various studies of complex formation and function (Figure 32.2) [38, 48]. Cloney and coworkers [48] demonstrated tetradecameric complex formation of chloroplast Cpn60s of *Brassica napus*, which were overexpressed in *E. coli*. Whereas expression of αCpn60 alone failed to form such complexes, expression of βCpn60 alone, as well as of both Cpn60 isoforms together, leads to complex formation (Figure 32.2). These results implicate that αCpn60 requires βCpn60 for complex formation but not vice versa. Functional analysis of Rubisco folding of these complexes revealed different requirements for GroES [49]. Thus, *E. coli* strains expressing βCpn60 are capable of Rubisco refolding independent of enhanced expression of GroES, whereas expression of both Cpn60s requires increased GroES expression for efficient Rubisco refolding. Due to missing complex formation, expression of αCpn60 did not lead to a refolding of Rubisco [49]. Similar observations were made for in vitro reconstitution of purified Cpn60s of pea chloroplasts [38]. αCpn60 failed to reconstitute in tetradecameric complexes, whereas βCpn60 and a mixture of both Cpn60 proteins lead to a successful reconstitution. It was found that the complex formation of a mixture of both Cpn60s is more enhanced by GroES proteins such as the bacterial GroES, mitochondrial Cpn10, or plastid Cpn20 in an ATP-dependent manner than is the complex formation by βCpn60 alone [38]. That the mixed Cpn60 complexes are approximately two times more

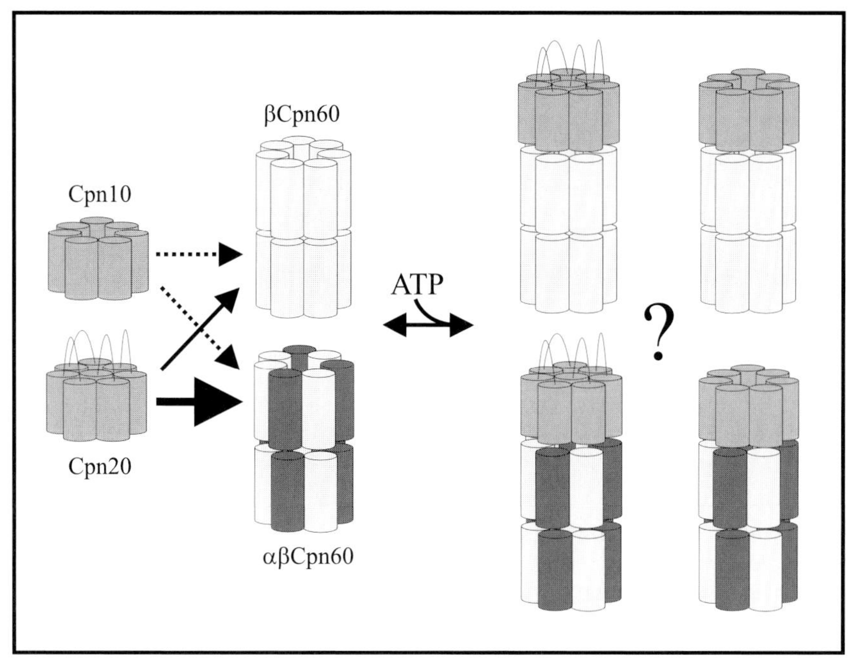

Fig. 32.2. The chaperonin system in chloroplasts. In line with the prokaryotic chaperonin structure of GroEL/GroES, the chaperonin of chloroplasts has to be assembled by a small and a large subunit. Within the chloroplasts, two homologues of the GroES, namely, cpn10 and cpn20, as well as two homologues to GroEL, namely, αCpn60 and βCpn60, exist. Whereas βcpn60 can assemble a homoheptameric complex, αCpn60 requires βCpn60 for complex formation. A heptameric formation of Cpn10 and a tetrameric structure of Cpn20 are assumed (left side). Therefore, there exist several possibilities for the ATP-dependent assembly of the chaperonin complex. However, assembly with Cpn60 homo- or heteroheptamers was observed only for the Cpn20 tetramer. For Cpn10, similar investigations are still missing.

dependent on the presence of GroES explains the observed complex formation in *E. coli* with overexpressed GroES [49]. Complex formation requires ATP hydrolysis, since the addition of the non-hydrolysable ATP analogue AMP-PNP inhibits complex formation in vitro [38]. In contrast, for βCpn60 the presence of endogenous GroES was sufficient for complex formation [49]. In electron micrographs no difference between the architecture of the two complexes was determined, suggesting that the refolding activity of both complex types is based on similar structures [38]. For refolding analysis the dimeric Rubisco from *Rhodospirillum rubrum* was used, since the refolding of the hexadecameric Rubisco of higher plants by chaperonins was not accomplished in vitro. The refolding of acid-denatured Rubisco of both complexes was dependent on the presence of diverse GroES homologues and ATP. The refolding activity was three to ten times more efficient when using the mixed complexes compared to the complexes composed only of βCpn60. The latter refolds Rubisco only in the presence of the mitochondrial Cpn10, but not in the presence of the chloroplast Cpn20 or the bacterial GroES [38]. Whether the assembly of a tetrameric complex by βCpn60 subunits alone reflects an in vitro artifact or is also present in the stroma remains to be clarified. The latter possibility provides an explanation for the presence of two different GroES homologues in chloroplasts and points to Cpn10 as a potential co-chaperone (Figure 32.2).

The molecular basics for the folding activity are well documented for the bacterial GroEL/ES machinery [31]. Much less is known about the plastid Cpn60. Electron micrographs and the tetrameric structure of Cpn60 complexes [35, 38, 48] suggest a mechanism similar to that described for GroEL/ES. This is further supported by the exchangeability of the GroES homologues within the bacterial, plastid, and mitochondrial chaperonin systems [38, 43]. The complex containing Cpn60s and Cpn20 is formed in the presence of ADP and forms asymmetric, bullet-shaped particles [35] that are remarkably similar to the structure of the bacterial GroEL/GroES system [5, 50]. It is proposed that such a complex consists of two rings of seven Cpn60 subunits and two rings formed by Cpn20 [35]. Due to the tandem organization of Cpn20, there has to be a difference from GroEL/ES complexes, which are composed of seven GroES subunits [11]. Accordingly, the stoichiometry of a complex composed of GroEL and Cpn20 was determined to be 14:3.6 [44]. Therefore, it seems likely that four Cpn20s form one ring. In the absence of nucleotides and co-chaperones, the Cpn60 complex reveals a symmetrical structure and is most likely composed of two heptameric rings (Figure 32.2) [35]. The folding activity of the bacterial homologues is dependent on the dynamic association and dissociation of the heptameric GroES ring to the barrel-like GroEL tetradecamer [11]. For refolding of acid-denatured Rubisco by purified chloroplast Cpn60 complexed with Cpn20, a half-time of 3 min was determined [35]. This rate is comparable to that of GroEL and GroES under similar conditions [51]. It was further observed that the assembly of Rubisco holo-enzyme proceeds through intermediates of different sizes in an ATP-dependent manner [6]. There were two particles of 7S and 29S described that contain both Cpn60 and the large subunit of Rubisco. The 29S particle is stabilized by ATP depletion and dissociates upon ATP addition, which leads to the 7S particles. At low ATP levels the 7S particles are competent for Rubisco assembly, which ends in the holo-enzyme formation of Rubisco [6]. It remains to be investigated whether the 7S particles present a building block of the 29S particles when substrate proteins are bound. Interestingly, the ATP-dependent refolding activity of Cpn60 complexes was potassium-independent [35]. This would be in contrast to other described chaperonin systems. However, the authors could not rule out that the purified proteins had already-bound potassium cations [35]. Further structural analysis of the chaperonin complexes and their structure might help to explain the specific features of chloroplast chaperonins.

32.2.3
The HSP100/Clp Protein Family in Chloroplasts

Hsp100 proteins fulfill multiple functions in the cell. In different prokaryotic and eukaryotic organisms [52], their expression is induced under extreme stress conditions such as starvation, heat shock, salt, oxidative stress, glucose/oxygen deprivation, and ethanol or puromycin treatment. Furthermore, Hsp100 homologues have been found to regulate gene expression and are involved in the virulence of *Leishmania* [53]. The functions of these proteins are based on their ATPase activity and formation of ring-structured hexamers, which are both mediated by the nucleotide-binding regions [53]. In this competent complex stage, Hsp100/Clp proteins resolubilize protein aggregates for proteolysis, reactivation, or regulation. The first biochemically characterized member was ClpA of *E. coli*, a component of an ATP-dependent protease complex, which was able to digest casein in vitro [54, 55]. The Hsp100/Clp family is divided into two classes: the feature of the proteins of the first class – ClpA, ClpB, ClpC, and ClpD – is the presence of two ATP-binding domains. ClpM, ClpN, ClpX, and ClpY of the second class comprise only one ATP-binding domain [53]. Structural organization and consensus features distinguish the subfamilies. Class 1 proteins are further characterized by the size of the region between the two nucleotide-binding regions [53].

In chloroplasts two Hsp100/Clp homologues were identified, namely, ClpC and ClpD, also named Erd1 [56, 57]. ClpD is unique for plants, whereas ClpC is also described for gram-positive bacteria and cyanobacteria [52]. In *A. thaliana* eight open reading frames for ClpA proteins with predicted plastid localization (Table 32.1) have been found. Two genes encode for ClpC, one for ClpD, and one for a ClpC-like protein (ClpS) [58, 59]. The presence of several Hsp100/Clp proteins in one organelle is concomitant with the expression of multiple chloroplast-localized ClpP proteins [60, 61]. Since ClpC and ClpD most likely form the regulatory subunit of a ClpP-protease complex, the presence of multiple hetero-complexes with distinct functions in the chloroplast stroma remains possible. ClpC was first identified as an inner-envelope membrane–associated stromal protein in pea. The 92-kDa protein is nuclear-encoded with an N-terminal transit sequence and imported into chloroplasts posttranslationally [56]. However, more detailed studies of the *A. thaliana* homologue revealed an exclusive stromal localization [62].

Although ClpC and ClpD are constitutively expressed [62, 63], the expression pattern of both chloroplastic Hsp100/Clp proteins reveals some differences. Thus, high mRNA and protein levels of ClpC were found in leaves, whereas rather low levels were described for roots, and for etiolated as well as senescing tissues [64–66]. Therefore, ClpC seems to be involved in protection of the plastid against stress conditions caused by the photosynthesis activity. Furthermore, the *Synechocystis* homologue is highly light- and CO_2-induced, but is not heat-induced [67]. The growth rate of *Synechocystis* correlates positively with the ClpC protein content, indicating an essential role of ClpC for photosynthetic growth [67]. The essential role of ClpC in cell survival is further documented by a nonviable knockout of the gene in *Synechocystis* [67]. In addition, it was not possible to generate a viable antisense plant of ClpC in *A. thaliana* [62].

For ClpD a higher transcript level was found in roots and in etiolated and senescing tissues of *A. thaliana* when compared to photosynthetic tissues [65]. Remarkably, the protein level of ClpD in leaves and stems is higher than in roots [61]. Therefore, the authors suggest a posttranslational regulation of ClpD expression. In agreement with this notion, the ClpD transcript level increased during senescence concomitant with a decrease in protein level [63]. The transcript level of ClpD is strongly enhanced under several stress conditions, including cold, light, or oxidative stress. To date, biochemical studies of ClpD, which could explain the altered protein requirements as suggested by the alteration of the expression levels, are still missing. Moreover, further work is required in order to underline the different requirement of the two Hsp100/Clp proteins in chloroplasts.

Only little is known about the biochemical features of the chloroplast Hsp100/Clp proteins. However, several studies on ClpC implicate their involvement in proteolysis. The *A. thaliana* ClpC together with ClpP in the presence of ATP is able to promote degradation of ^{3}H-methylcasein [62]. Furthermore, atClpC is able to functionally replace ClpA in *E. coli*, indicating a similar function of both proteins [62]. ClpC from spinach was found in several high-molecular-weight complexes when stromal extracts were fractionated by size-exclusion chromatography [68]. Fractions of higher molecular weight contain both ClpC and ClpP, indicating a complex formation of both proteins [68] that is ATP-dependent in the stroma [69–71]. In addition, in the presence of ATP, the ClpC-like ClpS1 is present in a 350-kDa complex together with different ClpP proteins [60]. The structure of the ClpP/ClpC complex reveals two heptameric rings similar to that described for the Clp protease of *E. coli* [72–74]. Therefore, for chloroplast Hsp100 proteins, an involvement in the Clp-protease complex is documented. However, evidence of a chaperone activity of these proteins is still missing.

32.2.4
The Small Heat Shock Proteins

The 15- to 42-kDa sHsps are produced in prokaryotic and eukaryotic cells upon heat stress [75]. In plants, sHsps are the most dominant proteins expressed during heat shock [76]. The important role of the small chaperones in plants is underlined by the presence of a larger variety of sHsps when compared to other eukaryotic organisms. Thus, in the *A. thaliana* genome 19 ORFs coding for sHsp-related proteins were identified [77]. The nuclear-encoded sHsps are divided into six classes. The proteins of the first three classes are distributed in the cytosol and nucleus. Furthermore, the small Hsps of the endoplasmatic reticulum, mitochondria, and plastids form individual classes [76]. Common for all sHsps is the C-terminal alpha-crystallin domain [77]. It consists of two consensus sequences separated by a hydrophilic domain of variable length and forms a beta-sandwich structure of two antiparallel β-sheets [77–79]. This domain distinguishes sHsps from other small heat-inducible proteins [80]. Furthermore, the proposed chaperone activity of sHsps is not ATP-dependent.

Plastid-localized sHsps can be identified by a third consensus motif in the N-terminus. It comprises an amphipathic helix with a 100% conserved hydrophilic site and conserved methionine residues on the hydrophobic site. This feature is characteristic for plastid sHsps of mono- and dicotyledons, whereas two Hsp21s of *Funaria hygrometria* do not share the conserved methionine bristle [81]. Therefore, the authors suggest a phylogenetic division of Hsp21 of monocotyledons and dicotyledons and then branching to mosses [81]. Comparison of the amino acid sequence of *Chlamydomonas reinhardtii* Hsp22 with sHsps of higher plants reveals even lower similarity in this region. Therefore, it is proposed that the methionine bristle evolved during land plant development. Since the amino acid sequences of plastid sHsps are distantly related to *Synechocystis* Hsp16, it is assumed that these proteins originated by gene duplication of a nuclear gene [81].

Pea Hsp21 mRNA is almost not detectable under non-stress conditions, whereas upon heat-stress treatment, the mRNA of Hsp21 increases up to 0.75% of total poly(A)RNA content [82]. Under non-stress conditions only a low protein level of pea Hsp21 was observed [83], whereas it accumulates upon heat stress to >200-fold in leaves and roots, with a half-life of 52 h [84]. Furthermore, overproduction of AtHsp25.3-P in *A. thaliana* provides an increased tolerance against heat and light stress [85]. The detailed mode of sHsps during stress conditions is not sufficiently clarified. Several studies indicate that the PSII complex is one target of sHsps activity. Hsp21 prevents heat-induced damage of PSII in tomato [86].

Pea and *A. thaliana* Hsp21 are localized in the stromal fraction [84], whereas for *Chlamydomonas reinhardtii* a thylakoid grana association and for *Chenopedium album* a thylakoid lumen localization was reported [87, 88]. Upon posttranslational import into chloroplasts and subsequent processing to the mature size, pea and *A. thaliana* Hsp21 forms homo-oligomers of 200–230 kDa and 300 kDa in native polyacrylamide gels, respectively [89, 90]. The size of the complexes is consistent with complexes formed by recombinant-expressed Hsp21 [90]. After heat shock treatment, the 300-kDa oligomer of atHsp21 undergoes a conformational change, which results in a lower mobility in a native PAGE [85]. Not much is known about the structure and function of plastid-localized sHsp complexes. A chaperone activity of recombinant-expressed *A. thaliana* Hsp21 oligomers was documented by the prevention of thermal-induced aggregation of citrate synthase in vitro [91]. Therefore, the chaperone activity is attributed to the complex structure, which is in line with proposed dissociation and reassembly of the complex during substrate protein interaction and binding (Figure 32.3a) [77]. The current knowledge about structure and function of the oligomers derives from the crystal structure of wheat cytosolic Hsp16.9 [92]. The building blocks of a homo-oligomer are dimers assembled in a dodecameric double disk, which is conserved in sHsps (Figure 32.3a) [77, 92]. Such dimers were also found for the chloroplast Hsp21 of pea directly after import [89]. In heat-treated chloroplasts such dimers were immediately used for complex formation, whereas in non-heat-treated chloroplasts, only the dimer form of Hsp21 was detected [89]. These results implicate a requirement of a specific concentration of Hsp21 dimers or Hsp21 substrates in chloroplasts in order to form oligomeric structures (Figure 32.3a). Therefore, the C-terminal heat shock domain is crucial for contact of other subunits and structural maintenance in the oligomer [93, 94]. Additional heat treatment of already heat-stressed tissues leads to the formation of large insoluble granules of MDa size [95]. To the same extent as heat stress granules (HSG) are formed, a decrement of the 230-kDa complexes was observed, suggesting the aggregation of the oligomers to the insoluble HSG (Figure 32.3a). The HSGs are insensitive to detergent, salt, and nucleotide treatment [95]. Such HSGs

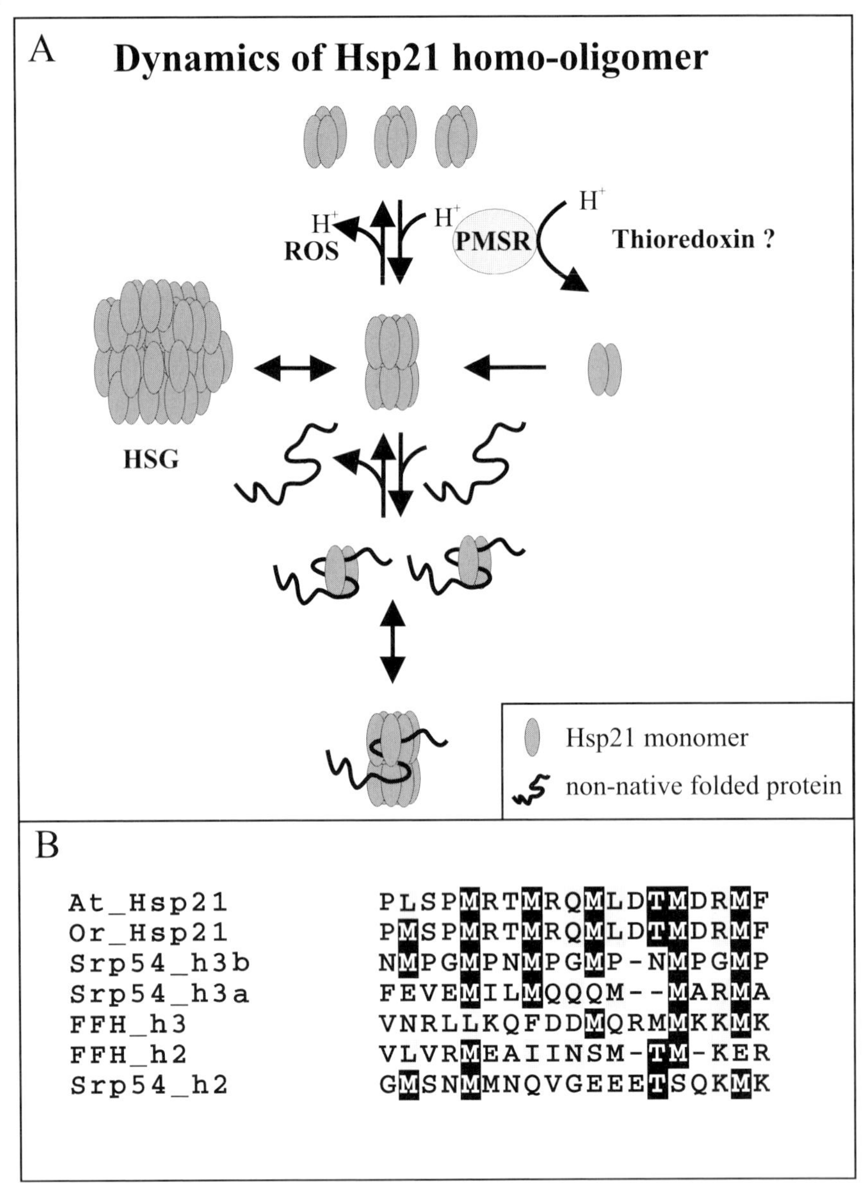

Fig. 32.3. The small heat shock proteins. (A) sHsps form several distinct complex structures in vitro and in situ. The dimeric building blocks assemble in a complex structure, which exhibits chaperone activity by binding of substrate proteins via a cycle of disassembly and assembly (left and lower part). Prolonged stress exposure leads to the formation of nonfunctional high-molecular-weight heat shock granules (HSG). The aggregation is reversible by subsequent cooling (right part). Finally, the active sHsp complex dissociates upon sulfo-oxidation by radical oxygen species (ROS) into tetramers, which lack chaperone activity. The functional complexes can be reformed upon reduction by a stromal thioredoxin-regulated peptide methionine sulfo reductase (PMSR). (B) The methionine-rich

are thought to be formed as long as the refolding capacity of the Hsp70 system is exceeded [96]. According to the documented presence of Hsp70 chaperones in HSG of cytoplasmic sHsps [96, 97], a presence of high-molecular-weight chaperones in chloroplastic HSG remains elusive. Finally, the oligomeric stage of plastid atHsp21 is also affected by sulfo-oxidation of the N-terminal methionine bristle. Sulfo-oxidation leads to a dissociation of the complex in 100-kDa oligomers, which do not exhibit chaperone activity (Figure 32.3a) [91]. Upon reduction of the inactive 100-kDa oligomers by a redox-regulated peptide methionine sulfo-oxide reductase (PMSR), the active complex structures are reassembled (Figure 32.3a) [98].

The chaperone activity is attributed to the N-terminal methionine-rich region [99]. Consistent with this conclusion is the homology of the methionine bristle to methionine-rich regions in Srp54 of yeast and its *E. coli* homologue Ffh (Figure 32.3b) [100], which forms the substrate-binding pocket [101].

32.2.5 Hsp90 Proteins of Chloroplasts

Hsp90 proteins have been described for some eubacteria, cyanobacteria, and eukaryotes. In eukaryotes Hsp90s were found in the cytosol, endoplasmatic reticulum, mitochondria, and chloroplasts [102]. Hsp90s participate in signal transduction, gene expression, and protein degradation [103]. Hsp90s generally are not involved in protein folding but exhibit a preferred recognition of proteins of signal transduction chains as substrate. In the cytosol the ATP-dependent chaperones Hsp70 and Hsp90 cooperate in the folding of nuclear receptors such as the glucocorticoid receptor. Thereby, Hsp90 is thought to receive substrate proteins with almost native conformation from Hsp70 [103]. In contrast, bacterial and endoplasmatic reticulum–localized Hsp90s function independently of co-chaperones [103]. For mitochondrial and chloroplast Hsp90s, analysis of the requirement of co-chaperones is still missing.

The first chloroplast Hsp90 was found in rye and was named, according to its molecular weight, cpHsp82 [104]. It was identified in a cDNA library made of ribosome-deficient plastids isolated from leaves of rye grown at 32 °C. The rye cpHsp82 is nuclear-encoded and synthesized with an N-terminal transit peptide to ensure targeting to chloroplasts. In *A. thaliana* one open reading frame coding for Hsp90 is described (Table 32.1), which contains an N-terminal transit peptide [102]. This protein can be imported in isolated chloroplasts in vitro [105]. Interestingly, for the Hsp90 of rye a higher protein content was found in etiolated and

region of the small heat shock proteins of *A. thaliana* (atHsp21) and *Oryza sativa* (orHsp21) are aligned to the predicted α-helix-forming regions of the M-domain of SRP from *Saccharomyces cerevisae* and its homologue from *E. coli* (FFH). Depicted is an alignment of the methionine-rich domain of *A. thaliana* Hsp21 (88–106); *Oryza sativa* Hsp21 (92–110); helix 2 (363–381), helix 3a (421–437), and helix 3b (459–476) of Srp54 from yeast; and helix 2 (372–388) and helix 3 (413–431) from ffh of *E. coli*. Conserved residues are marked by a black box and homologue residues by a gray box.

chlorotic 70S ribosome-deficient leaves than in photosynthetic active leaves. Furthermore, high Hsp90 protein content was observed in tomato pericarp and in all tomato fruits [104]. In contrast, the plastid-localized Hsp90 of *A. thaliana* reveals a more pronounced expression in leaves and flowers than in roots or stems [105]. High Hsp90 expression levels in reproductive organs and in early post-germination stages indicate a role of the chaperones in plant development. The Hsp90 mutant cr88 harbors a point mutation in a conserved motif, which is probably involved in dimerization and substrate recognition [106–108]. The chlorate-resistant mutant plants grow slower than wild-type plants and exhibit long hypocotyls in red light [105]. Micrograph images revealed less developed chloroplasts in cotyledons and young leaves. Furthermore, light-induced genes such as the gene encoding for the chlorophyll-binding protein show reduced expression. Interestingly, the expression of Hsp90 is light-induced. Based on these observations, the authors proposed an involvement of the plastid Hsp90 in photomorphogenesis [105]. The expression of rye and *A. thaliana* Hsp90 is heat-inducible [104]. However, the function of Hsp90 during heat stress and the role of Hsp90 during plant development or in cell survival against heat shock still are not understood.

The evolutionary origin of chloroplast-localized Hsp90 is still under debate. Based on sequence homology, Schmitz and coworkers [104] suggested a eubacterial origin of cpHsp82. Support comes from the identification of an Hsp90 in *Synechocystis* since it is assumed that chloroplasts derive from a cyanobacterial ancestor. In contrast, when performing signature search, traditional phylogenetic analysis, and statistical tests, Emelyanov [102] postulated that the chloroplast Hsp90 from rye and *A. thaliana* derives from endoplasmatic reticulum homologues.

32.2.6 Chaperone-like Proteins

32.2.6.1 The Protein Disulfide Isomerase (PDI)

PDIs are predominantly found in the ER lumen, where the proteins catalyze the formation, reduction, and isomerization of disulfide bonds during protein folding [109]. At high concentrations, the endoplasmatic reticulum PDI was reported to fulfill a chaperone-like function, distinct from its disulfide bond interaction, such as the inhibition of aggregation of hydrophobic proteins [110, 111]. The only known chloroplast-localized PDI, RB60, was identified in *Chlamydomonas reinhardtii* [112]. It exhibits high homology to endoplasmic reticulum–localized PDIs and contains a C-terminal ER retention signal, KDEL [112]. RB60 is translated as a precursor form and can be imported in isolated pea and *Chlamydomonas reinhardtii* chloroplasts [113]. RB60 partitions between stroma and thylakoids in chloroplasts [113]. The protein is part of a complex that regulates the expression of the psbA gene by binding to the 5′ untranslated region [114]. The complex is composed of four components: RB38, RB47, RB55, and RB60. The complex regulates the association of polysomes to psbA mRNA and thus translation of the gene [114–117]. Binding of the complex is regulated by two distinct mechanisms. ADP-dependent phosphorylation of RB60 inhibits complex binding at ATP levels attained in chloroplasts only in the dark [118]. Furthermore, the activity of the complex is regulated by dithiol reduction; in vivo, it is probably mediated by thioredoxin [114]. As a target site for redox regulation, vicinal dithiol groups of RB60 are assumed [119]. Thus, reduction of RB60 in the light – most likely by the ferredoxin-thioredoxin system – results in the binding of the complex to the 5′ untranslated region of psbA mRNA and leads to its translation [120]. In line with this proposal, RB60 was found to catalyze the binding of RB47, a protein with high homology to poly (A)-binding proteins [116, 117], to the 5′ untranslated region of psbA mRNA [121]. Thus, the known PDI of chloroplasts participates in the regulation of the expression of a central component of the PSII in response to light and dark conditions. Despite the presence of Rb60, a protein-folding function for PDIs in chloroplasts has not yet been identified.

32.2.6.2 The Peptidyl-prolyl *cis* Isomerase (PPIase)

PPIase catalyzes the interconversion of *cis*- and *trans*-rotamers of peptidyl-prolyl amide bonds [122, 123]. This rotamase activity accelerates the folding of proline-containing proteins in vitro and apparently in vivo [124]. The *cis-trans* isomerization is the rate-limiting step in the folding pathway of proteins. Protein-folding studies with carbonic anhydrase further indicate a function of PPIase as chaperone [125]. PPIases are also referred to as immunophilins in regard to their binding to immunosuppressive drugs. Immunophilins are divided into two classes corresponding to their interacting drug: cyclophilins bind to cyclosporin A (CsA), while FKBPs (FK506-binding protein) bind to FK506 and rapamycin. The cyclophilin A-CsA and FKPB-FK506 complexes interact and thus inhibit the activity of calcineurin and thereby block the Ca^{2+}-dependent signal transduction pathway in a variety of cells, including human T-lymphocytes [126–128]. PPIase are widely distributed from bacteria to mammals and have been found in all subcellular compartments [129, 130]. Although in the *A. thaliana* genome 51 open reading frames coding for immunophilins have been found [131], only little is known about the function of these proteins.

In chloroplasts of several higher plants including spinach (*Spinacea oleracea*), fava bean (*Vicia faba*), pea (*Pisum sativum*), and *A. thaliana*, PPIase activity has been described [131–135]. Low-molecular-weight immunophilins have been reported for chloroplasts, and cyclophilins have been found in the stroma and the thylakoid lumen. For the latter, FKBPs and cyclophilins are described [132]. In line with the diversity in the *A. thaliana* genome, four open reading frames for FKBPs and five for cyclophilins with predicted plastid localization have been found (Table 32.1). It was demonstrated that the stromal-localized PPIase of pea or fava bean chloroplasts can be blocked by CsA and therefore belongs to the cyclophilins. Subsequently, the protein of *Vicia faba* was named cyclophilin B [133, 134]. The rotamase activity of cyclophilin B exhibits a low substrate activity. The combination of CsA and cyclophilin B inhibits phosphatase activity of bovine calcineurin, confirming cyclophilin B as a member of the immunophilins [134]. The cDNA clone coding for cyclophilin B of 248 amino acids was isolated and exhibits an

N-terminal putative transit peptide of 65 amino acids [133, 134]. In *A. thaliana* a stromal-localized cyclophilin named Roc4 was reported [136]. A cDNA clone of Roc4 has 248 amino acid lengths with an N-terminal transit peptide predicted to be 82 amino acids in length [136]. Roc4 is synthesized in the cytosol and after import into chloroplasts is processed to the mature size. The presence of a CsA-sensitive PPIase in a stroma extract of *A. thaliana* chloroplasts has been documented. Whether this activity is based on Roc4 remains to be investigated [136]. Furthermore, Roc4 reveals only 66% sequence identity to human cytosolic cyclophilin, which is lower than for most plant cyclophilins. These features lead to the proposal that Roc4 belongs to a new class of cyclophilins [136]. Therefore, it can be assumed that at least one immunophilin is present in the stroma of chloroplasts of higher plants.

The expression of cyclophilin B and Roc4 is light-induced [133, 134, 136]. Cyclophilin B is also heat-inducible [133, 134]. Northern blot analysis revealed that Roc4 is expressed in leaves and stems but not in roots [136]. In line with this finding, the cyclophilin B protein content was significantly higher in leaves than in roots [133]. These data suggest a pronounced requirement of chloroplasts for PPIase in the stroma in photosynthetically active tissues. However, additional biochemical characterizations are required to identify the substrate proteins and therefore the function of stromal cyclophilins.

CsA-sensitive PPIase activity has been reported for a thylakoid fraction of isolated pea chloroplasts [132]. Since chlorophyll interferes with the PPIase assays, the PPIase activity of the thylakoid fraction was determined after removal of chlorophyll by solubilization and subsequent purification using anion-exchange chromatography [132]. Later studies identified two more cyclophilins in the thylakoid lumen of spinach chloroplasts, TLP40 and TLP20 (thylakoid lumen protein). Both were CsA-sensitive and therefore belong to the family of cyclophilins [135, 137]. Luan and coworkers [133] identified an immunophilin from the rapamycin-binding type in the thylakoid lumen of isolated chloroplasts of *Vicia faba*, which was subsequently named FKBP13. Investigation of its homologue in *A. thaliana* revealed the existence of a bipartite signal present in the cytosolic-synthesized precursor form. The protein was further identified to facilitate the ΔpH pathway [131].

In contrast to stromal cyclophilins, some functional details have been explored about thylakoid lumen–localized immunophilins [131, 135, 137]. The expression of FKBP13 in *Vicia faba* is strongly light-induced, and the expression pattern indicates a restriction to photosynthetic tissues since no FKBP13 was found in roots or in etiolated plants [133]. This observation argues for a requirement of FKBP13 in chloroplasts connected to photosynthesis. Indeed, when using a two-hybrid screen and in vitro protein-protein interaction assays, the cytosolic and intermediate stromal precursor forms of FKBP13 of *A. thaliana* were found to interact with both the precursor and mature form of the Rieske subunit of the cytochrome *b* complex [131]. However, further examinations are required to verify this notion. Furthermore, the function of the mature FKBP13 in the thylakoid lumen remains elusive.

For thylakoid fractions of spinach chloroplasts, the CsA-sensitive PPIase activity has been attributed to TLP20. The authors suggest that TLP20 is the major PPIase and that it catalyzes protein folding in the thylakoid lumen [137]. However, experimental data are missing to support this proposal. The spinach TLP40 was copurified with a strong phosphatase activity out of isolated thylakoids [135]. Tlp40 comprises a C-terminal cyclophilin domain and an N-terminal leucine zipper as well as a phosphatase-binding region. The cyclophilin domain shows only 25% sequence identity to the human cytosolic cyclophilin and exhibits a CyA-insensitive PPIase activity. The 40-kDa protein is nuclear-encoded, with an N-terminal bipartite transit peptide. TLP40 is more abundant in stroma than in grana of thylakoids [135]. The mRNA level is only moderately stimulated by illumination, which is in line with the detection of TLP40 in etiolated plant tissues [135]. According to the phosphatase-binding region of TLP40, an interaction with phosphatase activity was demonstrated by co-immunoprecipitation [135]. This interaction with the thylakoid membrane phosphatase is salt-sensitive [138]. On one site the activity of the phosphatase is inhibited by the interaction of TLP40 in the presence of a substrate peptide for the cyclophilin. On the other site binding of the PPIase inhibitor CyA leads to an activation of the phosphatase activity [138]. Therefore, the thylakoid lumen binding of TLP40 regulates the phosphatase on the stromal site of the thylakoids. Furthermore, high temperature induces a dissociation of TLP40 from the membrane to the thylakoid lumen, which is coincident with the activation of phosphatase [139]. The dephosphorylation of central components of PSII is thought to trigger repair of photodamaged PSII [139]. Thus, TLP40 might be involved in regulation of the turnover of central photosynthetic protein in the thylakoid membrane.

32.3 The Functional Chaperone Pathways in Chloroplasts

32.3.1 Chaperones Involved in Protein Translocation

The great majority of chloroplast proteins are nuclear-encoded, synthesized in the cytosol, and posttranslationally imported into the target organelle. The protein import comprises two major steps, which are also characterized by their specific chaperone requirements. The first part is the cytosolic synthesis of preproteins with an N-terminal transit peptide as well as the transport through the cytosol to the chloroplast surface, which involves cytosolic Hsp70s and a 14-3-3 protein. Subsequently, the second process of the translocation across the outer- and inner-envelope membrane of chloroplasts takes place. The ATP requirement of the translocation process can be divided into three major steps (Figure 32.4). First, the preproteins bind in an energy-independent manner to the chloroplast outer-envelope translocon, which asks for the unfolding activity of molecular chaperones to ensure translocation through the channel. Cytosolic Hsp70s and the outer-envelope-associated Hsp70, Com70, are likely candidates to fulfill this function (Figure 32.4, step 1). Subsequently, translocation across the outer envelope takes place. The low energy

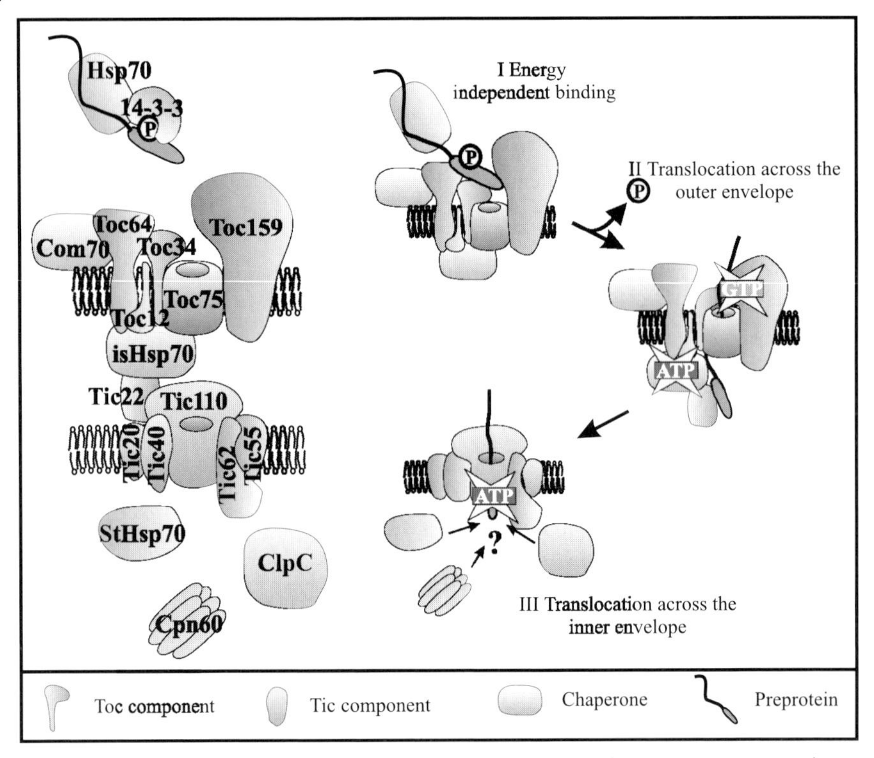

Fig. 32.4. Protein translocation across the envelope membranes. On the left side the molecular architecture of the translocon of the chloroplast envelope membranes is shown. Toc components are marked in blue, whereas Tic components are depicted in red. Chaperones involved in protein translocation are colored yellow. On the right side the translocation process is shown in a three-step model. First the preprotein is bound in an energy-independent way by the Toc receptors Toc34, Toc64, and Toc159 (step1). Upon dephosphorylation of the targeting signal and upon GTP hydrolysis by the GTPases Toc159 and Toc34, the preprotein is translocated across the outer-envelope membrane via the pore-forming Toc75. The translocation event is probably accompanied by the action of the intermembrane space facing Hsp70 via ATP hydrolysis (step 2). For the import in the stroma across the inner-envelope membrane via the Tic translocon, higher ATP levels are required. There are three chaperones suggested to be involved in this translocation process: Cpn60, Hsp70, and ClpC. Their specific role remains to be elucidated (step 3).

consumption for stable association with the Toc machinery is most likely based on the activity of two GTPases, Toc159 and Toc34, as well as of the intermembrane space facing Hsp70 (Figure 32.4, step 2). Finally, the translocation across the inner envelope into the stroma requires higher levels of ATP for the activity of stromal chaperones (Figure 32.4, step 3). However, their identity and docking site on the translocon remains to be clarified.

The involvement of chaperones of the Hsp70 type is most likely mediated by the nature of the transit peptide. Sequence analysis of a set of transit peptides with two different algorithms revealed the presence of at least one binding site for DnaK, the *E. coli* Hsp70 homologue. This feature is shared by >75% of the analyzed transit peptides [140–142]. There is still a debate about the location of the binding site within the transit peptide. Whereas Rial and coworkers [142] predicted binding sites within the central part, Ivey and coworkers [141] suggest a binding site at the N-terminus of the transit peptide. The presence of this predicted binding site is further supported by in vitro binding assays of Dnak and Css1, the major stromal Hsp70 in pea chloroplasts, to the two precursor proteins pSSU (precursor of the small subunit of the Rubisco) and pFNR (precursor of ferredoxin NADP oxidoreductase) in a transit peptide–dependent manner [140–142]. In vitro ATP application causes a dissociation of the Hsp70 homologues from pFNR, revealing characteristics of a substrate interaction of the chaperone [140, 141]. Rial and coworkers introduce mutations into the transit peptide of pFNR, which results in lower binding of Css1 [143]. However, such mutants are efficiently translocated into isolated chloroplasts in vitro. Based on this observation, the authors suggest only a minor role of Hsp70 homologues during protein translocation [143]. But it remains unknown whether Hsp70 binding is limited to the transit sequence or comprises further targets in the mature domain. When using different fusion proteins, Rial and coworkers [142] could not detect any binding of DnaK to the mature part of FNR unless an artificial binding site was introduced. However, studies using preproteins other than pFNR are still missing, and it remains elusive whether this observation can be generalized. On one hand, Pilon and coworkers [144] have demonstrated that an N-terminal deletion within the presequence of ferredoxin results in loss of import, whereas mutations in this region seemed not to affect protein translocation into the organelle. On the other hand, the studies of Rial were performed with a stromal-targeted preprotein, and therefore the binding of Hsp70 proteins to transmembrane regions of integral membrane proteins in order to prevent unspecific aggregation cannot be excluded. Finally, Rial and coworkers focused their investigation on the stromal Hsp70 Css1 and did not address the impact of the outer envelope Hsp70s of chloroplasts (see below).

Cytosolic targeting of precursor proteins to the chloroplast surface fulfils three major challenges. It prevents preprotein accumulation in the cytosol, it holds the precursors in an unfolded, import-competent state, and it warrants that the enzymatic activity of those proteins does not take place in a wrong cellular compartment. Since these demands are similar to chaperone function, the involvement of Hsp70 proteins is reasonable. Currently, three different pathways are proposed. These are (1) the direction of outer-envelope proteins to the chloroplast surface for membrane insertion, (2) the targeting pathway of non-phosphorylated precursor with the aid of unidentified proteins to the translocation machinery, and (3) the guiding to the chloroplast surface by the interaction to a hetero-oligomeric protein complex called guidance complex [145]. In all cases selective protein targeting through the cytosol most likely starts directly at the ribosome during the synthesis of preproteins. A probable common feature is Hsp70 binding during or after translation. Support for this idea comes from import studies of heterologously expressed and urea-denatured pLCHP (precursor of the light-harvesting protein).

The translocation across the envelope membranes is stimulated by leaf extract and by addition of purified Hsp70 and an additional unknown ATP-dependent factor [146]. However, the import of denatured pSSU and pFd was not stimulated by the addition of Hsp70 [147, 148], even though association of pSSU with HSP70 could be identified [149].

A subset of the precursor proteins contains a phosphorylation motif within the N-terminal transit peptide, which is specifically phosphorylated on serine or threonine by a cytosolic kinase [150]. The phosphorylation motif shares similarities to 14-3-3 binding sites. In line with this observation, size exclusion of the post-ribosomal supernatant of freshly wheat germ lysate–translated pSSU resulted in the identification of a so-called guidance complex [149]. This hetero-oligomeric complex consists of 14-3-3, Hsp70, and maybe other so far unknown components. Furthermore, May and Soll [149] demonstrated that the 14-3-3 of the cytosolic guidance complex is different from the mitochondrial import stimulating factor (MSF), another 14-3-3 protein postulated to promote import of a subset of mitochondrial precursor proteins [151, 152]. Without phosphorylation of the transit sequence, pSSU is found only in association with Hsp70 and imports with an efficiency about fourfold less compared to the guidance complex–bound precursor. Moreover, the guidance complex is not essential for pSSU import in vitro but provides a strong stimulating effect [149]. Whether the guidance complex is responsible for targeting of the preprotein to the receptor or whether the increase of the efficiency of the import process results from bypassing early steps in translocation is still unclear. Furthermore, the docking site of this hetero-oligomeric complex and the stage and mode of guidance complex dissociation remain to be investigated. In summary, the cytosolic targeting involves a 14-3-3 protein and Hsp70 homologues. Both proteins are thought to stimulate trafficking to the chloroplast surface. The Hsp70 system keeps preproteins in an unfolded state to ensure their transfer through the translocation channel. The binding of chaperones to integral membrane proteins is most likely based on the prevention of aggregation by the association of hydrophobic stretches.

The protein translocation process across the two envelope membranes comprises different steps, which are distinguished by their energy requirement. The guided preprotein is recognized by the chloroplast outer-envelope translocon in an energy-independent manner. Toc64, a dynamic associated receptor component, is suggested as a docking site for the guidance complex [153]. The consumption of low ATP levels results in a stable association of the preprotein with the translocon. The two outer-envelope membrane Hsp70 proteins are discussed as a possible source of ATP requirement. However, due to the presence of nucleotide kinases, a conversion of ATP to other nucleotide triphosphates remains possible. Therefore, the energy consumption for stable preprotein association might also reflect the activity of the two GTPases of the Toc translocon.

The translocation apparatus of the outer-envelope membrane is accompanied by two molecular chaperones, namely, Com70 on the cytosolic side [27] and another Hsp70 facing the intermembrane space [15, 154, 155]. Com70 is exposed to the cytosolic site of the outer envelope since it is degraded upon thermolysin treatment

[27]. Hints for its involvement in protein import come from its cross-links to preproteins at the early stage of import [27, 156]. The function of Com70 is still not understood. It might prevent back diffusion of preproteins into the cytosol or it might be involved in protein unfolding prior to translocation. Furthermore, Com70 could keep preproteins in an unfolded state to facilitate precursor passage through the translocation channel.

The second outer-envelope Hsp70 persists in the membrane upon high salt and alkali treatment, indicating an integral membrane character [15]. Thermolysin treatment of intact outer-envelope vesicles did not result in degradation of the chaperone. However, after solubilization of the membranes by Triton X-100 treatment, the protein became susceptible to thermolysin, indicating an orientation to the intermembrane space [15]. The similarity to other eukaryotic Hsp70 homologues shown by partial sequencing and the detection by Hsp70 antiserum against Hsp70 of human HeLa cells underline the identity of the protein as a molecular chaperone [155]. The association of Hsp70 with trapped precursor proteins and the binding to purified pSSU indicate an involvement in protein translocation [29, 155]. Furthermore, co-immunoprecipitation of Toc159 with antiserum against Hsp70 in the presence of a bound preprotein shows an interaction with the translocon of the outer-envelope membrane [154]. Interestingly, the intermembrane space Hsp70 was found as part of an intermembrane space complex. This complex is composed of Toc64, the inner-membrane-associated Tic22, and a J-domain-containing Toc12 [29]. However, a direct interaction with pSSU was demonstrated for Toc64 and Tic22 in vitro. Therefore, Toc12 probably recruits the Hsp70 protein to the complex, since no interaction between Toc12 and pSSU was observed in vitro [29]. The function of this complex remains elusive. However, it seems likely that the complex keeps preproteins in an unfolded state to facilitate translocation across the inner membrane and to prevent protein aggregation in the intermembrane space. It is argued that this pathway is specialized for a certain subset of preproteins, since the components of the intermembrane space complex are less abundant than the components of the core complex.

The translocation process across the inner-envelope membrane of chloroplasts is assisted by chaperones as well. Several chaperones are discussed for association of the inner-envelope translocon, such as ClpC, Cpn60, and a stromal Hsp70, named S78 [157–159]. Cpn60 was co-immunoprecipitated by antibodies against the major component of the inner-membrane translocon, Tic110, in the presence of preproteins [157]. Therefore, a transient interaction of Cpn60 to the translocon for folding of the incoming preprotein is proposed [157]. However, Cpn60 was identified as a Rubisco assembly factor (Section 32.3.3), which raises the question of whether the identified association of the chaperonin might be specific to the bait used in this experiment. In contrast, ClpC (Hsp 100) was also found in a cross-linked complex in the presence of a preprotein [158, 159]. This observation leads the authors to the proposal of ClpC being associated with the Toc-Tic super complex [159]. However, both Cpn60 and the Hsp100 molecule were identified after isolation of the translocon joint by a translocation intermediate of the small subunit of Rubisco, and in vitro analysis of interaction with recombinant proteins is still missing. Furthermore, one has to keep in mind that for the plastid ClpC no chaperone activity has been demonstrated so far. It is only known as a regulatory component of the Clp protease of the chloroplast stroma. Interestingly, the small subunit of Rubisco is rapidly degraded if it is not integrated into a functional complex, a process involving ClpC (Section 32.3.4). This raises the question of whether the association was specific or resulted from the prolonged experimental procedure. Therefore, the functional significance of these data remains to be clarified. In a high-molecular-weight complex after cross-linking and during translocation, the stromal Hsp70, termed S78, was also found. Furthermore, Tic40, a component of the translocon facing the stroma, comprises a C-terminal TPR domain, which is 32% identical to the Hop protein Sti1p. Hop proteins are involved in the formation of chaperone complexes by mediating the interaction between Hsp70 and Hsp90, and deletion of these proteins results in a drastic reduction of growth [160]. Tic40 was found by screening for a homologue of the Cim/Com44 of *Brassica napus* [161] in *Pisum sativum* [162]. Since a stromal exposition of the soluble C-terminal domain is proposed, Tic40 might act as a co-chaperone. However, evidence for such a role of Tic40 is still missing. Surprisingly, a null mutant of Tic40 is not lethal [160], which would have been expected in the case of such a crucial function in protein translocation.

32.3.2
Protein Transport Inside of Plastids

Proteins imported into chloroplasts (see Section 32.4.1) or synthesized in the organelle [163] have to be transported toward their place of function (Figure 32.1, pathway 6). This process is coupled either to the import of the preproteins or to their translation.

For the first case it is still under debate, which chaperone might be involved in the translocation event at the stromal side (Section 32.3.1). The ATP dependence of the translocation of proteins across the inner membrane [164–168] strongly suggests the action of a chaperone. Indeed, one stromal Hsp70, namely S78, was identified in high molecular Toc/Tic complexes [13, 169, 170]. Strikingly, this would assemble the current picture on the final translocation across membranes, since the mitochondrial import or bacterial export system also requires the action of an Hsp70 homologue. However, Hsp100 or Cpn60 were also found to be associated with the translocon (Section 32.3.1). Therefore, it remains under debate, which chaperone is generally involved in protein translocation across the inner envelope.

So far, not much is known about the transport systems within chloroplasts, since only few examples were studied. For the Rieske FeS protein inserted into the thylakoids by the TAT pathway [171] it was observed, that shortly after translocation into the stroma the protein is associated with the Cpn60. During further transport toward the TAT translocon the protein becomes associated with one of the stromal Hsp70 [172]. This would lead to the assumption that Cpn60 mediates the transport from the membrane into the stroma and Hsp70 is involved in the translocation into the thylakoids. However, knowing that Cpn60 and Hsp70 are also involved in folding (Section 32.3.3) and that TAT substrates are only translocated in a folded state it cannot be clearly distinguished between folding and targeting. Investigation of the translocation of a second protein, namely the precursor of the stromal ferredoxin-NADP reductase (FNR) revealed a different mode. pFNR first associates with Hsp70 and only subsequently with Cpn60 upon import into the stroma [173]. Furthermore, in vitro studies suggested that the association with Hsp70 is necessary for processing, whereas association with the Cpn60 homologue GroEL cannot restore processing [174]. These results indicate that Hsp70 might even be involved in the late translocation events of protein import. Even though more studies have to confirm this result, it can be suggested that one Hsp70 is involved in late steps of protein import or early steps of protein transport preceding the processing. Cpn60 might then be required for quality control and or folding and a second Hsp70 in further transport of thylakoid proteins toward the membrane or their storage in the stromal compartment (Section 32.3.5).

However, not only Hsp70 proteins are involved in transport of proteins toward thylakoids. One routing systems was found for the LHCP proteins, which require a signal recognition particle for their transport [175], which is essential and sufficient to target LHCP. However, the influence of an Hsp70 on this process was discussed controversially. On one hand it was suggested that Hsp70 is needed for the insertion process [176]. In contrast, it was demonstrated that addition of Hsp70 to unfolded protein could not restore translocation [177]. It might therefore be that Hsp70 is required for partly unfolding of the incoming protein to keep it competent for recognition by SRP. However, this again would suggest that an Hsp70 isoform is involved in the late translocation/early folding events.

32.3.3
Protein Folding and Complex Assembly Within Chloroplasts

Assembly of the functional form of Rubisco was found to be dependent on the Rubisco binding protein, which subsequently was identified as Cpn60. The first direct evidence for its function in assembly derived from the observation that antibodies against Cpn60 block the Rubisco holo-enzyme assembly [6]. The later identified chaperonin complex is not only essential but also sufficient to drive the formation of the active form of Rubisco in vitro [178]. However, investigations on the dimeric form of Rubisco of *Rhodospirillum rubrum* suggested that Cpn60 is actually involved in the folding of the subunits of the complex rather than in its assembly [178, 179]. In contrast to its bacterial homologue, Cpn60 does not require potassium for its action. An interaction of Cpn60 with a set of proteins after their translocation into the chloroplasts was demonstrated by non-denaturing gel electrophoresis. These proteins also include the monomeric β-lactamase, which does not form any functional oligomers [180]. Later studies further revealed that Cpn60 is involved in folding of ferredoxin and Rieske iron-sulfur protein [172, 173]. Therefore, Cpn60 is involved not only in the folding and assembly of components of multimeric complexes but also in the folding of monomeric proteins.

In addition, urea-denatured CF_1 subunits of the chloroplast F-type ATP synthetase require Cpn60 and Hsp70 for reconstitution in a functional CF_1 complex in vitro [181]. It is thought that first the trimeric complex of the α- and β-subunit is formed and then the γ-subunit is assembled [181]. However, the specific requirement of the chaperone systems in the specific steps remains unknown.

The integration of proteins into the thylakoids (for review see Ref. [182]) and the assembly of protein complexes require the action of chaperones or proteins with chaperone-like function. For example, the assembly of the photosystem II (PSII) requires HCF136. Even though it remains unknown how this protein influences the assembly of this complex, the deletion of HCF136 results in an inhibition of the assembly of the reaction center of PSII [183]. Similarly, the assembly of PSI is mediated by factors including the investigated Ycf3. Ycf3 contains a TPR motif at the amino terminus, which is essential for the assembly of the components of PSI. However, it remains elusive whether the TPR-containing region itself acts as assembly factor or if this region recruits other components of a larger assembly machinery.

32.3.4
Chloroplast Chaperones Involved in Proteolysis

Chloroplast development and biogenesis are strongly related to the adaptation of the proteome on the developmental condition and on removal of misfolded or unassembled proteins. Therefore, it is not surprising that a complex system for protein degradation was identified in the stroma of chloroplasts and that the expression of the components of the system is light- and developmental-dependent (see Section 32.2.3). It could be demonstrated that proteins of the stroma become degraded in an ATP-dependent manner [69, 184, 185]. Investigation of this degradation suggested that ClpP is involved in this process. When ClpP is downregulated, the adaptation to a high concentration of CO_2 is reduced, leading to the conclusion that these proteins are also involved in degradation of fully or partly assembled cytochrome b_6f complexes [185]. However, the proteolytic activity of ClpP is controlled by a variety of cofactors, and several distinct complexes have been identified (Section 32.2.3). The functional complementation of the ClpP from *E. coli* by ClpC cofactors of tobacco [62] suggests that the ClpP/ClpC system in plants has similar functional properties when compared to the *E. coli* system [52]. However, the extra amino acids on the N-terminus of ClpP suggest a more complex system than that found in the prokaryotic ancestor [65]. This is also supported by the observation that ClpP isolated from plastids can degrade the same peptide as its homologue from *E. coli*. One of the best-investigated co-factors for ClpP within chloroplasts is the nuclear-coded ClpC (Section 32.2.3). This protein is involved in the degradation of newly synthesized proteins within isolated chloroplasts [186]. ClpC is also involved in the ATP-dependent degradation of mistargeted and therefore unassembled proteins such as the 33-kDa subunit of the oxygen-evolving complex [69]. Furthermore, the data suggest that this protein is involved in the regulation of the Rubisco stoichiometry [52]. This is in line with the observation that the nuclear-encoded and subsequently translocated small subunit of Rubisco becomes rapidly degraded if not integrated in a Rubisco holo-complex [187]. Therefore, it becomes clear that the identified Clp system is involved in the housekeeping of the proteome. The Clp system could also be involved in protection against cellular damage caused by photo-inhibition by degradation of affected proteins. However, despite the stated observations, not much is known about protein degradation within chloroplasts and its regulation.

32.3.5
Protein Storage Within Plastids

Plastids undergo drastic changes during their transitions from one type to the other. This kind of transition requires a large change of the proteome. Not only the transition of plastids but also developmental or environmental changes require changes in the proteome. One way of adaptation is the increase of protein import for new proteins (Section 32.3.1) and protein degradation of proteins no longer required (Section 32.3.4). However, a further backup system might be the storage of proteins for further use. It could be demonstrated that the typically membrane-associated phytoene desaturase exists in a soluble form in forming chromoplasts. These "stored" proteins are complexed with Hsp70 [188, 189]. After association of the phytoene desaturase with FAD, the complex between the enzyme and the chaperone disassembles and the enzyme is targeted to the membrane surface. The authors discuss that this association prevents the degradation of the phytoene desaturase. Since chromoplasts derive from chloroplasts, massive changes of the ultrastructure are required [190], leading to the formation of up to 50 concentrically stacked membranes [188, 189]. This might require the preceding translocation of nuclear-encoded preproteins and their storage for further use. Even though the phytoene desaturase is the only studied example, it seems likely that chaperones, especially of the Hsp70 class, are involved in the storage of a subset of proteins during plastid transitions in order to prevent those pre-imported proteins from degradation. However, subsequent investigations will have to reveal whether this process is limited to a small subset of proteins with low expression levels, such as the studied phytoene desaturase [188], or whether this is a general mechanism within plastids.

It could also be demonstrated that immunophilins (Section 32.2.6.2) serve as molecular anchors for proteins. For example, the lumenal domain of the Rieske subunit of the cytochrome *b* complex was found to interact with the intermedial, stromal-localized form of FKBP13 [131]. In this way FKBP13 might regulate the content of the Rieske subunit in the thylakoid membrane and further serve as a reservoir of the Rieske subunit for conditions with pronounced requirements of proteins of the photosynthetic electron chain. This is further supported by the observation that gene silencing of FKBP13 by RNAi leads to a 70% increment of the Rieske subunit proteins in the thylakoid membranes [131]. Again, this demonstrates that stromal protein storage is essential for the adaptation to drastic developmental or environmental changes and that chaperones or chaperone-like proteins are involved in this storage.

32.3.6 Protein Protection and Repair

Photodamage of the photosystems can occur under high light conditions. The protection and/or repair of the photosystems are essential for the functionality of chloroplasts. A function of the proteins of the Hsp70 family could already be expected, since the expression of some isoforms was found to be light-inducible. Thus, studies with mutant strains exhibiting a lower expression of Hsp70-B render the photosystem II more sensitive to light-induced damage than the wild type. In contrast, overexpression leads to a higher protection of the PSII [191]. The higher expression leads especially to the stabilization of two proteins of the PSII, namely, D1 and CP43, two proteins highly sensitive to photodamage [192]. Further support comes from the observation that high light treatment induces the formation of a 320-kDa complex containing Hsp70 and the photodamaged but non-degraded D1, D2, and CP47 [193]. This complex was discussed to represent a photosystem II repair intermediate. In line with these observations, 5–25% of the total chloroplast content of Hsp70-B is membrane-associated [194]. Small Hsps were also proposed to be involved in protection from photodamage. It is proposed that small plastidial chaperones can act as scavengers for radical oxygen species (ROS) induced by heat or light stress or might be involved in prevention of protein aggregation. It was suggested that the surface-exposed amphipathic helix of the methionine bristle in the N-terminus can be sulfo-oxidized and might function as scavenger for ROS [91]. This conclusion is supported by the proposed function of methionine residues as antioxidant defense mechanism [195]. The sulfo-oxidation of methionine residues in the N-terminus results in drastic conformational changes of the *A. thaliana* Hsp21 in vitro. Therefore, sulfo-oxidation reduces the amount of surface-exposed amphipathic helices [91]. Sulfo-oxidation under cold stress leads to a formation of a 450-kDa oligomer. In contrast, under heat stress the functional oligomer dissociates upon sulfo-oxidation into nonfunctional 100-kDa complexes, which probably resemble tetramers [91]. These 100-kDa complexes do not possess chaperone activity. Similarly, phosphorylation of cytosolic Hsp27 of mammals leads to tetramer formation, which abolishes chaperone activity [196]. For plastid atHsp21, in vitro reduction of methionines of 100-kDa complexes by the peptide methionine sulfo-oxide reductase (PMSR) leads to the reassembly of functional oligomeric complexes [98]. The chloroplast-localized PMSR is highly expressed in photosynthetic tissues in *A. thaliana* [197]. Its activity is redox-regulated, which is shown by inactivation through sulfo-oxidation and reactivation through DTT treatment [98]. The authors suggest thioredoxin as a reductant of PMSR in vivo [98]. Thus, Hsp21 might play an essential role in protection of the photosynthetic apparatus as an ROS scavenger. However, data providing a direct connection between the in vitro data of sulfo-oxidation and the in vivo observed stabilization of PSII are still missing.

The second parameter that drastically influences the biogenesis of plastids is temperature. Temperature adaptation is largely achieved by alteration of the expression levels of the needed proteins. However, heat stress also induces oxidative stress, and the functionality of the complexes has to be warranted under high temperatures. In line with the observation that small Hsps are involved in protection of the photosystem II, heat stress leads to an association with the grana thylakoids as well [88]. As before, a function in protection of the photosystem II was suggested [87, 198]. Moreover, it could be demonstrated that small Hsps are involved in restoring electron transport in pre-heat-stressed plants [86]. Furthermore, biochemical evidence for chaperone activity of sHsps is provided by the prevention of thermal-induced aggregation of citrate synthase by Hsp21 in vitro [91]. Therefore, Hsp21 is thought to prevent substrate proteins from aggregation in order to keep them competent for the plastid refolding machinery [97]. Plastid sHsps seems to function not only under heat stress but also under cold stress. Thus, in tomato (*Lycopersicon esculentum*) plastid induction of Hsp21 provides cold-stress tolerance [199]. In addition, Cpn60 also seems to be involved in heat-stress adaptation, since a tDNA insertion in the gene of βCpn60 results in accelerated cell death in response to heat stress [200]. The molecular mechanism of this observation, however, remains to be investigated.

32.4 Experimental Protocols

32.4.1 Characterization of Cpn60 Binding to the Large Subunit of Rubisco via Native PAGE (adopted from Ref. [6])

1. Isolation of chloroplasts: Harvest leaves of 9- to 11-day-old peas. Disrupt the tissues with a blender in homogenization medium (20 mM Mops, 13 mM Tris, 0.1 mM $MgCl_2$, 330 mM sorbit, 0.02% BSA). Filtrate the homogenate through four layers mull and one layer gauze. Pellet chloroplasts (1500 g, 4 °C, 5 min) and apply the resuspended pellet on a Percoll (Amersham Bioscience, Upsalla, Sweden) gradient (10 min, 4°C, 7000 rpm). The Percoll gradient consists of a 40% and an 80% layer in 330 mM sorbit, 50 mM MOPS/KOH pH 7.9. Collect the intact chloroplasts on the 40% Percoll layer with a Pasteur pipette and transfer them in 12–14 mL of ice-cold resuspension buffer (50 μM EDTA, 0.2 mM $MgCl_2$, 375 mM sorbit, 35 mM HEPES/KOH pH 8.3, 0.96 mM DTT, 200 μM isoleucine, 200 μM threonine). Centrifuge the mixture (4000 g and brake immediately) and resuspend the pellet in a minimal volume of resuspension buffer. Determine the chlorophyll content according to Arnon [201].
2. *In Organello* protein synthesis: Illuminate isolated chloroplasts for 24 min in the presence of 400 mCi [^{35}S]methionine (> 1000 Ci $mmol^{-1}$). Resuspend the chloroplasts in a 10-fold excess of resuspension buffer and pellet the chloroplasts (4000 g and brake immediately).
3. Lyse the chloroplasts in 10 mM Tris/HCl pH 7.6, 1 mM benzamidine, 1 mM ε-aminocaproic acid, 7 mM β-mercaptoethanol, and 1 mM PMSF.

4. Remove the membranes by centrifugation (100 000 g, 30 min, 4 °C).
5. Divide the cleared lysate into 90 μL aliquots.
6. Bring the aliquot to 50 mM HEPES/KOH pH 7.6, 220 mM KCl, 6 mM $MgCl_2$, 20 mM DTT.
7. Incubate for 30 min at 0°C.
8. Incubate the sample for 60 min at 24°C.
9. Apply the sample on a non-denaturating polyacrylamide gel and run the gel overnight at 4°C.
10. Native PAGE: Prepare all buffers and gel solutions as described [202], but omit SDS. Pour a slab gel of 7.5% (w/v) polyacrylamide and a stacking gel of 4% (w/v) polyacrylamide concentration without the addition of an SDS cocktail. Add glycerol to the sample to a final concentration of 10% and apply it on the gel. Run the gel at a constant voltage of 2.5 mA until the voltage reaches 150–200 V. At this time switch the gel to a constant voltage of 150–200 V and run it for 18–24 h at 4 °C.
11. Soak the gel in EN^3HANCE, dry it, and expose it to X-ray film. Develop the film for analysis.

32.4.2
Purification of Chloroplast Cpn60 From Young Pea Plants (adopted from Ref. [203])

1. Plant material: Plant peas as a dense monolayer in trays with compost. The trays are subsequently incubated at low light (20–40 $\mu Em^2\ s^{-1}$) at 18–22 °C with a 12-h illumination cycle and rich watering. Cut leaves from 10-day-old plants (usually from four standard trays) for protein isolation.
2. Protein extract preparation: Perform all steps on ice or in the cold room. Grind the plants using Polytron homogenizer (usually in 10 portions) with 200 mL of ice-cold buffer (20 mM Tris/HCl pH 7.5, 10 mM $MgCl_2$, 10 mM EDTA, 50 mM β-mercaptoethanol, 50 g L^{-1} PVPP, 1 mM PMSF) per portion. Typical grinding conditions: four bursts of 10 s at the maximal speed. Add immediately prior grinding 50 mM β-mercaptoethanol, 50 g L^{-1} PVPP, and 1 mM PMSF (0.4 mL of 0.5 M solution in methanol per 200 mL). Centrifuge the homogenate in a high-volume rotor and filter the supernatant through eight layers of muslin.
3. Ammonium sulfate precipitation: Add solid $(NH_4)_2SO_4$ to the extract (2 × 1 L) to make 40% saturation. Allow the precipitate to form for 30 min in the cold room on a magnetic stirrer and subsequently centrifuge at 10 000 g for 30 min (Beckman JA10 rotor). Discard the pellets, add solid $(NH_4)_2SO_4$ to the supernatant to 60% saturation, and allow the precipitate to form for an hour. Collect the protein precipitate by centrifugation as described above (supernatant discarded) and dissolve in 1 L of cold column buffer (20 mM Tris/HCl pH 8, 0.1 mM EDTA, 10 β-mercaptoethanol, 0.5 mM PMSF [add before use]). Precipitate proteins by adding $(NH_4)_2SO_4$ to 60% saturation (usually left overnight).
4. Desalting on Sephadex G25: Centrifuge protein precipitate as above (supernatant discarded). Dissolve the pellets in a small volume of the column buffer ($<$ 130 mL resulting solution) and centrifuge at 18 000 g for 60 min (Sorvall

SS34 rotor). Discard green pellets, and pass the supernatant through a column filled with 400 mL Sephadex G-25 (medium). Combine all protein-containing fractions (eluted after the void volume of the column) and centrifuge as above; discard green pellets.

5. Q-Sepharose chromatography: Load the desalted material onto a column containing 300 mL of Q-Sepharose Fast Flow (Amersham Biosciences) equilibrated with the column buffer. Elute proteins in a gradient of 0–0.8 M NaCl (1-L gradient volume). Analyze the fractions by SDS-PAGE (using a gel composed of 15% acrylamide and 0.15% *bis*-acrylamide). Chloroplast Cpn60 elutes after Rubisco (the protein of the highest abundance) and is easily distinguished, as it exhibits characteristic dual bands due to the presence of two polypeptides of very similar size around 60 kDa (which could be resolved only on low cross-linked gels).
6. Fractogel TSK butyl 650 (S) chromatography: Combine fractions containing chloroplast Cpn60 and load onto a column (1.6×15 cm) with Fractogel TSK butyl 650(S) (Merck) equilibrated with buffer containing 20 mM Tric/HCl pH 8, 0.2 mM EDTA, 5 mM β-mercaptoethanol, and 20% of saturation of $(NH_4)_2SO_4$. Apply a decreasing (20% to 0%) $(NH_4)_2SO_4$ gradient followed by isocratic elution (no ammonium sulfate). Chloroplast Cpn60 elutes as a sharp peak soon after the end of the gradient.
7. MonoQ HR5/5 chromatography: Combine the Cpn60-containing fractions and apply the pooled sample onto a MonoQ HR5/5 column. Run a gradient of 0–0.5 M NaCl in 20 mM Tris/HCl pH 8, 5 mM β-mercaptoethanol. Collect the fractions containing Cpn60.
8. Superdex 200 HR10/30 chromatography: Combine the Cpn60-containing fractions from the MonoQ HR5/5 column and apply the sample on a Superdex 200 HR10/30 size-exclusion column. Use 40 mM triethanolamine acetate pH 7.5, 0.1 M potassium acetate, 0.1 mM EDTA, and 1 mM DTT for gel filtration. Concentrate the Cpn60-containing fractions and store them at -20 °C.

32.4.3 Purification of Chloroplast Hsp21 From Pea (*Pisum sativum*) (adopted from [90])

1. Plant material: Peas (*Pisum sativum* L. cv. Little Marvel) are planted on vermiculite and grown for 9 days at 22 °C using a 16-h photoperiod. Before harvest on the ninth day, the growth temperature has to be shifted from 22°C to 38 °C at a rate of 4°C per hour. The plants will then be kept for 4 h at 38°C before the temperature has to be decreased to 22°C at a rate of 4°C per hour.
2. Chloroplast isolation and stroma purification: Leaf tissues from 1 kg plant material (step I) are homogenized in a blender in isolation medium (330 mM sorbit, 20 mM Mops, 13 mM Tris, 0.1 mM $MgCl_2$, 0.02% BSA). The slurry is filtrated through four layers of mull and one layer of gauze. Chloroplasts are recovered by centrifugation at 1500 g for 5 min at 4 °C. The obtained pellet is resuspended in isolation media before loading on a Percoll step gradient composed of 13 mL 40% and 10 mL 80% Percoll in 50 mM HEPES/KOH pH 7.9; 330 mM sorbit. Intact chloroplasts on the interface between the 40% and 80%

Percoll layer are collected, resuspended in wash medium (330 mM sorbit, 50 mM HEPES/KOH pH 7.6, 3 mM $MgCl_2$) and pelleted by centrifugation at 1500 *g* for 5 min at 4°C. The washing step is then repeated. Washed chloroplasts have to be diluted in 10 mM HEPES/KOH pH 8, 10 mM ε-amino-*n*-caproic acid and 1 mM benzamidine to a final chlorophyll concentration of 5 mg mL^{-1}. Removal of sorbit results in lyses of chloroplasts. After incubation for 10 min at 4°C, chloroplast membranes are removed by centrifugation (100 000 *g*, 1 h). Subsequently the supernatant is collected and stored at −20°C for further use.

3. Ammonium sulfate precipitation: Add 0.1 M citrate buffer (one and half volume of the stromal extract) and ammonium sulfate to 40% final concentration to the stromal extract. After incubation for 1.5 h at 4°C protein precipitate is removed by centrifugation at 12 000 *g* for 45 min at 4°C. The supernatant contains Hsp21, which can be controlled by SDS-PAGE and immunoblotting.
4. Hydrophobic interaction column: In order to purify Hsp21, the supernatant will be filtered through a 0.2-μm filter and applied on a hydrophobic interaction column (21.4 × 100 mm, 300-Å pore size; model 83-B23-E, Dynamax Hydropore, Rainin, Woburn, MA). Hsp21 will be eluted by a gradient from 2 M to 0 M ammonium sulfate in 100 mM sodium phosphate, pH 7.0. A 50-mL gradient has to be applied with a flow rate of 5 mL min^{-1}. Five-milliliter fractions were collected and dialyzed against 10 mM Tris/HCL pH 8. The Hsp21-containing fractions are then identified by SDS-PAGE and immunoblotting.

32.4.4 Light-scattering Assays for Determination of the Chaperone Activity Using Citrate Synthase as Substrate (adopted from [196])

The oxidation of Hsp21 influences the functionality of the chaperone. First, oxidation induces the formation of larger oligomers of Hsp21, which can be assayed by native PAGE electrophoresis (see Section 32.1.1). Second, oxidation reduces the chaperone activity, which can be assayed as follows:

1. Oxidized Hsp21 is prepared by incubation of 0.4 mg mL^{-1} Hsp21 with 5 mM H_2O_2, 0.1 M ammonium bicarbonate buffer, pH 7.8 at 37 °C for 2 h.
2. For control, the oxidation of Hsp21 can be reverted by addition of 12 mM $MgCl_2$ (final), 34 mM KCl (final), 15 mM DTT, and 5 mM peptide methionine sulfoxide reductase (EC 1.8.4.6) to a final ratio of 1:50 (PMSR:Hsp21). This mixture has to be incubated for 30 min at 25 °C prior to use.
3. To assay the activity of Hsp21, the thermal aggregation/denaturation of citrate synthase can be determined. For this, 37.5 nM citrate synthase in 40 mM HEPES/KOH, pH 7, in the absence or presence of a 30-fold molar excess of Hsp21, Hsp21 oxidized, or Hsp21 oxidized and reduced is placed in a cuvette to determine the aggregation by light scattering. The sample will then be shifted to 43°C and the time-dependent light scattering will be recorded, for example, facilitating a RF-5301PC Shimadzu spectrofluorometer. The rate of aggregation

of citrate synthase can be monitored by a simple time scan where both the excitation and emission wavelengths are set to 500 nm with a spectral bandwidth of 2 nm.

32.4.5
The Use Of *Bis*-ANS to Assess Surface Exposure of Hydrophobic Domains of Hsp17 of *Synechocystis* (adopted from [202])

Bis-ANS (1,1′-*bis*(4-anilino)naphtalene-5,5′-disulfonic acid) is widely used for monitoring the surface exposure of hydrophobic domains to the protein surface. *Bis*-ANS is excited at 390 nm and shows a maximum fluorescence emission at 520 nm in aqueous buffer. *Bis*-ANS binds to accessible hydrophobic domains, which results in a shift of the maximum of the spectra to 495 nm. Therefore, by addition of a protein the hydrophobic exposure can be determined. Such assays can be used to determine the thermal-induced exposure of such regions as well as the interaction-induced exposure of hydrophobic surfaces induced by small heat shock proteins.

1. In order to monitor the thermal-induced exposure of hydrophobic surfaces in Hsp17 from *Synechocystis* and in malate dehydrogenase, 10 μM *bis*-ANS (final) is incubated with 1 μM Hsp17 or 0.8 μM malate dehydrogenase in 100 mM Tris/HCl pH 7.5; 150 mM KCl, and 20 mM Mg acetate at 25 °C. The increase of fluorescence at 495 nm in comparison to 10 μM *bis*-ANS can be determined. The protein-fluorophor mixture is then shifted to 47°C to determine the heat-induced fluorescence increase. The temperature can be controlled by a circulating water bath and directly measured in the cuvettes by a platinum probe.
2. To monitor the protein-induced exposure of hydrophobic surfaces, the increase of the fluorescence in the presence of 10 μM *bis*-ANS (final), 1 μM Hsp17, and 0.8 μM malate dehydrogenase in 100 mM Tris/HCl pH 7.5; 150 mM KCl, and 20 mM Mg acetate at 47°C is determined. The observed increase is then compared to the sum of the increase of the individual proteins. In this particular case, measurements were performed in a Quanta Master QM-1 fluorescence spectrometer (Photon Technology International, Princeton) [202].

32.4.6
Determination of Hsp17 Binding to Lipids (adopted from Refs. [204, 205])

Interaction with Lipids Determined by Steady-state Anisotropy Fluorescence

1. Lipid mixtures mimicking the membrane surface have to be created. In the cited manuscripts, DOPC (1,2-dioleolyl-sn-glycero-3-phosphochoiline), DOPG (1,2-dioleolyl-sn-glycero-3-phosphoglycerol), or total polar lipid extract of *Synechocystis* was used. Lipids are solubilized in chloroform.
2. Before preparation of large unilamellar vesicles (LUVs), 1,6-diphenyl-1,3,5-hexatriene (DPH) or 1-(4-trimethylammoniumphenyl)-6-phenyl-1,3,5-hexatridiene (TMA-DPH) is added to the lipid mixture in organic solvent before drying

of the lipid film by a constant nitrogen flow. The lipid-to-probe ratio should be between 1:1000 and 1:500 molar ratio (label to lipids total).

3. LUVs are created by resuspension of the lipid film in degassed appropriate buffer to the desired lipid concentration. Here, lipids were dissolved in 100 mM Tris pH 7.5, 150 mM KCl, and 20 mM Mg acetate to a final lipid concentration of 50 μM. The lipid film is then destroyed by five freeze/thaw cycles (using liquid N_2) to create multilamellar vesicles. The multilamellar vesicles have to be extruded 25 times through a 100-nm pore polycarbonate filter mounted in the mini-extruder (Liposofast, Avestin, Inc) to give unilamellar liposomes.
4. The effect of protein addition on the membrane physical state of the LUVs can now be determined. After determination of the fluorescence anisotropy of the label in LUVs without protein at 20 °C in 100 mM Tris/HCl pH 7.5; 150 mM KCl, and 20 mM Mg acetate for 300 s, purified Hsp17 is added to a final concentration of 1.5 μM. The measurement should be continued with constant stirring for the same time range after addition of the protein. In the cited manuscript, a T-format fluorescence spectrometer (Quanta Master QM-1, Photon Technology International, Princeton) was used. Excitation and emission wavelengths were set to 360 nm and 430 nm, respectively (5-nm slits). The temperature was measured directly in the cuvettes by a platinum probe and was controlled by a circulating water bath.

Interaction with Lipids Determined by Monolayer Experiments

The association of proteins with a lipid surface can also be determined using the monolayer technology as described below.

1. Monolayer experiments are carried out with a KSV3000 Langmuir Blodget instrument (KSV Instruments, Helsinki) in the work cited. A Teflon dish was used with a volume of 6.5 mL and a surface area of 9 cm^2 at 23 °C in buffer A (50 mM TEA·HCl, pH 7.5, 10 mM MgCl2, 100 mM KCl). Surface pressure is measured by the Wilhelmy method, using a platinum plate. Monomolecular lipid layers of 75% DOPE, 20% DOPG, and 5% ECCL (wt%) are spread from $CHCl_3$ lipid solution to give the desired initial surface pressure on a subphase of buffer A at the air-buffer interface. The subphase is continuously stirred with a magnetic bar. The subphase has to be washed in between the experiments by injecting and ejecting buffer solution at opposite sides of the dish, typically with a rate of about 10 mL min^{-1}.
2. Purified Hsp17 (0.5 μM final) is injected through a 0.5-cm^2 hole at an extended corner of the dish underneath a monolayer spread on 6.5 mL of 100 mM Tris/HCl pH 7.5; 150 mM KCl, 20 mM Mg acetate. The injected volume has to be kept below 1% of the total subphase volume. The increase of surface pressure by addition of the protein is a measure for insertion of hydrophobic protein segments into the lipid monolayer. With such a method the treatment-specific exposure of hydrophobic regions of proteins can be determined. The associated protein can be further analyzed by collecting the monolayer followed by Western blot analysis.

References

1 Laskey, R. A., Honda, B. M., Mills, A. D. & Finch, J. T. (1978). Nucleosomes are assembled by an acidic protein which binds histones and transfers them to DNA. *Nature* **275**, 416–420.

2 Pelham, H. R. (1986). Speculations on the functions of the major heat shock and glucose-regulated proteins. *Cell* **46**, 959–961.

3 Munro, S. & Pelham, H. R. (1986). An Hsp70-like protein in the ER: identity with the 78 kd glucose-regulated protein and immunoglobulin heavy chain binding protein. *Cell* **46**, 291–300.

4 Lewis, M. J. & Pelham, H. R. (1985). Involvement of ATP in the nuclear and nucleolar functions of the 70 kd heat shock protein. *EMBO J.* **4**, 3137–3143.

5 Pushkin, A. V., Tsuprun, V. L., Solovjeva, N. A., Shubin, V. V., Evstigneeva, Z. G. & Kretovich, W. L. (1982). High molcular weight pea leaf protein similar to the groE protein of escherichia coli. *Biochim. Biophys. Acta* **704**, 379–384.

6 Cannon, S., Wang, P. & Roy, H. (1986). Inhibition of ribulose bisphosphate carboxylase assembly by antibody to a binding protein. *J. Cell Biol.* **103**, 1327–1335.

7 Hemmingsen, S. M., Woolford, C., van der Vies, S. M., Tilly, K., Dennis, D. T., Georgopoulos, C. P., Hendrix, R. W. & Ellis, R. J. (1988). Homologous plant and bacterial proteins chaperone oligomeric protein assembly. *Nature* **333**, 330–334.

8 Cavalier-Smith, T. (1987). The simultaneous symbiotic origin of mitochondria, chloroplasts, and microbodies. *Ann. NY. Acad. Sci.* **503**, 55–71.

9 Martin, W. & Herrmann, R. G. (1998). Gene transfer from organelles to the nucleus: how much, what happens, and Why? *Plant Physiol.* **118**, 9–17.

10 Leister, D. (2003). Chloroplast research in the genomic age. *Trends Genet.* **19**, 47–56.

11 Bukau, B. & Horwich, A. L. (1998). The Hsp70 and Hsp60 chaperone machines. *Cell* **92**, 351–366.

12 Guy, C. L. & Li, Q. B. (1998). The organization and evolution of the spinach stress 70 molecular chaperone gene family. *Plant Cell* **10**, 539–556.

13 Lin, B. L., Wang, J. S., Liu, H. C., Chen, R. W., Meyer, Y., Barakat, A. & Delseny, M. (2001). Genomic analysis of the Hsp70 superfamily in Arabidopsis thaliana. *Cell Stress Chaperones* **6**, 201–208.

14 Scaramuzzi, C. D., Stokes, H. W. & Hiller, R. G. (1992). Heat shock Hsp70 protein is chloroplast-encoded in the chromophytic alga Pavlova lutherii. *Plant Mol. Biol.* **18**, 467–476.

15 Marshall, J. S., DeRocher, A. E., Keegstra, K. & Vierling, E. (1990). Identification of heat shock protein hsp70 homologues in chloroplasts. *Proc. Natl. Acad. Sci. USA.* **87**, 374–378.

16 Ko, K., Bornemisza, O., Kourtz, L., Ko, Z. W., Plaxton, W. C. & Cashmore, A. R. (1992). Isolation and characterization of a cDNA clone encoding a cognate 70-kDa heat shock protein of the chloroplast envelope. *J. Biol. Chem.* **267**, 2986–2993.

17 Drzymalla, C., Schroda, M. & Beck, C. F. (1996). Light-inducible gene HSP70B encodes a chloroplast-localized heat shock protein in Chlamydomonas reinhardtii. *Plant Mol. Biol.* **31**, 1185–1194.

18 Karlin, S. & Brocchieri, L. (1998). Heat shock protein 70 family: multiple sequence comparisons, function, and evolution. *J. Mol. Evol.* **47**, 565–577.

19 Li, Q. B., Haskell, D. W. & Guy, C. L. (1999). Coordinate and non-coordinate expression of the stress 70 family and other molecular chaperones at high and low temperature in spinach and tomato. *Plant Mol. Biol.* **39**, 21–34.

20 Marshall, J. S. & Keegstra, K.

(1992). Isolation and Characterization of a cDNA clone encoding the major Hsp70 of the peachloroplastic stroma. *Plant Physiol.* **100**, 1048–1054.

21 Schroda, M., Vallon, O., Whitelegge, J. P., Beck, C. F. & Wollman, F. A. (2001). The chloroplastic GrpE homolog of Chlamydomonas: two isoforms generated by differential splicing. *Plant Cell* **13**, 2823–2839.

22 von Gromoff, E. D., Treier, U. & Beck, C. F. (1989). Three light-inducible heat shock genes of Chlamydomonas reinhardtii. *Mol. Cell Biol.* **9**, 3911–3918.

23 Muller, F. W., Igloi, G. L. & Beck, C. F. (1992). Structure of a gene encoding heat-shock protein HSP70 from the unicellular alga Chlamydomonas reinhardtii. *Gene* **111**, 165–173.

24 Kropat, J., von Gromoff, E. D., Muller, F. W. & Beck, C. F. (1995). Heat shock and light activation of a Chlamydomonas HSP70 gene are mediated by independent regulatory pathways. *Mol. Gen. Genet.* **248**, 727–734.

25 Kropat, J., Oster, U., Rudiger, W. & Beck, C. F. (1997). Chlorophyll precursors are signals of chloroplast origin involved in light induction of nuclear heat-shock genes. *Proc. Natl. Acad. Sci. USA.* **94**, 14168–14172.

26 Li, Q. B., Haskell, D., Zhang, C., Sung, D. Y. & Guy, C. (2000). Diurnal regulation of Hsp70s in leaf tissue. *Plant J.* **21**, 373–378.

27 Kourtz, L. & Ko, K. (1997). The early stage of chloroplast protein import involves Com70. *J. Biol. Chem.* **272**, 2808–2813.

28 Schlicher, T. & Soll, J. (1997). Chloroplastic isoforms of DnaJ and GrpE in pea. *Plant Mol. Biol.* **33**, 181–185.

29 Becker, T., Hritz, J., Vogel, M., Caliebe, A., Bukau, B., Soll, J. & Schleiff, E. (2004). Toc 12, A Novel Subunit of the Intermembrane Space Preprotein Translocon of Chloroplasts. *Mol. Biol. Cell.* in press.

30 Orme, W., Walker, A. R., Gupta, R. & Gray, J. C. (2001). A novel plastid-targeted J-domain protein in Arabidopsis thaliana. *Plant Mol. Biol.* **46**, 615–626.

31 Hartl, F. U. & Hayer-Hartl, M. (2002). Molecular chaperones in the cytosol: from nascent chain to folded protein. *Science* **295**, 1852–1858.

32 Barraclough, R. & Ellis, R. J. (1980). Protein synthesis in chloroplasts. IX. Assembly of newly-synthesized large subunits into ribulose bisphosphate carboxylase in isolated intact pea chloroplasts. *Biochim. Biophys. Acta* **608**, 19–31.

33 Maier, U. G., Rensing, S. A., Igloi, G. L. & Maerz, M. (1995). Twintrons are not unique to the Euglena chloroplast genome: structure and evolution of a plastome cpn60 gene from a cryptomonad. *Mol. Gen. Genet.* **246**, 128–311.

34 Viale, A. (1995). GroEL (Hsp60)-based eubacterial and organellar phylogenies. *Mol. Microbiol.* **17**, 1013.

35 Viitanen, P. V., Schmidt, M., Buchner, J., Suzuki, T., Vierling, E., Dickson, R., Lorimer, G. H., Gatenby, A. & Soll, J. (1995). Functional characterization of the higher plant chloroplast chaperonins. *J Biol Chem* **270**, 18158–64.

36 Musgrove, J. E., Johnson, R. A. & Ellis, R. J. (1987). Dissociation of the ribulosebisphosphate-carboxylase large-subunit binding protein into dissimilar subunits. *Eur. J. Biochem.* **163**, 529–534.

37 Nishio, K., Hirohashi, T. & Nakai, M. (1999). Chloroplast chaperonins: evidence for heterogeneous assembly of alpha and beta Cpn60 polypeptides into a chaperonin oligomer. *Biochem. Biophys. Res. Commun.* **266**, 584–587.

38 Dickson, R., Weiss, C., Howard, R. J., Alldrick, S. P., Ellis, R. J., Lorimer, G., Azem, A. & Viitanen, P. V. (2000). Reconstitution of higher plant chloroplast chaperonin 60 tetradecamers active in protein folding. *J. Biol. Chem.* **275**, 11829–11835.

39 Martel, R., Cloney, L. P., Pelcher, L. E. & Hemmingsen, S. M. (1990). Unique composition of plastid

chaperonin-60: alpha and beta polypeptide-encoding genes are highly divergent. *Gene* **94**, 181–187.

40 Hartman, D. J., Dougan, D., Hoogenraad, N. J. & Høj, P. B. (1992). Heat shock proteins of barley mitochondria and chloroplasts Identification of organellar hsp 10 and 12: putative chaperonin 10 homologues. *FEBS-Let.* **305**, 147–150.

41 Zabaleta, E., Assad, N., Oropeza, A., Salerno, G. & Herrera-Estrella, L. (1994). Expression of one of the members of the Arabidopsis chaperonin 60 beta gene family is developmentally regulated and wound-repressible. *Plant Mol. Biol.* **24**, 195–202.

42 Bertsch, U., Soll, J., Seetharam, R. & Viitanen, P. V. (1992). Identification, characterization, and DNA sequence of a functional "double" groES-like chaperonin from chloroplasts of higher plants. *Proc. Natl. Acad. Sci. USA.* **89**, 8696–8700.

43 Baneyx, F., Bertsch, U., Kalbach, C. E., van der Vies, S. M., Soll, J. & Gatenby, A. A. (1995). Spinach chloroplast cpn21 co-chaperonin possesses two functional domains fused together in a toroidal structure and exhibits nucleotide-dependent binding to plastid chaperonin 60. *J. Biol. Chem.* **270**, 10695–10702.

44 Koumoto, Y., Shimada, T., Kondo, M., Takao, T., Shimonishi, Y., Hara-Nishimura, I. & Nishimura, M. (1999). Chloroplast Cpn20 forms a tetrameric structure in Arabidopsis thaliana. *Plant J.* **17**, 467–477.

45 Sharkia, R., Bonshtien, A. L., Mizrahi, I., Weiss, C., Niv, A., Lustig, A., Viitanen, P. V. & Azem, A. (2003). On the oligomeric state of chloroplast chaperonin 10 and chaperonin 20. *Biochim. Biophys. Acta* **1651**, 76–84.

46 Schlicher, T. & Soll, J. (1996). Molecular chaperones are present in the thylakoid lumen of pea chloroplasts. *FEBS Lett.* **379**, 302–304.

47 Koumoto, Y., Shimada, T., Kondo, M., Hara-Nishimura, I. & Nishimura, M. (2001). Chloroplasts have a novel Cpn10 in addition to Cpn20 as co-chaperonins in Arabidopsis thaliana. *J. Biol. Chem.* **276**, 29688–29694.

48 Cloney, L. P., Wu, H. B. & Hemmingsen, S. M. (1992). Expression of plant chaperonin-60 genes in Escherichia coli. *J. Biol. Chem.* **267**, 23327–23332.

49 Cloney, L. P., Bekkaoui, D. R., Wood, M. G. & Hemmingsen, S. M. (1992). Assessment of plant chaperonin-60 gene function in Escherichia coli. *J. Biol. Chem.* **267**, 23333–23336.

50 Saibil, H. R., Zheng, D., Roseman, A. M., Hunter, A. S., Watson, G. M. F., Chen, S., Auf der Mauer, A., O'Hara, B. P., Wood, S. P., Mann, N. H., Barnett, L. K. & Ellis, R. J. (1993). ATP induces large quaternary rearrngemants in a cage-like chaperonin structure. *Curr. Biol.* **3**, 265–273.

51 Dickson, R., Larsen, B., Viitanen, P. V., Tormey, M. B., Geske, J., Strange, R. & Bemis, L. T. (1994). Cloning, expression, and purification of a functional nonacetylated mammalian mitochondrial chaperonin 10. *J. Biol. Chem.* **269**, 26858–26864.

52 Porankiewicz, J., Wang, J. & Clarke, A. K. (1999). New insights into the ATP-dependent Clp protease: Escherichia coli and beyond. *Mol. Microbiol.* **32**, 449–458.

53 Schirmer, E. C., Glover, J. R., Singer, M. A. & Lindquist, S. (1996). HSP100/Clp proteins: a common mechanism explains diverse functions. *Trends. Biochem. Sci.* **21**, 289–296.

54 Hwang, B. J., Park, W. J., Chung, C. H. & Goldberg, A. L. (1987). Escherichia coli contains a soluble ATP-dependent protease (Ti) distinct from protease La. *Proc. Natl. Acad. Sci. USA.* **84**, 5550–5554.

55 Katayama, Y., Gottesman, S., Pumphrey, J., Rudikoff, S., Clark, W. P. & Maurizi, M. R. (1988). The two-component, ATP-dependent Clp protease of Escherichia coli. Purification, cloning, and mutational analysis of the ATP-binding

component. *J. Biol. Chem.* **263**, 15226–15236.

56 Moore, T. & Keegstra, K. (1993). Characterization of a cDNA clone encoding a chloroplast-targeted Clp homologue. *Plant Mol. Biol.* **21**, 525–537.

57 Kiyosue, T., Yamaguchi-Shinozaki, K. & Shinozaki, K. (1993). Characterization of cDNA for a dehydration-inducible gene that encodes a ClpA, B-like protein in Arabidopsis thaliana L. *Biochem. Biophys. Res. Commun.* **196**, 1214–1220.

58 Peltier, J. B., Friso, G., Kalume, D. E., Roepstorff, P., Nilsson, F., Adamska, I. & van Wijk, K. J. (2000). Proteomics of the chloroplast: systematic identification and targeting analysis of lumenal and peripheral thylakoid proteins. *Plant Cell* **12**, 319–341.

59 Adam, Z., Adamska, I., Nakabayashi, K., Ostersetzer, O., Haussuhl, K., Manuell, A., Zheng, B., Vallon, O., Rodermel, S. R., Shinozaki, K. & Clarke, A. K. (2001). Chloroplast and mitochondrial proteases in Arabidopsis. A proposed nomenclature. *Plant Physiol.* **125**, 1912–1918.

60 Peltier, J. B., Ytterberg, J., Liberles, D. A., Roepstorff, P. & van Wijk, K. J. (2001). Identification of a 350-kDa ClpP protease complex with 10 different Clp isoforms in chloroplasts of Arabidopsis thaliana. *J. Biol. Chem.* **276**, 16318–16327.

61 Zheng, B., Halperin, T., Hruskova-Heidingsfeldova, O., Adam, Z. & Clarke, A. K. (2002). Characterization of Chloroplast Clp proteins in Arabidopsis: Localization, tissue specificity and stress responses. *Physiol. Plant* **114**, 92–101.

62 Shanklin, J., DeWitt, N. D. & Flanagan, J. M. (1995). The stroma of higher plant plastids contain ClpP and ClpC, functional homologs of Escherichia coli ClpP and ClpA: an archetypal two-component ATP-dependent protease. *Plant Cell* **7**, 1713–1722.

63 Weaver, L. M., Froehlich, J. E. & Amasino, R. M. (1999). Chloroplast-targeted ERD1 protein declines but its mRNA increases during senescence in Arabidopsis. *Plant Physiol.* **119**, 1209–1216.

64 Ostersetzer, O. & Adam, Z. (1996). Effects of light and temperature on expression of ClpC, the regulatory subunit of chloroplastic Clp protease, in pea seedlings. *Plant Mol. Biol.* **31**, 673–676.

65 Nakabayashi, K., Ito, M., Kiyosue, T., Shinozaki, K. & Watanabe, A. (1999). Identification of clp genes expressed in senescing Arabidopsis leaves. *Plant Cell Physiol.* **40**, 504–514.

66 Nakashima, K., Kiyosue, T., Yamaguchi-Shinozaki, K. & Shinozaki, K. (1997). A nuclear gene, erd1, encoding a chloroplast-targeted Clp protease regulatory subunit homolog is not only induced by water stress but also developmentally up-regulated during senescence in Arabidopsis thaliana. *Plant J.* **12**, 851–861.

67 Clarke, A. K. & Eriksson, M. J. (1996). The cyanobacterium Synechococcus sp. PCC 7942 possesses a close homologue to the chloroplast ClpC protein of higher plants. *Plant Mol. Biol.* **31**, 721–730.

68 Sokolenko, A., Lerbs-Mache, S., Altschmied, L. & Herrmann, R. G. (1998). Clp protease complexes and their diversity in chloroplasts. *Planta* **207**, 286–295.

69 Halperin, T. & Adam, Z. (1996). Degradation of mistargeted OEE33 in the chloroplast stroma. *Plant Mol. Biol.* **30**, 925–933.

70 Desimone, M., Weiss-Wichert, W., Wagner, E., Altenfeld, U. & Johanningmeier, U. (1997). Immunochemical studies on the Clp-protease in chloroplasts: evidence for the formation of a ClpC/P complex. *Bot. Acta* **110**, 234–239.

71 Halperin, T., Ostersetzer, O. & Adam, Z. (2001). ATP-dependent association between subunits of Clp protease in pea chloroplasts. *Planta* **213**, 614–619.

72 Flanagan, J. M., Wall, J. S., Capel, M. S., Schneider, D. K. & Shanklin, J. (1995). Scanning transmission electron microscopy and small-angle scattering provide evidence that native Escherichia coli ClpP is a tetradecamer with an axial pore. *Biochem.* **34**, 10910–10917.

73 Shin, D. H., Lee, C. S., Chung, C. H. & Suh, S. W. (1996). Molecular symmetry of the ClpP component of the ATP-dependent Clp protease, an Escherichia coli homolog of 20 S proteasome. *J. Mol. Biol.* **262**, 71–76.

74 Wang, S. & Liu, X. Q. (1997). Identification of an unusual intein in chloroplast ClpP protease of Chlamydomonas eugametos. *J. Biol. Chem.* **272**, 11869–11873.

75 Vierling, E. (1997). The small heat shock protein in plants are members of an ancient family of heat induced proteins. *Acta Physiol. Plant.* **19**, 539–547.

76 Sun, W., Van Montagu, M. & Verbruggen, N. (2002). Small heat shock proteins and stress tolerance in plants. *Biochim. Biophys. Acta* **1577**, 1–9.

77 Scharf, K.-D., Siddique, M. & Vierling, E. (2001). The expanding family of Arabidopsis thaliana small heat stress proteins and a new family of proteins containing-crystallin domains (Acd proteins). *Cell Stress Chaperones* **6**, 225–237.

78 Vierling, E. (1991). The roles of heat shock proteinsin plants. *Rev. Plant Physiol. Plant Mol. Biol.* **42**, 579–620.

79 Waters, E. R., Lee, G. J. & Vierling, E. (1996). Evolution, structure and function of the small heat shock proteins in plants. *J. Exp. Bot.* **47**, 325–338.

80 Boston, R. S., Viitanen, P. V. & Vierling, E. (1996). Molecular chaperones and protein folding in plants. *Plant Mol. Biol.* **32**, 191–222.

81 Waters, E. R. & Vierling, E. (1999). Chloroplast small heat shock proteins: evidence for atypical evolution of an organelle-localized protein. *Proc. Natl. Acad. Sci. USA.* **96**, 14394–14399.

82 Vierling, E., Mishkind, M. L., Schmidt, G. W. & Key, J. L. (1986). Specific heat shock proteins are transported into chloroplasts. *Proc. Natl. Acad. Sci. USA.* **83**, 361–365.

83 Vierling, E., Harris, L. M. & Chen, Q. (1989). The major low-molecular-weight heat shock protein in chloroplasts shows antigenic conservation among diverse higher plant species. *Mol. Cell Biol.* **9**, 461–468.

84 Chen, Q., Lauzon, L. M., DeRocher, A. E. & Vierling, E. (1990). Accumulation, stability, and localization of a major chloroplast heat-shock protein. *J. Cell Biol.* **110**, 1873–1883.

85 Harndahl, U., Hall, R. B., Osteryoung, K. W., Vierling, E., Bornman, J. F. & Sundby, C. (1999). The chloroplast small heat shock protein undergoes oxidation-dependent conformational changes and may protect plants from oxidative stress. *Cell Stress Chaperones* **4**, 129–138.

86 Heckathorn, S. A., Downs, C. A., Sharkey, T. D. & Coleman, J. S. (1998). The small, methionine-rich chloroplast heat-shock protein protects photosystem II electron transport during heat stress. *Plant Physiol.* **116**, 439–444.

87 Adamska, I. & Kloppstech, K. (1991). Evidence for the localization of the nuclear-coded 22-kDa heat-shock protein in a subfraction of thylakoid membranes. *Eur. J. Biochem.* **198**, 375–381.

88 Downs, C. A., Coleman, J. S. & Heckathorn, S. A. (1999). The chloroplast 22-Ku heat-shock protein: a lumenal protein that associates with the oxygen evolving complex and protects photosystem II during heat stress. *J. Plant Physiol.* **155**, 477–487.

89 Chen, Q., Osteryoung, K. & Vierling, E. (1994). A 21-kDa chloroplast heat shock protein assembles into high molecular weight complexes in vivo and in Organelle. *J. Biol. Chem.* **269**, 13216–13223.

90 Suzuki, T. C., Krawitz, D. C. & Vierling, E. (1998). The chloroplast small heat-shock protein oligomer is

not phosphorylated and does not dissociate during heat stress in vivo. *Plant Physiol.* **116**, 1151–1161.

91 Harndahl, U., Kokke, B. P., Gustavsson, N., Linse, S., Berggren, K., Tjerneld, F., Boelens, W. C. & Sundby, C. (2001). The chaperone-like activity of a small heat shock protein is lost after sulfoxidation of conserved methionines in a surface-exposed amphipathic alpha-helix. *Biochim. Biophys. Acta* **1545**, 227–237.

92 van Montfort, R. L., Basha, E., Friedrich, K. L., Slingsby, C. & Vierling, E. (2001). Crystal structure and assembly of a eukaryotic small heat shock protein. *Nat. Struct. Biol.* **8**, 1025–1030.

93 Kim, K. K., Kim, R. & Kim, S.-H. (1998). Crystal structure of a small heat-shock protein. *Nature* **394**, 595–599.

94 Kirschner, M., Winkelhaus, S., Thierfelder, J. M. & Nover, L. (2000). Transient expression and heat-stress-induced co-aggregation of endogenous and heterologous small heat-stress proteins in tobacco protoplasts. *Plant J.* **24**, 397–411.

95 Osteryoung, K. W. & Vierling, E. (1994). Dynamics of small heat shock protein distribution within the chloroplasts of higher plants. *J. Biol. Chem.* **269**, 28676–28682.

96 Löw, D., Brändle, K., Nover, L. & Forreiter, C. (2000). Cytosolic heat-stress proteins Hsp17.7 class I and Hsp17.3 class II of tomato act as molecular chaperones in vivo. *Planta* **211**, 575–582.

97 Lee, G. J. & Vierling, E. (2000). A small heat shock protein cooperates with heat shock protein 70 systems to reactivate a heat-denatured protein. *Plant Physiol.* **122**, 189–198.

98 Gustavsson, N., Kokke, B. P., Harndahl, U., Silow, M., Bechtold, U., Poghosyan, Z., Murphy, D., Boelens, W. C. & Sundby, C. (2002). A peptide methionine sulfoxide reductase highly expressed in photosynthetic tissue in Arabidopsis thaliana can protect the chaperone-like activity of a chloroplast-localized small heat shock protein. *Plant J.* **29**, 545–553.

99 Plater, M. L., Goode, D. & Crabbe, M. J. (1996). Effects of site-directed mutations on the chaperone-like activity of alphaB-crystallin. *J. Biol. Chem.* **271**, 28558–28566.

100 Chen, Q. & Vierling, E. (1991). Analysis of conserved domains identifies a unique structural feature of a chloroplast heat shock protein. *Mol. Gen. Genet.* **226**, 425–431.

101 Keenan, R. J., Freymann, D. M., Stroud, R. M. & Walter, P. (2001). The signal recognition particle. *Annu. Rev. Biochem.* **70**, 755–775.

102 Emelyanov, V. V. (2002). Phylogenetic relationships of organellar Hsp90 homologs reveal fundamental differences to organellar Hsp70 and Hsp60 evolution. *Gene* **299**, 125–133.

103 Young, J. C., Moarefi, I. & Hartl, F. U. (2001). Hsp90: a specialized but essential protein-folding tool. *J. Cell Biol.* **154**, 267–273.

104 Schmitz, G., Schmidt, M. & Feierabend, J. (1996). Characterization of a plastid-specific HSP90 homologue: identification of a cDNA sequence, phylogenetic descendence and analysis of its mRNA and protein expression. *Plant Mol. Biol.* **30**, 479–492.

105 Cao, D., Froehlich, J. E., Zhang, H. & Cheng, C. L. (2003). The chlorate-resistant and photomorphogenesis-defective mutant cr88 encodes a chloroplast-targeted HSP90. *Plant J.* **33**, 107–118.

106 Nemoto, T., Ohara-Nemoto, Y., Ota, M., Takagi, T. & Yokoyama, K. (1995). Mechanism of dimer formation of the 90-kDa heat-shock protein. *Eur. J. Biochem.* **233**, 1–8.

107 Young, J. C., Schneider, C. & Hartl, F. U. (1997). In vitro evidence that hsp90 contains two independent chaperone sites. *FEBS Lett.* **418**, 139–143.

108 Scheibel, T., Weikl, T. & Buchner, J. (1998). Two chaperone sites in Hsp90 differing in substrate specificity and ATP dependence. *Proc. Natl. Acad. Sci. USA.* **95**, 1495–1499.

109 Freedman, R. B. (1989). Protein disulfide isomerase: multiple roles in the modification of nascent secretory proteins. *Cell* **57**, 1069–1072.

110 Cai, H., Wang, C. C. & Tsou, C. L. (1994). Chaperone-like activity of protein disulfide isomerase in the refolding of a protein with no disulfide bonds. *J. Biol. Chem.* **269**, 24550–24552.

111 Song, J. L. & Wang, C. C. (1995). Chaperone-like activity of protein disulfide-isomerase in the refolding of rhodanese. *Eur. J. Biochem.* **231**, 312–316.

112 Kim, J. & Mayfield, S. P. (1997). Protein disulfide isomerase as a regulator of chloroplast translational activation. *Science* **278**, 1954–1957.

113 Trebitsh, T., Meiri, E., Ostersetzer, O., Adam, Z. & Danon, A. (2001). The protein disulfide isomerase-like RB60 is partitioned between stroma and thylakoids in Chlamydomonas reinhardtii chloroplasts. *J. Biol. Chem.* **276**, 4564–4569.

114 Danon, A. & Mayfield, S. P. (1991). Light regulated translational activators: identification of chloroplast gene specific mRNA binding proteins. *EMBO J.* **10**, 3993–4001.

115 Yohn, C. B., Cohen, A., Danon, A. & Mayfield, S. P. (1996). Altered mRNA binding activity and decreased translational initiation in a nuclear mutant lacking translation of the chloroplast psbA mRNA. *Mol. Cell Biol.* **16**, 3560–3566.

116 Yohn, C. B., Cohen, A., Danon, A. & Mayfield, S. P. (1998). A poly(A) binding protein functions in the chloroplast as a message-specific translation factor. *Proc. Natl. Acad. Sci. USA.* **95**, 2238–2243.

117 Yohn, C. B., Cohen, A., Rosch, C., Kuchka, M. R. & Mayfield, S. P. (1998). Translation of the chloroplast psbA mRNA requires the nuclear-encoded poly(A)-binding protein, RB47. *J. Cell Biol.* **142**, 435–442.

118 Danon, A. & Mayfield, S. P. (1994). ADP-dependent phosphorylation regulates RNA-binding in vitro: implications in light-modulated translation. *EMBO J.* **13**, 2227–2235.

119 Trebitsh, T., Levitan, A., Sofer, A. & Danon, A. (2000). Translation of chloroplast psbA mRNA is modulated in the light by counteracting oxidizing and reducing activities. *Mol. Cell Biol.* **20**, 1116–1123.

120 Trebitsh, T. & Danon, A. (2001). Translation of chloroplast psbA mRNA is regulated by signals initiated by both photosystems II and I. *Proc. Natl. Acad. Sci. USA.* **98**, 12289–12294.

121 Kim, J. & Mayfield, S. P. (2002). The active site of the thioredoxin-like domain of chloroplast protein disulfide isomerase, RB60, catalyzes the redox-regulated binding of chloroplast poly(A)-binding protein, RB47, to the 5′ untranslated region of psbA mRNA. *Plant Cell Physiol.* **43**, 1238–1243.

122 Fischer, G., Wittmann-Liebold, B., Lang, K., Kiefhaber, T. & Schmid, F. X. (1989). Cyclophilin and peptidyl-prolyl *cis-trans* isomerase are probably identical proteins. *Nature* **337**, 476–478.

123 Harding, M. W., Galat, A., Uehling, D. E. & Schreiber, S. L. (1989). A receptor for the immuno-suppressant FK506 is a *cis-trans* peptidyl-prolyl isomerase. *Nature* **341**, 758–760.

124 Gething, M. J. (1991). Molecular chaperones: individualists or groupies? *Curr. Opin. Cell Biol.* **3**, 610–614.

125 Freskgard, P. O., Bergenhem, N., Jonsson, B. H., Svensson, M. & Carlsson, U. (1992). Isomerase and chaperone activity of prolyl isomerase in the folding of carbonic anhydrase. *Science* **258**, 466–468.

126 Liu, J., Farmer, J. D. Jr, Lane, W. S., Friedman, J., Weissman, I. & Schreiber, S. L. (1991). Calcineurin is a common target of cyclophilin-cyclosporin A and FKBP-FK506 complexes. *Cell* **66**, 807–815.

127 Clipstone, N. A. & Crabtree, G. R. (1992). Identification of calcineurin as a key signalling enzyme in T-lymphocyte activation. *Nature* **357**, 695–697.

128 O'Keefe, S. J., Tamura, J., Kincaid, R. L., Tocci, M. J. & O'Neill, E. A. (1992). FK-506- and CsA-sensitive activation of the interleukin-2 promoter by calcineurin. *Nature* **357**, 692–694.

129 Schreiber, S. L. (1991). Chemistry and Biology of the Immunophilins and Their Immunosuppressive Ligands. *Science* **251**, 283–287.

130 Fink, A. L. (1999). Chaperone-mediated protein folding. *Physiol. Rev.* **79**, 425–449.

131 Gupta, R., Mould, R. M., He, Z. & Luan, S. (2002). A chloroplast FKBP interacts with and affects the accumulation of Rieske subunit of cytochrome bf complex. *Proc. Natl. Acad. Sci. USA.* **99**, 15806–15811.

132 Breiman, A., Fawcett, T. W., Ghirardi, M. L. & Mattoo, A. K. (1992). Plant organelles contain distinct peptidylprolyl *cis,trans*-isomerases. *J. Biol. Chem.* **267**, 21293–21296.

133 Luan, S., Albers, M. W. & Schreiber, S. L. (1994). Light-regulated, tissue-specific immunophilins in a higher plant. *Proc. Natl. Acad. Sci. USA.* **91**, 984–988.

134 Luan, S., Lane, W. S. & Schreiber, S. L. (1994). pCyP B: a chloroplast-localized, heat shock-responsive cyclophilin from fava bean. *Plant Cell* **6**, 885–892.

135 Fulgosi, H., Vener, A. V., Altschmied, L., Herrmann, R. G. & Andersson, B. (1998). A novel multi-functional chloroplast protein: identification of a 40 kDa immunophilin-like protein located in the thylakoid lumen. *EMBO J.* **17**, 1577–1587.

136 Lippuner, V., Chou, I. T., Scott, S. V., Ettinger, W. F., Theg, S. M. & Gasser, C. S. (1994). Cloning and characterization of chloroplast and cytosolic forms of cyclophilin from Arabidopsis thaliana. *J. Biol. Chem.* **269**, 7863–7868.

137 Edvardsson, A., Eshaghi, S., Vener, A. V. & Andersson, B. (2003). The major peptidyl-prolyl isomerase activity in thylakoid lumen of plant chloroplasts belongs to a novel cyclophilin TLP20. *FEBS Lett.* **542**, 137–141.

138 Vener, A. V., Rokka, A., Fulgosi, H., Andersson, B. & Herrmann, R. G. (1999). A cyclophilin-regulated PP2A-like protein phosphatase in thylakoid membranes of plant chloroplasts. *Biochem.* **38**, 14955–14965.

139 Rokka, A., Aro, E. M., Herrmann, R. G., Andersson, B. & Vener, A. V. (2000). Dephosphorylation of photosystem II reaction center proteins in plant photosynthetic membranes as an immediate response to abrupt elevation of temperature. *Plant Physiol.* **123**, 1525–1536.

140 Ivey, R. A. 3rd & Bruce, B. D. (2000). In vivo and in vitro interaction of DnaK and a chloroplast transit peptide. *Cell Stress Chaperones* **5**, 62–71.

141 Ivey, R. A. 3rd, Subramanian, C. & Bruce, B. D. (2000). Identification of a Hsp70 recognition domain within the rubisco small subunit transit peptide. *Plant Physiol.* **122**, 1289–1299.

142 Rial, D. V., Arakaki, A. K. & Ceccarelli, E. A. (2000). Interaction of the targeting sequence of chloroplast precursors with Hsp70 molecular chaperones. *Eur. J. Biochem.* **267**, 6239–6248.

143 Rial, D. V., Ottado, J. & Ceccarelli, E. A. (2003). Precursors with altered affinity for Hsp70 in their transit peptides are efficiently imported into chloroplasts. *J. Biol. Chem.* [Epub ahead of print].

144 Pilon, M., Wienk, H., Sips, W., de Swaaf, M., Talboom, I., van't Hof, R., de Korte-Kool, G., Demel, R., Weisbeek, P. & de Kruijff, B. (1995). Functional domains of the ferredoxin transit sequence involved in chloroplast import. *J. Biol. Chem.* **270**, 3882–3893.

145 Schleiff, E. & Soll, J. (2000). Travelling of proteins through membranes: translocation into chloroplasts. *Planta* **211**, 449–456.

146 Waegemann, K., Paulsen, H., Soll, J. (1990) Translocation of proteins into chloroplasts requires cytosolic factors

to obtain import competence. *FEBS-Lett.* **161**, 89–92.

147 Pilon, M., de Boer, A. D., Knols, S. L., Koppelman, M. H., van der Graaf, R. M., de Kruijff, B. & Weisbeek, P. J. (1990). Expression in Escherichia coli and purification of a translocation-competent precursor of the chloroplast protein ferredoxin. *J. Biol. Chem.* **265**, 3358–3361.

148 Pilon, M., de Kruijff, B. & Weisbeek, P. J. (1992). New insights into the import mechanism of the ferredoxin precursor into chloroplasts. *J. Biol. Chem.* **267**, 2548–2556.

149 May, T. & Soll, J. (2000). 14-3-3 proteins form a guidance complex with chloroplast precursor proteins in plants. *Plant Cell* **12**, 53–64.

150 Waegemann, K. & Soll, J. (1996). Phosphorylation of the transit sequence of chloroplast precursor proteins. *J. Biol. Chem.* **271**, 6545–6554.

151 Hachiya, N., Mihara, K., Suda, K., Horst, M., Schatz, G. & Lithgow, T. (1995). Reconstitution of the initial steps of mitochondrial protein import. *Nature* **376**, 705–709.

152 Pfanner, N. & Geissler, A. (2001). Versatility of the mitochondrial protein import machinery. *Nat. Rev. Mol. Cell. Biol.* **2**, 339–349.

153 Sohrt, K. & Soll, J. (2000). Toc64, a new component of the protein translocon of chloroplasts. *J. Cell Biol.* **148**, 1213–1221.

154 Waegemann, K., Soll, J. (1991) Characterization of the protein import apparatus in isolated outer enevlopes of chloroplasts. *Plant Journal* **1**, 149–158.

155 Schnell, D. J., Kessler, F. & Blobel, G. (1994). Isolation of components of the chloroplast protein import machinery. *Science* **266**, 1007–1012.

156 Wu, C., Seibert, F. S. & Ko, K. (1994). Identification of chloroplast envelope proteins in close physical proximity to a partially translocated chimeric precursor protein. *J. Biol. Chem.* **269**, 32264–32271.

157 Kessler, F., Blobel, G. (1996) Interaction of the protein import and folding machineries of the chloroplast. *Proc. Natl. Acad. Sci. USA.* **93**, 7684–7689.

158 Akita, M., Nielsen, E. & Keegstra, K. (1997). Identification of protein transport complexes in the chloroplastic envelope membranes via chemical cross-linking. *J. Cell Biol.* **136**, 983–994.

159 Nielsen, E., Akita, M., Davila-Aponte, J. & Keegstra, K. (1997). Stable association of chloroplastic precursors with protein translocation complexes that contain proteins from both envelope membranes and a stromal Hsp100 molecular chaperone. *EMBO J.* **16**, 935–946.

160 Chou, M. L., Fitzpatrick, L. M., Tu, S. L., Budziszewski, G., Potter-Lewis, S., Akita, M., Levin, J. Z., Keegstra, K. & Li, H. M. (2003). Tic40, a membrane-anchored co-chaperone homolog in the chloroplast protein translocon. *EMBO J.* **22**, 2970–2980.

161 Ko, K., Budd, D., Wu, C., Seibert, F., Kourtz, L. & Ko, Z. W. (1995). Isolation and characterization of a cDNA clone encoding a member of the Com44/Cim44 envelope components of the chloroplast protein import apparatus. *J. Biol. Chem.* **270**, 28601–28608.

162 Stahl, T., Glockmann, C., Soll, J. & Heins, L. (1999). Tic40, a new "old" subunit of the chloroplast protein import translocon. *J. Biol. Chem.* **274**, 37467–37472.

163 Choquet, Y. & Wollman, F. A. (2002). Translational regulations as specific traits of chloroplast gene expression. *FEBS Lett.* **529**, 39–42.

164 Flügge, U. I. & Hinz, G. (1986). Energy dependence of protein translocation into chloroplasts. *Eur. J. Biochem.* **160**, 563–570.

165 Schindler, C., Hracky, R., Soll, J. (1987) Protein transport in chloroplasts: ATP is prerequisit. *Zeitschr. Naturfrosch.* **42c**, 103–108.

166 Pain, D. & Blobel, G. (1987). Protein import into chloroplasts requires a chloroplast ATPase. *Proc. Natl. Acad. Sci. USA.* **84**, 3288–3292.

167 Olsen, L. J., Theg, S. M., Selman, B. R. & Keegstra, K. (1989). ATP is required for the binding of precursor proteins to chloroplasts. *J. Biol. Chem.* **264**, 6724–6729.
168 Olsen, L. J. & Keegstra, K. (1992). The binding of precursor proteins to chloroplasts requires nucleoside triphosphates in the intermembrane space. *J. Biol. Chem.* **267**, 433–439.
169 Breton, G., Danyluk, J., Charron, J. B. & Sarhan, F. (2003). Expression profiling and bioinformatic analyses of a novel stress-regulated multispanning transmembrane protein family from cereals and Arabidopsis. *Plant Physiol.* **132**, 64–74.
170 Leszczynski, D., Pitsillides, C. M., Pastila, R. K., Rox Anderson, R. & Lin, C. P. (2001). Laser-beam-triggered microcavitation: a novel method for selective cell destruction. *Radiat. Res.* **156**, 399–407.
171 Molik, S., Karnauchov, I., Weidlich, C., Herrmann, R. G. & Klosgen, R. B. (2001). The Rieske Fe/S protein of the cytochrome b6/f complex in chloroplasts: missing link in the evolution of protein transport pathways in chloroplasts? *J. Biol. Chem.* **276**, 42761–42766.
172 Madueno, F., Napier, J. A. & Gray, J. C. (1993). Newly Imported Rieske Iron-Sulfur Protein Associates with Both Cpn60 and Hsp70 in the Chloroplast Stroma. *Plant Cell* **5**, 1865–1876.
173 Tsugeki, R. & Nishimura, M. (1993). Interaction of homologues of Hsp70 and Cpn60 with ferredoxin-NADP+ reductase upon its import into chloroplasts. *FEBS Lett.* **320**, 198–202.
174 Dionisi, H. M., Checa, S. K., Krapp, A. R., Arakaki, A. K., Ceccarelli, E. A., Carrillo, N. & Viale, A. M. (1998). Cooperation of the DnaK and GroE chaperone systems in the folding pathway of plant ferredoxin-NADP+ reductase expressed in Escherichia coli. *Eur. J. Biochem.* **251**, 724–728.
175 Li, X., Henry, R., Yuan, J., Cline, K. & Hoffman, N. E. (1995). A chloroplast homologue of the signal recognition particle subunit SRP54 is involved in the posttranslational integration of a protein into thylakoid membranes. *Proc Natl Acad Sci USA.* **92**, 3789–93.
176 Yalovsky, S., Paulsen, H., Michaeli, D., Chitnis, P. R. & Nechushtai, R. (1992). Involvement of a chloroplast HSP70 heat shock protein in the integration of a protein (light-harvesting complex protein precursor) into the thylakoid membrane. *Proc. Natl. Acad. Sci. USA.* **89**, 5616–5619.
177 Yuan, J., Henry, R. & Cline, K. (1993). Stromal factor plays an essential role in protein integration into thylakoids that cannot be replaced by unfolding or by heat shock protein Hsp70. *Proc. Natl. Acad. Sci. USA.* **90**, 8552–8556.
178 Goloubinoff, P., Christeller, J. T., Gatenby, A. A. & Lorimer, G. H. (1989). Reconstitution of active dimeric ribulose bisphosphate carboxylase from an unfoleded state depends on two chaperonin proteins and Mg-ATP. *Nature* **342**, 884–889.
179 Viitanen, P. V., Lubben, T. H., Reed, J., Goloubinoff, P., O'Keefe, D. P. & Lorimer, G. H. (1990). Chaperonin-facilitated refolding of ribulosebis-phosphate carboxylase and ATP hydrolysis by chaperonin 60 (groEL) are K+ dependent. *Biochem.* **29**, 5665–5671.
180 Lubben, T. H., Donaldson, G. K., Viitanen, P. V. & Gatenby, A. A. (1989). Several proteins imported into chloroplasts form stable complexes with the GroEL-related chloroplast molecular chaperone. *Plant Cell* **1**, 1223–1230.
181 Chen, G. G. & Jagendorf, A. T. (1994). Chloroplast molecular chaperone-assisted refolding and reconstitution of an active multisubunit coupling factor CF1 core. *Proc. Natl. Acad. Sci. USA.* **91**, 11497–11501.
182 Choquet, Y. & Vallon, O. (2000). Synthesis, assembly and degradation of thylakoid membrane proteins. *Biochimie* **82**, 615–634.
183 Plucken, H., Muller, B.,

Grohmann, D., Westhoff, P. & Eichacker, L. A. (2002). The HCF136 protein is essential for assembly of the photosystem II reaction center in Arabidopsis thaliana. *FEBS Lett.* **532**, 85–90.

184 Liu, X.-Q. & Jagendorf, A. T. (1984). ATP-dependent proteolysis in pea chloroplasts. *FEBS Lett.* **166**, 248–252.

185 Majeran, W., Wollman, F. A. & Vallon, O. (2000). Evidence for a role of ClpP in the degradation of the chloroplast cytochrome b(6)f complex. *Plant Cell* **12**, 137–150.

186 Malek, L., Bogorad, L., Ayers, A. R. & Goldberg, A. L. (1984). Newly synthesized proteins are degraded by an ATP-stimulated proteolytic process in isolated pea chloroplasts. *FEBS Lett.* **166**, 253–257.

187 Schmidt, G. W. & Mishkind, M. L. (1983). Rapid degradation of unassembled ribulose-1,5-bisphosphate carboxylase small subunit in chloroplasts. *Proc. Natl. Acad. Sci. USA.* **80**, 2632–2636.

188 Al-Babili, S., von Lintig, J., Haubruck, H. & Beyer, P. (1996). A novel, soluble form of phytoene desaturase from Narcissus pseudonarcissus chromoplasts is Hsp70-complexed and competent for flavinylation, membrane association and enzymatic activation. *Plant J.* **9**, 601–612.

189 Bonk, M., Tadros, M., Vandekerckhove, J., Al-Babili, S. & Beyer, P. (1996). Purification and characterization of chaperonin 60 and heat-shock protein 70 from chromoplasts of Narcissus pseudonarcissus. *Plant Physiol.* **111**, 931–939.

190 Liedvogel, B., Sitte, P. & Falk, H. (1976). Chromoplasts in the daffodil: fine structure and chemistry. *Cytobiology* **12**, 155–174.

191 Schroda, M., Vallon, O., Wollman, F. A. & Beck, C. F. (1999). A chloroplast-targeted heat shock protein 70 (HSP70) contributes to the photoprotection and repair of photosystem II during and after photoinhibition. *Plant Cell* **11**, 1165–1178.

192 Schuster, G., Timberg, R. & Ohad, I. (1988). Turnover of thylakoid photosystem II proteins during photoinhibition of Chlamydomonas reinhardtii. *Eur. J. Biochem.* **177**, 403–410.

193 Yokthongwattana, K., Chrost, B., Behrman, S., Casper-Lindley, C. & Melis, A. (2001). Photosystem II damage and repair cycle in the green alga Dunaliella salina: involvement of a chloroplast-localized HSP70. *Plant Cell Physiol.* **42**, 1389–1397.

194 Schroda, M., Kropat, J., Oster, U., Rudiger, W., Vallon, O., Wollman, F. A. & Beck, C. F. (2001). Possible role for molecular chaperones in assembly and repair of photosystem II. *Biochem. Soc. Trans.* **29**, 413–418.

195 Levine, R. L., Mosoni, L., Berlett, B. S. & Stadtman, E. R. (1996). Methionine residues as endogenous antioxidants in proteins. *Proc. Natl. Acad. Sci. USA.* **93**, 15036–15040.

196 Rogalla, T., Ehrnsperger, M., Preville, X., Kotlyarov, A., Lutsch, G., Ducasse, C., Paul, C., Wieske, M., Arrigo, A. P., Buchner, J. & Gaestel, M. (1999). Regulation of Hsp27 oligomerization, chaperone function, and protective activity against oxidative stress/tumor necrosis factor alpha by phosphorylation. *J. Biol. Chem.* **274**, 18947–18956.

197 Sadanandom, A., Poghosyan, Z., Fairbairn, D. J. & Murphy, D. J. (2000). Differential regulation of plastidial and cytosolic isoforms of peptide methionine sulfoxide reductase in Arabidopsis. *Plant Physiol.* **123**, 255–264.

198 Glaczinski, H. & Kloppstech, K. (1988). Temperature-dependent binding to the thylakoid membranes of nuclear-coded chloroplast heat-shock proteins. *Eur. J. Biochem.* **173**, 579–583.

199 Sabehat, A., Lurie, S. & Weiss, D. (1998). Expression of small heat-shock proteins at low temperatures. A possible role in protecting against chilling injuries. *Plant Physiol.* **117**, 651–658.

200 Ishikawa, A., Tanaka, H., Nakai, M. & Asahi, T. (2003). Deletion of a

chaperonin 60 beta gene leads to cell death in the Arabidopsis lesion initiation 1 mutant. *Plant Cell Physiol.* **44**, 255–261.

201 Arnon, D. (1949). Copper enzyme in isolated chloroplasts. Polyphend oxidase in Beta vulgaris. *Plant. Physiol.* **24**, 1–15.

202 Laemmli, U. K. (1970). Cleavage of structural proteins during assembly of phage T4 head. *Nature* **227**, 680–685.

203 Lissin, N. M. (1995). In vitro dissociation of self-assembly of three chaperonin 60s: the role of ATP. *FEBS Lett.* **361**, 55–60.

204 Török, Z., Goloubinoff, P., Horváth, I., Tsvetkova, N. M., Glatz, A., Balogh, G., Varvasovski, V., Los, D. A., Vierling, E., Crowe, J. h., and Vigh, L. (2001) Synechocystis Hsp17 is an amphitropic protein that stabililizes heat-stressed membranes and binds denaturated proteins for subsequent chaperone-mediated refolding. *Proc. Natl. Acad. Sci. USA.* **98**, 3098–3103.

205 Török, Z., Horváth, I., Goloubinoff, P., Kovács, E., Glatz, A., Balogh, G., and Vigh, L. (1997) Evidence for a lipochaperonin: association of active protein-folding GroESL oligomers with lipids can stabilize membranes under heat shock conditions. *Proc. Natl. Acad. Sci. USA.* **94**, 2192–2197.

33
An Overview of Protein Misfolding Diseases

Christopher M. Dobson

33.1
Introduction

The ability of even the most complex protein molecules to fold to their biologically functional states is perhaps the most fundamental example of biological self-assembly, one of the defining characteristics of living systems [1]. In recent years, considerable progress has been made in understanding both the general principles of protein folding and the specific structural transitions that are involved in the folding of individual proteins [2, 3]. In addition to the detailed mechanistic studies that have largely been carried out in vitro, considerable efforts have been directed at understanding the manner in which folding occurs in vivo. Although the fundamental principles underlying the mechanism of folding are unlikely to differ in any significant manner from those elucidated from in vitro studies, many details of the way in which individual proteins fold undoubtedly depend on the environment in which they are located. In particular, given the complexity and stochastic nature of the folding process, it is inevitable that misfolding events will occur under some circumstances [4]. As misfolded or even incompletely folded chains inevitably expose to the outside world many regions of the polypeptide molecule that are buried in the native state, such species are prone to aberrant interactions with other molecules within the complex and crowded cellular environment [5]. Such interactions can result both in the disruption of normal cellular processes and in self-association or aggregation. In order to cope with the inevitable issues of misfolding and aggregation, therefore, living systems have evolved a range of elaborate strategies, including molecular chaperones, folding catalysts, and quality-control and degradation mechanisms [6, 7].

The fact that proteins can misfold and interact inappropriately with one another is well established. For example, one of the major problems in generating recombinant protein for research purposes or for use in biotechnology is recovering significant quantities of soluble and fully active protein from in vitro refolding procedures carried out in the laboratory [8]. The fact that misfolding, and its consequences such as aggregation, can be an important feature of the behavior of proteins in vivo is clearly evident from observations that the levels of certain pro-

Protein Folding Handbook. Part II. Edited by J. Buchner and T. Kiefhaber

ISBN: 3-527-30784-2

teins known to be able to deal with such problems are increased substantially during cellular stress [9]. Indeed, many molecular chaperones were first recognized in prokaryotic (bacterial) systems that had been subjected to stress generated by exposure to elevated temperatures, as their nomenclature as heat shock proteins (Hsps) indicates. As well as assisting folding and preventing incompletely folded chains from forming misfolded species, it is clear that some molecular chaperones are able to rescue proteins that have misfolded and allow them a second chance to fold correctly. There are also examples of molecular chaperones that are known to be able to solubilize misfolded aggregates under at least some circumstances. Such active intervention requires energy, and, not surprisingly, ATP is required for many of the molecular chaperones to function correctly [5]. Despite the fact that molecular chaperones are often expressed at high levels only in stressed systems, it is clear that they have a critical role to play in all organisms even when present at lower levels under normal physiological conditions.

In eukaryotic systems, many of the proteins that are synthesized in a cell are destined for secretion to the extracellular environment. These proteins are translocated into the ER, where folding takes place prior to secretion through the Golgi apparatus. The ER contains a wide range of molecular chaperones and folding catalysts to promote efficient folding; in addition, the proteins involved are subject to stringent quality control [6]. The quality-control mechanism in the ER can involve a complex series of glycosylation and deglycosylation processes and acts to prevent misfolded proteins from being secreted from the cell. Once recognized, unfolded and misfolded proteins are targeted for degradation through the ubiquitin-proteasome pathway [7]. The details of how these regulatory systems operate provide remarkable evidence of the stringent mechanisms that biology has established to ensure that misfolding and its consequences are minimized. The importance of such a process is underlined by the fact that recent experiments suggest that up to half of all polypeptide chains fail to satisfy the quality-control mechanism in the ER, and for some proteins the success rate is even lower [10]. Like the "heat shock response" in the cytoplasm, the "unfolded protein response" in the ER is upregulated during stress and, as we shall see below, is strongly linked to the avoidance of misfolding diseases.

33.2 Protein Misfolding and Its Consequences for Disease

Folding and unfolding are the ultimate ways of generating and abolishing specific cellular activities, and unfolding is also the precursor to the ready degradation of a protein [11]. Moreover, it is increasingly apparent that some events in the cell, such as translocation across membranes, can require proteins to be in unfolded or partially folded states. Processes as apparently diverse as trafficking, secretion, the immune response, and the regulation of the cell cycle are in fact now recognized to be directly dependent on folding and unfolding [12]. It is not surprising, therefore, that failure to fold correctly, or to remain correctly folded, gives rise to the malfunc-

Tab. 33.1. Representative protein-folding diseases (adapted from Ref. [14]).

Disease	*Protein*	*Site of folding*
Hypercholesterolemia	Low-density lipoprotein receptor	ER
Cystic fibrosis	Cystic fibrosis transmembrane regulator	ER
Phenylketonuria	Phenylalanine hydroxylase	Cytosol
Huntington's disease	Huntingtin	Cytosol
Marfan syndrome	Fibrillin	ER
Osteogenesis imperfecta	Procollagen	ER
α1-Antitrypsin deficiency	α1-Antitrypsin	ER
Tay-Sachs disease	β-Hexosaminidase	ER
Scurvy	Collagen	ER
Alzheimer's disease	Amyloid β-peptide/tau	ER
Parkinson's disease	α-Synuclein	Cytosol
Scrapie/Creutzfeldt-Jakob disease	Prion protein	ER
Familial amyloidoses	Transthyretin/lysozyme	ER
Retinitis pigmentosa	Rhodopsin	ER
Cataract	Crystallins	Cytosol
Cancer	p53	Cytosol

tioning of living systems and therefore to disease. Indeed, it is becoming increasingly evident that a wide range of human diseases is associated with aberrations in the folding process (Table 33.1) [13, 14]. Some of these diseases (e.g., cystic fibrosis) can be attributed to the simple fact that if proteins do not fold correctly, they will not able to exercise their proper function; such misfolded species are often degraded rapidly within the cell. Others (e.g., disorders associated with α1-antitrypsin) result at least in part from the failure of proteins to be trafficked to the appropriate organs in the body [15, 16]. In other cases, misfolded proteins escape all the protective mechanisms discussed above and form intractable aggregates within cells or in extracellular space. An increasing number of pathologies – including Alzheimer's and Parkinson's diseases, the spongiform encephalopathies, and late-onset diabetes – are known to be directly associated with the deposition of such aggregates in tissue (Table 33.2) [13, 14, 17–19]. Diseases of this type are among the most debilitating, socially disruptive, and costly in the modern world, and they are becoming increasingly prevalent as our societies age and as new agricultural, dietary, and medical practices are adopted [20].

One of the most characteristic features of many of the aggregation diseases is that they give rise to the deposition of proteins in the form of amyloid fibrils and plaques [17, 18]. Such deposits can form in the brain, in vital organs such as the liver and spleen, or in skeletal tissue, depending on the disease involved. In the case of neurodegenerative diseases, the quantity of such aggregates can be almost undetectable in some cases, while in systemic diseases, kilograms of protein can be found in such deposits. Each amyloid disease primarily involves the aggregation of a specific peptide or protein, although a range of additional components, including other proteins and carbohydrates, are also incorporated into the deposits when they

Tab. 33.2. Examples of diseases associated with amyloid deposition (adapted from Refs. [14] and [18]).

Clinical syndrome	*Fibril component*	*Type*
Alzheimer's disease	Aβ peptide, 1-42, 1-43	Organ limited
Spongiform encephalopathies	Full-length prion or fragments	Organ limited
Primary systemic amyloidosis	Intact light chain or fragments	Systemic
Secondary systemic amyloidosis	76-residue fragment of amyloid A protein	Systemic
Familial amyloidotic polyneuropathy I	Transthyretin variants and fragments	Systemic
Senile systemic amyloidosis	Wild-type transthyretin and fragments	Systemic
Hereditary cerebral amyloid angiopathy	Fragment of cystatin-C	Organ limited
Hemodialysis-related amyloidosis	β2-microglobulin	Systemic
Familial amyloidotic polyneuropathy II	Fragments of apolipoprotein A-I	Systemic
Finnish hereditary amyloidosis	71-residue fragment of gelsolin	Systemic
Type II diabetes	Islet-associated polypeptide	Organ limited
Medullary carcinoma of the thyroid	Fragments of calcitonin	Organ limited
Atrial amyloidosis	Atrial natriuretic factor	Organ limited
Lysozyme amyloidosis	Full-length lysozyme variants	Systemic
Insulin-related amyloidosis	Full-length insulin	Systemic
Fibrinogen α-chain amyloidosis	Fibrinogen α-chain variants	Systemic

form in vivo. The characteristics of the soluble forms of the 20 or so proteins involved in the well-defined amyloidoses vary greatly, ranging from intact globular proteins to largely unstructured peptide molecules, but the aggregated forms have many common characteristics [21]. Thus, amyloid deposits all show specific optical properties (such as birefringence) on binding certain dye molecules, notably Congo red; these properties have been used in diagnosis for over a century. The fibrillar structures that are characteristic of many of these types of aggregates have very similar morphologies (long, unbranched, and often twisted structures a few nanometers in diameter) and a characteristic "cross-beta" X-ray fiber diffraction pattern [21]. The latter reveals that the organized core structure is composed of β-sheets with the strands running perpendicular to the fibril axis (Figure 33.1, see p. 1100) [22]. Fibrils having the essential characteristics of those found in ex vivo deposits can be reproduced in vitro from the component proteins under appropriate conditions, showing that they can self-assemble without the need for other components.

For many years it was generally assumed that the ability to form amyloid fibrils with the characteristics described above was limited to a relatively small number of proteins, largely those seen in disease states, and that these proteins must possess specific sequence motifs encoding the amyloid core structure. However, recent studies have suggested that the ability of polypeptide chains to form such structures is common and indeed can be considered a generic feature of polypeptide chains [4, 23–25]. The most direct evidence for the latter statement is that fibrils can be formed under appropriate conditions by many different proteins that are not associated with disease, including such well-known proteins as myoglobin [26], as well as by homopolymers such as polythreonine or polylysine [27]. Remarkably, fibrils of similar appearance to those containing large proteins can be formed by peptides with just a handful of residues [28]. One can consider that amyloid fibrils are highly organized structures (effectively one-dimensional crystals) adopted by an unfolded polypeptide chain when it behaves as a typical polymer molecule; similar types of structures can be formed, for example, by synthetic polymers. The essential features of such structures are therefore determined by the physicochemical properties of the polymer chain. As with other highly organized materials (including crystals) whose structures are based on repetitive long-range interactions, the most stable structures are usually those consisting of a single type of molecular species (e.g., a specific peptide sequence) where such interactions can be optimized [25].

33.3 The Structure and Mechanism of Amyloid Formation

Studies of amyloid fibrils formed by both disease-associated and other peptides and proteins have enabled many of the features of these structures to be defined [21, 29–33], although no complete structure has yet been determined in atomic detail. It is clear that the core structure of the fibrils is stabilized primarily by interactions, particularly hydrogen bonds, involving the atoms of the extended polypeptide main chain. As the main chain is common to all polypeptides, this observation explains why fibrils formed from polypeptides of very different amino acid sequences are similar in appearance. The side chains are likely to be incorporated in whatever manner is most favorable for a given sequence within the amyloid structures; they appear to affect the details of the fibrillar assembly but not the fundamental structure of the fibril [34]. In addition, the proportion of a polypeptide chain that is incorporated in the core structure can vary substantially; in some cases only a handful of residues may be involved in such structure, with the remainder of the chain associated in some other manner with the fibrillar assembly. This generic type of structure contrasts strongly with the globular structures of most natural proteins that result from the uniquely favorable packing of a given set of side chains into a particular fold. On the conceptual model outlined here, structures such as amyloid fibrils occur because interactions associated with the highly specific packing of the side chains can sometimes override the conformational preferences of the main chain [24–27]. The strands and helices so familiar in the structures of native proteins are then the most stable structures that the main chain can adopt in the folds that are primarily defined by the side chain interactions. However, if the solution environment (pH, temperature, concentration, etc.) in which the molecules are found is such that these side chain interactions are insufficiently

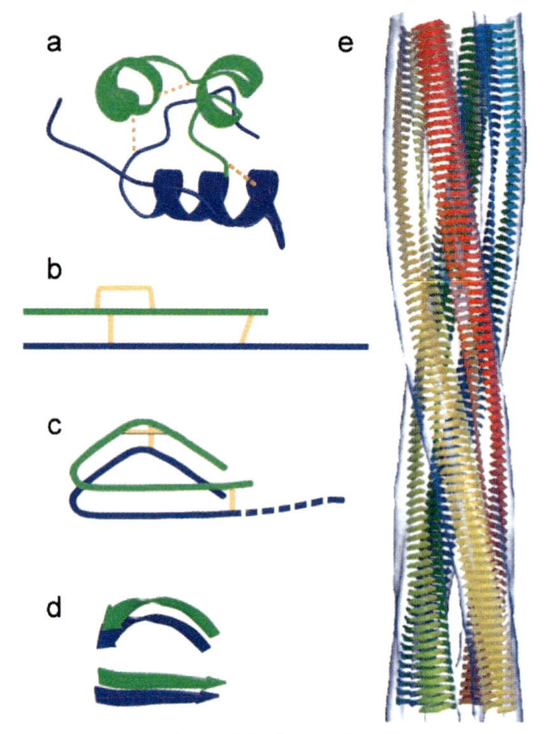

Fig. 33.1. Molecular model of an amyloid fibril. This model is derived from cryo-EM analysis of fibrils grown from bovine insulin whose native topology is indicated in (a). The two chains (A, green, and B, blue) are connected by three disulfide bridges (b). A possible topology for insulin in the fibrils is illustrated in (c), while (d) indicates how strands could be assembled in a complete fibril, a model of which is shown in (e). The fibril consists of four "protofilaments" that twist around one another to form a regular pattern with a diameter of approximately 10 nm (from Ref. [31]).

stable, the structures can unfold and may then, at least under some circumstances, reassemble in the form of amyloid fibrils.

In order to understand the occurrence and significance of amyloid formation in biological systems, it is essential to establish the mechanism by which such structures are assembled from the soluble precursor species. In globular proteins the polypeptide main chain is largely buried within the folded structure. Conditions that favor formation of any form of aggregate by such species are therefore ones in which the molecules involved are at least partially unfolded, as occurs, for example, at low pH or elevated temperature [23, 35]. Because of the importance of the globular fold in preventing aggregation, the fragmentation of proteins, through proteolysis or other means, is another ready mechanism to stimulate amyloid formation. Indeed, many amyloid disorders, including Alzheimer's disease, involve aggregation of fragments of larger precursor proteins that are unable to fold in the absence of the remainder of the protein structure (see Table 33.2). Experiments in vitro indicate that the formation of fibrils, by appropriately destabilized or fragmented proteins, is then generally characterized by a lag phase followed by a period of relatively rapid growth [36]. Behavior of this type is typical of nucleated processes such as crystallization; as with crystallization, the lag phase in amyloid formation can generally be eliminated by addition of preformed fibrils to fresh solutions, a process known as seeding [36]. Although the specific events taking place during fibril growth are not yet elucidated in any detail, it is becoming possible to simulate the overall kinetic profiles using relatively simple models that incorporate well-established principles of nucleated processes [37].

One of the key findings of studies of the formation of amyloid fibrils is that there are many common features in the behavior of the different systems that have been examined [19, 36, 38, 39]. The first phase of the aggregation process generally appears to involve the formation of more or less disordered oligomeric species as a result of relatively nonspecific interactions, although in some cases specific structural transitions, such as domain swapping [40], may be involved if such processes increase the rate of aggregation. The earliest species visible by electron or atomic force microscopy often resemble small bead-like structures, sometimes described as amorphous aggregates or as micelles. These early "pre-fibrillar aggregates" then appear to transform into species with more distinctive morphologies, sometimes described as "protofibrils" or "protofilaments" [39, 41]. These structures are commonly short, thin, sometimes curly, fibrillar species that are thought to be able to assemble into mature fibrils, perhaps by lateral association accompanied by some degree of structural reorganization [42] (Figure 33.2). The extent to which aggregates dissolve and reassemble into more regular structures involved at the different stages of fibril assembly is not clear, but it could well be important under the slow growth conditions in which the most highly structured fibrils are formed. The earliest aggregates are likely in most situations to be relatively disorganized structures that expose to the outside world a variety of segments of the protein that are normally buried in the globular state. In other cases, however, such species appear to adopt distinctive structures, including the well-defined "doughnut-shaped" species that have been seen for a number of systems (Figure 33.2) [39, 44].

Although the ability of polypeptides to form amyloid fibrils appears generic, the propensity to do so varies dramatically with amino acid composition and sequence. At the most fundamental level, some types of amino acids are much more soluble than others, such that the concentration required for aggregation to occur will be much greater for some polypeptides than for others. In addition, the aggregation process, like crystallization, needs to be nucleated and the rates at which this process takes place can be highly dependent on many different factors. It is clear that even single changes of amino acids in protein sequences can change the rates at which the unfolded polypeptide chains aggregate by an order of magnitude or more [45]. Moreover, it has recently proved possible to correlate changes in aggregation rates caused by mutations with changes in simple properties that result from substitutions, such as charge, secondary structure propensities, and hydrophobicity [45]. As this correlation has been found to hold for a wide range of different sequences (see Figure 33.3), it strongly endorses the concept of a common

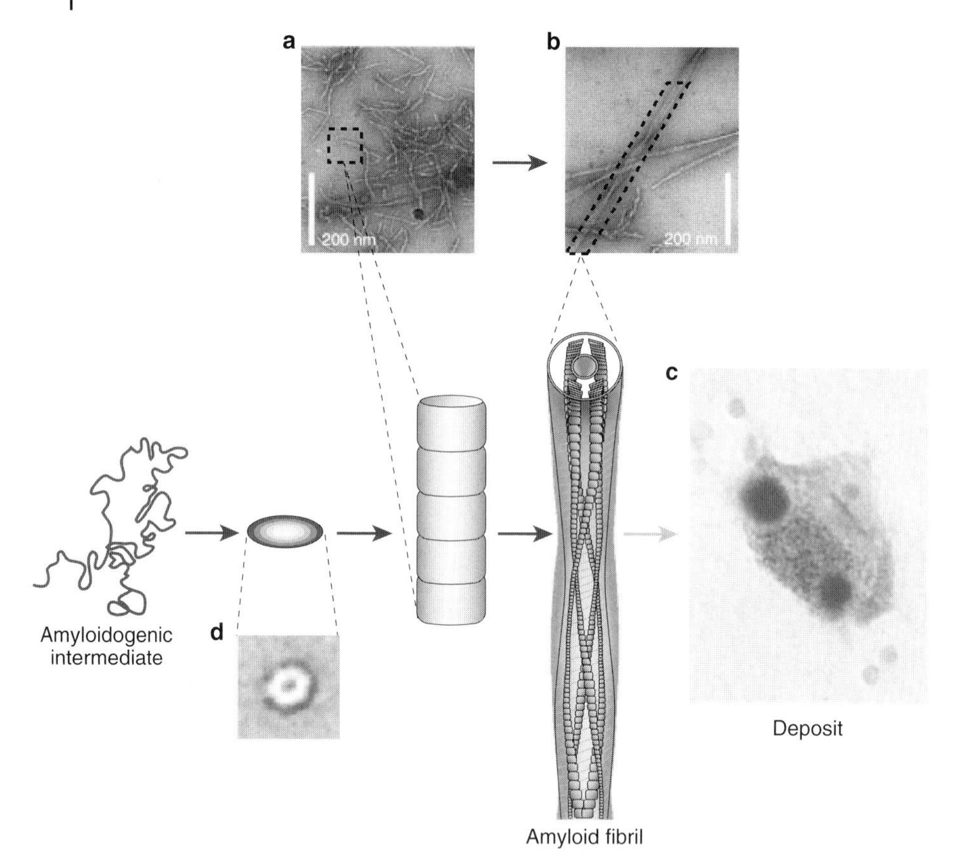

Fig. 33.2. Schematic representation of the formation of amyloid fibrils and pathological deposits. An unfolded or partially unfolded peptide or protein initially forms small soluble aggregates that then assemble to form a variety of "pre-fibrillar" species, some or all of which appear to be toxic to living cells. These species undergo a series of additional assembly and reorganizational steps to give ordered amyloid fibrils of the type illustrated in Figure 33.3. Pathological deposits such as the Lewy bodies associated with Parkinson's disease [43] frequently contain large assemblies of fibrillar aggregates, along with other components including molecular chaperones (from Ref. [4]).

mechanism of amyloid formation. In accordance with such ideas, those proteins that are completely or partially unfolded under normal conditions in the cell have sequences that are likely to have very low propensities to aggregate. An interesting and potentially important additional observation is that specific regions of the polypeptide chain appear to be responsible for nucleating the aggregation process [46]. Interestingly, the residues that nucleate the folding of a globular protein seem to be distinct from those that nucleate its aggregation into amyloid fibrils. Such a characteristic, which may reflect the different nature of the partially folded species that initiate the assembly processes, offers the opportunity for evolutionary pressure to select sequences that favor folding over aggregation [46, 47].

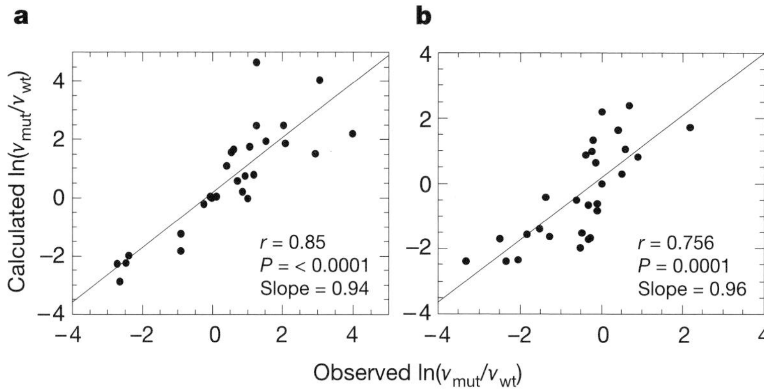

Fig. 33.3. Rationalization of the effects of mutations on the aggregation rates of peptides and proteins. The experimental aggregation rates of a variety of short peptides or natively unfolded proteins, including amylin, amyloid β-peptide, tau, and α-synuclein (see Tables 33.1 and 33.2 for a summary of the diseases with which these are associated), are shown plotted against rates calculated from an algorithm derived from extensive mutational studies of the protein acylphosphatase; the analogous plot of the data for acylphosphatase is shown in (b). The correlation shown in (a) argues strongly for a common mechanism for amyloid formation and provides a platform for both predicting the effects of natural mutations and designing polypeptides with altered aggregation properties (from Ref. [45]).

33.4
A Generic Description of Amyloid Formation

It is clear that proteins can adopt different conformational states under different conditions and that there can be considerable similarities in this regard between different proteins. This situation can be summarized by using a schematic representation of the different states that are in principle accessible to a polypeptide chain (see Figure 33.4) [4, 25, 48]. In such a situation, the state that a given system populates under specific conditions will depend on the relative thermodynamic stabilities of the different states (in the case of oligomers and aggregates, the concentration will be a critical parameter) and on the kinetics of the various interconversion processes. In this diagram, amyloid fibrils are included as just one of the types of aggregates that can be formed by proteins, although they have particular significance in that their highly organized, hydrogen-bonded structure gives them unique kinetic stability. This type of representation emphasizes that complex biological systems have become robust by controlling and regulating the various states accessible to a given polypeptide chain at given times and under given conditions, just as they regulate and control the various chemical transformations that take place in the cell [4, 24]. The latter is achieved primarily through enzymes and the former by means of the molecular chaperones and degradation mechanisms mentioned above. And just as the aberrant behavior of enzymes can cause metabolic disease, the aberrant behavior of the chaperone and other machinery regulating polypeptide conformations can contribute to misfolding and aggregation diseases [49].

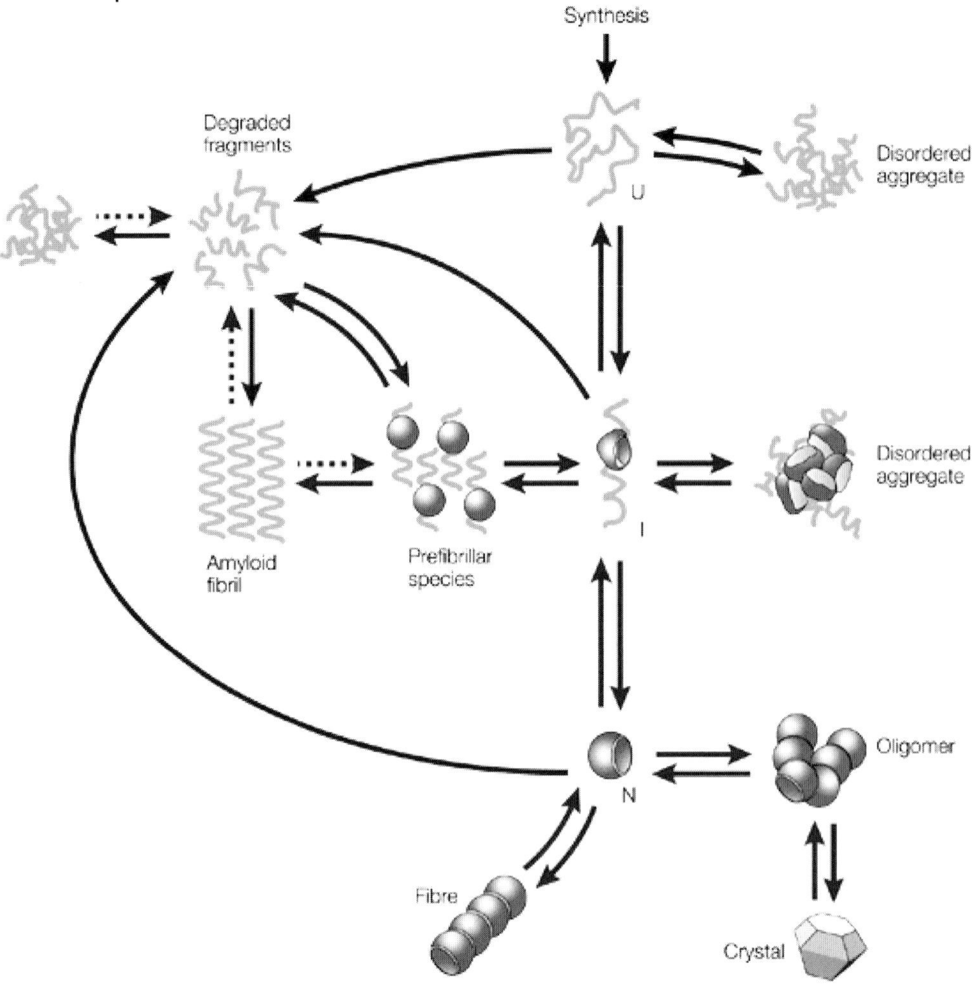

Fig. 33.4. Schematic representation of some of the states accessible to a polypeptide chain following its synthesis on a ribosome. The relative populations of the different states depend on the kinetics and thermodynamics of the various equilibria shown in the diagram. In living systems the fate of a given molecule is closely regulated, using mechanisms such as those illustrated in Figures 33.1 and 33.2, rather like metabolic pathways in cells are controlled by enzymes and associated molecules such as cofactors (from Ref. [25]).

The type of diagram shown in Figure 33.4 serves as a framework for understanding the fundamental molecular events that underlie the regulatory and quality-control mechanisms and the origins of the amyloid diseases that can result when these fail or are overwhelmed. As we have discussed above, partially or completely unfolded polypeptides are highly aggregation-prone and represent the species that trigger amyloid formation. Such species are, however, inherent in the folding process, and for this reason a variety of molecular chaperones is present in abundance in the cellular compartments wherever such processes occur. It is possible, however, that chaperones also exist in environments where *ab initio* folding does not take place. Indeed, an extracellular chaperone (clusterin) has been identified recently [50]. Nevertheless, it is undoubtedly important in the context of avoiding aggregation that proteins are correctly folded prior to their secretion from the cell, hence the need for a highly effective system of quality control in the ER. In this context it is interesting that the majority of the diseases associated with amyloid formation involve deposits that are extracellular in nature, although it is not yet clear exactly where their formation in fact takes place [19]. As biology is a dynamic process, there is a continuous need to degrade as well as synthesize proteins, and the degradative mechanisms target misfolded as well as redundant proteins. It is during such processes, which require unfolding and proteolysis of polypeptide chains, that aggregation may be particularly likely. Degradation pathways, such as those of the ubiquitin-proteasome system, are therefore highly regulated in order to avoid the occurrence of such events [7, 51].

In order to understand misfolding and aggregation diseases, we need to know not only how such systems function efficiently but also why they fail to do so under some circumstances [51–53]. The effects of many pathogenic mutations can be particularly well understood from the schematic representation given in Figure 33.4. Many of the mutations associated with the familial deposition diseases increase the population of partially unfolded states either within or outside the cell by decreasing the stability or cooperativity of the native state [54–56]. Cooperativity is in fact a crucial factor in enabling proteins to remain soluble, as it ensures that, even for a protein that is marginally stable, the equilibrium population of unfolded molecules or of unfolded regions of the polypeptide chain is minimal [24]. Other familial diseases are associated with the accumulation of fragments of native proteins, which are often produced by aberrant processing or incomplete degradation; such species are usually unable to fold into aggregation-resistant states. Other pathogenic mutations act by enhancing the propensities of such species to aggregate, for example, by increasing their hydrophobicity or decreasing their charge [45]. In the case of the transmissible encephalopathies, it is likely that ingestion of pre-aggregated states of an identical protein (e.g., by cannibalistic means or by contamination of surgical instruments) dramatically increases the rate of aggregation within the individual concerned and hence underlies the mechanism of transmission [36, 57, 58]. Such seeding phenomena may also at least partly explain why some deposition conditions such as Alzheimer's disease progress so rapidly once the initial symptoms are evident [36, 59].

33.5
The Fundamental Origins of Amyloid Disease

Despite our increasing knowledge of the general principles that underlie protein misfolding diseases, the manner in which improperly folded proteins and aggregated proteins can generate pathological behavior is not yet understood in detail. In the case of systemic disease, the sheer mass of protein that can be deposited may physically disrupt the functioning of specific organs [17]. In other cases it may be that the loss of functional protein results in the failure of some crucial cellular process [39]. For neurodegenerative disorders such as Alzheimer's disease, however, it appears that the primary symptoms arise from the destruction of cells such as neurons by a "toxic gain of function" that results from the misfolding and aggregation process [19, 39]. It has recently become apparent that the early pre-fibrillar aggregates of proteins associated with neurological disorders can be highly damaging to cells; by contrast, the mature fibrils are relatively benign [39, 60]. There is evidence, however, that the toxic nature of protein aggregates is not restricted to species formed from the peptides and proteins associated with pathological conditions. Experiments have recently indicated that pre-fibrillar aggregates of several proteins that are not connected with any known diseases have a degree of toxicity comparable to that of similar species formed from the Aβ-peptide [61]. The concept that the effects of such aggregates on cells are more strongly related to their generic nature than to their specific sequence has recently been reinforced through experiments with polyclonal antibodies that cross-react with early aggregates of different peptides and proteins and, moreover, inhibit their toxicity in cellular assays [62]. It is possible that there are specific mechanisms for such toxicity, for example, through the doughnut-shaped aggregates that resemble the toxins produced by bacteria that form pores in membranes and disrupt the ion balance in cells [44]. It is also possible that disorganized early aggregates, or even misfolded monomers, are toxic through a less specific mechanism, e.g., because exposed, nonnative hydrophobic surfaces may stimulate aberrant interactions with cell membranes or other cellular components [53, 63, 64].

Such findings raise the question as to how cellular systems are able to tolerate the intrinsic tendency of incompletely folded proteins to aggregate. The answer is almost certainly that under normal circumstances the molecular chaperones and other "housekeeping" mechanisms are remarkably efficient in ensuring that such potentially toxic species are neutralized [5, 61]. Molecular chaperones of various types are able to shield hydrophobic regions, to unfold some forms of aggregates, or to alter the partitioning between different forms of aggregates. The latter mechanism, for example, could convert the precursors of amyloid fibrils into less intractable species, allowing them to be refolded or disposed of by the cellular degradation systems. Indeed, evidence has been obtained that such a situation occurs with polyglutamine sequences associated with disorders such as Huntington's disease [65]. In this case the precursors of amyloid fibrils appear to be diverted into amorphous, and hence more readily degradable, species by the action of molecular chaperones. If such protective processes fail, it may be possible for potentially harmful species to be sequestered in relatively harmless forms such as inclusion bodies in bacteria or aggresomes in eukaryotic systems. Indeed, it has been suggested that the formation of mature amyloid fibrils, whose toxicity appears to be much lower than that of their precursors (as we have discussed above), may itself represent a protective mechanism in some cases [19, 36].

Most of the aggregation diseases, however, do not result from genetic mutations or infectious agents but rather are sporadic or associated with old age. The ideas summarized in this chapter offer a qualitative explanation of why such a situation arises. We have seen that all proteins have an inherent tendency to aggregate unless they are maintained in a highly regulated environment. Selective pressure during evolution has therefore resulted in protein molecules that are normally able to resist aggregation during our normal lifespan, enabling us to develop, to pass on genes, and to give appropriate protection to our offspring. Evolution, however, can only generate sequences and cellular environments that are adequate for those conditions under which selective pressure exists [20]. There is no reason to suppose that random mutations will in general further reduce the propensity to aggregate; indeed they will generally increase it – just as random mutations generally reduce the stability of native proteins. We can see, therefore, that our recent ability to prolong life [66] is leading to the proliferation of these diseases as the limitations in their ability to resist reversion to aggregates, including the intractable and damaging amyloid structures, become evident. However, it is intriguing to speculate that favorable mutations in aggregation-prone proteins might be the reason that some in the population do not readily succumb to diseases such as Alzheimer's even in extreme old age [45].

The link with aging is likely to involve more than just a greater probability of aggregation taking place as we get older. It is likely to be linked more fundamentally to the failure of the "housekeeping" mechanisms in our bodies under such circumstances [20, 67]. This failure may in part be a result of the need for greater protective capacity in old age as aggregation becomes more prevalent, perhaps as a consequence of the increasing accumulation of misfolded and damaged proteins, leading to chaperone overload [68]. But as we age, it is likely that the activity of our chaperone response and degradative mechanisms declines, resulting in the increasing probability that the protective mechanisms are overwhelmed. One reason for this decline, although there may be many, is that cells become less efficient in producing the ATP that is needed for the effective functioning of many chaperones. Similarly, the rapidity with which we have introduced practices that have not been experienced previously in history – including new agricultural practices (associated with the emergence of BSE and variant CJD [57]), a changing diet and lifestyle (associated with the prevalence of type II diabetes [69]), and new medical procedures (e.g., resulting in iatrogenic diseases such as CJD [57] and amyloid deposition in hemodialysis, during which the concentration of β-2 microglobulin in serum increases [70]) – means that we have not had time to evolve effective protective mechanisms [20]. Thus, we are now increasingly observing the limitations of our proteins and their environments to resist aggregation under conditions that differ from those under which natural selection has taken place.

33.6
Approaches to Therapeutic Intervention in Amyloid Disease

The progress made in understanding the underlying causes of the amyloid diseases is also leading to new approaches to their prevention and treatment [71, 72]. For proteins whose functional state is a tightly packed globular fold, we have seen that an essential first step in fibril formation is undoubtedly the partial or complete unfolding of the native structure that otherwise protects the aggregation-prone polypeptide backbone. Thus, many of the familial forms of amyloid disease are associated with genetic mutations that decrease protein stability and thereby promote unfolding. In such cases, one approach to therapy is to find ways of stabilizing the native states of amyloidogenic protein variants. In accordance with this proposal, a recent study has reported a series of small-molecule analogues of the hormone thyroxine, the natural ligand of transthyretin, that act in this manner. Transthyretin is the protein associated with the most common type of systemic amyloidosis, and these potential drugs have been found to dramatically inhibit the rate at which the disease-associated variants aggregate, at least in vitro [73]. In a similar approach, specific antibodies raised against lysozyme have been found to prevent amyloid fibril formation by pathogenic forms of this well-know antibacterial protein whose aggregation is linked to another systemic amyloidosis [74]. Moreover, quinacrine, a drug previously used to combat malaria, has been found to limit dramatically the replication in cell cultures of the pathogenic form of the prion protein associated with CJD [75]. The finding that this molecule interacts with the soluble form of the protein [76] suggests that a similar mechanism could be operating here. Quinacrine is already entering clinical trials, and determined efforts are now underway to find more potent forms of this compound [75].

A possible therapeutic strategy for those amyloid diseases that result from the aggregation of proteolytic fragments rather than intact proteins would be to reduce the levels of the aggregation-prone peptides by inhibiting the enzymes through whose action the fragments are generated. An extremely important example of this approach involves the 40- or 42-residue Aβ-peptide derived from the amyloid precursor protein (APP), as the aggregation of this species is linked to Alzheimer's disease. Much is now known about the complex secretase enzymes whose role in processing APP gives rise to the Aβ-peptide, and a number of potent inhibitors have already been developed as potential therapeutics for this debilitating neurodegenerative disease [77, 78]. Another way of reducing the levels of aggregation-prone species in an organism is to enhance their clearance from the body. A number of strategies based on this general idea have emerged recently, some of the most exciting of which involve the action of antibodies raised against specific amyloid-forming polypeptides. Of particular interest is the finding that active immunization with Aβ-peptides can result in the extensive clearance of amyloid deposits in transgenic mice that overexpress the Aβ-peptide. Although the first set of clinical trials with human Alzheimer victims was terminated as a result of an inflammatory response in a significant number of patients, the potential of this type of approach has been clearly demonstrated [79]. One possible variation on the original procedure is to use passive immunotherapy, in which antibodies are infused into, rather than generated within, the patient. Additionally, there are means of stimulating clearance other than by the direct action of antibodies. In a study of the effects of a ligand designed to bind to serum amyloid P (SAP), a protein that is associated with amyloid deposits in vivo and is known to inhibit natural clearance mechanisms, the levels of SAP in serum were reduced dramatically [80]. It remains to be established how such molecules fare in ongoing clinical trials, but the general principle of using small molecules to target specific proteins for degradation is an extremely interesting development.

Enhancing the clearance of the amyloidogenic species may well be of particular importance in other forms of amyloid disorders that involve the deposition of peptides and proteins that, although intact, appear not to fold to globular structures even under normal physiological conditions. Examples of such cases include α-synuclein, whose aggregation is associated with Parkinson's disease, and IAPP (amylin), which is involved in type II diabetes [14, 18]. Another strategy that has been proposed is to intervene in the aggregation process directly by means of small peptides or peptide analogues designed to bind tightly to fibrils as they form, hence blocking their further growth. A variety of such species is being explored, particularly in the context of Alzheimer's disease [81]. A potential problem with this approach is that inhibition of the formation of amyloid fibrils might occur at the expense of the small aggregates that are their precursors. Such an effect could be highly counterproductive, as the latter appear to be the primary pathogenic agents, at least in the neurodegenerative forms of amyloid disease, as we discuss above. But provided that problems of this type can be avoided, this approach could potentially be rather generally applicable.

It is a particularly satisfying aspect of research directed at understanding the origins of the amyloid disorders to find that so many different therapeutic approaches can be rationalized using the relatively simple framework illustrated in Figure 33.4 [72]. Particularly, as there are a number of different stages in the aggregation process where intervention is potentially possible, one can be optimistic that novel and efficacious forms of treatment will emerge in the not-too-distant future. Moreover, with the increasing evidence for the generic nature of the amyloid conditions, it is interesting to speculate that there might be generic approaches to inhibiting a variety, perhaps even all, of the amyloid diseases with a single drug [25, 72]. That such an idea could in principle be realized is suggested by the observation that antibodies raised against small aggregates of the Aβ-peptide not only recognize but also suppress the toxicity of similar aggregates formed from other proteins [62]. An approach of this type would be particularly dramatic in light of the number of different types of amyloid disorders that increasingly place the aging populations of the world at risk. Alternatively, one might even speculate that a means could be found to extend the period of time that our natural defenses are able to combat aggregation, and hence defer the onset of these disorders. And looking perhaps even further into the future, our ability to design peptides and proteins whose tendencies to aggregate are reduced, often substantially, provides the basis for a range of exciting approaches involving gene therapy and stem cell techniques [45].

33.7 Concluding Remarks

This chapter has discussed the state of our developing knowledge of the origin of amyloid disease from the perspective of our increasing understanding of the fundamental properties of proteins. We have stressed the fact that the structure and biological effects of amyloid deposition can be considered to be generic and that this fact can be rationalized, at least in outline, through the ideas of polymer science. The propensities of different sequences to aggregate under specific conditions, however, can vary significantly, and it is becoming possible to explain the origins of pathogenic behavior in terms of the factors that affect such propensities. The amyloid diseases can be thought of as fundamentally resulting from the reversion of normally soluble proteins to the generic amyloid structure as a result of the failures of the mechanisms that have evolved to prevent such behavior. A range of novel approaches is now being developed to combat the various amyloid diseases and can to a large extent be rationalized in terms of intervention at different steps in the process of aggregation as represented in Figure 33.4. Many of these approaches show considerable promise for the treatment of specific diseases, and the generic nature of amyloid formation raises the exciting possibility that there could be generic approaches to preventing some or indeed all of the various members of the family of diseases. We can therefore be optimistic that the better understanding of protein misfolding and aggregation that is developing from recent research of the type discussed in this article will enable us to devise increasingly effective strategies for drug discovery based on rational arguments.

Acknowledgements

The ideas in this article have emerged from extensive discussions with outstanding students, research fellows, and colleagues over many years. I am most grateful to all of them; they are too numerous to mention here, but the names of many appear in the list of references. The content of this chapter is derived in part from a review article published in *Nature* in 2003 [4] and a perspective published in *Science* in 2004 [72]. The research of C.M.D. is supported by Programme Grants from the Wellcome Trust and the Leverhulme Trust, as well as by the BBSRC, EPSRC, and MRC.

References

1 M. Vendruscolo, J. Zurdo, C. E. MacPhee and C. M. Dobson, "Protein Folding and Misfolding: A Paradigm of Self-Assembly and Regulation in Complex Biological Systems", *Phil. Trans. R. Soc. Lond.*, **A 361**, 1205–1222 (2003).

2 A. R. Fersht, *Structure and Mechanism in Protein Science; A Guide to Enzyme Catalysis and Protein Folding*, W.H. Freeman, New York, 1999.

3 R. H. Pain (Ed.), *Protein Folding* (2nd ed.) Oxford University Press, Oxford, 2000.

4 C. M. Dobson, "Protein Folding and Misfolding", *Nature*, **426**, 884–890 (2003).

5 F. U. Hartl and M. Hayer-Hartl, "Molecular Chaperones in the Cytosol: From Nascent Chain to Folded Protein", *Science*, **295**, 1852–1858 (2002).

6 R. Sitia and I. Braakman, "Quality Control in the Endoplasmic Reticulum Protein Factory", *Nature*, **426**, 891–894 (2003).

7 A. L. Goldberg, "Protein Degradation and Protection against Misfolded or Damaged Proteins", *Nature*, **426**, 895–899 (2003).

8 R. R. Kopito, "Aggresomes, Inclusion Bodies and Protein Aggregates", *Trends Cell Biol.*, **10**, 524–530 (2000).

9 H. R. Pelham, "Speculations on the Functions of the Major Heat Shock and Glucose-Regulated Proteins", *Cell*, **46**, 959–961 (1968).

10 U. Schubert, L. C. Anton, J. Gibbs, C. C. Orbyry, J. W. Yewdell and J. R. Bennink, "Rapid Degradation of a Large Fraction of Newly Synthesized Proteins by Proteasomes, *Nature*, **404**, 770–774 (2000).

11 A. Matouschek, "Protein Unfolding – an Important Process in Vivo?", *Curr. Opin. Struct. Biol.*, **13**, 98–109 (2001).

12 S. E. Radford and C. M. Dobson, "From Computer Simulations to Human Disease: Emerging Themes in Protein Folding", *Cell*, **97**, 291–298 (1999).

13 P. J. Thomas, B. H. Qu and P. L. Pedersen, "Defective Protein Folding as a Basis of Human Disease", *Trends Biochem. Sci.*, **20**, 456–459 (1995).

14 C. M. Dobson, "The Structural Basis of Protein Folding and its Links with Human Disease", *Phil. Trans. R. Soc. Lond.*, **B 356**, 133–145 (2001).

15 K. Powell and P. L. Zeitlin, "Therapeutic Approaches to Repair Defects in ΔF508 CFTR Folding and Cellular Targeting", *Adv. Drug Deliv. Rev.*, **54**, 1395–1408 (2002).

16 R. W. Carrell and D. A. Lomas, "Mechanisms of Disease. α1-Antitrypsin Deficiency – A Model for Conformational Diseases", *N. Engl. J. Med*, **346**, 45–53 (2002).

17 M. B. Pepys, "Amyloidoses" in The Oxford Textbook of Medicine (eds D. J. Weatherall, J. G. Ledingham and D. A. Warrel), 3rd Edition Oxford University Press, Oxford, 1512–1524 (1995).

18 D. J. Selkoe, "Folding Proteins in Fatal Ways", *Nature*, **426**, 900–904 (2003).

19 E. H. Koo, P. T. Lansbury Jr., and J. W. Kelly, "Amyloid Diseases: Abnormal Protein Aggregation in Neurodegeneration", *Proc. Natl. Acad. Sci. USA*, **96**, 9989–9990 (1999).

20 C. M. Dobson, "Getting Out of Shape – Protein Misfolding Diseases", *Nature*, **418**, 729–730 (2002).

21 M. Sunde and C. C. F. Blake, "The Structure of Amyloid Fibrils by Electron Microscopy and X-ray Diffraction", *Adv. Protein Chem.*, **50**, 123–159 (1997).

22 J. L. Jiménez, J. I. Guijarro, E. Orlova, J. Zurdo, C. M. Dobson, M. Sunde and H. R. Saibil, "Cryo-electron Microscopy Structure of an SH3 Amyloid Fibril and Model of the Molecular Packing", *EMBO J.*, **18**, 815–821 (1999).

23 F. Chiti, P. Webster, N. Taddei, A.

Clark, M. Stefani, G. Ramponi and C. M. Dobson, "Designing Conditions for in vitro Formation of Amyloid Protofilaments and Fibrils", *Proc. Natl. Acad. Sci. USA*, **96**, 3590–3594 (1999).
24 C. M. Dobson, "Protein Misfolding, Evolution and Disease", *Trends Biochem. Sci.*, **24**, 329–332 (1999).
25 C. M. Dobson, "Protein Folding and Disease: A View from the First Horizon Symposium", *Nature Rev. Drug Disc.*, **2**, 154–160 (2003).
26 M. Fändrich, M. A. Fletcher and C. M. Dobson, "Amyloid Fibrils from Muscle Myoglobin", *Nature*, **410**, 165–166 (2001).
27 M. Fändrich and C. M. Dobson, "The Behavior of Polyamino Acids Reveals an Inverse Side-chain Effect in Amyloid Structure Formation", *EMBO J.*, **21**, 5682–5690 (2002).
28 M. Lopez de la Paz, K. Goldie, J. Zurdo, E. Lacrois, C. M. Dobson, A. Hoenger and L. Serrano, "De Novo Designed Peptide-based Amyloid Fibrils", *Proc. Natl. Acad. Sci. USA*, **99**, 16052–16057 (2002).
29 L. C. Serpell, C. C. F. Blake and P. E. Fraser, "Molecular Structure of a Fibrillar Alzheimer's Aβ Fragment", *Biochemistry*, **39**, 13269–13275 (2000).
30 H. Wille, M. D. Michelitsch, V. Guenebaut, S. Supattapone, A. Serban, F. E. Cohen, D. A. Agard and S. B. Prusiner, "Structural Studies of the Scrapie Prion Protein by Electron Crystallography", *Proc. Natl. Acad. Sci, USA*, **99**, 3563–3568 (1999).
31 J. L. Jiménez, E. J. Nettleton, M. Bouchard, C. V. Robinson, C. M. Dobson and H. R. Saibil, "The Protofilament Structure of Insulin Amyloid Fibrils", *Proc. Natl. Acad. Sci. USA*, **99**, 9196–9201 (2002).
32 A. T. Petkova, Y. Ishii, J. J. Balbach, O. N. Antzutkin, R. D. Leapman, F. Delaglio and R. Tycko, "A Structural Model for Alzheimer's β-amyloid Fibrils Based on Experimental Constraints from Solid State NMR", *Proc. Natl. Acad. Sci. USA*, **99**, 16742–16747 (2002).
33 C. P. Jaroniec, C. E. MacPhee, V. S. Bajaj, M. T. McMahon, C. M. Dobson and R. G. Griffin, "High Resolution Molecular Structure of a Peptide in an Amyloid Fibril Determined by Magic Angle Spinning NMR Spectroscopy", *Proc. Natl. Acad. Sci. USA*, **101**, 711–716 (2004).
34 A. Chamberlain, C. E. MacPhee, J. Zurdo, L. A. Morozova-Roche, H. A. O. Hill, C. M. Dobson and J. Davis, "Ultrastructural Organisation of Amyloid Fibrils by Atomic Force Microscopy", *Biophys. J.*, **79**, 3282–3293 (2000).
35 J. Kelly, "Alternative Conformation of Amyloidogenic Proteins and their Multi-step Assembly Pathways", *Curr. Opin. Struct. Biol.*, **8**, 101–106 (1998).
36 J. D. Harper and P. T. Lansbury Jr., "Models of Amyloid Seeding in Alzheimer's Disease and Scrapie: Mechanistic Truths and Physiological Consequences of the Time-dependent Solubility of Amyloid Proteins", *Annu. Rev. Biochem.*, **66**, 385–407 (1997).
37 S. B. Padrick and A. D. Miranker, "Islet Amyloid" Phase Partitioning and Secondary Nucleation are Central to the Mechanism of Fibrillogenesis", *Biochemistry*, **41**, 4694–4703 (2002).
38 E. J. Nettleton, P. Tito, M. Sunde, M. Bouchard, C. M. Dobson and C. V. Robinson, "Characterization of the Oligomeric States of Insulin in Self-assembly and Amyloid Fibril Formation by Mass Spectrometry", *Biophys. J.*, **79**, 1053–1065 (2000).
39 B. Caughey and P. T. Lansbury Jr., "Protofibrils, Pores, Fibrils, and Neurodegeneration: Separating the Responsible Protein Aggregates from the Innocent Bystanders", *Annu. Rev. Neurosci.*, **26**, 267–298 (2003).
40 M. P. Schlunegger, M. J. Bennett and D. Eisenberg, "Oligomer Formation by 3D Domain Swapping: A Model for Protein Assembly and Misassembly", *Adv. Protein Chem.*, **50**, 61–122 (1997).
41 G. Bitan, M. D. Kirkitadze, A. Lomakin, S. S. Vollers, G. B. Benedek and D. B. Teplow. "Amyloid β-protein assembly: Aβ40 and Aβ42

oligomerize through distinct pathways," *Proc. Natl. Acad. Sci. USA*, **100**, 330–335 (2003).

42 M. Bouchard, J. Zurdo, E. J. Nettleton, C. M. Dobson and C. V. Robinson, "Formation of Insulin Amyloid Fibrils Followed by FTIR Simultaneously with CD and Electron Microscopy", *Protein Sci.*, **9**, 1960–1967 (2000).

43 M. G. Spillantini, M. L. Schmidt, V. M.-Y. Lee, J. Q. Trojanowski, R. Jakes and M. Goedert, "α-Synuclein in Lewy Bodies", *Nature*, **388**, 839–840 (1997).

44 H. A. Lashuel, D. Hartley, B. M. Petre, T. Walz, P. T. Lansbury Jr., "Neurodegenerative Disease: Amyloid Pores from Pathogenic Mutations", *Nature*, **418**, 291 (2002).

45 F. Chiti, M. Stefani, N. Taddei, G. Ramponi and C. M. Dobson, "Rationalisation of Mutational Effects on Peptide and Protein Aggregation Rates", *Nature*, **424**, 805–808 (2003).

46 F. Chiti, N. Taddei, F. Baroni, C. Capanni, M. Stefani, G. Ramponi and C. M. Dobson, "Kinetic Partitioning of Protein Folding and Aggregation", *Nature Struct. Biol.*, **9**, 137–143 (2002).

47 S. W. Raso and J. King, "Protein Folding and Human Disease", in **Protein Folding**, R. H. Pain (Ed.), (2nd ed) Oxford University Press, Oxford, 2000, pp 406–428.

48 C. M. Dobson, "How Do We Explore the Energy Landscape for Folding?", *Simplicity and Complexity in Proteins and Nucleic Acids* (eds. H. Fraunfelder, J. Deisenhofer and P. G. Wolynes), Dahlem University Press, Berlin, pp. 15–37 (1999).

49 A. J. L. Macario and E. C. de Macario, "Sick Chaperones and Ageing: A Perspective", *Ageing Res. Rev.*, **1**, 295–311 (2002).

50 M. R. Wilson and S. B. Easterbrook-Smith, "Clusterin is a Secreted Mammalian Chaperone", *Trends Biochem Sci.*, **25**, 95–98 (2000).

51 N. F. Bence, R. M. Sampat and R. R. Kopito, "Impairment of the Ubiquitin-proteosome System by Protein Aggregation", *Science*, **292**, 1552–1555 (2001).

52 A. Horwich, "Protein Aggregation in Disease: A Role for Folding Intermediates Forming Specific Multimeric Interactions", *J. Clin Invest.*, **110**, 1221–1232 (2002).

53 M. Stefani and C. M. Dobson, "Protein Aggregation and Aggregate Toxicity: New Insights into Protein Folding, Misfolding Diseases and Biological Evolution", *J. Mol. Med.*, **81**, 678–699 (2003).

54 D. R. Booth, M. Sunde, V. Bellotti, C. V. Robinson, W. L. Hutchinson, P. E. Fraser, P. N. Hawkins, C. M. Dobson, S. E. Radford, C. C. F. Blake and M. B. Pepys, "Instability, Unfolding and Aggregation of Human Lysozyme Variants Underlying Amyloid Fibrillogenesis", *Nature*, **385**, 787–793 (1997).

55 M. Ramirez-Alvarado, J. S. Merkel and L. Regan, "A Systematic Exploration of the Influence of Protein Stability on Amyloid Fibril Formation in vitro", *Proc. Natl. Acad. Sci. USA*, **97**, 8979–8984 (2000).

56 M. Dumoulin, A. M. Last, A. Desmyter, K. Decanniere, D. Canet, A. Spencer, D. B. Archer, S. Muyldermans, L. Wyns, A. Matagne, C. Redfield, C. V. Robinson and C. M. Dobson, "A Camelid Antibody Fragment Inhibits Amyloid Fibril Formation by Human Lysozyme", *Nature*, **424**, 783–788 (2003).

57 S. Prusiner, "Prion Diseases and the BSE Crisis", *Science*, **278**, 245–251 (1997).

58 P. Chien, A. H. DePace, S. R. Collins and J. S. Weissman, "Generation of Prion Transmission Barriers by Mutational Control of Amyloid Conformations", *Nature*, **424**, 948–951 (2003).

59 M. F. Perutz and A. H. Windle, "Cause of Neuronal Death in Neuro-degenerative Diseases Attributable to Expansion of Glutamine Repeats", *Nature*, **412**, 143–144 (2001).

60 D. M. Walsh, I. Klyubin, J. V. Fadeeva, W. K. Cullen, R. Anwyl, M. S. Wolfe, M. J. Rowan and D. J.

Selkoe, "Naturally Secreted Oligomers of Amyloid Beta Protein Potently Inhibit Hippocampal Long-term Potentiation in vivo", *Nature*, **416**, 535–539 (2002).

61 M. Bucciantini, E. Giannoni, F. Chiti, F. Baroni, L. Formigli, J. Zurdo, N. Taddei, G. Ramponi, C. M. Dobson and M. Stefani, "Inherent Cytotoxicity of Aggregates Implies a Common Origin for Protein Misfolding Diseases", *Nature*, **416**, 507–511 (2002).

62 R. Kayed, E. Head, J. L. Thompson, T. M. McIntire, S. C. Milton, C. W. Cotman and C. G. Glabe, "Common Structure of Soluble Amyloid Oligomers Implies Common Mechanisms of Pathogenesis", *Science*, **300**, 486–489 (2003).

63 M. Svensson, H. Sabharwal, A. Hakansson, A. K. Mossberg, P. Lipniunas, H. Leffler, C. Svanborg, S. Linset, "Molecular Characterization of α-Lactalbumin Folding Variants that Induce Apoptosis in Tumor Cells", *J. Biol. Chem.*, **274**, 6388–6396 (1999).

64 P. Polverino de Laureto, N. Taddei, E. Frare, C. Capanni, S. Constantini, J. Zurdo, F. Chiti, C. M. Dobson and A. Fontana, "Protein Aggregation and Amyloid Fibril Formation by an SH3 Domain Probed by Limited Proteolysis", *J. Mol. Biol.*, **334**, 129–141 (2003).

65 P. J. Muchowski, G. Schaffar, A. Sittler, E. E. Wanker, M. K. Hayer-Hartl and F. U. Hartl, "Hsp70 and Hsp40 Chaperones Can Inhibit Self-assembly of Polyglutamine Proteins into Amyloid-like Fibrils", *Proc. Natl. Acad. Sci. USA*, **97**, 7841–7846 (2000).

66 J. Oeppen and J. W. Vaupel, "Broken Limits to Life Expectancy", *Science*, **296**, 1029–31 (2002).

67 P. Csermely, "Chaperone Overload is a Possible Contributor to 'Civilization Diseases'", *Trends Gen.*, **17**, 701–704 (2001).

68 A. J. L. Macario and E. C. Macario, "Sick Chaperones and Ageing: A Perspective", *Ageing Res.*, **Rev. 1**, 295–311 (2002).

69 J. W. M. Höppener, M. G. Nieuwenhuis, T. M. Vroom, B. Ahrén and C. J. M. Lips, "Role of Islet Amyloid in Type 2 Diabetes Mellitus: Consequence or Cause?", *Mol. Cell. Endocrin.*, **197**, 205–212 (2002).

70 F. Gejyo, N. Homma, Y. Suzuki and M. Arakawa, "Serum Levels of β2-microglobulin as a New Form of Amyloid Protein in Patients Undergoing Long-term Hemodialysis", *N. Engl. J. Med.*, **314**, 585–586 (1986).

71 F. E. Cohen and J. W. Kelly, "Therapeutic Approaches to Protein-misfolding Diseases", *Nature*, **426**, 905–909 (2003).

72 C. M. Dobson, "In the Footsteps of Alchemists", *Science*, **304**, 1259–1262 (2004).

73 P. Hammarstrom, R. L. Wiseman, E. T. Powers and J. W. Kelly, "Prevention of Transthyretin Amyloid Disease by Changing Protein Misfolding Energetics", *Science*, **299**, 713–716 (2003).

74 M. Dumoulin, A. M. Last, A. Desmyter, K. Decanniere, D. Canet, G. Larsson, A. Spencer, D. B. Archer, J. Sasse, S. Muyldermans, L. Wyns, C. Redfield, A. Matagne, C. V. Robinson and C. M. Dobson, "A Camelid Antibody Fragment Inhibits the Formation of Amyloid Fibrils by Human Lysozyme", *Nature*, **424**, 783–788 (2003).

75 B. C. May, A. T. Fafarman, S. B. Hong, M. Rogers, L. W. Deady, S. B. Prusiner, F. E. Cohen, "Potent Inhibition of Scrapie Prion Replication in Cultured Cells by Bis-acridines", *Proc. Natl. Acad. Sci. USA*, **100**, 3416–3421 (2003).

76 M. Vogtherr, S. Grimme, B. Elshorst, D. M. Jacobs, K. Fiebig, C. Griesinger and R. Zahn, "Antimalarial Drug Quinacrine Binds to C-terminal Helix of Cellular Prion Protein", *J. Med. Chem.*, **46**, 3563–3564 (2003).

77 M. S. Wolfe, "γ-Secretase as a Target for Alzheimer's Disease", *Curr. Top. Med. Chem.*, **2**, 371–383 (2002).

78 R. Vassar, "Beta-secretase (BACE) as a Drug Target for Alzheimer's

Disease", *Adv. Drug Deliv. Rev.*, **54**, 1589–1602 (2002).

79 J. A. R. Nicoll, D. Wilkinson, C. Holmes, P. Steart, H. Markham and R. O. Weller, "Neuropathology of Human Alzheimer Disease after Immunization with Amyloid-beta Peptide: A Case Report", *Nature Med.*, **9**, 448–452 (2003).

80 M. B. Pepys, J. Herbert, W. L. Hutchinson, G. A. Tennent, H. J. Lachmann, J. R. Gallimore, L. B. Lovat, T. Bartfai, A. Alanine, C. Hertel, T. Hoffmann, R. Jakob-Roetner, R. D. Norcross, J. A. Kemp, K. Yamamura, M. Suzuki, G. W. Taylor, S. Murray, D. Thompson, A. Purvis, S. Kolstoe, S. P. Wood, P. N. Hawkins, "Targeted Pharmacological Depletion of Serum Amyloid P Component for Treatment of Human Amyloidosis", *Nature*, **417**, 254–259 (2002).

81 C. Soto, "Unfolding the Role of Protein Misfolding in Neurodegenerative Diseases", *Nature Rev. Neurosci*, **4**, 49–60 (2003).

34
Biochemistry and Structural Biology of Mammalian Prion Disease

Rudi Glockshuber

34.1
Introduction

34.1.1
Prions and the "Protein-Only" Hypothesis

Transmissible spongiform encephalopathies (TSEs) or prion diseases in humans and mammalians have been the subject of intensive research during the last decade, and there are excellent reviews available dealing with all aspects of TSEs [1–8]. This chapter focuses on biophysical and structural studies on the recombinant prion protein (PrP) and their implications for understanding the mechanism of prion propagation in mammals.

In contrast to infectious diseases caused by microorganisms or viruses, mammalian transmissible spongiform encephalopathies (TSEs) are caused by an infectious agent, the prion (for prion terminology, see Table 34.2), which appears to be devoid of informational nucleic acid [9, 10]. According to the "protein-only" hypothesis [11–13], the prion consists mainly, if not entirely, of the insoluble, oligomeric scrapie isoform (PrP^{Sc}) of the monomeric, cellular prion protein (PrP^{C}) that is produced by the infected host [14]. The protein-only hypothesis further assumes that the subunits in PrP^{Sc} oligomers have the same covalent structure as PrP^{C} [15–17], but a different tertiary structure [18], and that PrP^{Sc} oligomers propagate by triggering the conversion of PrP^{C} (e.g., from a non-infected cell) into new PrP^{Sc} molecules. These basic assumptions are supported by the fact that highly enriched prions isolated from the brain of infected animals essentially consist of PrP^{Sc} oligomers [9, 10, 19] and, unlike nucleic acid–containing infectious agents, are resistant to strong UV irradiation [11, 20]. The strongest evidence in favor of the protein-only hypothesis comes from the fact that knockout mice devoid of the gene encoding the murine prion protein are resistant to prion infections, and transgenic mice overexpressing PrP^{C} are particularly susceptible to prions [21]. This proves that PrP is indeed an indispensable component in prion propagation. In addition, all inherited prion diseases in humans described to date are associated with mutations in the gene encoding human PrP [1, 22], arguing that these mutations favor

Protein Folding Handbook. Part II. Edited by J. Buchner and T. Kiefhaber

ISBN: 3-527-30784-2

spontaneous formation of PrP^{Sc} in the corresponding individuals. Despite these facts, the protein-only hypothesis on the nature of TSE prions still cannot be considered proven [23].

At first view, it appears that the prion phenomenon is not in accordance with one of the basic principles in protein biochemistry, namely, the dogma that polypeptide chains adopt only one specific conformation, which is the thermodynamically most stable state in aqueous solution [24]. The prion hypothesis now suggests that there are at least two conformational states of the prion protein (PrP^{C} and PrP^{Sc}) and implies that PrP^{Sc} is more stable than the functional, cellular form of the prion protein. A simple way to resolve this apparent contradiction is to restrict the central dogma of protein folding to the physiologically relevant association state of proteins in vivo. Under these prerequisites, monomeric PrP^{C} would be more stable than monomeric PrP^{Sc}, which agrees well with the fact that monomeric PrP^{Sc} has indeed never been observed. Moreover, it is practically impossible to assess the thermodynamic stability of PrP^{Sc} because it cannot be reversibly dissociated and reassembled in vitro [25]. The tertiary structure of PrP^{Sc} thus appears to be stable only in the context of the quaternary structure of the PrP^{Sc} oligomer. Overall, the thermodynamic stability of the PrP^{Sc} oligomer is not accessible experimentally, and there is no principle theoretical difficulty in combining the protein-only hypothesis with Anfinsen's central dogmas of protein folding [24]. Table 34.1 summarizes the known forms of human and mammalian TSEs.

34.1.2 Models of PrP^{Sc} Propagation

A key feature of any infectious agent is its ability to replicate. In contrast to amyloids associated with amyloid diseases in humans, including Alzheimer's and Parkinson diseases, PrP^{Sc} aggregates are the only known protein aggregates in mammalians that can be *transmitted* and cause disease by propagation in the newly infected host. Within the framework of the protein-only hypothesis, there are mainly two potential mechanisms that are discussed for replication of PrP^{Sc} aggregates, i.e., formation of new PrP^{Sc} oligomers from PrP^{C} of the host. These models also account for the de novo formation of prions in inherited or sporadic prion diseases. The template assistance or heterodimer model [26] proposes that the rate-limiting step in the de novo generation of PrP^{Sc} is formation of a PrP^{Sc} monomer from PrP^{C}. A PrP^{Sc} monomer would then form a heterodimer with a benign PrP^{C} molecule and convert this PrP^{C} into PrP^{Sc}. This requires direct or indirect specific recognition of PrP^{C} by PrP^{Sc}. Newly converted PrP^{Sc} molecules would then further convert PrP^{C} molecules into PrP^{Sc} (Figure 34.1A). The nucleation-polymerization model [27] (Figure 34.1B) assumes that PrP^{C} and PrP^{Sc} are always in a dynamic equilibrium, but PrP^{C} monomers are by far more stable than PrP^{Sc} monomers, such that the small fraction of PrP^{Sc} monomers is not observed experimentally. Here, the rate-limiting step in spontaneous PrP^{Sc} formation is the oligomerization of the small fraction of PrP^{Sc} monomers to an oligomer of critical size, which is

Tab. 34.1. Transmissible spongiform encephalopathies in humans and mammals.

Disease	***Origin/Mechanism***
	Human prion diseases
Sporadic Creutzfeld-Jakob disease (CJD)	Spontaneous formation of prions from PrP^C Most common human prion disease; nevertheless extremely rare (1 case per one million per year) Affected individuals are almost exclusively homozygous with respect to the polymorphism at residue 129 (Met/Met or Val/Val) Average age of onset: >60 years
Variant CJD	Infection from bovine prions Infected individuals are generally Met/Met homozygous with respect to the polymorphism at residue 129, which is also Met in bovine PrP
Familial CJD	Germ-line mutations in the PrP gene
Gerstmann-Sträussler-Scheinker disease	Germ-line mutations in the PrP gene
Fatal familial insomnia	Germ-line mutations in the PrP gene
Iatrogenic CJD	Infection from prion-contaminated human growth hormone, dura mater grafts, etc.
Kuru	Infection through ritual cannibalism (Fore people)
	Prion diseases in mammals
Bovine spongiform encephalopathy	BSE-contaminated food
Scrapie (sheep and goats)	Genetic susceptibility to spontaneous formation of scrapie prions. Transmission among sheep?
Transmissible mink encephalopathy	Infection with prions from cattle or sheep
Chronic wasting disease (elk, deer)	Origin and transmission unknown
Feline spongiform encephalopathy (cats)	Infection with prion-contaminated beef

sufficiently stabilized by quaternary structure contacts such that it can no longer dissociate. This nucleus is capable of growing further, by pulling additional PrP^{Sc} monomers from the equilibrium into the PrP^{Sc} oligomer. An important difference between the models is that the nucleation-polymerization model does not necessarily require a direct contact and specific recognition between PrP^C and PrP^{Sc}. Moreover, it much better accounts for the prion strain phenomenon, as will be discussed below (see Section 34.5).

In addition to these models for the generation of PrP^{Sc}, another mechanism of amyloid formation has been postulated for the aggregation of human lysozyme variants in hereditary systemic amyloidosis. Here, single amino acid replacements in lysozyme destabilize the native state relative to partially structured intermedi-

A) Template assistance model

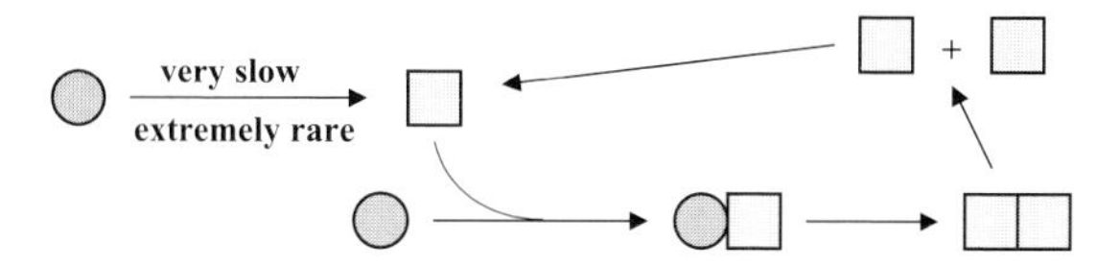

B) Nucleation-polymerization model

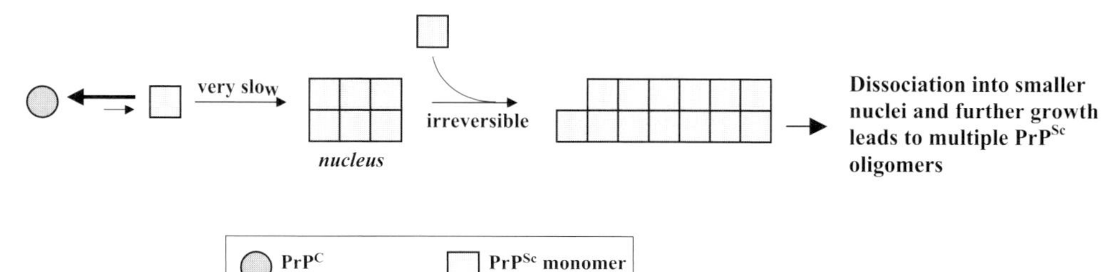

Fig. 34.1. Schematic representation of the template-assistance model (A) and nucleation-polymerization model (B) of PrP^{Sc} propagation.

ates. This leads to an increase of the fraction of partially folded intermediates that act as amyloid precursors and are pulled from the normal folding equilibrium during amyloid propagation [28]. A further mechanism is observed in the formation of amyloid fibrils of transthyretin (TTR), which are the putative cause of senile systemic amyloidosis (SSA) and familial amyloidotic polyneuropathy (FAP). Here, dissociation of the native TTR homotetramer into monomers is the rate-limiting step in amyloid formation, which is followed by partial monomer denaturation and misassembly into fibrils [29].

34.2 Properties of PrP^C and PrP^{Sc}

Mammalian PrP^C is a cell surface glycoprotein of about 210 amino acids that is strongly expressed in neurons but is also present in most other tissues. PrP^C belongs to the most conserved proteins known, and pairwise alignment of PrPs from different species generally reveals more than 90% sequence identity [30, 31]. In addition, there are three types of invariant posttranslational modifications found in mature PrP^C: two *N*-glycosylation sites at Asn181 and Asn197 [34], a single disulfide bond between Cys179 and Cys214 [33], and a C-terminal glycosyl-phosphatidyl-inositol (GPI) anchor at residue 231 (amino acid numbering according to human PrP) [16, 17, 32]. Another typical feature of mammalian PrP sequences is a segment of five octapeptide repeats in the N-terminal region (Figure

34.3A). Although the strong conservation of PrP in mammals suggests an important cellular role of the protein, the biological function of PrP^C remains unknown to date. This is mainly due to the fact that PrP knockout mice are essentially healthy [35] and show only subtle phenotypes that are not completely preserved in different knockout strains. Proposed functions for PrP include roles in signal transduction [36], copper storage [37], circadian rhythm regulation [38], and maintenance of synaptic function [39] and Purkinje cell survival [40] (also reviewed in Ref. [8]). PrP^C is monomeric, is soluble in non-denaturing detergents, and contains a high fraction of α-helical secondary structure [18]. In vivo, PrP^C is enriched in cholesterol and sphingolipid-rich membrane rafts, subject to caveolae-mediated endocytosis [41], and has a half-life of about 3–6 h [42]. There is evidence that raft localization of PrP^C and co-localization with PrP^{Sc} in the same membrane are required for its conversion into PrP^{Sc} [43]. Most PrP molecules synthesized in the cell reach the cell surface and are anchored to the cell membrane via their GPI anchor, but two intracellular transmembrane forms of PrP have been identified that span the ER membrane in different orientations, of which the form with the C-terminus towards the ER lumen (^{Ctm}PrP) may be associated with neurodegeneration in prion-infected individuals [44]. Whether aberrant membrane topology of PrP at the ER has general functional implications still has to be determined. Moreover, retrograde transport of misfolded PrP from the ER to the cytosol normally leads to PrP degradation by the proteasome, but if this degradation is compromised, the accumulation of cytosolic PrP has severe toxic effects on neurons and has been proposed to be the general mechanism underlying the death of neurons during prion pathogenesis [45].

The insoluble, oligomeric scrapie form PrP^{Sc} has so far been isolated only from the brain of prion-infected individuals or prion-infected cell cultures. Although the subunits in PrP^{Sc} have the same covalent structure and posttranslational modifications as PrP^C, it differs from PrP^C with respect to all its biochemical properties: it is insoluble in non-denaturing detergents and shows an increased β-sheet content and a decreased α-helix content compared to PrP^C [18]. Another important property of PrP^{Sc} is its partial resistance to proteinase K (PK). While PrP^C is readily degraded by PK, the polypeptide chains in PrP^{Sc} exhibit a protease-resistant core ranging from approximately residue 90 to the C-terminal residue 231. In contrast, the N-terminal PrP^{Sc} segment from residue 23 to approximately residue 90 is degraded (Figure 34.2) [46, 47]. As will be discussed in Section 34.5, this observation has important implications, as it shows that PrP^{Sc} is an *ordered* oligomer in which all subunits appear to have a very similar structural environment that protects residues 90–231 in *every* subunit from access by PK. This is a strong hint that PrP^{Sc} does not represent a nonspecific, inclusion body–like protein aggregate but rather a regular array of subunits with identical tertiary structures in which both the fold of the subunits and their quaternary structure contacts confer protease resistance. This is in good agreement with the nucleation-polymerization model of PrP^{Sc} propagation [27], which assumes specific subunit-subunit contacts in the nucleation and further growth of the PrP^{Sc} oligomer. The importance of the protease-resistant core of PrP^{Sc}, termed PrP27–30, is further underlined by the

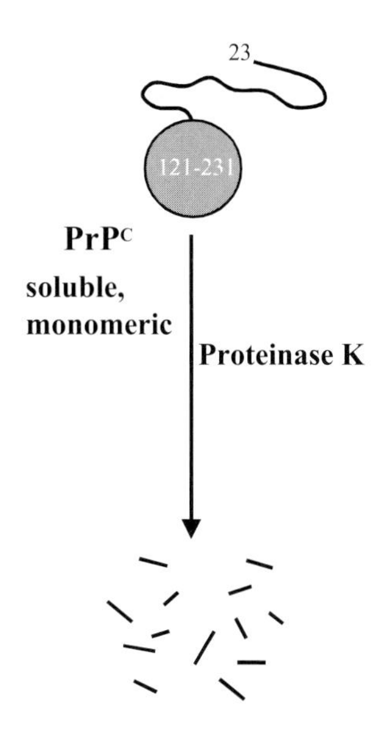

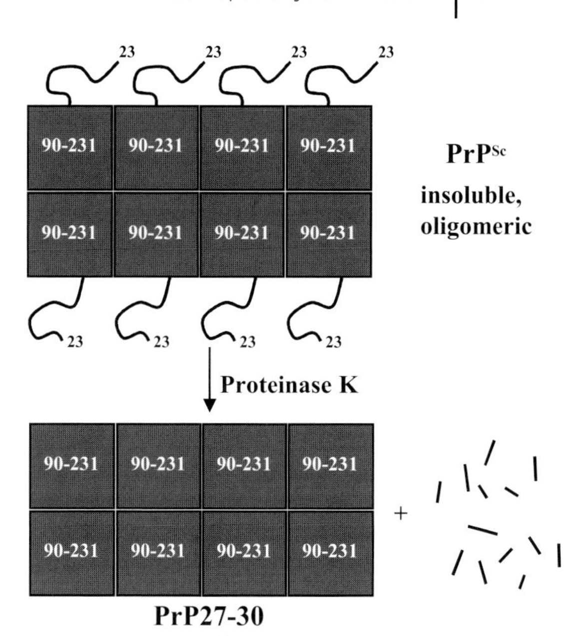

Fig. 34.2. Protease resistance of PrP^{Sc}. While PrP^C is readily degraded by proteinase K into peptide fragments, the subunits of PrP^{Sc} contain a protease-resistant core (PrP27–30) from about residue 90 to the C-terminal residue 231 (amino acid numbering according to human PrP). Thus, only the N-terminal segment 23–90 in PrP^{Sc} subunits is accessible to proteinase K degradation. The structure of PrP^C, comprised of the flexibly disordered N-terminal tail 23–120, and the structured C-terminal domain PrP(121–231) (cf. Figure 34.3) are also indicated.

fact that it retains prion infectivity and forms amyloid-like fibers [19]. In addition, transgenic mice exclusively expressing N-terminally truncated forms of PrP that still contain segment 90–231 are susceptible to prions and able to replicate prions [48]. This shows that the octapeptide repeats in the N-terminal segment of PrP^C are not required for prion propagation. The identification of PrP27–30 after proteinase K treatment of brain extracts, protein separation by SDS-PAGE, and immunospecific detection in Western blots represents the basis for both post-mortem and pre-clinical detection of prions, e.g., in BSE-infected cattle, and, as outlined in Section 34.5, biochemical characterization of prion strains [6, 47]. Interestingly, a high-molecular-weight polysaccharide consisting of 1,4-, 1,6-, and 1,4,6-linked glucose units was found to be associated with infectious preparations of PrP27–30, and it was postulated that this polysaccharide is a scaffold that promotes formation of PrP^{Sc} and contributes to its extraordinary physical and chemical stability [49]. Additional structural information on PrP^{Sc} has become available from fiber diffraction studies and atomic force microscopy studies indicating that PrP^{Sc} contains β-sheets that are arranged perpendicular to the fiber axis [50] and may form parallel β-helices [51].

34.3
Three-dimensional Structure and Folding of Recombinant PrP

34.3.1
Expression of the Recombinant Prion Protein for Structural and Biophysical Studies

The protein-only hypothesis and all theoretical models of PrP^{C} to PrP^{Sc} conversion imply that knowledge of the three-dimensional structures of PrP^{C} and PrP^{Sc} is the prerequisite for understanding prion propagation at a molecular level. As PrP^{Sc} is insoluble, and because no soluble, low-molecular-weight oligomers with defined stoichiometry have been isolated so far, PrP^{Sc} has not been accessible to atomic structure determination by X-ray crystallography or nuclear magnetic resonance (NMR) spectroscopy.

Because it is extremely difficult to purify milligram quantities of PrP^{C} from mammalian brain, production of recombinant PrP in *Escherichia coli* has proved to be necessary to determine the three-dimensional structure of the monomeric protein. Since there is neither *N*-linked glycosylation nor GPI anchor biogenesis in *E. coli*, bacterially produced PrP lacks these modifications. Nevertheless, there is no evidence at present that these modifications significantly influence the tertiary structure of the prion protein. To allow formation of the invariant disulfide bond between Cys179 and 214 in the recombinant prion protein, two strategies have been pursued. As disulfide bonds cannot form in the reducing environment of the bacterial cytoplasm, the prion protein and N-terminally truncated fragments were secreted into periplasm with the *E. coli* OmpA signal sequence [52]. The disulfide bond was formed, but the protein was N-terminally degraded by periplasmic proteases such that only the fragment from residues 121–231 stayed stable in the periplasm. The fragment was termed PrP(121–231) [52]. This was the first evidence that the entire N-terminal segment 23–120 of recombinant PrP does not adopt a defined structure in solution (see below). Because any attempt to produce full-length PrP in the bacterial periplasm has failed so far, the full-length protein was expressed in the cytoplasm where the protein accumulates in the reduced form and nonspecifically aggregates into inclusion bodies. Oxidative refolding from the inclusion body fraction in vitro and further purification yielded large quantities of homogeneous, recombinant PrP for structure analysis and biochemical characterrization [53, 54]. Details on different refolding protocols are given in the Experimental Section.

34.3.2
Three-dimensional Structures of Recombinant Prion Proteins from Different Species and Their Implications for the Species Barrier of Prion Transmission

34.3.2.1 Solution Structure of Murine PrP

The first three-dimensional structure of a recombinant prion protein was reported in 1996 for the fragment PrP(121–231) of the murine prion protein, which was determined by nuclear magnetic resonance spectroscopy (NMR) [56, 57] (Figure

34.3B). PrP(121–231) adopts a unique fold that has so far not been found in other proteins. The fold consists of a short, two-stranded, antiparallel β-sheet (residues 128–131 and 161–164) and three α-helices (residues 144–154, 175–194, and 200–226). The fold is additionally characterized by a tightly packed hydrophobic core of 20 amino acids (residues 134, 137, 139, 141, 158, 161, 175, 176, 179, 180, 184, 198, 203, 205, 206, 209, 210, 213, 214, and 215). The invariant disulfide bond between Cys179 and Cys214 is part of this hydrophobic core and connects the second with the third α-helix. The hydrophobic core is surrounded by a shell of hydrogen-bonded residues that further stabilize the fold [58]. Overall, the structure of PrP(121–231) resembles a flat ellipsoid that is characterized by an uneven distribution of surface charges between the two flat surfaces. The two glycosylation sites at Asn181 and Asn189 are located on the negatively charged side of the molecule, while the opposite surface is positively charged. It has been proposed that the uneven charge distribution may contribute to the orientation of PrP^C relative to the cell membrane. One would expect that the protein associates via its positively charged surface with the membrane, such that the two glycosylation sites would be oriented towards the extracellular space [56].

The NMR structure of full-length murine PrP, which was reported in 1997, revealed that residues 23–120 (including the octapeptide repeats), in contrast to the compact domain PrP(121–231), are flexibly disordered in solution (Figure 34.3B), with all residues in this segment being flexible on a sub-nanosecond timescale [59]. The structure of residues 121–231 in the context of the full-length protein proved to be indistinguishable from that in isolated PrP(121–231). The functional role of the flexible tail is presently unclear. However, a plausible model has been proposed for structure formation within the octapeptide repeat regions upon cooperative binding of Cu^{2+} ions [60]. As there is increasing evidence that a substantial fraction of cellular proteins are "natively unfolded" in isolation and adopt a defined tertiary structure only in complex with target molecules [61], it appears likely that the flexible tail of PrP is required for the interaction of PrP^C with its natural ligands.

Overall, the structure of recombinant, non-glycosylated, full-length PrP produced in *E. coli* is in full agreement with the known properties of PrP^C isolated from mammalian cells. As natural PrP^C, it predominantly contains α-helical secondary structure and is soluble and monomeric. It is therefore generally accepted that the structure of recombinant PrP corresponds to that of the benign cellular prion protein. As will be discussed in Section 34.3.2.2, the NMR structure of murine PrP is also very similar to the structures of other mammalian prion proteins. Because the segment 90–231 is protease-resistant in PrP^{Sc}, an important conclusion from the structure of full-length PrP(23–231) is that residues 90–120, which are required for prion propagation [36], must adopt a defined three-dimensional structure in PrP^{Sc}. This corresponds to the minimal structural change that has to be postulated for a PrP polypeptide when it is converted from PrP^C to a subunit of PrP^{Sc} (Figure 34.3B).

As one might expect from the flexibly disordered N-terminal tail, attempts to crystallize recombinant prion proteins have failed so far. The only exception is the X-ray structure of human PrP(90–231), which crystallized as a covalently linked

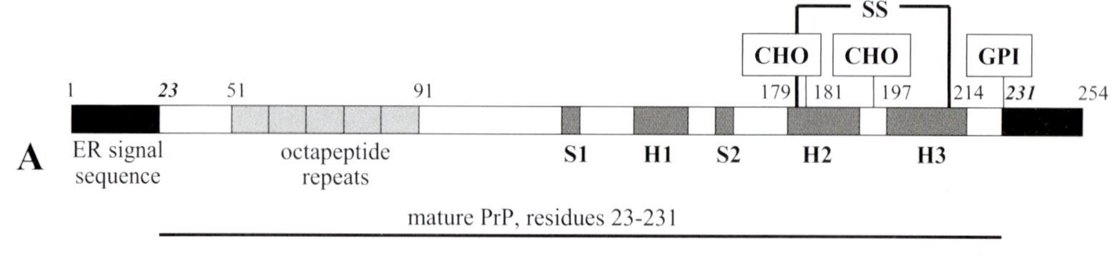

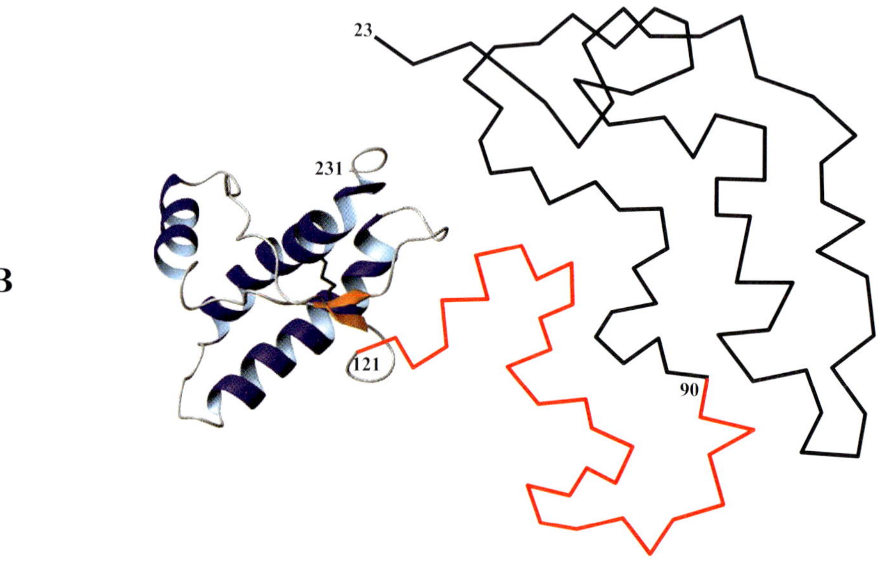

Fig. 34.3. Posttranslational modifications of PrP^C and three-dimensional structure of recombinant PrP. (A) Mammalian prion proteins contain an N-terminal signal sequence (residues 1–22) that directs the protein to the endoplasmic reticulum, where glycans (CHO) at Asn181 and Asn197 are attached, the C-terminal peptide 232–254 is replaced by a GPI anchor at the carboxylate of residue 231, and the single disulfide bridge between Cys179 and Cys214 is formed (amino acid numbering according to human PrP). Positions of β-strands (S1, S2) and α-helices (H1–H3) in the NMR structure of the recombinant PrP are indicated, as well the length of the protease-resistant core of PrP^{Sc} and the structured C-terminal domain PrP(121–231). (B) Ribbon diagram of the NMR structure of recombinant murine PrP. The protein is composed of a flexibly disordered N-terminal tail (residues 23–120, represented as random coil C^{α} trace) and a structured domain (residues 121–231), which consists of a two-stranded, antiparallel β-sheet and three α-helices. The single disulfide bond (black lines) connects the second and third α-helix and is completely shielded from the solvent. The solvent-accessible residues 90–120, which become protease-resistant in PrP^{Sc}, are indicated as red lines.

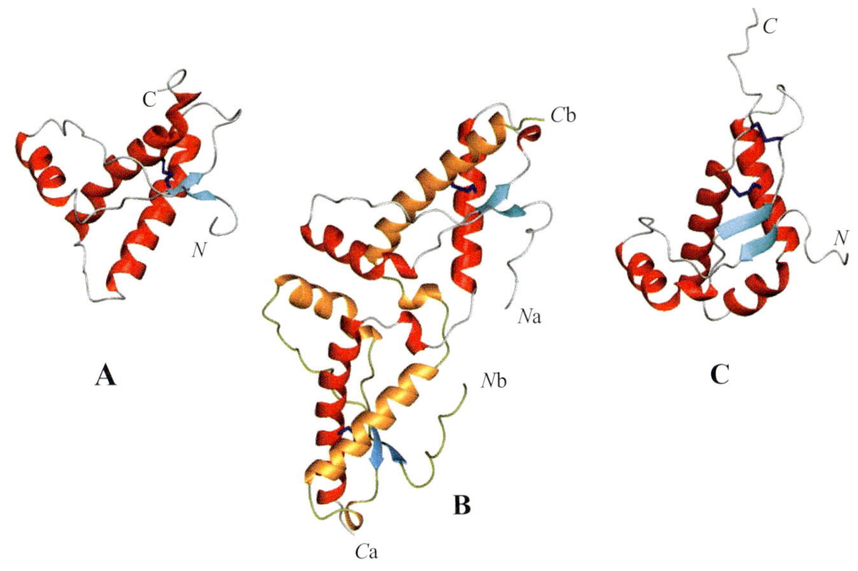

Fig. 34.4. Comparison of the NMR structure of recombinant murine PrP(121–231) (pdb code: 1AG2) (A) with the X-ray structure of a dimer of recombinant human PrP(90–231) (B) (code1I4M) and the NMR structure of the murine Doppel protein (C) (code 1I17). (B) In the dimer of human PrP(90–231), the C-terminal α-helices are swapped and disulfide is cross-linked with the second α-helix of the other subunit. The molecule may be an artifact resulting from oxidative refolding of the recombinant protein from *E. coli* inclusion bodies under nonnative conditions. (C) Note that the Doppel protein is stabilized by two intramolecular disulfide bonds (dark blue lines).

homodimer in which α-helices 3 are swapped between the subunits such that α-helix 2 in each subunit is now disulfide-bonded to α-helix 3 from the other subunit (Figure 34.4B) [62]. This structure may be an experimental artifact that was produced during oxidative in vitro refolding of the recombinant protein under conditions that favored intermolecular disulfide bond formation. Disulfide-bonded homodimers have not been observed for PrP^{C}, nor have intermolecular disulfide bonds been reported for PrP^{Sc} oligomers [33].

As mentioned above, the structure of the C-terminal domain of the mammalian prion protein represents a new fold that has not been observed in other proteins. In this context, it is interesting to discuss the recently solved NMR structure of the mammalian protein with closest homology to the prion protein, termed Doppel (Dpl) [63] (Figure 34.4C). Dpl lacks the N-terminal octapeptide repeats found in PrP and is most likely not linked with prion diseases, as it is normally not expressed in the central nervous system. In contrast, it was shown to be required for male fertility in mice by controlling male gametogenesis and sperm-egg interaction [64]. Figure 34.4C shows the NMR structure of murine Dpl [65], which shares 21% amino acid sequence identity with murine PrP and is very similar to the structure of human Dpl [66]. The Dpl fold is similar to that of PrP(121–231) in that the secondary structures fall in the same positions in the two proteins. Differences are observed mainly in the lengths of the α-helices and the positions of the β-strands (in Dpl, the strands are displaced by one to two residues relative to those in the PrP structure). Moreover, Dpl contains two intramolecular disulfide bonds. The disulfide bond 109–143 corresponds to that of PrP(121–231), and the second disulfide 95–148 connects the segment after β-strand 1 with the C-terminal segment following α-helix 3. The strongest structural difference between the proteins is that the plane of the β-sheet is parallel to the axis of α-helix 2 in Dpl, while it is perpendicular to this axis in the structure of PrP(121–231). Moreover, there is a marked kink in α-helix 2 of Dpl, which is not observed in PrP(121–231). In summary, the structured domains of Dpl and PrP appear to have evolved from the same ancestor by gene duplication but now fulfill different functions.

34.3.2.2 Comparison of Mammalian Prion Protein Structures and the Species Barrier of Prion Transmission

In addition to the structure of murine PrP, the NMR structures of the recombinant prion proteins from hamster [67, 68], human [69], and cattle [70] have been determined. All proteins consist of very similar global architectures, i.e., the flexible N-terminal tail of about 100 residues, followed by the folded C-terminal domain. The folds of C-terminal domains are again very similar but exhibit interesting local differences [57]. For example, α-helix 3 is kinked in the structure of murine PrP but straight in the structures of hamster, bovine, and human PrP. Comparison of the structures of mouse and hamster PrP revealed a better-defined loop segment between β-strand 1 and α-helix 2 and slight local differences between the structures in the segment between α-helix 1 and β-strand 2. In the structure of human PrP, α-helix 3 is also straight and well defined and extends to residue 228. Interestingly, the C-terminal turns of α-helices 2 and 4 appear to be in a local unfolding equilibrium. In addition, short, transient contacts between the flexible tail and the folded C-terminal domain were detected for human PrP, which may slightly contribute to a stabilization of the C-terminal residues of α-helices 2 and 3 against unfolding [57]. The structure of bovine PrP is most closely related to that of human PrP but shows the differences in local backbone conformation from murine and hamster PrP described above (Figure 34.5).

Comparison of the structure of bovine PrP(121–231) with that from humans, mice, and hamsters revealed RMSD values of 0.98 Å, 1.66 Å, and 1.68 Å, respectively [57, 70]. The striking structural similarity between bovine and human PrP^C, in conjunction with the fact that BSE prions can be transmitted to humans and cause new variant Creutzfeld-Jakob disease (CJD) (Table 34.1), suggests that the structural relationship at the level of PrP^C inversely correlates with the species barrier of prion transmission. The species-barrier phenomenon has so far been discussed mainly at the level of sequence similarity between PrPs from different species with the view that the transmission barrier decreases with increasing sequence identity [30, 31]. A detailed structural comparison of all available PrP solution structures suggests that the most relevant structural feature with regard to the species-barrier phenomenon, besides the conformational differences described above, are differences in local surface charges, even though the uneven surface charge distribution on both flat surfaces of the C-terminal PrP domain is preserved in all structures [57].

In the context of the nucleation-polymerization mechanism of prion propagation, an inverse correlation between structural similarity of PrP^Cs from different species and species barrier could mean that those parts of the PrP^C structure that are similar between a pair of species but different in other PrP^C structures are preserved in the structure of the PrP^{Sc} subunits and possibly are involved in subunit-subunit contacts. This could mean that substantial parts of the structure of the PrP^C segment 121–231 are still contained in the tertiary structure of PrP^{Sc} subunits. In addition, a host-specific factor, called "protein X," may be responsible for the species-barrier phenomenon. "Protein X" is believed to be contained in every mammalian species and to be required for prion propagation [71].

34.3.3 Biophysical Characterization of the Recombinant Prion Protein

34.3.3.1 Folding and Stability of Recombinant PrP

Studies on the folding and thermodynamic stability of murine PrP(121–231) at neutral pH and 25 °C revealed that the C-terminal PrP domain behaves perfectly according to the two-state model as one would expect for a one-domain domain protein. It shows entirely reversible unfolding and refolding transitions in denaturant-induced equilibrium experiments, with a free energy of folding of -30 kJ mol^{-1} and a cooperativity expected for a 111-residue protein ($m = 4.8$ kJ mol^{-1}/M urea) [52]. In the context of full-length PrP(23–231), the stability of the domain is slightly decreased (-26 kJ mol^{-1}), which is due to a somewhat lower cooperativity, while the transition midpoint is the same as that of PrP(121–231) (6.3 M urea). The lower folding cooperativity observed for full-length PrP may be due to formation of residual structure in the unfolded state in which residues 23–120 interact with the parts of segment 121–231 [55]. Reversible unfolding transitions at neutral pH were also confirmed for human PrP(90–231) [72–75]. Hydrogen-exchange experiments on human PrP(90–231) revealed that only a small segment of about 10 residues around the disulfide bond retains residual structure in the unfolded state [74]. Further studies revealed that the stability of recombinant PrP decreases with decreasing pH [87, 96] and shows an unusual salt dependence in that salts destabilize the protein at concentrations below 50 mM increase the stability at high concentrations according to the Hofmeister series [96].

Assuming that folding of recombinant PrP corresponds to that of PrP^C, the complete reversibility of PrP^C folding nicely explains within the framework of the protein-only hypothesis why treatment of prions with denaturants such as urea or guanidinium chloride (GdmCl) completely abolishes infectivity, even after subsequent removal of the denaturant [76]: high denaturant concentrations dissociate

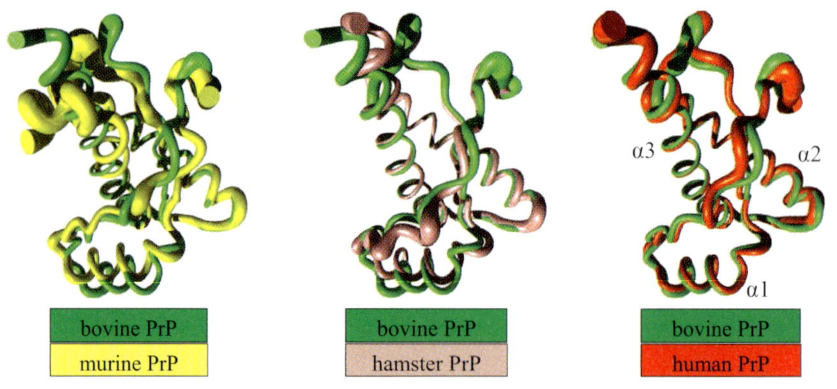

Fig. 34.5. Comparison of the NMR structure of bovine PrP(121–231) (green, pdb code 1DX0) with that of recombinant PrP(121–231) from mice (yellow, code 1AG2), hamsters (pink, code 1B10), and humans (red, code 1E1J). The C^{α} traces are represented as spline functions, and the thickness of the cylindrical rods represents the mean global displacement per residue after superposition of the backbone atoms in the set of energy-minimized conformers that represents each NMR structure (picture from Dr. R. Riek, ETH Zurich).

PrP^{Sc} into monomers and denature its subunits, so that a solution of dissociated and unfolded PrP^{Sc} is identical to a solution of unfolded PrP^{C}. Consequently, unfolded PrP^{Sc} subunits will always refold to PrP^{C}.

Analysis of the kinetics of folding of murine PrP(121–231) showed that the C-terminal domain belongs to the fastest folding proteins described so far. It folds with a half-time of about 200 μs at neutral pH and 4 °C and unfolds with a half-time of 4.6 min under these conditions [77]. Rapid folding of PrP(121–231) is consistent with the absence of *cis*-proline peptide bonds in the three-dimensional structure [57, 67–70]. Analysis of the dependence of the rate constants of unfolding and refolding on denaturant concentration showed that the compactness of the transition state of folding is closer to the unfolded state than to the native state. No transient folding intermediates have been detected so far in the kinetics of folding of PrP(121–231). However, a kinetic folding intermediate has been identified in the folding of recombinant human PrP(90–231), which nevertheless also folds extremely rapidly [78, 79]. Whether this difference is due to the additional segment 90–120 in human PrP(90–231) or reflects intrinsic differences in the folding of human and murine PrP(121–231) still has to be established. Overall, the very rapid folding of PrP seems to exclude that kinetic folding intermediates can serve as a source of PrP^{Sc} at neutral pH. However, the population of kinetic intermediates is uniformly higher in all recombinant human PrP(90–231) variants that bear mutations linked with inherited human TSEs [79].

PrP^{Sc} has been shown to accumulate in lysosomes of prion-infected cells [80–82] where the pH varies between 4 and 6 [83]. This opens the possibility that conversion of PrP^C into PrP^{Sc} does not occur at the cell surface but rather in the acidic environment of lysosomes. For this reason, folding of recombinant PrP has also been analyzed in detail under acidic conditions, where the properties of the protein indeed change dramatically. At pH 4, the folding mechanism ($U \leftrightarrow N$) of murine PrP(121–231) changes from a two-state to a three-state equilibrium in which an acid-induced unfolding intermediate is observed by a pronounced plateau phase in the circular dichroism signal at 222 nm [72, 84–86]. At pH 4, the intermediate is maximally populated at 3 M urea and still significantly populated in the absence of urea [85]. In contrast to the α-helical CD spectrum of native PrP(121–231) at pH 4, the intermediate shows far-UV CD spectra of a β-sheet protein with a single minimum at 215 nm that are reminiscent of PrP^{Sc}. The transition from two-state to three-state equilibrium transitions has an apparent pK_a of 4.5, which indicates that protonation of acidic side chains is required for stabilization of the intermediate [85]. Very similar results were obtained for the folding of the full-length murine PrP(23–231) and human PrP(90–231). Consequently, formation of the acid-induced unfolding intermediate is an intrinsic property of the domain PrP(121–231). Further investigations demonstrated that formation of the intermediate is dependent on protein concentration and that the intermediate is most likely is a homodimer. In addition, ionic strengths above 50 mM are required for the intermediate to be observed. Equilibrium folding of recombinant PrP at neutral and acidic pH can thus by described by the following scheme [87]:

pH 4–8, low ionic strength at pH 4–5: $U \leftrightarrow N$
pH 4, ionic strength > 50 μM, [PrP] > 20 mM: $2\ U \leftrightarrow I_2 \leftrightarrow 2\ N$

An attractive hypothesis resulting from these data is that the intermediate represents an early dimeric intermediate in the formation of oligomeric PrP^{Sc}. Indeed, starting with conditions that favor the population of the intermediate, fibrillar aggregates of recombinant PrP(90–231) could be obtained that showed a somewhat increased resistance to proteinase K [88]. Nevertheless, as in all attempts to generate prions de novo from recombinant PrP performed so far, it could not be demonstrated that these aggregates are infectious and relevant for understanding prion propagation. The same holds true for a β-sheet-rich conformation of human PrP(91–231), which is observed after thermal unfolding and cooling of the protein [86].

34.3.3.2 **Role of the Disulfide Bond in PrP**

The disulfide bond between the only cysteine residues of PrP, Cys179, and Cys214 is invariant in all mammalian prion protein sequences [31]. It is formed quantitatively in PrP^C, and it has been shown unequivocally that free thiol groups are also absent in PrP^{Sc} [33]. Thus, there is primarily no indication that reduced PrP is involved in the mechanism of prion propagation. Nevertheless, a series of interesting recent observations argue in favor of a transient reduction of the disulfide bond during PrP^{Sc} formation. For example, it could be that the disulfide bond of

PrP^C, after caveolin-dependent endocytosis into lysosomes, becomes reduced and oligomerizes into reduced PrP^{Sc} in which the subunits again become re-oxidized when PrP^{Sc} is released into the extracellular space after death of the infected cell. Alternatively, it could be that reduced PrP is required for formation of PrP^{Sc} nuclei but that further incorporation of oxidized PrP into these nuclei eventually leads to a PrP^{Sc} oligomer in which the vast majority of subunits are oxidized, such that the small fraction of reduced PrP that was initially present is no longer detectable.

The characterization of reduced, recombinant prion proteins has indeed revealed that the prion protein has the property of adopting two entirely different tertiary structures, depending on whether or not the Cys179-Cys214 disulfide is formed. This was first observed for recombinant PrP(91–231) [89, 90]. Reduction of the disulfide bond in human PrP(90–231) at pH 4 and low ionic strength leads to transition of the predominantly α-helical protein to a monomeric polypeptide that shows β-sheet-like CD spectra [90]. Loss of chemical shift dispersion in ^1H–^{15}N heteronuclear NMR spectra, however, indicates that the tertiary structure content in this conformational state is significantly lower compared to that of the oxidized protein. Increase in ionic strength at acidic pH then leads to formation of amyloid-like fibers of reduced PrP(91–231), which show increased protease resistance [90]. In addition to these in vitro experiments, several studies on the expression of full-length PrP in the reducing environment of the yeast cytoplasm have revealed that self-replicating amyloid-like fibers of reduced PrP are formed in vivo and show a protease-resistance pattern similar to that of the non-glycosylated PrP from PrP^{Sc} [91, 92]. In this context, it is very interesting that an in vitro protocol for transformation of recombinant hamster PrP(90–231) into amyloids could be developed that requires reduction and re-oxidation of the recombinant protein. The subunits in the resulting oligomers were connected by intermolecular disulfide bonds, which can be explained by a sub-domain swapping mechanism according to which each PrP molecule is covalently connected to two other PrPs via its cysteine residues and each oligomer has two non-satisfied ends with a free thiol group [93]. Here, the domain contacts and the structures of the monomers must be different compared to the X-ray structure of homodimeric human PrP(90–231), which lacks free thiols, as each Cys179 is disulfide-bonded with Cys214 of the other subunit of the dimer [62].

In summary, amyloid-like oligomers can be formed with both oxidized and reduced recombinant PrP in vitro. In none of these cases, however, could it be shown that infectious prions were generated. In this context, it should be emphasized that strong evidence has been accumulated during the past few years that nearly any protein, including purely α-helical proteins such as myoglobin that are not related to known amyloid diseases, can be converted into β-sheet-rich amyloid fibrils in vitro [94, 95]. This suggests that protein amyloids are primarily stabilized by β-sheet hydrogen bonds between main chain atoms. Thus, amyloid formation appears to be an intrinsic property of any polypeptide chain, independent of the amino acid sequence [95]. Therefore, it could well be that none of the aggregates of recombinant PrP reported so far are related to infectious PrP^{Sc}.

34.3.3.3 Influence of Point Mutations Linked With Inherited TSEs on the Stability of Recombinant PrP

All known inherited prion diseases in humans are associated with mutations in the gene encoding human PrP. There are three different familial human TSEs, all of which are autosomal dominant: the familial Creutzfeld-Jakob diseases (CJD), Gerstmann-Sträussler-Scheinker syndrome (GSS), and fatal familial insomnia (FFI) (Table 34.1) [1]. Figure 34.6A summarizes the location of amino acid replacements in human PrP linked with the three different disease phenotypes. Interestingly, the mutation D178N causes either CJD or FFI, depending on the polymorphism at position 129 (Val or a Met, respectively). Figure 34.6B shows that disease-related mutations are mainly located in the folded C-terminal domain of PrP. There is, however, no structural clustering of amino acid replacements related to a specific disease phenotype. As the affected individuals with a point mutation in one of the two PrP alleles spontaneously develop prions without having had contact with external prions, an obvious mechanism underlying this phenomenon could be a destabilization of PrP^C due to the mutation and, consequently, a facilitated conversion to PrP^{Sc}. Inspection of the NMR structure of recombinant PrP indeed predicts a destabilization in the case of the replacements D178N (loss of a salt bridge to Arg164), T183A (loss of two hydrogen bonds from the hydroxyl group of Thr183 to the H^N of Tyr163 and the carbonyl oxygen of Cys179), F198S (generation of a cavity), and Q2017R (loss of a hydrogen bond to the carbonyl atom of Ala133) [57, 58]. However, there are no clear-cut predictions for the other amino acid replacements, and the GSS mutations P102L, P105L, and A117V are located in the flexible tail and should not affect the stability of PrP^C at all.

Therefore, all disease-related point mutations were introduced into recombinant PrP(121–231), and the thermodynamic stabilities of the variants were analyzed. All variants showed far-UV CD spectra that were indistinguishable from that of the wild-type protein, demonstrating that none of these mutations a priori induces a PrP^{Sc}-like, β-sheet-rich conformation [55]. The thermodynamic stabilities of the PrP(121–231) variants with single disease-related amino acid replacements are summarized in Figure 34.6C and reveal that some but not all mutations destabilize PrP^C. A significant destabilization was indeed observed only for the above replacements that were predicted to be destabilizing from analysis of the three-dimensional structure. The most destabilizing mutation proved to be T183A, which decreases the free energy of folding relative to the wild-type protein from -30 to -10 kJ mol^{-1} [55]. Other mutations such as E220K or V210I were completely neutral with respect to PrP^C stability. Analogous results were obtained for variants of recombinant PrP(90–231) [62]. Consequently, destabilization of PrP^C cannot be the only mechanism underlying the spontaneous formation of prions in inherited human TSEs, and destabilization may trigger disease only in the case of mutations D178N, T183A, F198S, R208H, and Q217R (Figure 34.6C). Another result from this analysis is that there is no correlation between the thermodynamic stability of a TSE-related PrP^C variant and the disease phenotype. For example, both the most destabilizing mutation T183A and the "neutral" mutation E200K are linked with inherited CJD (Figure 34.6). It follows that there are most likely

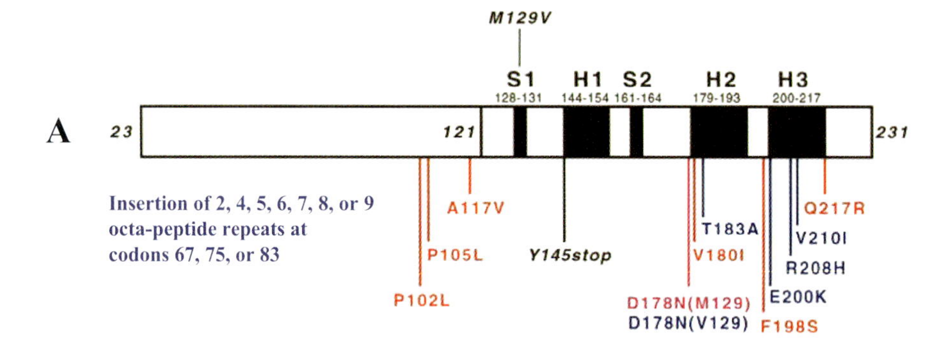

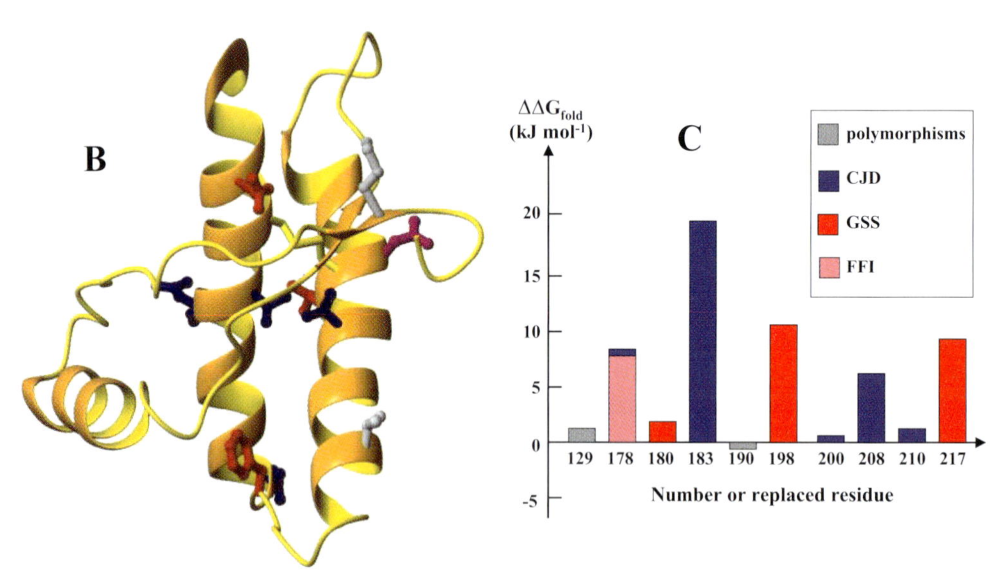

Fig. 34.6. Point mutations in human PrP that are associated with inherited prion diseases and their effect on the intrinsic stability of recombinant PrP(121–231). (A) Scheme of the primary structure of mature human PrP^C, with the amino acid replacements that are linked with the three inherited human TSEs: CJD (blue), GSS (red), and FFI (pink). The Met/Val polymorphism at residue 129 (gray) determines the phenotype of the disease related to the replacement D178N (FFI or CJD, respectively). (B) Location of residues that are replaced in inherited TSE in the structure of PrP(121–231). The polymorphism in murine PrP at residue 190 is also shown (same color code as in (A)). (C) Thermodynamic stabilities of recombinant variants of PrP(121–231), with disease-related, single amino acid replacements, relative to PrP(121–231) wild type. Positive $\Delta\Delta G^\circ$ values indicate that the variant is less stable than wild-type PrP(121–231) (same color code as in (A)). Wild-type PrP(121–231) has a free energy of folding of -30 kJ mol^{-1}. The least stable variant T183A is thus destabilized threefold relative to the wild-type protein.

multiple and independent mechanisms that can lead to spontaneous generation of prions as a consequence of a mutation in PrP. Such mechanisms may, e.g., be an increased stability of PrP^{Sc} ("wild-type" PrP^{Sc} has a half-life of about two days in vivo) or faster nucleation kinetics of PrP^{Sc}. The latter may be caused by the above-mentioned higher fraction of kinetic folding intermediates that has been reported for variants of human PrP(90–231) bearing amino acid replacements associated with inherited TSEs [78, 79].

34.4 Generation of Infectious Prions in Vitro: Principal Difficulties in Proving the Protein-Only Hypothesis

Most scientists would consider the in vitro generation of prions from purified PrP^{C} the final proof of the protein-only hypothesis [8]. In principle, there are several ways to perform such an experiment. One approach is the propagation of prions by inoculation of purified PrP^{C} (or recombinant PrP) with catalytic amounts of PrP^{Sc} isolated from the brain of prion-infected animals. Another (and possibly even more convincing) experiment is the de novo generation of prions from purified PrP^{C} or recombinant PrP by applying in vitro conditions that favor both spontaneous formation and propagation of PrP^{Sc}. The only way to monitor the success of these experiments is to test newly converted PrP^{Sc} molecules for infectivity. This is done by injection of prions into the brain of transgenic mice that overexpress PrP^{C}, and are therefore particularly susceptible to prions, and determination of the incubation time until the onset of the disease. Unfortunately, there is only a very rough, inverse linear correlation between the logarithm of the concentration of infectious units and incubation time [48, 97]. In practice, this means that differences in prion concentration between two samples (e.g., before and after an in vitro conversion experiment) can be detected reliably only when the concentrations differ by two to three orders of magnitude. If we assume that a difference in concentration of infectious units by a factor of 100 is minimally required to distinguish two prion preparations, a conversion experiment in which an infectious PrP^{Sc} preparation is used as an inoculum has to be designed such that PrP^{C} is used at 100-fold excess over PrP^{Sc} and that the conversion would be 100%. It is, however, obvious that the conditions for an in vitro conversion experiment will not be optimal. Thus, if the conversion efficiency was, e.g., only 1%, PrP^{C} would have to be used at an excess of at least 10 000-fold over PrP^{Sc}. In summary, the low sensitivity of the only available assay for infectious prions requires very high in vitro conversion efficiencies for detection of newly generated prions.

The most promising attempts to convert PrP^{C} to the protease-resistant, oligomeric form (which is also termed PrP^{res}) are experiments in which ^{35}S-labeled PrP^{C} isolated from mammalian cells is incubated with PrP^{Sc} in vitro [47, 98]. Numerous experiments performed according to the following reaction scheme revealed that ^{35}S-labeled PrP^{C} can indeed be converted in vitro into a protease-resistant form, PrP^{res}, which has the same biochemical properties as proteinase K–treated PrP^{Sc} [47].

^{35}S-labeled PrP^{C} from mammalian cells + PrP^{Sc} from prion-infected individuals

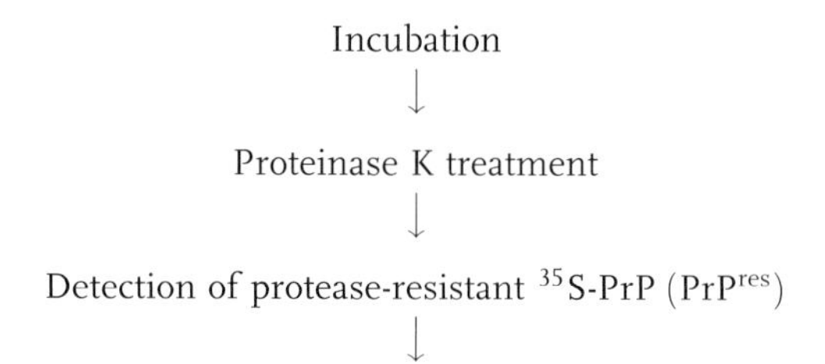

Further mechanistic studies demonstrated that this in vitro conversion of PrP^{C} to PrP^{res} is characterized by a rapid and specific binding step of PrP^{C} to PrP^{Sc} oligomers, followed by a slow conformational transition to PrP^{res} in the PrP^{Sc}-bound state [47]. Strikingly, this assay reproduces experimentally determined species barriers of prion diseases, in that PrP^{Sc} preparations from one species convert PrP^{C} from another species with similar efficiency, as the disease can be transmitted between the two species [99]. Unfortunately, the seeded in vitro conversion of PrP^{C} to PrP^{sen} generally does not generate more than stoichiometric amounts of PrP^{res} relative to the PrP^{Sc} molecules used as seeds [47], which prevents detection of newly generated infectivity. However, the seeded in vitro conversion has been significantly improved recently by a protocol in which growing PrP^{Sc}/PrP^{res} oligomers were subjected to repeated rounds of sonification and re-incubation with PrP^{C}, with the idea of breaking growing oligomers into smaller nuclei by the sonification step and thereby increasing the concentration of "growing ends" [100]. Indeed, this PCR-like approach strongly increased the efficiency of the in vitro conversion process, such that a 50-fold molar excess of converted PrP^{sen} over the initial PrP^{Sc} seeds was achieved. Despite this significant improvement, it could still not be shown that the concentration of infectivity increased and new prions were generated. In any case, this new method is predicted to improve the sensitivity of prion assays based on the detection of protease-resistant PrP by almost two orders of magnitude. Regarding the above described principle difficulties in proving generation of prions from PrP^{C} when PrP^{Sc} is used as an inoculum, the de novo generation of prions from recombinant PrP, which requires only a qualitative proof of existence of infectivity, may be the more promising approach towards the final proof of the protein-only hypothesis. However, the seeded in vitro conversion method has the potential of being further improved through the finding that mammalian RNA preparations stimulate the in vitro amplification of PrP^{res} [101]. This hints at a role for host-encoded stimulatory RNA molecules in prion pathogenesis, which is, e.g., principally possible after retrograde transport of PrP into the cytosol.

34.5
Understanding the Strain Phenomenon in the Context of the Protein-Only Hypothesis: Are Prions Crystals?

The most frequently raised argument against the protein-only hypothesis is the occurrence of prion strains. This relates to the fact that different prions can be isolated from a single species that differ in incubation time, clinical signs, vacuolar

Tab. 34.2. Glossary of prion terminology.

Prion	Infectious agent of transmissible spongiform encephalopathies. According to the protein-only hypothesis, the prion is identical with PrP^{Sc}.
PrP^{C}	Benign, cellular form of the prion protein.
PrP^{Sc}	Insoluble, oligomeric form of the prion protein that can so far be isolated only from the brains of prion-infected individuals or from prion-infected cell cultures.
PrP27–30	Oligomeric, protease-resistant core of the prion protein comprised of PrP subunits containing residues 90–231.
PrP^{res}	Protease-resistant form of the prion protein. PrP^{res} can be generated in vitro from PrP^{C} by incubation of PrP^{C} with PrP^{Sc} and is biochemically indistinguishable from PrP27–30.
Recombinant PrP	Prion protein produced in bacteria, composed of amino acids 23–231 but lacking the C-terminal GPI anchor at residue 231 and the two glycans at Asn181 and Asn197.
PrP(121–231)	Structured C-terminal fragment of the recombinant prion protein, comprising residues 121–231.
"Protein X"	Host-encoded factor that has been postulated to be required for prion replication and to regulate in the species barrier of prion transmission.

brain pathology, and PrP^{Sc} deposition in the brain [8, 23]. These specific phenotypic properties of a given prion strain are reproducibly preserved during consecutive rounds of experimental infection and re-isolation of the strain from the brain of the infected individual. Strain phenomena are typically observed for viruses, which are subject to high mutation rates of the viral genome. The prion strain phenomenon can, however, be explained in the context of the protein-only hypothesis on the basis of the nucleation-polymerization model of prion propagation [47]. This is because there is very convincing evidence that prion strains differ not only with respect to the above phenotypic criteria but also in the biochemical properties of their PrP^{Sc} deposits [6, 47, 102]. Figure 34.7A schematically depicts a Western blot analysis of human PrP^{Sc} from the brain of different patients after proteinase K digestion and SDS-PAGE separation [102]. It shows that the proteinase K–resistant core of the PrP^{Sc} oligomers that are associated with a given strain differs both in the fractions of the incorporated PrP glycoforms (in mammalian cells, there is always a mixture of non-glycosylated, mono-glycosylated, and di-glycosylated PrP^{C}) and in the exact length of the proteinase K–resistant polypeptide fragments. Like the phenotypic properties described above, the characteristic strain-specific banding patterns are always preserved, even after crossing the species barrier, and thus are diagnostic for each prion strain. For example, a characteristic feature of BSE prions from cattle is the high fraction of di-glycosylated PrP contained in the PrP^{Sc} oligomers, which is also observed in PrP^{Sc} isolated from patients with variant CJD [102] (Figure 34.7A, Table 34.1). The simplest model that explains these similarities is that PrP^{Sc} is a crystal-like oligomer. Similar to the fact that proteins frequently form different crystal forms, even under identical crystallization conditions, PrP^{Sc} appears to be capable of forming multiple macromolecular assemblies that differ in subunit composition and arrangement in their quaternary structures and possibly also in the tertiary structures of their subunits (Figure 34.7B). Each prion strain thus would correspond to a different PrP^{Sc} "crystal form," and this single crystal form then acts as a seed for formation of new PrP^{Sc} oligomers that grow further in the same crystal form. This model can also explain why humans who are heterozygous with respect to the polymorphism at amino acid 129 (Met or Val in humans) are essentially protected from sporadic CJD: although about 51% of the European population is Met/Met or Val/Val homozygous and 49% is Met/Val heterozygous, 95% of all sporadic CJD patients are homozygous [103, 104]. If we assume that residue 129 is solvent-exposed in PrP^{Sc} subunits (as it is in PrP^{C}; Figure 34.6B) and involved in subunit-subunit contacts in PrP^{Sc}, it is

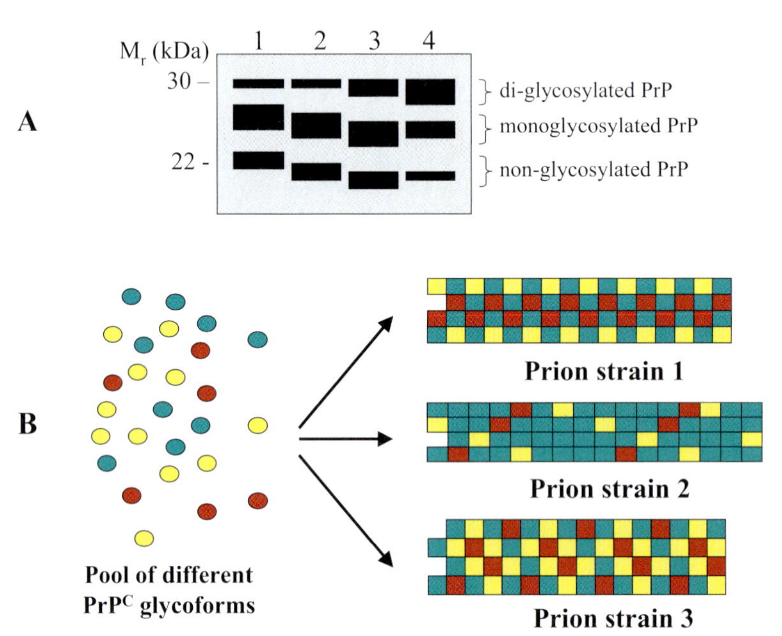

Fig. 34.7. Explanation of different prion strains within the framework of the protein-only hypothesis: (A) Besides disease symptoms and incubation time, prion strains isolated from mammalians and humans can be distinguished on the basis of the biochemical properties of PrP^{Sc} subunits after proteinase K treatment and separation by SDS-PAGE and Western blotting. Specifically, prion strains differ in the fractions of non-glycosylated, mono-glycosylated, and doubly glycosylated PrP contained in PrP^{Sc} oligomers, as well as the N-terminus of the protease-resistant core of the individual subunits. The figure schematically shows the Western blot band patterns of four different human prion strains after proteinase K treatment of PrP^{Sc}: (1) sporadic CJD, Met/Met 129 genotype; (2) sporadic CJD, Val/Val 129 genotype; (3) iatrogenic CJD, Val/Val129 genotype; (4) variant CJD, Met/Met129 genotype. (B) The simplest explanation for the occurrence of different prion strains are crystal-like PrP^{Sc} oligomers that differ in the fractions of the incorporated PrP glycoforms, the arrangement of subunits in the quaternary structure, and/or slightly different tertiary structures of incorporated subunits. In this model, each prion strain corresponds to a specific crystal form. Only one crystal form of PrP^{Sc} would then occur in a prion-infected individual, and only this crystal form would be further propagated when the disease is transmitted to another individual.

easy to imagine that a PrP^{Sc} oligomer composed of identical polypeptide chains could more easily assemble spontaneously into a quaternary structure with a regular array of subunits, compared to a 1:1 mixture of different polypeptides that are incorporated statistically into the oligomer. Another remarkable finding in this context is that residue 129 is exclusively Met in cattle PrP, and variant CJD patients are exclusively Met/Met homozygous [105].

34.6 Conclusions and Outlook

Although essentially all available experimental data on TSE prions and their propagation are either in agreement with or can be explained within the framework of the protein-only hypothesis, the most convincing proof of the prion hypothesis, i.e., the experimental replication of TSE prions in vitro, has not been reported to date. Some of the principal difficulties in proving the generation of infectious prions in vitro by seeding experiments have already been addressed in Section 34.4. In addition, if the "crystal model" of PrP^{Sc} that explains the strain-specific fractions of the different PrP glycoforms in PrP^{Sc} is correct, it follows that the PrP glycans are critically involved in subunit-subunit contacts that determine the quaternary structure of PrP^{Sc}. In the worst case, this means that it will not be possible to produce a PrP^{Sc} oligomer that is infectious and identical to PrP^{Sc} isolated from prion-infected brains from non-glycosylated, recombinant PrP. Moreover, experiments on the de novo formation of prions from recombinant PrP have so far essentially been performed in solution, while the conversion process in vivo most likely occurs at the surface of membranes. The covalent linkage to the GPI anchor and a membrane localization of PrP may therefore be additional prerequisites for generation of prions in vitro. A new protocol for the preparation of recombinant PrP covalently attached to liposomes has recently been established [106] that will allow in vitro conversion experiments with membrane-bound, recombinant PrP. In addition, like cellular RNAs [101], the glucose-based polyglycans found to be associated with PrP^{Sc} [49] may be another critical cofactor of prion replication.

A more general question is which experimental strategies should be pursued to gain mechanistic insights into prion propagation at the molecular level. Because there appears to be an inverse correlation between structural similarity at the level of recombinant PrP and species barrier, the determination of additional structures of recombinant PrPs is certainly a promising approach towards a further understanding of the species barrier. Even more important, however, appears to be the investigation of the oligomeric structure of PrP^{Sc}. For example, PrP^{Sc} oligomers that are spontaneously formed in inherited human prion diseases can either be homo-oligomers of the mutated prion protein or hetero-oligomers of mutated and non-mutated PrP. Both cases have been reported (see references in [55]), but no comprehensive mechanistic investigation on these phenomena is available. As PrP^{Sc} is insoluble and not accessible to structure determination by NMR spectroscopy and X-ray crystallography, the combination of electron microscopy, electron crystallography, and biochemical and genetic studies will be required to obtain further insights into the molecular details of the PrP^{Sc} quaternary structure.

In contrast to mammalian TSEs, the protein-only hypothesis can be considered proven in the case of the yeast prions Ure2p and Sup35p, which are responsible for the non-genetically transmissible phenotypes [*URE3*] and [*PSI*], respectively [107–109]. Specifically, infectious aggregates of Sup35p have been generated from recombinant, bacterial protein in vitro and then successfully transmitted to yeast cells that further propagated the [*PSI*] phenotype [110]. In addition, the species-barrier phenomenon was unequivocally shown to be caused by *specific* aggregates of Sup35p from different yeasts that replicate independently in the same test tube [111]. Thus, there are no principle reasons that the protein-only hypothesis should not be valid for mammalian TSEs.

34.7 Experimental Protocols

As outlined in Section 34.3.1, milligram quantities of correctly folded, disulfide-bonded recombinant prion proteins for biochemical and structural studies can be obtained from *E. coli* by either refolding cytoplasmically expressed proteins from inclusion bodies [53–55] or, in the case of PrP(121–231), by purification from the bacterial periplasm after secretory expression [52]. The more efficient method is the production of recombinant full-length PrP(23–231) and the N-terminally truncated forms PrP(90–231) and PrP(121–231) as cytoplasmic inclusion bodies under control of the T7 promoter, followed by oxidative refolding and purification to homogeneity. For this purpose, essentially two protocols are available, which are briefly described below.

34.7.1 Protocol 1 [53, 55]

Reduced, recombinant PrP is solubilized by treatment of the inclusion bodies with 8 M urea and purification of PrP under denaturing conditions by cation-exchange chromatography and gel filtration in 8 M deionized urea. Disulfide bond formation at pH 8.5 is then initiated by addition of 1 mM $CuSO_4$ as a catalyst of air oxidation for 2 h. The oxidation reaction is then stopped by addition of EDTA and a shift to pH 6, oxidized, and unfolded prion protein is concentrated to 10 mg mL^{-1} and refolded by 1:50 dilution at pH 6. After a final cation-exchange chromatography step at pH 6 under native conditions, homogeneous, recombinant PrP is obtained with yields of about 20 mg per liter of bacterial culture.

34.7.2 Protocol 2 [54]

The full-length prion protein or fragments thereof are expressed with the N-terminal oligo-histidine tag NH_3^+-MRGSHHHHHHGLVPR↓GS ..., which can be cleaved specifically by trypsin (cleavage site indicated by the arrow). Cytoplasmic inclusion bodies of the tagged prion proteins are solubilized by 6 M guanidinium chloride (GdmCl) and applied to a Ni^{2+}-NTA column at pH 8. The prion proteins are then re-oxidized by air and refolded by using a gradient from 6 M to 0 M GdmCl while bound to the column. On-column refolding prevents nonspecific aggregation and intermolecular disulfide bond formation. After cleavage of the histidine tag with thrombin and removal of thrombin by an anion-exchange resin, pure recombinant prion proteins are obtained after a last cation-exchange chromatography step with yields of 10–50 mg per liter of bacterial culture.

For periplasmic production of PrP(121–231) in *E. coli*, the protein is secreted through fusion of the bacterial OmpA signal sequence to its N-terminus under control of the *lac* promoter [52]. The signal sequence is correctly cleaved by *E. coli* signal peptidase. Expression is performed at 25 °C in order to obtain maximum yields of soluble PrP(121–231) in the periplasmic fraction. PrP(121–231) can then be purified to homogeneity from the periplasmic extract by anion-exchange chromatography at pH 7, hydrophobic chromatography at pH 6, and gel filtration at pH 6, with yields of about 3 mg pure PrP(121–231) per liter of bacterial culture.

Purified recombinant prion proteins can be stored at −20 °C after dialysis against distilled water. If recombinant PrP(23–231) or PrP(90–231) is stored at 4 °C, addition of protease inhibitor tablets ("COMPLETE," Roche) is recommended. Very detailed protocols for cytoplasmic expression of recombinant PrP(23–231), PrP(90–231), and PrP(121–231) as inclusion bodies, on-column refolding, and isolation of periplasmically expressed PrP(121–231), including information on expression vectors and bacterial growth conditions, are given in Refs. [52–55].

Finally, a protocol for covalent attachment of recombinant, full-length PrP to liposomes has been established. For this purpose, murine PrP(23–231), elongated by the hexapeptide $(Gly)_5$-Cys-CO_2^- at its C-terminus, is cytoplasmically expressed in *E. coli*, refolded, and purified from inclusion bodies essentially as described under Protocol 2. The recombinant PrP(23–231) variant with its free C-terminal thiol is then covalently coupled to the thiol-reactive lipid *N*-((2-pyridyldithio)-propionyl)-1,2-dihexdecanoyl-sn-glycero-3-phosphoethanolamine (PDP-DHPE; Molecular Probes) that has previously been incorporated into liposomes (100-nm diameter) composed of phosphatidylcholine, cholesterol (molar ratio 1:1), and 1% (w/w) PDP-DHPE. Liposome-bound PrP can then be separated from soluble PrP and nonspecific aggregates of PrP by ultracentrifugation. Typically, liposomes bearing 200 molecules of PrP(23–231) at their surface are obtained. Circular dichroism spectra and limited proteolysis experiments revealed that liposome-bound PrP has the same structure and stability as recombinant PrP(23–231) in solution. A very detailed protocol is given in Ref. [106].

Note added in the proof:

While this article was in press, the group of Stanley B. Prusiner reported the first generation of synthetic prions from recombinant PrP (Lagname G. et al., *Science* **305**, 673–676). Recombinant murine PrP(89-231) was polymerized into amyloid fibrils with high β-sheet content in vitro. When these fibrils were used for intracerebral inoculation of transgenic mice exclusively overexpressing murine PrP(89-231), all infected animals developed neurologic disease between 380 and 660 days after inoculation, while control animals inoculated with phosphate-buffered saline did not show signs of disease after 660 days. Prions isolated from brains of mice inoculated with synthetic PrP(89-231) fibrils could be successfully transmitted to wild type mice and mice overexpressing full-length PrP(23-231) with incubation times of 150 and 90 days, respectively, and analysis of brain extracts revealed accumulation of protease K resistant PrP^{Sc}. The authors state that their findings provide compelling evidence that TSE prions are infectious proteins.

References

1 Prusiner, S. B. (1997). Prion diseases and the BSE crisis. *Science* **278**, 245–51.
2 Cohen, F. E. (1999). Protein misfolding and prion diseases. *Journal of Molecular Biology* **293**, 313–320.
3 Weissmann, C. (1999). Molecular Genetics of Transmissible Spongiform Encephalopathies. *J. Biol. Chem.* **274**, 3–6.
4 Aguzzi, A., Montrasio, F. & Kaeser, P. S. (2001). Prions: health scare and biological challenge. *Nat Rev Mol Cell Biol* **2**, 118–26.
5 Weissmann, C., Enari, M., Klohn, P. C., Rossi, D. & Flechsig, E. (2002). Transmission of prions. *J Infect Dis* **186 Suppl 2**, S157–65.
6 Wadsworth, J. D., Hill, A. F., Beck, J. A. & Collinge, J. (2003). Molecular and clinical classification of human prion disease. *Br Med Bull* **66**, 241–54.
7 Osherovich, L. Z. & Weissman, J. S. (2002). The utility of prions. *Dev Cell* **2**, 143–51.
8 Aguzzi, A. & Polymenidou, M. (2004). Mammalian prion biology: one century of evolving concepts. *Cell* **116**, 313–27.
9 Bolton, D. C., McKinley, M. P. & Prusiner, S. B. (1982). Identification of a protein that purifies with the scrapie prion. *Science* **218**, 1309–11.
10 Riesner, D., Kellings, K., Wiese, U., Wulfert, M., Mirenda, C. & Prusiner, S. B. (1993). Prions and nucleic acids: search for "residual" nucleic acids and screening for mutations in the PrP-gene. *Dev Biol Stand* **80**, 173–81.
11 Alper, T., Cramp, W. A., Haig, D. A. & Clarke, M. C. (1967). Does the agent of scrapie replicate without nucleic acid? *Nature* **214**, 764–6.
12 Griffith, J. S. (1967). Self-replication and scrapie. *Nature* **215**, 1043–4.
13 Prusiner, S. B. (1982). Novel proteinaceous infectious particles cause scrapie. *Science* **216**, 136–44.
14 Basler, K., Oesch, B., Scott, M., Westaway, D., Walchli, M., Groth, D. F., McKinley, M. P., Prusiner, S. B. & Weissmann, C. (1986). Scrapie and cellular PrP isoforms are encoded by the same chromosomal gene. *Cell* **46**, 417–28.
15 Hope, J., Morton, L. J., Farquhar, C. F., Multhaup, G., Beyreuther, K. & Kimberlin, R. H. (1986). The major polypeptide of scrapie-associated fibrils (SAF) has the same size, charge distribution and N-terminal protein sequence as predicted for the normal brain protein (PrP). *Embo J* **5**, 2591–7.
16 Stahl, N. & Prusiner, S. B. (1991). Prions and prion proteins. *Faseb J* **5**, 2799–807.
17 Stahl, N., Baldwin, M. A., Teplow, D. B., Hood, L., Gibson, B. W., Burlingame, A. L. & Prusiner, S. B. (1993). Structural studies of the scrapie prion protein using mass spectrometry and amino acid sequencing. *Biochemistry* **32**, 1991–2002.
18 Pan, K. M., Baldwin, M., Nguyen, J., Gasset, M., Serban, A., Groth, D., Mehlhorn, I., Huang, Z., Fletterick, R. J., Cohen, F. E. & et al. (1993). Conversion of alpha-helices into beta-sheets features in the formation of the scrapie prion proteins. *Proc Natl Acad Sci USA* **90**, 10962–6.
19 Prusiner, S. B., McKinley, M. P., Bowman, K. A., Bolton, D. C., Bendheim, P. E., Groth, D. F. & Glenner, G. G. (1983). Scrapie prions aggregate to form amyloid-like birefringent rods. *Cell* **35**, 349–58.
20 Alper, T. (1985). Scrapie agent unlike

viruses in size and susceptibility to inactivation by ionizing or ultraviolet radiation. *Nature* **317**, 750.

21 Bueler, H., Aguzzi, A., Sailer, A., Greiner, R. A., Autenried, P., Aguet, M. & Weissmann, C. (1993). Mice devoid of PrP are resistant to scrapie. *Cell* **73**, 1339–47.

22 Hsiao, K., Baker, H. F., Crow, T. J., Poulter, M., Owen, F., Terwilliger, J. D., Westaway, D., Ott, J. & Prusiner, S. B. (1989). Linkage of a prion protein missense variant to Gerstmann-Straussler syndrome. *Nature* **338**, 342–5.

23 Chesebro, B. (1998). BSE and prions: uncertainties about the agent. *Science* **279**, 42–3.

24 Anfinsen, C. B. (1973). Principles that govern the folding of protein chains. *Science* **181**, 223–30.

25 Prusiner, S. B., Groth, D., Serban, A., Stahl, N. & Gabizon, R. (1993). Attempts to restore scrapie prion infectivity after exposure to protein denaturants. *Proc Natl Acad Sci USA* **90**, 2793–7.

26 Prusiner, S. B. (1991). Molecular biology of prion diseases. *Science* **252**, 1515–22.

27 Jarrett, J. T. & Lansbury, P. T., Jr. (1993). Seeding "one-dimensional crystallization" of amyloid: a pathogenic mechanism in Alzheimer's disease and scrapie? *Cell* **73**, 1055–8.

28 Booth, D. R., Sunde, M., Bellotti, V., Robinson, C. V., Hutchinson, W. L., Fraser, P. E., Hawkins, P. N., Dobson, C. M., Radford, S. E., Blake, C. C. & Pepys, M. B. (1997). Instability, unfolding and aggregation of human lysozyme variants underlying amyloid fibrillogenesis. *Nature* **385**, 787–93.

29 Hammarstrom, P., Wiseman, R. L., Powers, E. T. & Kelly, J. W. (2003). Prevention of transthyretin amyloid disease by changing protein misfolding energetics. *Science* **299**, 713–6.

30 Schatzl, H. M., Da Costa, M., Taylor, L., Cohen, F. E. & Prusiner, S. B. (1995). Prion protein gene variation among primates. *J Mol Biol* **245**, 362–74.

31 Wopfner, F., Weidenhofer, G., Schneider, R., von Brunn, A., Gilch, S., Schwarz, T. F., Werner, T. & Schatzl, H. M. (1999). Analysis of 27 mammalian and 9 avian PrPs reveals high conservation of flexible regions of the prion protein. *J Mol Biol* **289**, 1163–78.

32 Hope, J., Morton, L. J., Farquhar, C. F., Multhaup, G., Beyreuther, K. & Kimberlin, R. H. (1986). The major polypeptide of scrapie-associated fibrils (SAF) has the same size, charge distribution and N-terminal protein sequence as predicted for the normal brain protein (PrP). *Embo J* **5**, 2591–7.

33 Turk, E., Teplow, D. B., Hood, L. E. & Prusiner, S. B. (1988). Purification and properties of the cellular and scrapie hamster prion proteins. *Eur J Biochem* **176**, 21–30.

34 Caughey, B., Race, R. E., Ernst, D., Buchmeier, M. J. & Chesebro, B. (1989). Prion protein biosynthesis in scrapie-infected and uninfected neuroblastoma cells. *J Virol* **63**, 175–81.

35 Bueler, H., Fischer, M., Lang, Y., Bluethmann, H., Lipp, H. P., DeArmond, S. J., Prusiner, S. B., Aguet, M. & Weissmann, C. (1992). Normal development and behaviour of mice lacking the neuronal cell-surface PrP protein. *Nature* **356**, 577–82.

36 Shmerling, D., Hegyi, I., Fischer, M., Blattler, T., Brandner, S., Gotz, J., Rulicke, T., Flechsig, E., Cozzio, A., von Mering, C., Hangartner, C., Aguzzi, A. & Weissmann, C. (1998). Expression of amino-terminally truncated PrP in the mouse leading to ataxia and specific cerebellar lesions. *Cell* **93**, 203–14.

37 Brown, D. R., Qin, K., Herms, J. W., Madlung, A., Manson, J., Strome, R., Fraser, P. E., Kruck, T., von Bohlen, A., Schulz-Schaeffer, W., Giese, A., Westaway, D. & Kretzschmar, H. (1997). The cellular prion protein binds copper in vivo. *Nature* **390**, 684–7.

38 Tobler, I., Gaus, S. E., Deboer, T., Achermann, P., Fischer, M., Rulicke, T., Moser, M., Oesch, B., McBride, P. A. & Manson, J. C. (1996). Altered circadian activity rhythms and sleep in mice devoid of prion protein. *Nature* **380**, 639–42.

39 Collinge, J., Whittington, M. A., Sidle, K. C., Smith, C. J., Palmer, M. S., Clarke, A. R. & Jefferys, J. G. (1994). Prion protein is necessary for normal synaptic function. *Nature* **370**, 295–7.

40 Sakaguchi, S., Katamine, S., Nishida, N., Moriuchi, R., Shigematsu, K., Sugimoto, T., Nakatani, A., Kataoka, Y., Houtani, T., Shirabe, S., Okada, H., Hasegawa, S., Miyamoto, T. & Noda, T. (1996). Loss of cerebellar Purkinje cells in aged mice homozygous for a disrupted PrP gene. *Nature* **380**, 528–31.

41 Baron, G. S., Wehrly, K., Dorward, D. W., Chesebro, B. & Caughey, B. (2002). Conversion of raft associated prion protein to the protease-resistant state requires insertion of PrP-res (PrP(Sc)) into contiguous membranes. *Embo J* **21**, 1031–40.

42 Peters, P. J., Mironov, A., Jr., Peretz, D., van Donselaar, E., Leclerc, E., Erpel, S., DeArmond, S. J., Burton, D. R., Williamson, R. A., Vey, M. & Prusiner, S. B. (2003). Trafficking of prion proteins through a caveolae-mediated endosomal pathway. *J Cell Biol* **162**, 703–17.

43 Caughey, B. & Raymond, G. J. (1991). The scrapie-associated form of PrP is made from a cell surface precursor that is both protease- and phospholipase-sensitive. *J Biol Chem* **266**, 18217–23.

44 Hegde, R. S., Mastrianni, J. A., Scott, M. R., DeFea, K. A., Tremblay, P., Torchia, M., DeArmond, S. J., Prusiner, S. B. & Lingappa, V. R. (1998). A transmembrane form of the prion protein in neurodegenerative disease. *Science* **279**, 827–34.

45 Ma, J., Wollmann, R. & Lindquist, S. (2002). Neurotoxicity and neurodegeneration when PrP accumulates in the cytosol. *Science* **298**, 1781–5.

46 McKinley, M. P., Bolton, D. C. & Prusiner, S. B. (1983). A protease-resistant protein is a structural component of the scrapie prion. *Cell* **35**, 57–62.

47 Caughey, B., Raymond, G. J., Callahan, M. A., Wong, C., Baron, G. S. & Xiong, L. W. (2001). Interactions and conversions of prion protein isoforms. *Adv Protein Chem* **57**, 139–69.

48 Fischer, M., Rulicke, T., Raeber, A., Sailer, A., Moser, M., Oesch, B., Brandner, S., Aguzzi, A. & Weissmann, C. (1996). Prion protein (PrP) with amino-proximal deletions restoring susceptibility of PrP knockout mice to scrapie. *Embo J* **15**, 1255–64.

49 Appel, T. R., Dumpitak, C., Matthiesen, U. & Riesner, D. (1999). Prion rods contain an inert polysaccharide scaffold. *Biol Chem* **380**, 1295–306.

50 Nguyen, J. T., Inouye, H., Baldwin, M. A., Fletterick, R. J., Cohen, F. E., Prusiner, S. B. & Kirschner, D. A. (1995). X-ray diffraction of scrapie prion rods and PrP peptides. *J Mol Biol* **252**, 412–22.

51 Wille, H., Michelitsch, M. D., Guenebaut, V., Supattapone, S., Serban, A., Cohen, F. E., Agard, D. A. & Prusiner, S. B. (2002). Structural studies of the scrapie prion protein by electron crystallography. *Proc Natl Acad Sci USA* **99**, 3563–8.

52 Hornemann, S. & Glockshuber, R. (1996). Autonomous and reversible folding of a soluble amino-terminally truncated segment of the mouse prion protein. *J Mol Biol* **261**, 614–9.

53 Hornemann, S., Korth, C., Oesch, B., Riek, R., Wider, G., Wuthrich, K. & Glockshuber, R. (1997). Recombinant full-length murine prion protein, mPrP(23–231): purification and spectroscopic characterization. *FEBS Lett* **413**, 277–81.

54 Zahn, R., von Schroetter, C. & Wuthrich, K. (1997). Human prion proteins expressed in Escherichia coli and purified by high-affinity column refolding. *FEBS Lett* **417**, 400–4.

55 Liemann, S. & Glockshuber, R. (1999). Influence of amino acid substitutions related to inherited human prion diseases on the thermodynamic stability of the cellular prion protein. *Biochemistry* **38**, 3258–67.

56 Riek, R., Hornemann, S., Wider, G., Billeter, M., Glockshuber, R. &

Wuthrich, K. (1996). NMR structure of the mouse prion protein domain PrP(121–321). *Nature* **382**, 180–2.

57 Wuthrich, K. & Riek, R. (2001). Three-dimensional structures of prion proteins. *Adv Protein Chem* **57**, 55–82.

58 Riek, R., Wider, G., Billeter, M., Hornemann, S., Glockshuber, R. & Wuthrich, K. (1998). Prion protein NMR structure and familial human spongiform encephalopathies. *Proc Natl Acad Sci USA* **95**, 11667–72.

59 Riek, R., Hornemann, S., Wider, G., Glockshuber, R. & Wuthrich, K. (1997). NMR characterization of the full-length recombinant murine prion protein, mPrP(23–231). *FEBS Lett* **413**, 282–8.

60 Viles, J. H., Cohen, F. E., Prusiner, S. B., Goodin, D. B., Wright, P. E. & Dyson, H. J. (1999). Copper binding to the prion protein: structural implications of four identical cooperative binding sites. *Proc Natl Acad Sci USA* **96**, 2042–7.

61 Dyson, H. J. & Wright, P. E. (2002). Coupling of folding and binding for unstructured proteins. *Curr Opin Struct Biol* **12**, 54–60.

62 Knaus, K. J., Morillas, M., Swietnicki, W., Malone, M., Surewicz, W. K. & Yee, V. C. (2001). Crystal structure of the human prion protein reveals a mechanism for oligomerization. *Nat Struct Biol* **8**, 770–4.

63 Moore, R. C., Lee, I. Y., Silverman, G. L., Harrison, P. M., Strome, R., Heinrich, C., Karunaratne, A., Pasternak, S. H., Chishti, M. A., Liang, Y., Mastrangelo, P., Wang, K., Smit, A. F., Katamine, S., Carlson, G. A., Cohen, F. E., Prusiner, S. B., Melton, D. W., Tremblay, P., Hood, L. E. & Westaway, D. (1999). Ataxia in prion protein (PrP)-deficient mice is associated with upregulation of the novel PrP-like protein doppel. *J Mol Biol* **292**, 797–817.

64 Behrens, A., Genoud, N., Naumann, H., Rulicke, T., Janett, F., Heppner, F. L., Ledermann, B. & Aguzzi, A. (2002). Absence of the prion protein homologue Doppel causes male sterility. *Embo J* **21**, 3652–8.

65 Mo, H., Moore, R. C., Cohen, F. E., Westaway, D., Prusiner, S. B., Wright, P. E. & Dyson, H. J. (2001). Two different neurodegenerative diseases caused by proteins with similar structures. *Proc Natl Acad Sci USA* **98**, 2352–7.

66 Luhrs, T., Riek, R., Guntert, P. & Wuthrich, K. (2003). NMR structure of the human doppel protein. *J Mol Biol* **326**, 1549–57.

67 James, T. L., Liu, H., Ulyanov, N. B., Farr-Jones, S., Zhang, H., Donne, D. G., Kaneko, K., Groth, D., Mehlhorn, I., Prusiner, S. B. & Cohen, F. E. (1997). Solution structure of a 142-residue recombinant prion protein corresponding to the infectious fragment of the scrapie isoform. *Proc Natl Acad Sci USA* **94**, 10086–91.

68 Donne, D. G., Viles, J. H., Groth, D., Mehlhorn, I., James, T. L., Cohen, F. E., Prusiner, S. B., Wright, P. E. & Dyson, H. J. (1997). Structure of the recombinant full-length hamster prion protein PrP(29–231): the N terminus is highly flexible. *Proc Natl Acad Sci USA* **94**, 13452–7.

69 Zahn, R., Liu, A., Luhrs, T., Riek, R., von Schroetter, C., Lopez Garcia, F., Billeter, M., Calzolai, L., Wider, G. & Wuthrich, K. (2000). NMR solution structure of the human prion protein. *Proc Natl Acad Sci USA* **97**, 145–50.

70 Lopez Garcia, F., Zahn, R., Riek, R. & Wuthrich, K. (2000). NMR structure of the bovine prion protein. *Proc Natl Acad Sci USA* **97**, 8334–9.

71 Telling, G. C., Scott, M., Mastrianni, J., Gabizon, R., Torchia, M., Cohen, F. E., DeArmond, S. J. & Prusiner, S. B. (1995). Prion propagation in mice expressing human and chimeric PrP transgenes implicates the interaction of cellular PrP with another protein. *Cell* **83**, 79–90.

72 Swietnicki, W., Petersen, R., Gambetti, P. & Surewicz, W. K. (1997). pH-dependent stability and conformation of the recombinant human prion protein PrP(90–231). *J Biol Chem* **272**, 27517–20.

73 Swietnicki, W., Petersen, R. B., Gambetti, P. & Surewicz, W. K. (1998). Familial mutations and the thermodynamic stability of the recombinant human prion protein. *J Biol Chem* **273**, 31048–52.

74 Hosszu, L. L., Baxter, N. J., Jackson, G. S., Power, A., Clarke, A. R., Waltho, J. P., Craven, C. J. & Collinge, J. (1999). Structural mobility of the human prion protein probed by backbone hydrogen exchange. *Nat Struct Biol* **6**, 740–3.

75 Jackson, G. S., Hosszu, L. L., Power, A., Hill, A. F., Kenney, J., Saibil, H., Craven, C. J., Waltho, J. P., Clarke, A. R. & Collinge, J. (1999a). Reversible conversion of monomeric human prion protein between native and fibrilogenic conformations. *Science* **283**, 1935–7.

76 Prusiner, S. B., Groth, D., Serban, A., Stahl, N. & Gabizon, R. (1993). Attempts to restore scrapie prion infectivity after exposure to protein denaturants. *Proc Natl Acad Sci USA* **90**, 2793–7.

77 Wildegger, G., Liemann, S. & Glockshuber, R. (1999). Extremely rapid folding of the C-terminal domain of the prion protein without kinetic intermediates. *Nat Struct Biol* **6**, 550–3.

78 Apetri, A. C. & Surewicz, W. K. (2002). Kinetic intermediate in the folding of human prion protein. *J Biol Chem* **277**, 44589–92.

79 Apetri, A. C., Surewicz, K. & Surewicz, W. K. (2004). The effect of disease-associated mutations on the folding pathway of human prion protein. *J Biol Chem* **279**, 18008–14.

80 Vey, M., Pilkuhn, S., Wille, H., Nixon, R., DeArmond, S. J., Smart, E. J., Anderson, R. G., Taraboulos, A. & Prusiner, S. B. (1996). Subcellular colocalization of the cellular and scrapie prion proteins in caveolae-like membranous domains. *Proc Natl Acad Sci USA* **93**, 14945–9.

81 Arnold, J. E., Tipler, C., Laszlo, L., Hope, J., Landon, M. & Mayer, R. J. (1995). The abnormal isoform of the prion protein accumulates in late-endosome-like organelles in scrapie-infected mouse brain. *J Pathol* **176**, 403–11.

82 Borchelt, D. R., Taraboulos, A. & Prusiner, S. B. (1992). Evidence for synthesis of scrapie prion proteins in the endocytic pathway. *J Biol Chem* **267**, 16188–99.

83 Lee, R. J., Wang, S. & Low, P. S. (1996). Measurement of endosome pH following folate receptor-mediated endocytosis. *Biochim Biophys Acta* **1312**, 237–42.

84 Zhang, H., Stockel, J., Mehlhorn, I., Groth, D., Baldwin, M. A., Prusiner, S. B., James, T. L. & Cohen, F. E. (1997). Physical studies of conformational plasticity in a recombinant prion protein. *Biochemistry* **36**, 3543–53.

85 Hornemann, S. & Glockshuber, R. (1998). A scrapie-like unfolding intermediate of the prion protein domain PrP(121–231) induced by acidic pH. *Proc Natl Acad Sci USA* **95**, 6010–4.

86 Jackson, G. S., Hill, A. F., Joseph, C., Hosszu, L., Power, A., Waltho, J. P., Clarke, A. R. & Collinge, J. (1999b). Multiple folding pathways for heterologously expressed human prion protein. *Biochim Biophys Acta* **1431**, 1–13.

87 Glockshuber, R. (2001). Folding dynamics and energetics of recombinant prion proteins. *Adv Protein Chem* **57**, 83–105.

88 Swietnicki, W., Morillas, M., Chen, S. G., Gambetti, P. & Surewicz, W. K. (2000). Aggregation and fibrillization of the recombinant human prion protein huPrP90-231. *Biochemistry* **39**, 424–31.

89 Mehlhorn, I., Groth, D., Stockel, J., Moffat, B., Reilly, D., Yansura, D., Willett, W. S., Baldwin, M., Fletterick, R., Cohen, F. E., Vandlen, R., Henner, D. & Prusiner, S. B. (1996). High-level expression and characterization of a purified 142-residue polypeptide of the prion protein. *Biochemistry* **35**, 5528–37.

90 Jackson, G. S., Hosszu, L. L., Power, A., Hill, A. F., Kenney, J., Saibil, H., Craven, C. J., Waltho, J. P., Clarke, A. R. & Collinge, J. (1999a).

Reversible conversion of monomeric human prion protein between native and fibrilogenic conformations. *Science* **283**, 1935–7.

91 Ma, J. & Lindquist, S. (1999). De novo generation of a PrPSc-like conformation in living cells. *Nat Cell Biol* **1**, 358–61.

92 Ma, J., Wollmann, R. & Lindquist, S. (2002). Neurotoxicity and neurodegeneration when PrP accumulates in the cytosol. *Science* **298**, 1781–5.

93 Lee, S. & Eisenberg, D. (2003). Seeded conversion of recombinant prion protein to a disulfide-bonded oligomer by a reduction-oxidation process. *Nat Struct Biol* **10**, 725–30.

94 Fandrich, M., Fletcher, M. A. & Dobson, C. M. (2001). Amyloid fibrils from muscle myoglobin. *Nature* **410**, 165–6.

95 Dobson, C. M. (2003). Protein folding and misfolding. *Nature* **426**, 884–90.

96 Apetri, A. C. & Surewicz, W. K. (2003). Atypical effect of salts on the thermodynamic stability of human prion protein. *J Biol Chem* **278**, 22187–92.

97 Brandner, S., Isenmann, S., Raeber, A., Fischer, M., Sailer, A., Kobayashi, Y., Marino, S., Weissmann, C. & Aguzzi, A. (1996). Normal host prion protein necessary for scrapie-induced neurotoxicity. *Nature* **379**, 339–43.

98 Kocisko, D. A., Come, J. H., Priola, S. A., Chesebro, B., Raymond, G. J., Lansbury, P. T. & Caughey, B. (1994). Cell-free formation of protease-resistant prion protein. *Nature* **370**, 471–4.

99 Raymond, G. J., Hope, J., Kocisko, D. A., Priola, S. A., Raymond, L. D., Bossers, A., Ironside, J., Will, R. G., Chen, S. G., Petersen, R. B., Gambetti, P., Rubenstein, R., Smits, M. A., Lansbury, P. T., Jr. & Caughey, B. (1997). Molecular assessment of the potential transmissibilities of BSE and scrapie to humans. *Nature* **388**, 285–8.

100 Saborio, G. P., Permanne, B. & Soto, C. (2001). Sensitive detection of pathological prion protein by cyclic amplification of protein misfolding. *Nature* **411**, 810–3.

101 Deleault, N. R., Lucassen, R. W. & Supattapone, S. (2003). RNA molecules stimulate prion protein conversion. *Nature* **425**, 717–20.

102 Collinge, J., Sidle, K. C., Meads, J., Ironside, J. & Hill, A. F. (1996a). Molecular analysis of prion strain variation and the aetiology of 'new variant' CJD. *Nature* **383**, 685–90.

103 Palmer, M. S., Dryden, A. J., Hughes, J. T. & Collinge, J. (1991). Homozygous prion protein genotype predisposes to sporadic Creutzfeldt-Jakob disease. *Nature* **352**, 340–2.

104 Collinge, J., Palmer, M. S. & Dryden, A. J. (1991). Genetic predisposition to iatrogenic Creutzfeldt-Jakob disease. *Lancet* **337**, 1441–2.

105 Collinge, J., Beck, J., Campbell, T., Estibeiro, K. & Will, R. G. (1996b). Prion protein gene analysis in new variant cases of Creutzfeldt-Jakob disease. *Lancet* **348**, 56.

106 Eberl, H., Tittmann, P. & Glockshuber, R. (2004). Characterization of recombinant, membrane-attached full-length prion protein. *J Biol Chem* **279**, 25058–65.

107 Uptain, S. M. & Lindquist, S. (2002). Prions as protein-based genetic elements. *Annu Rev Microbiol* **56**, 703–41.

108 Osherovich, L. Z. & Weissman, J. S. (2002). The utility of prions. *Dev Cell* **2**, 143–51.

109 Wickner, R. B., Edskes, H. K., Roberts, B. T., Pierce, M. & Baxa, U. (2002). Prions of yeast as epigenetic phenomena: high protein "copy number" inducing protein "silencing". *Adv Genet* **46**, 485–525.

110 Sparrer, H. E., Santoso, A., Szoka, F. C., Jr. & Weissman, J. S. (2000). Evidence for the prion hypothesis: induction of the yeast [PSI+] factor by in vitro – converted Sup35 protein. *Science* **289**, 595–9.

111 Santoso, A., Chien, P., Osherovich, L. Z. & Weissman, J. S. (2000). Molecular basis of a yeast prion species barrier. *Cell* **100**, 277–88.

35
Insights into the Nature of Yeast Prions

Lev Z. Osherovich and Jonathan S. Weissman

35.1
Introduction

The concept of an infectious protein (prion) was first put forth to explain the properties of a set of related transmissible spongiform encephalopathies (TSEs) including scrapie in sheep and Cruetzfeld-Jakob disease (CJD) in man (for reviews, see Prusiner 1998; Weissmann 1999). Purification of the infectious agent revealed that it was composed largely (if not entirely) of a single endogenous protein called PrP. Remarkably, PrP can interconvert between two states: the normal form (termed PrP^{c}) and its infectious variant (PrP^{Sc}). PrP^{c} has high α-helical content, whereas PrP^{Sc} adopts a β-sheet-rich conformation and in some cases aggregates to form long polymers termed amyloid fibers. These findings led to the "protein only" hypothesis in which replication of the infectious agent results from the ability of aggregated PrP^{Sc} to bind and catalyze conversion of PrP^{c} to PrP^{Sc}.

TSEs such as mad cow disease continue to imperil domesticated mammals and potentially threaten human health. Since the advent of the prion concept, a great deal of effort has been put into understanding the mechanisms for the appearance and transmission of protein-based infectious diseases, with considerable success. However, the inability to reconstitute critical aspects of mammalian prion diseases in vitro, as well as the difficulty of working with animal models, has greatly slowed efforts to investigate many key aspects of prion biology. In particular, the origin and limits of the barriers that inhibit the transmission of prion diseases between different species as well as the molecular basis of the strain phenomenon, wherein prions composed of the same infectious protein can cause distinct disease states, have remained largely mysterious.

Wickner (1994) extended the prion concept to explain the inheritance of two enigmatic non-Mendelian elements in the yeast *Saccharomyces cerevisiae*, called [*URE3*] and [PSI^{+}] (for review, see Uptain and Lindquist 2002). In contrast to the mammalian case, yeast prions do not cause cell death but rather act as epigenetic modulators of protein function. The phenotypes associated with [*URE3*] or [PSI^{+}] elements are not particularly remarkable: [*URE3*] alters nitrogen catabolite uptake (Lacroute 1971), while [PSI^{+}] allows for the suppression of some nonsense

Protein Folding Handbook. Part II. Edited by J. Buchner and T. Kiefhaber

ISBN: 3-527-30784-2

mutations (Cox 1965). Indeed, traditional loss-of-function mutations in the chromosomally encoded nitrogen catabolism repressor Ure2p and the essential translation termination factor Sup35p mimic the [*URE3*] and [PSI^+] states, respectively. What makes [*URE3*] and [PSI^+] remarkable are their epigenetic properties; for example, they are inherited by all of the meiotic progeny of "heterozygous" diploid cells and can be transmitted by transfer of cytoplasm from one cell to another without the exchange of genetic material.

To explain the unusual inheritance of [*URE3*] and [PSI^+], Wickner (1994) proposed that these states result from the presence of self-propagating (prion) forms of the Ure2p and Sup35p proteins, respectively (see Figure 35.1). This model provides a mechanism for the non-nuclear inheritance and faithful propagation of the [*URE3*] and [PSI^+] states; the prion forms can convert free molecules of the affectted protein and are themselves distributed along with the cytoplasm to all of the cell's progeny. The prion model also explains why the phenotypes of [*URE3*] and [PSI^+] mimic the loss of function of Ure2p and Sup35p, as conversion inactivates the affected protein. Since this time, new fungal prions and prion-like states have been identified, including the naturally occurring yeast prion [RNQ^+] (Sondheimer and Lindquist 2000) (also known as [PIN^+]) (Derkatch et al. 2001), the artificial prion [NU^+] (Santoso et al. 2000), and the *Podospora anserina* prion [*Het-s*] (Coustou et al. 1997). These have been comprehensively reviewed elsewhere (Uptain and Lindquist 2002; Osherovich and Weissman 2002). Importantly, formal proof of the prion hypothesis has come from studies with [PSI^+] and [Het-s]. When introduced into cells, prion seeds generated in vitro from purified recombinant Sup35p (Sparrer et al. 2000; Tanaka et al. 2004; King and Diaz-Avalos 2004) or HET-s protein (Maddelein et al. 2002) are able to cause de novo formation of the [PSI^+] and [Het-s] states, respectively. The powerful genetic tools of the yeast *S. cerevisae* together with the ability to create infectious material from recombinant protein have greatly facilitated efforts to understand the requirements and properties of prion inheritance. Drawing on our detailed understanding of the yeast prions, we examine the general principles of prion formation, replication, and transmission. In addition we discuss the possibility that, by allowing for stable self-organizing structures, prion-like conformations may play a broader role in normal biology (for reviews of the mammalian prion literature, see Prusiner 1998; Weissmann 1999).

35.2
Prions as Heritable Amyloidoses

Prion-based inheritance requires that a protein be capable of adopting at least one alternate, infectious protein state that is able to bind and convert the normal cellular form of the protein. A range of studies indicate that both mammalian and yeast prions accomplish this by directing the conformational change of a normal cellular – or, in the case of PrP, cell surface – host protein into an alternate prion conformation. These alternate conformations are β-sheet-rich multimers and re-

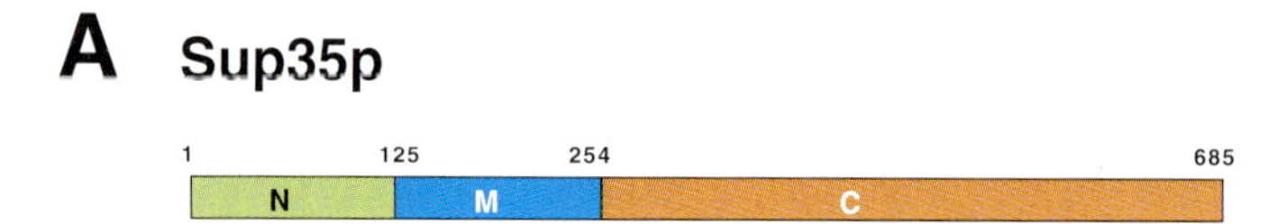

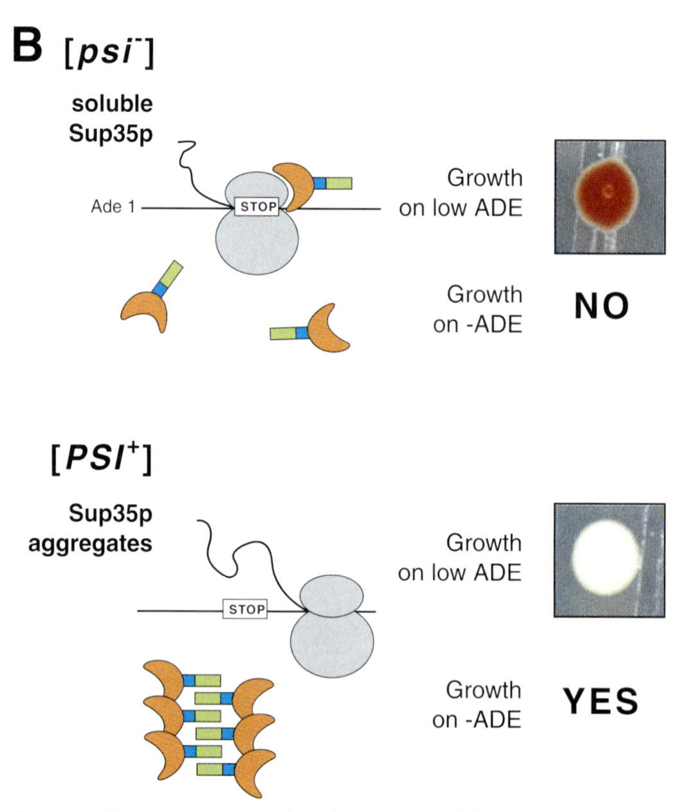

Fig. 35.1. The yeast prions [PSI^+] and [*URE3*] are due to self-propagating protein conformations. Adapted from (Chien et al. 2004). (A) Sup35p is an essential modular protein involved in translation termination whose self-propagating aggregation is responsible for the [PSI^+] phenotype. The N-terminus, (N), residues 1–125 (green), is glutamine- and asparagine-rich and mediates prion behavior. The middle domain (M), residues 126–254 (blue), is rich in charged residues. Commonly, purified NM fusions are used in vitro. The C-terminal domain (C), residues 254–685 (orange), is the essential part of the protein and is responsible for faithful translation termination at stop codons. (B) In [psi^-] yeast, Sup35p is soluble and able to interact with the ribosome to facilitate translation termination. In [PSI^+] yeast, Sup35p is aggregated, sequestering it from the ribosome and therefore allowing nonsense suppression. Typically, the [PSI^+] nonsense-suppression phenotype is monitored by measuring read-through of a reporter gene carrying a premature stop, such as the *ade*-1 gene, which leads to a convenient color change on low-adenine media and to a differential growth phenotype on media lacking adenine.

C Ure2p

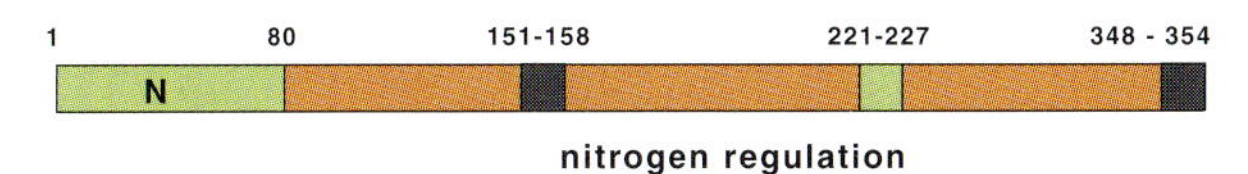

D [*ure-o*]

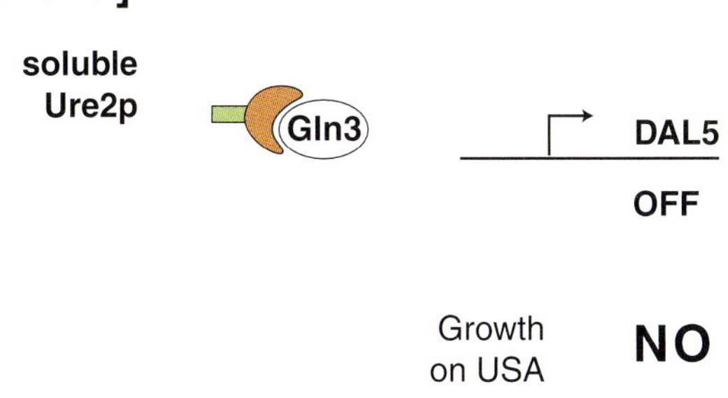

[*URE3*]

Ure2p
aggregates

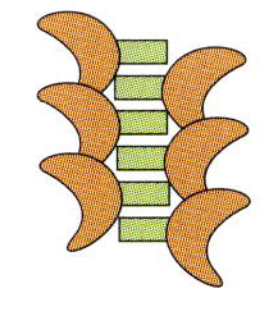

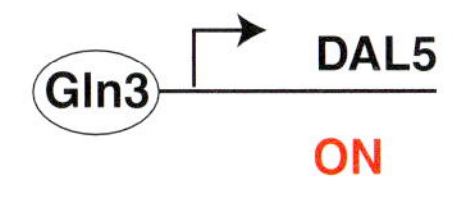

Growth
on USA

YES

Fig. 35.1. (C) Ure2p is a modular protein involved in regulation of nitrogen catabolism whose self-propagating aggregation is responsible for the [*URE3*] phenotype. In addition to the glutamine/asparagine-rich amino terminus (green), Ure2p also contains another region that facilitates prion behavior (green) and portions that antagonize prion formation (black). The C-terminus (C), residues 81–354 (orange), resembles glutathione-S-transferase, and signals the presence of high-quality nitrogen sources through Gln3. (D) Normally, Ure2p binds the transcription factor Gln3p in the cytoplasm, keeping Gln3p from turning on a host of genes required for uptake of poor nitrogen sources, such as *DAL5*, a plasma membrane import protein. Serendipitously, Dal5p imports not only the poor nitrogen source allantoate but also USA (*N*-carbamyl-aspartate), an intermediate in uracil biosynthesis. This allows selection for [URE3] on USA medium in a uracil auxotroph. In a [ure-o] yeast, Ure2p is complexed with Gln3, and DAL5 is turned off, prohibiting growth on USA medium. In a [URE3] yeast, Ure2p is aggregated, releasing Gln3 to the nucleus, where it activates transcription of *DAL5* and allows USA uptake and growth on USA medium.

semble a broader class of ordered protein aggregates often referred to as amyloids (Dobson 2001). In the past few years, amyloids have received a great deal of attention due to their involvement in a wide variety of protein-misfolding disorders, including neuropathies such as Alzheimer's, Parkinson's, and Huntington's diseases (Taylor et al. 2002) as well as a number of systemic amyloidoses. A number of non-disease-associated proteins, such as acylphosphatase and the SH3 domain from PIP_3 kinase, have been shown to form amyloid under mildly denaturing conditions (Chiti et al. 1999; Guijarro et al. 1998). The ability of so many unrelated proteins to form amyloids argues that this fold is generally accessible to polymers of amino acids, perhaps because it is stabilized by main-chain rather than side-chain interactions (Dobson 1999). Polyglutamine amyloids of the kind implicated in trinucleotide expansion diseases may gain additional stability due to hydrogen bonding between main-chain and side-chain amide moieties (Perutz et al. 1994).

Like mammalian prions, fungal prions appear to be amyloid conformational variants of normal cellular proteins. All fungal prion proteins identified have been shown to form high-molecular-weight complexes in prion-containing cells (Coustou et al. 1997; Masison and Wickner 1995; Patino et al. 1996; Paushkin et al. 1996; Santoso et al. 2000; Sondheimer and Lindquist 2000), although the extent of aggregation can vary from strain to strain (Ripaud et al. 2003). These prion aggregates can in some conditions be visualized in cells by expressing GFP fusions of the prion protein. These reporter fusions are soluble and are distributed evenly throughout the cytoplasm in prion-free cells but coalesce into punctate foci in prion-containing cells (Osherovich and Weissman 2001; Patino et al. 1996; Sondheimer and Lindquist 2000; Zhou et al. 2001). Ure2p has also been visualized by thin-section EM followed by immunogold staining and has been shown to form short cytoplasmic fibrils specifically in [*URE3*] yeast (Speransky et al. 2001). These prion aggregates are typically stable and have altered resistance to protease digestion (Coustou et al. 1997; Masison and Wickner 1995; Patino et al. 1996; Paushkin et al. 1996). Finally, de novo formation of these aggregates is slow, but once formed they are stably inherited by daughter cells during mitosis.

Self-propagating amyloid aggregation is only one of a number of possible mechanisms of protein-based epigenetic inheritance (Griffith 1967). Indeed, Roberts and Wickner (2003) have used a zymogen, or self-activating enzyme, to create a heritable positive feedback loop that resembles a prion in its inheritance and transmissibility but does not involve protein aggregation. However, all hitherto described naturally occurring prions are based on self-propagating amyloid aggregates.

The fungal prions have proven to be far more amenable to reconstitution in vitro than the mammalian prion system. Extracts from [PSI^+] yeast can induce conversion of soluble Sup35p in extracts from [psi^-] yeast (Paushkin et al. 1997). Moreover, Sup35p, Ure2p, HET-s, and Rnq1p, the prion proteins responsible for [PSI^+], [*URE3*], [Het-s], and [RNQ^+]/[PIN^+], respectively, have all been shown to form self-seeding amyloids in vitro (Baxa et al. 2003; Dos Reis et al. 2002; Glover et al. 1997; Jiang et al. 2004; King et al. 1997; Taylor et al. 1999). Non-amyloid fibrillar forms of Ure2p have also been suggested to underlie [*URE3*] (Bousset et al. 2002; Fay et al. 2003). Several lines of evidence argue that self-seeded aggregation drives prion inheritance in vivo. For example, mutations in Sup35 that affect aggregation in vivo have parallel effects on in vitro polymerization (DePace et al. 1998; Liu and Lindquist 1999; Patino et al. 1996). More directly for Sup35p and HET-s, it has been possible to create amyloids in vitro from recombinant protein and to use these to convert wild-type cells to the prion state when introduced into the cytoplasm of prion-free cells (Maddelein et al. 2002; Sparrer et al. 2000; Tanaka et al. 2004; King and Diaz-Avalos 2004). These experiments have provided the first direct demonstration that a pure protein can indeed act as a conformationally based infectious agent. Although the self-seeding nature of amyloid aggregates constitutes the basis of prion-based inheritance, not all self-seeding amyloids can support prion inheritance (see below).

35.3 Prion Strains and Species Barriers: Universal Features of Amyloid-based Prion Elements

Studies of the yeast prion [PSI^+] have shed light on two of the most perplexing features of prion biology. One of these is the existence of transmission barriers that inhibit the passage of prions between related species (Chien et al. 2004). The primary structure of the prion protein is a critical determinant of the specificity of propagation, as susceptibility to cross-species infection is intimately related to the degree of similarity between the sequences of the two prion proteins (Collinge 2001; Prusiner et al. 1990; Scott et al. 1989). Indeed, even point mutations or allelic variants can have dramatic effects on the specificity of prion propagation (Bossers et al. 1997; Bruce et al. 1994; Manson et al. 1999, 2000; Mastrianni et al. 2001).

The second remarkable feature is the existence of multiple prion strains, wherein infectious particles composed of the same protein give rise to a range of prion states that vary in incubation time, pathology, and other phenotypic aspects (Chien et al. 2004). Strain variability had been observed long before the prion hypothesis; in fact, it was originally interpreted as evidence for the existence of a nucleic acid genome in the infectious particle, with strain variation arising from mutations in this genome. To reconcile the presence of prion strain with the protein-only hypothesis, it has been proposed that a single polypeptide can misfold into multiple infectious conformations, at least one for each phenotypic variant (Aguzzi and Haass 2003; Weissmann 1991). Rather than being peculiarities of mammalian prions, it is now clear that both transmission barriers and strain variation are common features of amyloid-based prion elements and that both arise from the general ability of prion proteins to adopt multiple amyloid conformations.

Barriers inhibiting yeast prion transmission have been extensively studied using the yeast prion [PSI^+]. The cloning of *SUP35* genes from a broad range of budding yeast revealed that although the sequence of the amino-terminal domain varies, these domains can support prion states when expressed in a heterologous *S. cerevisiae* system (Chernoff et al. 2000; Kushnirov et al. 2000a; Santoso et al. 2000) and, in one case, in the original yeast species (*Kluyveromyces lactis*) from which it was derived (Nakayashiki et al. 2001). As with mammalian prions, the induction and transmission of [*PSI*$^+$] prions are typically confined to proteins from the same species (Chernoff et al. 2000; Crist et al. 2003; Kushnirov et al. 2000a; Nakayashiki et al. 2001; Santoso et al. 2000). An example of a particularly robust barrier is that between prions of *S. cerevisiae*- and *Candida albicans*-derived SUP35 prion domains. Although these organisms would not naturally interact, specificity can be studied using genetically manipulated yeast. Overexpression of *S. cerevisiae* Sup35p induces [*PSI*$^+$] in wild-type *S. cerevisiae* but not in yeast where the *SUP35* gene encodes the *C. albicans* prion domain and vice versa (Santoso et al. 2000). Even a single point mutation within the *S. cerevisiae SUP35* sequence is sufficient to alter specificity (DePace et al. 1998). However, in some cases, cross-transmission of a prion state between different *SUP35* proteins is possible, albeit with reduced efficiency compared to homotypic transmission (Chernoff et al. 2000; Chien et al. 2003).

Although cellular factors such as chaperones are needed for the inheritance of prions (see below), they do not determine specificity of prion propagation, as the species barrier can be reconstituted solely from recombinant prion proteins (Chien et al. 2003; Glover et al. 1997; Santoso et al. 2000). For example, both *S. cerevisiae*- and *C. albicans*-derived prion domains form amyloid fibers in vitro after characteristic lag times; addition of preformed fibers of *S. cerevisiae* Sup35p prion domains efficiently seeds polymerization of *S. cerevisiae* Sup35p prion domains but not of domains derived from *C. albicans* and vice versa. Even when present together in the same reaction, these two species of prion domains form homopolymeric fibers (Santoso et al. 2000). Specificity seems to be a common feature of amyloids (including those of noninfectious proteins) and can be explained by a requirement for conformational compatibility between amyloid fibril ends and soluble monomers during growth (Chien et al. 2004; O'Nuallain et al. 2004). Nevertheless, this does not preclude a role for host factors in modulating the host specificity of certain prions.

Like mammalian prions (Prusiner 1998) and other yeast prions (Bradley et al. 2002; Schlumpberger et al. 2001), [*PSI*$^+$] exhibits a range of heritable phenotypic strain variants (Derkatch et al. 1996), which are linked to differences in the conformation of the infectious prion protein. [*PSI*$^+$] strain variants differ in mitotic stability (Derkatch et al. 1996), in their interaction with cellular chaperone machinery (Kushnirov et al. 2000b), and in the solubility and activity of Sup35p protein (Derkatch et al. 1996; Zhou et al. 1999). Sup35p aggregates purified from different [*PSI*$^+$] strains also differ in their ability to seed polymerization of purified Sup35 in vitro (Uptain et al. 2001). In addition, strains can play a major role in determining specificity of prion transmission: [*PSI*$^+$] prion strains differ greatly in their ability to recruit various mutant forms of Sup35p (Chien et al. 2003; Chien and Weissman 2001).

The strain phenomenon is in fact intimately related to the alternative conformational states available to prion aggregates. For example, a chimeric prion domain containing the first 40 amino acids of *S. cerevisiae* SUP35 fused to the remainder of the *C. albicans* SUP35 prion domain can adopt at least two amyloid conformations, one that seeds *S. cerevisiae* protein and another that seeds *C. albicans* protein (Chien and Weissman 2001). Each conformation results in a phenotypically distinctive prion state that recapitulates the variation observed between naturally occurring prion strains (Chien et al. 2003). Moreover, the infection of yeast with recombinant Sup35p in distinct amyloid conformations generated in vitro leads to heritable differences in [PSI^+] prion strains (Tanaka et al. 2004; King and Diaz-Avalos 2004). These studies provide the first proof that prion strain difference can be "encoded" within the conformation of the infectious protein.

A synthesis of a wide range of experimental observations points to the following findings linking prion strains, species barriers, and the physical principles that govern protein misfolding to form amyloid aggregates (Chien et al. 2004).

1. Self-propagation of amyloid-like protein aggregates underlies prion growth and specificity.
2. A single protein can often misfold into multiple different amyloid conformations (Figure 35.2).
3. The phenotypic consequences resulting from an aggregated protein are highly dependent on the specific amyloid conformation.
4. The particular amyloid conformation that a protein adopts determines the specificity of growth.
5. Changes in protein sequence can modulate the spectrum of favored amyloid conformations.

Based on these observations, a model emerges in which prion strains and transmission barriers are in large part two different manifestations of the same phenomenon: the ability of a protein to misfold into multiple amyloid conformations (Figure 35.2). These conformations determine both the specificity of growth and the phenotypic consequences of harboring a prion. Changes in sequence in turn alter the range of preferred amyloid conformations, thereby modulating transmission barriers and strain phenotypes. The ability of mutations to alter the conformation and thus the specificity of the prion forms preferred by a given protein helps to explain the ubiquitous presence of species barriers.

35.4 Prediction and Identification of Novel Prion Elements

[*URE3*] and [PSI^+] (but not [Het-s] or the mammalian prions) are examples of a broader phenomenon of glutamine- and asparagine-rich (Gln/Asn-rich) protein aggregation of the kind implicated in human polyGln expansion disorders such as Huntington's disease (Cummings and Zoghbi 2000). Ure2p and the various Sup35 homologues from budding yeast all contain prion domains with a very high glutamine and asparagine (Q/N) content. High Q/N content is also a conserved feature in Ure2p, but whether Ure2p proteins from yeasts other than *S. cerevisiae* can support prion inheritance remains an open question (Baudin-Baillieu

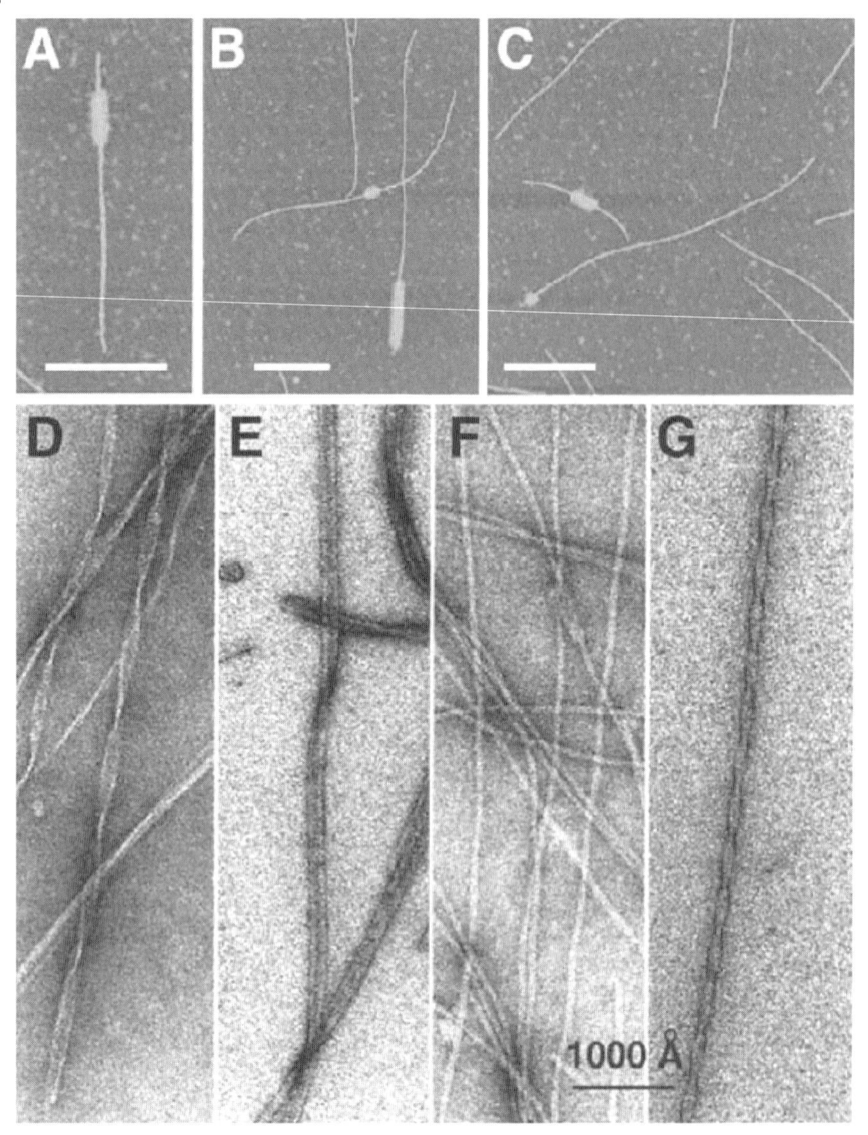

Fig. 35.2. Amyloid fibers adopt multiple distinguishable structures (adapted from Chien et al. 2004). (A–C). Amyloid fibers formed spontaneously by Sup35p vary in their growth patterns, including overall rate and polarity of growth (from DePace and Weissman 2002). Four kinetic fiber types visualized by an AFM single-fiber growth assay are shown. The original seed is labeled with antibody and therefore is wider than the new growth extending from its ends. Note the presence of long and short symmetric and asymmetric fibers. Scale bar is 500 nm. Polarity in growth has also been seen by EM (Inoue et al. 2001) and fluorescence-based assays (Scheibel et al. 2001). (D–G) Negative stain EM of amyloid fibers formed spontaneously by the SH3 domain from PIP_3 kinase illustrates that they vary in the number of protofilaments and helical pitch. Scale bar is 100 nm.

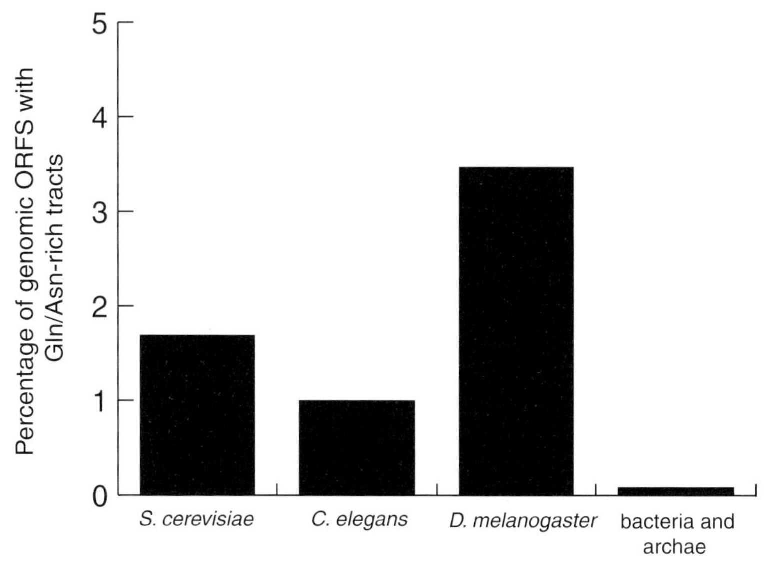

Fig. 35.3. Gln/Asn-rich tracts are abundant in certain eukaryotic proteomes (adapted from Michelitsch and Weissman 2000). The complete genomic sequences of three eukaryotes as well as of 20 bacteria and archaea were searched for Gln/Asn-rich tracts, defined as regions of open reading frames (ORFs) with 30 or more Gln and Asn residues per 80-residue window. The frequency of such sequences in the eukaryotes is much greater than expected if amino acid composition were due to chance. Interestingly, the recently sequenced human genome has a lower abundance (~0.3%) of predicted Gln/Asn-rich tracts than do other metazoans.

et al. 2003; Edskes and Wickner 2002). Moreover, mutational studies on Sup35 have established that this high glutamine/asparagine content plays a critical role in mediating the formation of amyloid aggregates (DePace et al. 1998). The structural basis of Q/N-rich aggregation has been suggested to involve formation of "polar zippers" in which the β-sheets are stabilized by a network of hydrogen bonds involving the glutamine and asparagine side chains (Perutz et al. 1994).

A surprisingly large number of Q/N-rich sequences reminiscent of the prion domains of Ure2p and Sup35p are found in many eukaryotic genomes, including those of *S. cerevisiae*, *C. elegans*, and *D. melanogaster* (Figure 35.3) (Harrison and Gerstein 2003; Michelitsch and Weissman 2000). Efforts to test prion formation by these proteins have largely focused on yeast because of the ease of manipulating this organism (Santoso et al. 2000; Si et al. 2003b; Sondheimer and Lindquist 2000). As it is difficult to anticipate the prion-associated phenotypes for a given protein, such studies have typically started by fusing putative prion domains to reporter proteins whose activity can be readily monitored.

Using this "artificial prion" method, two previously uncharacterized proteins (New1p and Rnq1p) were found to contain prion domains. A fusion of the first 153 residues of New1p and the translation domain of Sup35p could interconvert between two states, termed [nu^-] and [NU^+], which respectively resulted from soluble and aggregated forms of the reporter protein (Santoso et al. 2000). Similarly, a Sup35p fusion with the carboxy-terminal portion of Rnq1p could reversibly form an aggregated state (Sondheimer and Lindquist 2000). The full-length Rnq1p also formed a genuine endogenous prion in a number of wild-type yeast strains.

Another prion phenomenon was described in yeast some years earlier, although initial evidence was strictly genetic. Derkatch et al. (1997) uncovered a cryptic epigenetic state called [PIN^+] (PSI inducibility) that influenced the cell's susceptibility to the de novo induction of [PSI^+]; only strains that possessed the [PIN^+] trait could readily convert to [PSI^+] upon overexpression of Sup35p. [PIN^+] had all the hallmarks of a prion, including cytoplasmic inheritance, reversible curing, and dependence on cellular chaperones (Derkatch et al. 1997, 2000). Fortuitously, the protein responsible for the naturally occurring [PIN^+] state turned out to be Rnq1p (Derkatch et al. 2001; Osherovich and Weissman 2001). However, several other proteins containing Q/N-rich domains were also found to confer a PIN phenotype when overexpressed in a manner that correlated with the aggregation of the overexpressed protein (Derkatch et al. 2001; Osherovich and Weissman 2001). Whereas heterologous Q/N-rich aggregates have a stimulatory effect on de novo prion induction and the aggregation of polyglutamine proteins, in another context such aggregates can interfere with prion inheritance (Bradley and Liebman 2003). The mechanism of this cross-interaction between different prions remains a mystery.

A recent study (Si et al. 2003b) suggests that prion-like aggregation may play an important role in the normal (non-pathological) biology of metazoans including man. Si et al. identified a neuronal member of the cytoplasmic polyadenylation element-binding protein (CPEB) family that has an amino-terminal extension that resembles yeast prion proteins in its Q/N content. When fused to a reporter protein in yeast, this Gln-rich region displayed certain features of an epigenetically heritable state, including interconversion between soluble and aggregated forms. The full-length protein underwent similar interconversion, but it was the prion-like form that was most active in stimulating translation of CPEB-regulated mRNA. These authors suggest that conversion of CPEB to a prion-like state in stimulated synapses helps to maintain long-term synaptic changes associated with memory storage (Si et al. 2003a). It will thus be important to explore the prion-like properties of CPEB in its natural context.

35.5
Requirements for Prion Inheritance beyond Amyloid-mediated Growth

Although a broad range of unrelated proteins can form amyloid fibrils (Dobson 2001), very few such aggregates are infectious and/or heritable prions. The distinction between prions and more common amyloids appears to lie in the way in which they interact with cellular factors. These interactions involve features of the prion distinct from those required for amyloid formation. The complete prion replication cycle requires the periodic alternation of growth, division, and transmission (Tuite and Cox 2003) (Figure 35.4a). While the self-seeding nature of amyloids

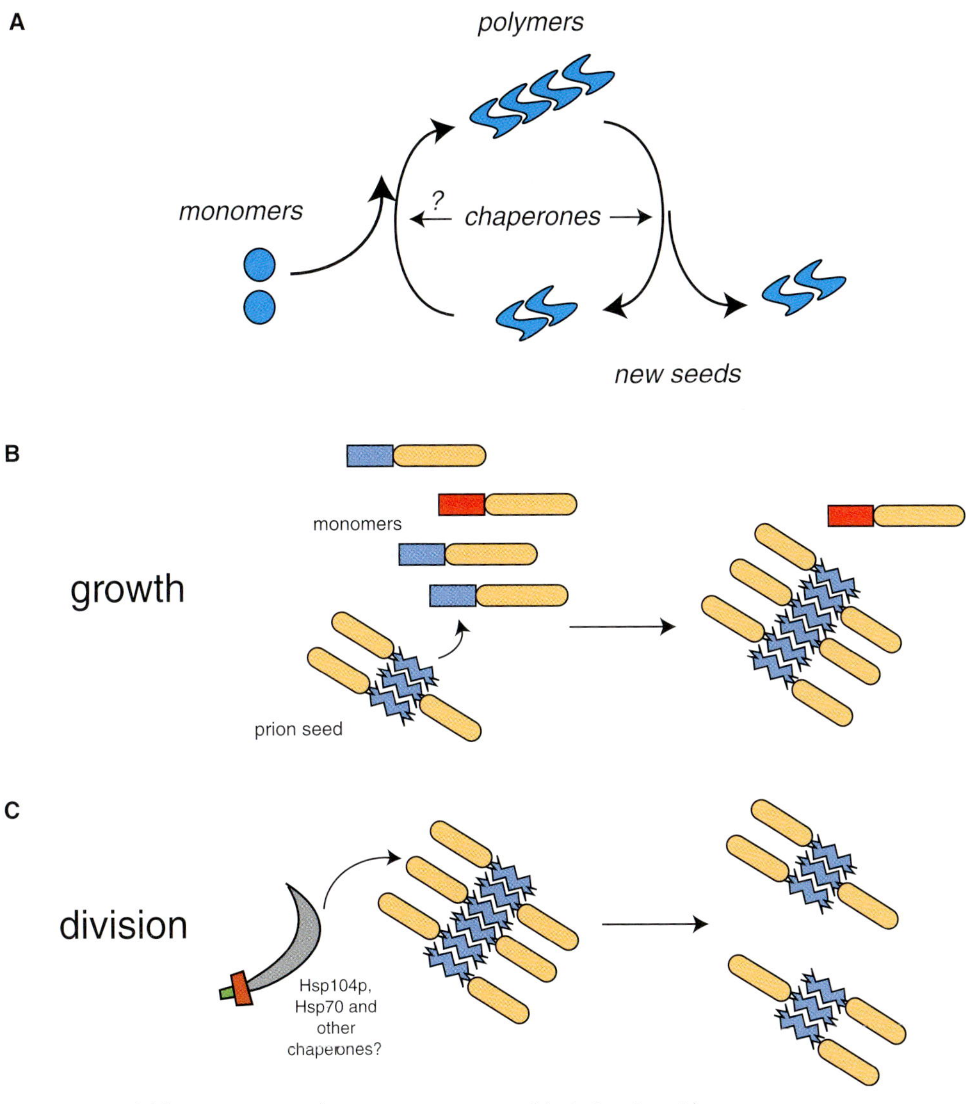

Fig. 35.4. Model for yeast prion replication (adapted from Osherovich et al. 2004). (A) A general schematic of the prion replication cycle. Monomer addition to polymer ends occurs spontaneously in vitro but may involve the action of chaperones in vivo. New seeds, consisting of small polymers, are generated by the chaperone-mediated severing of large polymers (see Kushnirov and Ter-Avanesyan 1998). (B) During prion growth, polymers seed the incorporation of monomers through interactions between Q/N-rich aggregation sequences (blue). Proteins with non-cognate aggregation sequences (red) are excluded. (C) The division phase of prion replication requires replication elements (orange), such as the oligopeptide repeats of Sup35p and New1p. These may serve as binding sites for chaperones such as the Hsp104p/Hsp70/Hsp40 complex (scimitar) or may alter the structure of the polymer in a way that facilitates chaperone action. New seeds are then partitioned between mother and daughter cells.

accounts for the growth phase of the prion replication cycle, in yeast prion division, and perhaps transmission, is dependent on the cell's chaperone machinery (Osherovich and Weissman 2002). A range of studies has now begun to define the sequence elements within the prion protein and the chaperone requirements for the replication and transmission steps.

Mutational and functional analyses argue that the Sup35p and New1p prion domains have a largely modular structure (Osherovich et al. 2004). In both proteins, a Q/N-rich tract mediates sequence-specific aggregation (Balbirnie et al. 2001; Hara et al. 2003; Osherovich et al. 2004). These Q/N-rich tracts may form the core of the amyloid structure, thereby allowing the self-specific growth of the prion state (Figure 35.4B). In contrast, the replication and transmission (Figure 35.4C) of both Sup35p and New1p prions are mediated by a stretch of oligopeptide repeats with the consensus sequence (P/Q)QGGYQ(Q/S)YN (Osherovich et al. 2004), which loosely resembles a repeating sequence (PHGGGWGQ) in mammalian PrP (Liu and Lindquist 1999; Parham et al. 2001).

The sequence features required for prion replication and transmission in other prion-forming proteins remain poorly understood. For example, Ure2p and Rnq1p lack obvious oligopeptide repeats, although a region (residues 218–405) of the Rnq1p prion domain has an amino acid content reminiscent of the oligopeptide repeat sequence (i.e., numerous Q, N, S, Y, and G residues) (Resende et al. 2003). Although mammalian PrP contains a sequence resembling the oligopeptide repeat that can functionally replace one of the Sup35p repeats (Parham et al. 2001), it is unclear what role this sequence plays in the replication of the PrP^{Sc} state. A better understanding of the sequence requirements for prion replication and transmission could greatly aid in the identification of novel prion elements. Of the 100 or so Q/N-rich domains encoded in the *S. cerevisiae* genome, only two other proteins (YDR210W and YBR016W) have clearly recognizable oligopeptide repeats as well as Q/N-rich regions. YBR016W forms aggregates when overexpressed (Sondheimer et al. 2001), but it is not known whether either protein can maintain a heritable aggregated state.

Based on this modular model for prion domains, we have designed two novel artificial prions by fusing the replication element of Sup35p to aggregation sequences from other proteins, including pathogenically expanded polyglutamine (Osherovich et al. 2004). The first such artificial prion, termed [F^+], consists of the aggregation sequence of New1p and the oligopeptide repeat of Sup35p. Despite a sequence derived primarily from Sup35p, the F chimera showed induction and growth properties similar to those of New1p, consistent with the notion that the Q/N-rich region forms the amyloid core of the prion and dictates self-specificity. A second prion, termed [Q^+], consists of an aggregation-prone sequence derived from pathogenically expanded glutamine fused to the Sup35 oligopeptide repeat sequence. The [Q^+] prion demonstrates that it is possible to turn a generic aggregating sequence into a heritable prion element. Such artificial prions could potentially serve as models for aggregate-chaperone interactions in metazoans as well as a genetic system for the high-throughput screening of modulators of human aggregation diseases (Li and Lindquist 2000; Meriin et al. 2003; Muchowski et al. 2000).

35.6
Chaperones and Prion Replication

The division of prion seeds into heritable units depends on *trans*-acting cellular chaperones, among which HSP104 plays a central role (Chernoff et al. 1995). Genetic and pharmacological analysis places Hsp104p in the division phase of prion replication (Eaglestone et al. 2000; Jung and Masison 2001), consistent with its biochemical activity as a general disaggregase capable of untangling large complexes of misfolded proteins (Glover and Lindquist 1998). Hsp104p is not required for the de novo formation of prion seeds in vivo or of amyloid fibrils in vitro (Glover et al. 1997), although it may contribute to prion growth in vitro (Shorter and Lindquist, 2004). However, diminished HSP104 activity through mutation, deletion, or pharmacological inhibition by guanidinium chloride (GdmCl) results in the arrest of prion replication, evident through the appearance of [*psi*$^-$] colonies as [*PSI*$^+$] cells continue to divide without replenishing their prion titer (Eaglestone et al. 2000; Tuite et al. 1981). Consistent with this model, HSP104 activity is needed to maintain [*PSI*$^+$] only so long as the cells are dividing (Eaglestone et al. 2000). A mutation within the oligopeptide repeats of Sup35p mimics the effect of Hsp104p inhibition (Osherovich et al. 2004), suggesting that Sup35p and Hsp104p may interact through this region, although perhaps indirectly.

A mechanistic understanding of how Hsp104p interacts with amyloid aggregates to promote prion inheritance is now the major challenge to workers in this field. A recent report (Shorter and Lindquist, 2004) demonstrates that purified recombinant Hsp104p can act on Sup35p amyloids to facilitate prion replication. However, evidence suggests that Hsp104p does not act alone in the disaggregation of misfolded proteins and the replication of prion aggregates. Rather, it is likely that Hsp104p works together with Hsp70 and Hsp40 cofactors to recognize prion proteins as substrates and remodel them in a way that generates new prion seeds and/or promotes prion transmission (Chernoff et al. 1999; Jones and Masison 2003; Jung et al. 2000; Moriyama et al. 2000a; Newnam et al. 1999; Sondheimer et al. 2001).

Hsp70s are ubiquitous ATPases involved in the folding and localization of proteins during translation and after protein folding stresses such as heat shock (Bukau and Horwich 1998; Hartl and Hayer-Hartl 2002). In yeast, two cytosolic Hsp70s (SSB1 and SSB2) facilitate co-translational protein folding, while four highly homologous Hsp70s (SSA1–SSA4) act independently of the ribosome to maintain protein solubility under stressful conditions (Craig et al. 1995). All Hsp70s require the action of co-chaperones, which modulate the cycle of ATP binding and hydrolysis needed for substrate recognition and release. These include the DnaJ/Hsp40 proteins, marked by the presence of a J-domain, as well as other structurally distinct co-chaperone families, such the tetratricopeptide repeat (TPR) "clamp" proteins and so-called Bag domain proteins. The yeast genome encodes a large number of known and putative co-chaperones, many of which are involved in the subcellular localization of Hsp70 for specific tasks (Young et al. 2003).

Biochemical and genetic evidence points to a concerted role for Hsp104p and the Hsp70/co-chaperone system in prion replication. In vitro refolding of model substrates such as heat-denatured luciferase requires the joint action of Hsp104p with Ssa1p and Ydj1p, a cytoplasmic Hsp40 protein (Glover et al. 1997). In vivo, Hsp104p is needed for the inheritance of all naturally occurring yeast prions (Chernoff et al. 1999; Moriyama et al. 2000b; Sondheimer and Lindquist 2000), suggesting that it performs a common biochemical activity needed for prion maintenance. Whereas Hsp70s and Hsp40 co-chaperones have also been implicated in prion maintenance, their effects are highly specific to each prion protein (summarized in Osherovich and Weissman 2002). At least one Hsp70 or Hsp40 is required for the stable inheritance of each type of prion, and many prions (both natural and artificial) are sensitive to the dosage of various Hsp70 and Hsp40 genes. For example, the inheritance of [*RNQ*$^{+}$] depends on the interaction of Rnq1p with a region of the Hsp40 protein Sis1p (Lopez et al. 2003; Sondheimer et al. 2001; Yang Fan et al. 2003), while two Hsp70s (Ssa1p and Ssa2p) are involved in the maintenance of [*PSI*$^{+}$] and [*URE3*], respectively (Jung et al. 2000; Roberts et al. 2004).

The general role of Hsp104p in prion maintenance compared with the highly specific genetic interaction of Hsp70 and co-chaperones with prions suggests that cytoplasmic Hsp70 and its co-chaperones act as coupling factors to recruit Hsp104p to prion aggregates. Electron microscopy of Sup35p yeast prion amyloids has shown that fibrils are laterally decorated with non-amyloidogenic portions of the protein, which protrude at regular intervals from the amyloid core like bristles from a bottle brush (Glover et al. 1997). The specific features of these potentially nonnative protein surfaces could attract Hsp70 and co-chaperones, which recognize and attempt to refold these proteins by working together with the more general disaggregation activity of Hsp104p. An appealing model for the chaperone-dependent replication of prion particles involves the liberation of partially solubilized but still amyloidogenic "seeds" from large prion aggregates by the combined action of Hsp104p, Hsp70, and Hsp40 co-chaperones.

35.7
The Structure of Prion Particles

Two recent studies have helped to better define the genetic and physical characteristics of yeast prions in vivo. Through the clever manipulation of chaperone activity, Cox and coworkers have devised a means of counting the number of heritable prion units or propagons in [*PSI*$^{+}$] cells and have used this method to probe the temporal and spatial regulation of prion replication (Cox et al. 2003). Two observations from this study point to unexpected interplay between the prion replication cycle and the broader cell division cycle. First, propagons are replicated discontinuously, with a sharp increase during the G2 phase, perhaps because of variation in chaperone activity during the cell cycle. Second, the partitioning of propagons displays a strong maternal bias, suggesting that the physical segregation of propagons that is proposed to take place after seed division may require an active transport process, perhaps involving interactions with cytoskeletal components.

What are propagons physically? From the number of genetically defined propagons and the number of Sup35p molecules present in typical cells, it can been estimated that propagons consist of 10–200 molecules of Sup35p (Cox et al. 2003). The minimum number of molecules needed to form a physically stable amyloid seed is unknown. However, it is unlikely that large amyloid fibrils of the kind observed in vitro in the absence of shearing activity can be acted upon by cellular chaperones; indeed, a mutant of Sup35p lacking much of the oligopeptide repeat region forms large, linear aggregates in vivo that can be visualized with GFP staining and that correlate with the loss of [PSI^+] (Osherovich, unpublished observations). A hint at the macromolecular nature of the propagon comes from a recent study in which Sup35p from [psi^-] and [PSI^+] strains could be distinguished using non-denaturing gel electrophoresis (Kryndushkin et al. 2003). Whereas Sup35p from [psi^-] migrates as a monomer, the material from [PSI^+] cells migrates as a broad smear between 700 and 4000 kDa. This large complex, termed a prion polymer by the authors, corresponds to 9–50 Sup35p monomers by molecular mass. As a number of proteins other than Sup35p have previously been found in association with prion aggregates, an estimate based solely on non-dentaturing gel migration may in fact overestimate the number of Sup35p molecules needed to make a minimal amyloid "seed." Nonetheless, it is possible that heritable prion units (propagons) consist of bundles of coalesced amyloid polymers.

35.8 Prion-like Structures as Protein Interaction Modules

The sequence requirements for prion inheritance are now understood to be more stringent than simply high Q/N content. For this reason, despite the general propensity of Q/N-rich proteins to misfold into amyloid aggregates, very few of these are likely to adopt heritable prion states. Why, then, are there so many Q/N-rich sequences (Michelitsch and Weissman 2000)? One possibility is that Q/N-rich domains act as protein-protein interaction modules that organize stable multicomponent complexes in a manner that may be structurally similar to amyloid aggregates (Perutz et al. 2002). Two examples of amyloid formation in the context of normal biological processes have been reported recently (Berson et al. 2003; Chapman et al. 2002). Non-prion "pseudo-amyloid" complexes may feature discrete structures held together by the same cross-pleated β-sheet interactions found in prions, but with limited opportunities for growth.

Although this model is speculative in general, studies in yeast hint at a role for Q/N-rich proteins in organizing multicomponent complexes. Perhaps the best-studied example occurs at the interface of the actin cytoskeleton and the endocytic apparatus (Jeng and Welch 2001). Pan1p is an essential EH domain protein that, together with Sla1p and End3p, recruits and activates the Arp2/3 complex to the surface of nascent endocytic vesicles, generating the cortical actin structures needed for endocytosis (Duncan et al. 2001; Tang et al. 2000). The amino terminal portion of Pan1p is extremely Q/N-rich and interacts with the comparably Q/N-rich carboxy-terminal region of Sla1p; this interaction is needed for the recruitment of the Arp2/3-activating C-terminus of Pan1p to clathrin-coated endocytic vesicles. Interestingly, the C-terminus of Sla1p can itself interact with the prion domain of Sup35p in a yeast two-hybrid assay (Bailleul et al. 1999), suggesting that this Q/N-rich domain can bind promiscuously to other Q/N-rich proteins. When overexpressed as a GFP fusion, the Q/N-rich portion of Pan1p forms visible aggregates similar to those of Sup35p (Osherovich and Weissman 2001). Unlike proper prions, these aggregates are not heritable and vanish after overexpression is stopped.

Another possible example comes from the study of the nonsense-mediated mRNA decay pathway (Gonzalez et al. 2001). Here, a complex of proteins coalesces into punctate structures termed P bodies that are the site of mRNA decapping and degradation (Sheth and Parker 2003). Several P body components, including Lsm4p and Dhh1p, have Q/N-rich regions, although whether these prion-like sequences are needed for P body formation or activity is unknown. However, Lsm4p had been identified as a PIN factor (Derkatch et al. 2001), forming [PSI^+]-promoting aggregates when overexpressed. Likewise, several Q/N-rich members of the SWI/SNF complex, which is involved in heterochromatin remodeling in yeast (Martens and Winston 2003), also came out of the screen for PIN factors.

The potential for Q/N-rich sequences to act as modular interaction domains may well explain their abundance in eukaryotic genomes. In turn, these relatively common Q/N-rich domains may have provided the raw material from which the true prions (those of Sup35p, Ure2p, and Rnq1p) arose. In this view, prions may be an accidental elaboration of a broader class of "pseudo-amyloid" protein complexes.

35.9 Experimental Protocols

35.9.1 Generation of Sup35 Amyloid Fibers in Vitro

Production of Sup35 Prion Domain (NM)

The prion domain of Sup35 is produced by overexpression in *E. coli* using a plasmid in which the T7 promoter drives expression of a gene encoding the NM domains (residues 1–254) or NM tagged with either three repeats of the intact hemagglutinin (HA) epitope (NM-HA_3) or a derivative (NM-$HA^*{}_3$) containing three conservative point mutations that prevent antibody recognition (the plasmids are termed, pAED4-NM, pAED4-NM-HA_3, and pAED4-NM-$HA_3{}^*$, respectively). In all cases the NM domains also contain a histidine tag (six to seven histidines long) at the C-terminus to facilitate purification. *E. coli* BL21(DE3) freshly transformed with the appropriate plasmid is grown to $OD_{650} = 0.5$ in LB broth supplemented with 100 μg mL^{-1} carbenicillin. Following 3 h induction, cell pellets from 1–2 L of growth are lysed in 30 mL of buffer A (25 mM Tris pH 7.8, 300 mM NaCl, 8 M urea). Subsequent to centrifugation at 25 000 g (20 min), the supernatant is filtered by a 0.22-μm filter (Millipore) and applied to a 20-mL Ni-NTA agarose (Qiagen) column. The column is washed with five column volumes of buffer A and eluted with a pH gradient (pH 7.8 to pH 4.5) in the same buffer without NaCl. Pooled fractions are applied to a 6-mL Resource S column (Pharmacia) equilibrated in 50 mM MES pH 6.0, 8 M urea, and eluted with 0–1 M NaCl gradient in the same buffer. Pure NM is concentrated using a microcon-10 (Amicon), filtered by microcon-100 (Amicon), yielding a final concentration of at least 500 μM, aliquoted, and stored at −80 °C. Protein concentration is determined by UV absorption at 275 nm using an extinction coefficient of 24.2 mM^{-1} cm^{-1}.

Preparation of NM Fibers

To initiate amyloid formation, concentrated NM is diluted at least 100-fold into buffer B (5 mM potassium phosphate, 150 mM NaCl) in 2-mL microcentrifuge tubes at room temperature to a final concentration of 2.5 μM. The solution is subjected to constant rotation (7.5 rpm) using a RKVS Rotamix (ATR, Inc.) for at least 2 h. The extent of amyloid formation can be monitored by thioflavin T, a dye that specifically binds amyloid fibers, leading to a large increase in fluorescence. This is accomplished by removing an aliquot of the polymerization reaction, adding an equal volume of 25 μM thioflavin T (pH 8.5 in 50 mM glycine), and monitoring fluorescence using an excitation wavelength of 442 nm and an emission wavelength of 483 nm. In order to obtain consistent size distributions easily visualized by AFM, it is important to generate new NM amyloid seeds for each AFM elongation reaction. To generate short fibers convenient for AFM measurements, prior to use, fibers are sheared by pulling through a 25-gauge needle 10 times, resulting in seeds that vary in length between 50 and 500 nm.

35.9.2
Thioflavin T–based Amyloid Seeding Efficacy Assay (Adapted from Chien et al. 2003)

Conversion of a soluble protein to an amyloid form is typically a multi-step process involving de novo amyloid seed formation, breakage of preexisting seed to generate new seed, and growth of soluble monomers onto the ends of existing seeds. Here we describe an assay that makes it possible to specifically measure the ability of a given preparation of preformed amyloid seeds to polymerize soluble protein. This is accomplished by measuring the initial rate of amyloid growth upon addition of preformed amyloid seeds to soluble protein. By examining initial rates of polymerization, complications due to the breakage of preexisting fibers or to the de novo formation of amyloid seeds can be minimized. Differences in seeding efficacy measured in this assay reflect differences in elongation rates of fibers and/or differences in the number of fiber ends. This assay is performed in a 96-well microplate using a microplate spectrofluorometer, so that many samples can be analyzed in parallel. It is a continuous assay and thus samples need be prepared only once and then monitored over time by the instrument. To control for the possible effects of thioflavin T on amyloid formation, it is important to confirm the results by an alternative means (e.g., a discontinuous assay in which thioflavin T is added after polymerization).

Preparation of Samples

In a microplate well, mix 100 μL of 25-μM thioflavin T (pH 8.5 in 50 mM glycine) and 90 μL of the soluble protein whose polymerization you wish to monitor at 2.22 times the desired final concentration (we usually use 2.5 μM final soluble NM). Seed should be added immediately before reading in the microplate fluorometer because polymerization will begin as soon as the seed is added. To each well add 10 μL of the desired seed preparation. Insert the plate into the microplate spectrophotometer and read every 30 s for 2 h. Fluorescence of thioflavin T is measured by excitation at 442 nm and emission at 483 nm.

Analysis of Seeding Efficacy

The initial rate of amyloid formation is determined by measuring the rate of increase in thioflavin T fluorescence at early time points (thioflavin T binds amyloid fibers, leading to a large increase in fluorescence). The initial rate is determined by taking the slope of the initial linear phase of the reaction (often the first 10 or 20 min), before the rate starts to decline as soluble protein starts to be depleted. Care should be taken in comparing rates from experiments performed on different days since the age or degree of exposure to light can affect the fluorescence of a thioflavin T solution. Additionally, care should be taken in comparing rates of polymerization from seeds made from different proteins, since fibers of different proteins may stimulate fluorescence of thioflavin T to varying degrees.

35.9.3 AFM-based Single-fiber Growth Assay

The following protocols describe a nanometer-resolution single-fiber growth assay (DePace and Weissman 2002) based on two variants of the Sup35 prion domain (NM) that can be differentially labeled with antibodies and distinguished by atomic force microscopy (AFM). The two Sup35 variants used in our assay consist of the N and M domains (residues 1–254) tagged with either three repeats of the intact hemagglutinin (HA) epitope (NM-HA_3) or a derivative (NM-$HA^*{}_3$) containing three conservative point mutations that prevent antibody recognition. Because C-terminal fusions do not interfere with prion activity, these epitope tags were introduced at the C-terminus, followed by a six-histidine tag to facilitate purification. The strategy of this assay is to produce short fibers of one variant and to use these to seed polymerization of the other variant in solution for a defined period of time. The elongation reactions are then rapidly quenched (10–20 s) by deposition onto freshly cleaved mica, and unpolymerized monomer is removed by washing. Fibers are subsequently labeled by incubation with an HA antibody directly on the mica. The fibers are then visualized by AFM, which can readily distinguish regions of the fibers that are bound by antibody from those that are not, allowing measurement of the growth of new NM monomers onto preformed amyloid seed (Figure 35.5). While the assay described below has been optimized for the analysis of Sup35 fibers, in principle it should be straightforward to use a similar method to monitor the growth of any amyloid fiber the can be epitope-labeled.

The assay has the following desirable properties. First, because the antibody la-

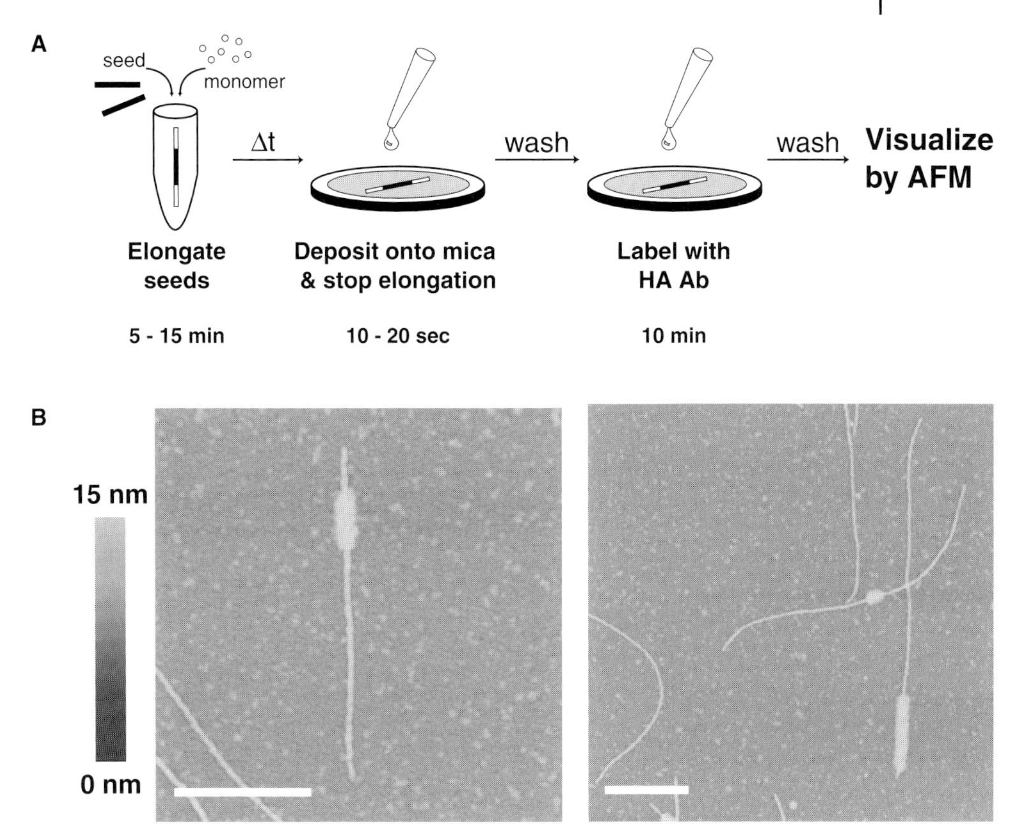

Fig. 35.5. Visualization of growth from single Sup35-NM fibers. (A) Schematic representation of the single-fiber elongation assay protocol. (B) Representative images of NM-HA_3 seeds elongated with NM-HA^*_3 monomer. Note the presence of both long and short symmetric and asymmetric fibers.

beling is dense and compact and provides a well-defined boundary between the seeds, new growth is unambiguously distinguishable from the seed. Second, elongation is measured in solution, avoiding complications arising from assembling fibers directly on a substrate, which can strongly alter the spectrum of fibrillar species observed. Third, the elongation reaction is rapidly quenched to allow precise time measurements. Fourth, the labeling step takes place after the reaction has been quenched, which avoids possible complications due to steric interference from the label. Finally, the assay is simple and rapid. Multiple samples can prepared in parallel and then can be stored indefinitely prior to analysis by AFM, and hundreds of fibers can be measured in a day.

Preparing Samples for Single-round Growth Assays

For visualization of growth from single fibers composed of NM-HA_3, seeds are elongated by incubation with 2.5 μM monomeric NM-$HA_3{}^*$ in buffer B; these reactions are not rotated to avoid shearing forces. At the appropriate time point 20 μL of the elongation reaction is removed and deposited onto freshly cleaved mica. The reaction is allowed to incubate on the mica for 30 s, and then 200 μL of water is added to the disc and immediately aspirated away. This wash step is repeated once. Twenty microliters of FAB-fragmented 16B12 (Babco) diluted to 20 ng μL^{-1} in buffer B is added directly to the mica surface. Antibody is incubated on the mica for 10 min and then washed twice with 200 μL of water as above. Samples are allowed to air dry for at least 1 h before imaging. To avoid shearing during all manipulations, wide-bore pipette tips are used; 200-μL yellow tips are clipped, removing 2 cm from the end to create an opening 3 mm in diameter.

Preparing Fibers for Two-round Growth Assays

In the first reaction, NM-HA^*_3 seeds, generated as described above, are used at 25% (v/v) to seed 2.5 μM NM-HA_3 monomer. This reaction is allowed to proceed for 7.5 min and then is diluted 1:10 into a new polymerization reaction containing 2.5 μM monomeric NM-HA^*_3. The second reaction is allowed to proceed for 7.5 min before it is deposited directly onto mica (in some cases it is helpful to dilute the reactions 1:2 or 1:5 in polymerization buffer before deposition). The remainder of the sample preparation proceeds as described above. High seed concentration (25% v/v) is used in the initial polymerization reaction to allow for a convenient fiber concentration following dilution into the second elongation reaction. Additionally, higher seed concentrations help to deplete NM-HA_3 monomer, which would otherwise interfere with visualization of growth in the second round.

Imaging Fiber Growth by Atomic Force Microscopy

We collect our images using a Digital Instruments MultiMode AFM, equipped with a Nanoscope IIIa controller. Tapping mode is employed with all samples; either Digital Instrument's TESP or Micromasch's non-contact NSC15 tips are used. To measure growth from individual seeds, 25-μm^2 scans are taken. To measure fiber height, 1.0-μm^2 scans of representative fibers are taken using the same TESP tip.

Image Analysis/Measurement of Fiber Dimensions

Fiber lengths are determined using the point-and-click measurement tool available in the Nanoscope software. When fibers are curved, a series of lines is used to trace the fiber axis, and all individual line segments are added. Height is determined using the section tool of the Nanoscope software.

35.9.4
Prion Infection Protocol (Adapted from Tanaka et al. 2004)

Preparation of Infectious Material From Pure Protein or Yeast Extracts

Amyloid fibers of NM are formed as described above. Fibers are collected by centrifugation at 20 000 g for 20 min, resuspended with buffer B, and sonicated (10 s 20% power on a Sonic Dismembrator Model 500 [Fisher Scientific] equipped with a micro-tip probe). The final concentration of amyloid fiber in infection experiments is typically 2.5 μM calculated in terms of Sup35 monomers. For preparation of yeast extracts containing prions, [*PSI*$^+$] yeast cells were spheroplasted with lyticase and sonicated. Concentration of total protein in the yeast extracts was measured by Bradford assay (Pierce). For infections involving extracts, the final concentration of total protein of the crude yeast extract is 200 μg mL^{-1}.

Prion Infection

[*psi*$^-$] yeast strains (W303 [*ade1-14, his3-11,15, leu2-3, trp1-1, ura3-1*] or 74D-694 [MATa *ade1-14 his3 leu2 trp1 ura3* [*PIN*$^+$] [Chernoff et al. 1995]) are grown in YEPD media to an OD_{600} of 0.5 and successively washed with sterile H_2O, 1 M sorbitol, and SCE-buffer (1 M sorbitol, 10 mM EDTA, 10 mM DTT, 100 mM citrate, pH 5.8). Cells are spheroplasted with lyticase (~250 μg for yeast cells of 50 mL YEPD) in SCE buffer at 30 °C for 30 min (see below for description of lyticase preparation). Spheroplasts are washed with 1 M sorbitol and STC buffer (1 M sorbitol, 10 mM $CaCl_2$, 10 mM Tris, pH 7.5). Pelleted cells are resuspended in STC buffer and mixed with sonicated amyloid fibers, *URA3* marked plasmid (pRS316) (20 μg mL^{-1}), and salmon sperm DNA (100 μg mL^{-1}). Fusion is induced by addition of nine volumes of PEG buffer (20% [w/v] PEG 8000, 10 mM $CaCl_2$, 10 mM Tris, pH 7.5) for 30 min. Cells are centrifuged, resuspended in SOS buffer (1 M sorbitol, 7 mM $CaCl_2$, 0.25% yeast extract, 0.5% bacto-peptone), incubated at 30 °C for 30 min, and plated on synthetic media lacking uracil overlaid with top agar (2.5% agar).

Identification of [*PSI*$^+$]-infected Yeast

Individual Ura+ colonies are re-streaked on 1/4 YEPD (2.5 g yeast extract, 5 g bacto-peptone, 20 g agar, 1 L H_2O plus 40 mL 50% filter-sterilized glucose added after autoclaving above solution). Strong [*PSI*$^+$] and [*psi*$^-$] strains are included as controls on all plates to aid in judging color phenotype. Plates are examined after 3–5 days of incubation at 30 °C. [*PSI*$^+$] colonies are confirmed by GdmCl curing: colonies are streaked onto YEPD plates supplemented with 3 mM GdmCl and re-streaked on 1/4YEPD to look for loss of [*PSI*$^+$].

35.9.5
Preparation of Lyticase

An *E coli.* bacterial plasmid that express periplasmically localized lyticase under control of T7 promoter (pUV5-lyticase; Scott and Schekman 1980) was transformed into *E coli.* BL21(DE3). A single colony was inoculated into 100 mL of LB media including 100 μg mL^{-1} ampicillin and cultured at 37 °C until OD_{600} reached ~0.5. Twenty milliliters of the culture media was transferred to 1 L of LB media and cultured again at 37 °C at 200 rpm (Innova 4400, New Brunswick Scientific). Isopropyl-β-thiogalactoside (final concentration of 0.5 mM) was added when OD_{600} reached ~0.5 to initiate protein expression. After 3 h, the culture media was centrifuged (5000 rpm) and the pellet was washed with distilled H_2O. Cells were resuspended in 20 mL of 25-mM Tris-HCl (pH 7.4), incubated at room temperature with gentle agitation for 30 min using a nutator, and centrifuged at 7500 rpm (Sorvall SS-34 Rotor, Du Pont Instruments) for 10 min. The pellet was resuspended in 20 mL of 5-mM $MgCl_2$, incubated at 4 °C for 30 min, and centrifuged at 15 000 rpm for 30 min (Sorvall SS-34 Rotor, Du Pont Instruments) to separate periplasmic components. This supernatant was dialyzed against 2 L of 50-mM sodium citrate buffer (pH 5.8) overnight and was stored at −80 °C. Concentration of lyticase was determined by the Bradford method, and ~250 μg of lyticase was used to spheroplast pelleted yeast cells obtained from 50 mL of YEPD media (OD_{600} = ~0.5).

35.9.6
Protocol for Counting Heritable Prion Units (Adapted from Cox et al. 2003)

Principle

The replication of prion particles depends on the action of Hsp104p, a chaperone that can be inhibited by the small molecule GdmCl. To count the number of heritable prion particles (propagons) in [*PSI*$^+$] cells, we perform a pulsed shutoff of Hsp104p activity using GdmCl. The transient arrest of prion replication while cells continue to divide eventually dilutes the prion particles to one or fewer per cell. At this point, the prion replication arrest is lifted by the removal of GdmCl, and the cells are recovered in a medium that selects for [*PSI*$^+$] (SD-ade with a dash of YEPD to aid initial growth). Thus, the total number of [*PSI*$^+$] progeny of a single cell that has undergone this transient prion replication arrest reports on the number of propagons the cell harbored at the start of the experiment.

Materials

- YEPD + GdmCl agar plates (1% yeast extract, 1% peptone, 2% dextrose supplemented to 3 mM guanidine hydrochloride from a sterile 3-M stock)
- SD-ade + 5% YEPD agar plates (synthetic defined medium lacking adenine to which a 5% volume of YEPD liquid medium has been added prior to pouring)
- Yeast micromanipulator microscope such as Singer MSM system or Zeiss Axioskop 40 Tetrad system
- Low-powered stereo dissection microscope with a 10× objective and illuminated stage
- 1000 microliter micropipette tips, with ~5 mm of tip trimmed away with a sterile razor blade

Procedure

Day 1: Inoculate a [*PSI*$^+$] culture into YEPD or appropriate selective medium and grow overnight until mid-logarithmic phase. It is important to use actively dividing cells.

Day 2: Spot 30 μL of the overnight culture onto a YEPD + GdmCl plate. Using the micromanipulator, move >40 individual cells away from the spot onto a widely spaced grid. Be consistent in choosing the cell cycle stage of cells; propagon numbers can vary throughout the cell cycle. We prefer cells that are showing a hint of a bud (G1 phase). Incubate overnight at 30 °C.

Day 3: In the morning, check for colony formation by individual picked cells. It is vital to allow the cells to have doubled at least 10 times but not to have grown to such a size as to make picking the entire colony difficult. Ideally, proceed to the next step when microcolonies are ~0.5 mM in diameter, consisting of 2000–4000 cells. Note that not all cells with have grown and that there will be some heterogeneity in the colony size. Allow colonies to grow to an appropriate size by incubating for longer if necessary.

Pick the Colonies

1. When microcolonies are ready, aliquot 500 μL of sterile water into microcentrifuge tubes for each colony that you intend to pick.
2. Using the dissection microscope for magnification, punch out an entire colony using a trimmed micropipette tip. It is critical to encapsulate the entire colony in the pipette tip; practice until you get it down.
3. Resuspend the colony and the agar plug in a water aliquot and plate the entire volume onto an SD-Ade + 5% YEPD plate.
4. Repeat as necessary using fresh tips and tubes.
5. Incubate at 30 °C for 5–7 days, depending on the yeast strain used. W303-derived [PSI^+] strains grow more quickly (5 days) than 74D694 and other [PSI^+] strains.

Count [PSI^+] Colonies

After 5–7 days, you should see several thousand tiny red microcolonies that will be [psi^-] due to the effect of GdmCl. Interspersed among them will be [PSI^+] colonies that have grown past the microcolony stage as a result of their adenine prototrophy. Genuine [PSI^+] colonies will be pink and will turn red when re-streaked onto YEPD + GdmCl. There will also be a small number of Ade+ revertants that typically form larger colonies than [PSI^+] cells and do not turn red upon YEPD + GdmCl re-streaking. Wild-type [PSI^+] strains typically yield 50–200 [PSI^+] colonies, with a small number of "jackpot" cells yielding > 500 propagons (see Cox et al. 2003).

Acknowledgements

L.Z.O was supported by a Howard Hughes Medical Institute Predoctoral Fellowship and the Jane Coffin Childs Foundation. J.S.W. is funded through the Howard Hughes Medical Institute, the Packard Foundation, and the NIH. We thank Angela DePace, Sean Collins, Peter Chien and Motomasa Tanaka for contributing protocols. J.S.W. thanks Strai-Kin Lee for critical insights.

References

1 Aguzzi, A., and Haass, C. (2003). Games played by rogue proteins in prion disorders and Alzheimer's disease. *Science 302*, 814–818.

2 Bailleul, P. A., Newnam, G. P., Steenbergen, J. N., and Chernoff, Y. O. (1999). Genetic study of interactions between the cytoskeletal assembly protein sla1 and prion-forming domain of the release factor Sup35 (eRF3) in Saccharomyces cerevisiae. *Genetics 153*, 81–94.

3 Balbirnie, M., Grothe, R., and Eisenberg, D. S. (2001). An amyloid-forming peptide from the yeast prion Sup35 reveals a dehydrated beta-sheet structure for amyloid. *Proc Natl Acad Sci USA* 98, 2375–2380.

4 Baudin-Baillieu, A., Fernandez-Bellot, E., Reine, F., Coissac, E., and Cullin, C. (2003). Conservation of the prion properties of Ure2p through evolution. *Mol Biol Cell 14*, 3449–3458.

5 Baxa, U., Taylor, K. L., Wall, J. S., Simon, M. N., Cheng, N., Wickner, R. B., and Steven, A. C. (2003). Architecture of Ure2p prion filaments: the N-terminal domains form a central core fiber. *J Biol Chem 278*, 43717–43727.

6 Berson, J. F., Theos, A. C., Harper, D. C., Tenza, D., Raposo, G., and Marks, M. S. (2003). Proprotein convertase cleavage liberates a fibrillogenic fragment of a resident glycoprotein to initiate melanosome biogenesis. *J Cell Biol 161*, 521–533.

7 Bossers, A., Belt, P., Raymond, G. J., Caughey, B., de Vries, R., and Smits, M. A. (1997). Scrapie susceptibility-linked polymorphisms modulate the in vitro conversion of sheep prion protein to protease-resistant forms. *Proc Natl Acad Sci USA* 94, 4931–4936.

8 Bousset, L., Thomson, N. H., Radford, S. E., and Melki, R. (2002). The yeast prion Ure2p retains its native alpha-helical conformation upon assembly into protein fibrils in vitro. *EMBO J 21*, 2903–2911.

9 Bradley, M. E., Edskes, H. K., Hong, J. Y., Wickner, R. B., and Liebman, S. W. (2002). Interactions among prions and prion "strains" in yeast. *Proc Natl Acad Sci USA 99 Suppl 4*, 16392–16399.

10 Bradley, M. E., and Liebman, S. W. (2003). Destabilizing interactions among [PSI(+)] and [PIN(+)] yeast prion variants. *Genetics 165*, 1675–1685.

11 Bruce, M., Chree, A., McConnell, I., Foster, J., Pearson, G., and Fraser, H. (1994). Transmission of bovine spongiform encephalopathy and scrapie to mice: strain variation and the species barrier. *Philos Trans R Soc Lond B Biol Sci 343*, 405–411.

12 Bukau, B., and Horwich, A. L. (1998). The Hsp70 and Hsp60 chaperone machines. *Cell 92*, 351–366.

13 Chapman, M. R., Robinson, L. S., Pinkner, J. S., Roth, R., Heuser, J., Hammar, M., Normark, S., and Hultgren, S. J. (2002). Role of Escherichia coli curli operons in directing amyloid fiber formation. *Science 295*, 851–855.

14 Chernoff, Y. O., Galkin, A. P., Lewitin, E., Chernova, T. A., Newnam, G. P., and Belenkiy, S. M. (2000). Evolutionary conservation of prion-forming abilities of the yeast Sup35 protein. *Molecular Microbiology 35*, 865–876.

15 Chernoff, Y. O., Lindquist, S. L., Ono, B., Inge-Vechtomov, S. G., and Liebman, S. W. (1995). Role of the chaperone protein Hsp104 in propagation of the yeast prion-like factor [psi+]. *Science 268*, 880–884.

16 Chernoff, Y. O., Newnam, G. P., Kumar, J., Allen, K., and Zink, A. D. (1999). Evidence for a protein mutator in yeast: role of the Hsp70-related chaperone ssb in formation, stability, and toxicity of the [PSI] prion. *Molecular and Cellular Biology 19*, 8103–8112.

17 Chien, P., DePace, A. H., Collins, S. R., and Weissman, J. S. (2003).

Generation of prion transmission barriers by mutational control of amyloid conformations. *Nature 424*, 948–951.

18 Chien, P., and Weissman, J. S. (2001). Conformational diversity in a yeast prion dictates its seeding specificity. *Nature 410*, 223–227.

19 Chien, P., Weissman, J. S., and DePace, A. H. (2004). Emerging Principles of Conformation-based Prion Inheritance. *Annu Rev Biochem in press.*

20 Chiti, F., Webster, P., Taddei, N., Clark, A., Stefani, M., Ramponi, G., and Dobson, C. M. (1999). Designing conditions for in vitro formation of amyloid protofilaments and fibrils. *Proc Natl Acad Sci USA 96*, 3590–3594.

21 Collinge, J. (2001). Prion diseases of humans and animals: their causes and molecular basis. *Annu Rev Neurosci 24*, 519–550.

22 Coustou, V., Deleu, C., Saupe, S., and Begueret, J. (1997). The protein product of the het-s heterokaryon incompatibility gene of the fungus Podospora anserina behaves as a prion analog. *Proc Natl Acad Sci USA 94*, 9773–9778.

23 Cox, B. S. (1965). y, a cytoplasmic suppressor of super-suppressor in yeast. *Heredity 20*, 505–521.

24 Cox, B. S., Ness, F., and Tuite, M. F. (2003). Analysis of the Generation and Segregation of Propagons: Entities That Propagate the [PSI+] Prion in Yeast. *Genetics 165*, 23–33.

25 Craig, E., Ziegelhoffer, T., Nelson, J., Laloraya, S., and Halladay, J. (1995). Complex multigene family of functionally distinct Hsp70s of yeast. *Cold Spring Harb Symp Quant Biol 60*, 441–449.

26 Crist, C. G., Nakayashiki, T., Kurahashi, H., and Nakamura, Y. (2003). [PHI+], a novel Sup35-prion variant propagated with non-Gln/Asn oligopeptide repeats in the absence of the chaperone protein Hsp104. *Genes Cells 8*, 603–618.

27 Cummings, C. J., and Zoghbi, H. Y. (2000). Trinucleotide repeats: mechanisms and pathophysiology. *Annu Rev Genomics Hum Genet 1*, 281–328.

28 DePace, A. H., Santoso, A., Hillner, P., and Weissman, J. S. (1998). A critical role for amino-terminal glutamine/asparagine repeats in the formation and propagation of a yeast prion. *Cell 93*, 1241–1252.

29 DePace, A. H., and Weissman, J. S. (2002). Origins and kinetic consequences of diversity in Sup35 yeast prion fibers. *Nat Struct Biol 9*, 389–396.

30 Derkatch, I. L., Bradley, M. E., Hong, J. Y., and Liebman, S. W. (2001). Prions affect the appearance of other prions: the story of [PIN(+)]. *Cell 106*, 171–182.

31 Derkatch, I. L., Bradley, M. E., Masse, S. V., Zadorsky, S. P., Polozkov, G. V., Inge-Vechtomov, S. G., and Liebman, S. W. (2000). Dependence and independence of [PSI(+)] and [PIN(+)]: a two-prion system in yeast? *EMBO J 19* 1942–1952.

32 Derkatch, I. L., Bradley, M. E., Zhou, P., Chernoff, Y. O., and Liebman, S. W. (1997). Genetic and environmental factors affecting the de novo appearance of the [PSI+] prion in Saccharomyces cerevisiae. *Genetics 147*, 507–519.

33 Derkatch, I. L., Chernoff, Y. O., Kushnirov, V. V., Inge-Vechtomov, S. G., and Liebman, S. W. (1996). Genesis and variability of [PSI] prion factors in Saccharomyces cerevisiae. *Genetics 144*, 1375–1386.

34 Dobson, C. M. (1999). Protein misfolding, evolution and disease. *Trends Biochem Sci 24*, 329–332.

35 Dobson, C. M. (2001). The structural basis of protein folding and its links with human disease. *Philos Trans R Soc Lond B Biol Sci 356*, 133–145.

36 Dos Reis, S., Coulary-Salin, B., Forge, V., Lascu, I., Begueret, J., and Saupe, S. J. (2002). The HET-s prion protein of the filamentous fungus Podospora anserina aggregates in vitro into amyloid-like fibrils. *J Biol Chem 277*, 5703–5706.

37 Duncan, M. C., Cope, M. J., Goode, B. L., Wendland, B., and Drubin, D. G. (2001). Yeast Eps15-like endocytic protein, Pan1p, activates the Arp2/3 complex. *Nat Cell Biol 3*, 687–690.

38 Eaglestone, S. S., Ruddock, L. W., Cox, B. S., and Tuite, M. F. (2000). Guanidine hydrochloride blocks a critical step in the propagation of the prion-like determinant [PSI(+)] of Saccharomyces cerevisiae. *Proc Natl Acad Sci USA 97*, 240–244.

39 Edskes, H. K., and Wickner, R. B. (2002). Conservation of a portion of the S. cerevisiae Ure2p prion domain that interacts with the full-length protein. *Proc Natl Acad Sci USA 99 Suppl 4*, 16384–16391.

40 Fay, N., Inoue, Y., Bousset, L., Taguchi, H., and Melki, R. (2003). Assembly of the yeast prion Ure2p into protein fibrils. Thermodynamic and kinetic characterization. *J Biol Chem 278*, 30199–30205.

41 Glover, J. R., Kowal, A. S., Schirmer, E. C., Patino, M. M., Liu, J. J., and Lindquist, S. (1997). Self-seeded fibers formed by Sup35, the protein determinant of [PSI+], a heritable prion-like factor of S. cerevisiae. *Cell 89*, 811–819.

42 Glover, J. R., and Lindquist, S. (1998). Hsp104, Hsp70, and Hsp40: a novel chaperone system that rescues previously aggregated proteins. *Cell 94*, 73–82.

43 Gonzalez, C. I., Wang, W., and Peltz, S. W. (2001). Nonsense-mediated mRNA decay in Saccharomyces cerevisiae: a quality control mechanism that degrades transcripts harboring premature termination codons. *Cold Spring Harb Symp Quant Biol 66*, 321–328.

44 Griffith, J. S. (1967). Self-replication and scrapie. *Nature 215*, 1043–1044.

45 Guijarro, J. I., Sunde, M., Jones, J. A., Campbell, I. D., and Dobson, C. M. (1998). Amyloid fibril formation by an SH3 domain. *Proc Natl Acad Sci USA 95*, 4224–4228.

46 Hara, H., Nakayashiki, T., Crist, C. G., and Nakamura, Y. (2003). Prion domain interaction responsible for species discrimination in yeast [PSI+] transmission. *Genes Cells 8*, 925–939.

47 Harrison, P. M., and Gerstein, M. (2003). A method to assess compositional bias in biological sequences and its application to prion-like glutamine/asparagine-rich domains in eukaryotic proteomes. *Genome Biol 4*, R40.

48 Hartl, F. U., and Hayer-Hartl, M. (2002). Molecular chaperones in the cytosol: from nascent chain to folded protein. *Science 295*, 1852–1858.

49 Inoue, Y., Kishimoto, A., Hirao, J., Yoshida, M. & Taguchi, H. (2001). Strong growth polarity of yeast prion fiber revealed by single fiber imaging. *J. Biol. Chem. 276*, 35227–35230.

50 Jeng, R. L., and Welch, M. D. (2001). Cytoskeleton: actin and endocytosis – no longer the weakest link. *Curr Biol 11*, R691–694.

51 Jiang, Y., Li, H., Zhu, L., Zhou, J. M., and Perrett, S. (2004). Amyloid nucleation and hierarchical assembly of Ure2p fibrils. Role of asparagine/glutamine repeat and nonrepeat regions of the prion domains. *J Biol Chem 279*, 3361–3369.

52 Jones, G. W., and Masison, D. C. (2003). Saccharomyces cerevisiae Hsp70 mutations affect [PSI+] prion propagation and cell growth differently and implicate Hsp40 and tetratricopeptide repeat cochaperones in impairment of [PSI+]. *Genetics 163*, 495–506.

53 Jung, G., Jones, G., Wegrzyn, R. D., and Masison, D. C. (2000). A role for cytosolic hsp70 in yeast [PSI(+)] prion propagation and [PSI(+)] as a cellular stress. *Genetics 156*, 559–570.

54 Jung, G., and Masison, D. C. (2001). Guanidine Hydrochloride Inhibits Hsp104 Activity In Vivo: A Possible Explanation for Its Effect in Curing Yeast Prions. *Current Microbiology 43*, 7–10.

55 King, C. Y., Tittmann, P., Gross, H., Gebert, R., Aebi, M., and Wüthrich, K. (1997). Prion-inducing domain 2-114 of yeast Sup35 protein transforms in vitro into amyloid-like filaments.

Proc Natl Acad Sci USA 94, 6618–6622.

56 King, C. Y. and Diaz-Avalos, R. (2004) Protein only transmission of three prion strains: Sup35 amyloid causes [PSI^+]. *Nature 428*, 319–323.

57 Kryndushkin, D. S., Alexandrov, I. M., Ter-Avanesyan, M. D., and Kushnirov, V. V. (2003). Yeast [PSI+] prion aggregates are formed by small Sup35 polymers fragmented by Hsp104. *J Biol Chem 278*, 49636–43.

58 Kushnirov, V. V., Kochneva-Pervukhova, N. V., Chechenova, M. B., Frolova, N. S., and Ter-Avanesyan, M. D. (2000a). Prion properties of the Sup35 protein of yeast Pichia methanolica. *EMBO J 19*, 324–331.

59 Kushnirov, V. V., Kryndushkin, D. S., Boguta, M., Smirnov, V. N., and Ter-Avanesyan, M. D. (2000b). Chaperones that cure yeast artificial [PSI+] and their prion-specific effects. *Current Biology 10*, 1443–1446.

60 Kushnirov, V. V., and Ter-Avanesyan, M. D. (1998). Structure and replication of yeast prions. *Cell 94*, 13–16.

61 Lacroute, F. (1971). Non-Mendelian mutation allowing ureidosuccinic acid uptake in yeast. *J Bacter 106*, 519–522.

62 Li, L., and Lindquist, S. (2000). Creating a protein-based element of inheritance. *Science 287*, 661–664.

63 Lindquist, S. (1997). Mad cows meet psi-chotic yeast: the expansion of the prion hypothesis. *Cell 89*, 495–498.

64 Liu, J. J., and Lindquist, S. (1999). Oligopeptide-repeat expansions modulate 'protein-only' inheritance in yeast. *Nature 400*, 573–576.

65 Lopez, N., Aron, R., and Craig, E. A. (2003). Specificity of class II Hsp40 Sis1 in maintenance of yeast prion [RNQ+]. *Mol Biol Cell 14*, 1172–1181.

66 Maddelein, M. L., Dos Reis, S., Duvezin-Caubet, S., Coulary-Salin, B., and Saupe, S. J. (2002). Amyloid aggregates of the HET-s prion protein are infectious. *Proc Natl Acad Sci USA 99*, 7402–7407.

67 Manson, J. C., Barron, R., Jamieson, E., Baybutt, H., Tuzi, N., McConnell, I., Melton, D., Hope, J., and Bostock, C. (2000). A single amino acid alteration in murine PrP dramatically alters TSE incubation time. *Arch Virol Suppl*, 95–102.

68 Manson, J. C., Jamieson, E., Baybutt, H., Tuzi, N. L., Barron, R., McConnell, I., Somerville, R., Ironside, J., Will, R., Sy, M. S., et al. (1999). A single amino acid alteration (101L) introduced into murine PrP dramatically alters incubation time of transmissible spongiform encephalopathy. *EMBO J 18*, 6855–6864.

69 Martens, J. A., and Winston, F. (2003). Recent advances in understanding chromatin remodeling by Swi/Snf complexes. *Curr Opin Genet Dev 13*, 136–142.

70 Masison, D. C., and Wickner, R. B. (1995). Prion-inducing domain of yeast Ure2p and protease resistan of Ure2p in prion-containing cells. *Science 270*, 93–95.

71 Mastrianni, J. A., Capellari, S., Telling, G. C., Han, D., Bosque, P., Prusiner, S. B., and DeArmond, S. J. (2001). Inherited prion disease caused by the V210I mutation: transmission to transgenic mice. *Neurology 57*, 2198–2205.

72 Meriin, A. B., Zhang, X., Miliaras, N. B., Kazantsev, A., Chernoff, Y. O., McCaffery, J. M., Wendland, B., and Sherman, M. Y. (2003). Aggregation of expanded polyglutamine domain in yeast leads to defects in endocytosis. *Mol Cell Biol 23*, 7554–7565.

73 Michelitsch, M. D., and Weissman, J. S. (2000). A census of glutamine/asparagine-rich regions: implications for their conserved function and the prediction of novel prions. *Proc Natl Acad Sci USA 97*, 11910–11915.

74 Moriyama, H., Edskes, H. K., and Wickner, R. B. (2000a). [URE3] prion propagation in Saccharomyces cerevisiae: requirement for chaperone Hsp104 and curing by overexpressed chaperone Ydj1p. *Mol Cell Biol 20*, 8916–8922.

75 Moriyama, H., Edskes, H. K., and

Wickner, R. B. (2000b). [URE3] prion propagation in Saccharomyces cerevisiae: requirement for chaperone Hsp104 and curing by overexpressed chaperone Ydj1p. *Mol Cell Biol 20*, 8916–8922.

76 Muchowski, P. J., Schaffar, G., Sittler, A., Wanker, E. E., Hayer-Hartl, M. K., and Hartl, F. U. (2000). Hsp70 and hsp40 chaperones can inhibit self-assembly of polyglutamine proteins into amyloid-like fibrils. *Proc Natl Acad Sci USA 97*, 7841–7846.

77 Nakayashiki, T., Ebihara, K., Bannai, H., and Nakamura, Y. (2001). Yeast [PSI+] "prions" that are crosstransmissible and susceptible beyond a species barrier through a quasi-prion state. *Mol Cell 7*, 1121–1130.

78 Newnam, G. P., Wegrzyn, R. D., Lindquist, S. L., and Chernoff, Y. O. (1999). Antagonistic interacttions between yeast chaperones Hsp104 and Hsp70 in prion curing. *Mol Cell Biology 19*, 1325–1333.

79 O'Nuallain, B., Williams, A. D., Westermark, P., and Wetzel, R. (2004). Seeding specificity in amyloid growth induced by heterologous fibrils. *J Biol Chem.*

80 Osherovich, L. Z., Cox, B. S., Tuite, M. F., and Weissman, J. S. (2004). Dissection and Design of Yeast Prions. *PLoS Biology 2.*

81 Osherovich, L. Z., and Weissman, J. S. (2001). Multiple Gln/Asn-rich prion domains confer susceptibility to induction of the yeast [PSI(+)] prion. *Cell 106*, 183–194.

82 Osherovich, L. Z., and Weissman, J. S. (2002). The utility of prions. *Dev Cell 2*, 143–151.

83 Parham, S. N., Resende, C. G., and Tuite, M. F. (2001). Oligopeptide repeats in the yeast protein Sup35p stabilize intermolecular prion interactions. *EMBO J 20*, 2111–2119.

84 Patino, M. M., Liu, J. J., Glover, J. R., and Lindquist, S. (1996). Support for the prion hypothesis for inheritance of a phenotypic trait in yeast. *Science 273*, 622–626.

85 Paushkin, S. V., Kushnirov, V. V., Smirnov, V. N., and Ter-Avanesyan, M. D. (1996). Propagation of the yeast prion-like [psi+] determinant is mediated by oligomerization of the SUP35-encoded polypeptide chain release factor. *EMBO J 15*, 3127–3134.

86 Paushkin, S. V., Kushnirov, V. V., Smirnov, V. N., and Ter-Avanesyan, M. D. (1997). In vitro propagation of the prion-like state of yeast Sup35 protein. *Science 277*, 381–383.

87 Perutz, M. F., Johnson, T., Suzuki, M., and Finch, J. T. (1994). Glutamine repeats as polar zippers: their possible role in inherited neurodegenerative diseases. *Proc Natl Acad Sci USA 91*, 5355–5358.

88 Perutz, M. F., Pope, B. J., Owen, D., Wanker, E. E., and Scherzinger, E. (2002). Aggregation of proteins with expanded glutamine and alanine repeats of the glutamine-rich and asparagine-rich domains of Sup35 and of the amyloid beta-peptide of amyloid plaques. *Proc Natl Acad Sci USA 99*, 5596–5600.

89 Prusiner, S. B. (1998). Prions. *Proc Natl Acad Sci USA 95*, 13363–13383.

90 Prusiner, S. B., Scott, M., Foster, D., Pan, K. M., Groth, D., Mirenda, C., Torchia, M., Yang, S. L., Serban, D., Carlson, G. A., and et al. (1990). Transgenetic studies implicate interactions between homologous PrP isoforms in scrapie prion replication. *Cell 63*, 673–686.

91 Resende, C. G., Outeiro, T. F., Sands, L., Lindquist, S., and Tuite, M. F. (2003). Prion protein gene polymorphisms in Saccharomyces cerevisiae. *Mol Microbiol 49*, 1005–1017.

92 Ripaud, L., Maillet, L., and Cullin, C. (2003). The mechanisms of [URE3] prion elimination demonstrate that large aggregates of Ure2p are dead-end products. *EMBO J 22*, 5251–5259.

93 Roberts, B. T., Moriyama, H., and Wickner, R. B. (2004). [URE3] prion propagation is abolished by a mutation of the primary cytosolic Hsp70 of budding yeast. *Yeast 21*, 107–117.

94 Roberts, B. T., and Wickner, R. B. (2003). Heritable activity: a prion that propagates by covalent autoactivation. *Genes Dev 17*, 2083–2087.

95 Santoso, A., Chien, P., Osherovich, L. Z., and Weissman, J. S. (2000). Molecular basis of a yeast prion species barrier. *Cell 100*, 277–288.

96 Scheibel, T., Kowal, A. S., Bloom, J. D., and Lindquist, S. L. (2001). Bidirectional amyloid fiber growth for a yeast prion determinant. *Curr Biol 11*, 366–369.

97 Schlumpberger, M., Prusiner, S. B., and Herskowitz, I. (2001). Induction of distinct [URE3] yeast prion strains. *Mol Cell Biol 21*, 7035–7046.

98 Scott, J. H., and R. Schekman. (1980). Lyticase: endoglucanase and protease activities that act together in yeast cell lysis. *J Bacteriol 142*, 414–423.

99 Scott, M., Foster, D., Mirenda, C., Serban, D., Coufal, F., Walchli, M., Torchia, M., Groth, D., Carlson, G., DeArmond, S. J., and et al. (1989). Transgenic mice expressing hamster prion protein produce species-specific scrapie infectivity and amyloid plaques. *Cell 59*, 847–857.

100 Sheth, U., and Parker, R. (2003). Decapping and decay of messenger RNA occur in cytoplasmic processing bodies. *Science 300*, 805–808.

100a Shorter, J. and Lindquist, S. (2004). Hsp104 catalyzes formation and elimination of self-replicating Sup35 prion conformers. *Science 304*, 1793–1797.

101 Si, K., Giustetto, M., Etkin, A., Hsu, R., Janisiewicz, A. M., Miniaci, M. C., Kim, J. H., Zhu, H., and Kandel, E. R. (2003a). A neuronal isoform of CPEB regulates local protein synthesis and stabilizes synapse-specific long-term facilitation in aplysia. *Cell 115*, 893–904.

102 Si, K., Lindquist, S., and Kandel, E. R. (2003b). A neuronal isoform of the aplysia CPEB has prion-like properties. *Cell 115*, 879–891.

103 Sondheimer, N., and Lindquist, S. (2000). Rnq1: an epigenetic modifier of protein function in yeast. *Mol Cell 5*, 163–172.

104 Sondheimer, N., Lopez, N., Craig, E. A., and Lindquist, S. (2001). The role of Sis1 in the maintenance of the [RNQ+] prion. *EMBO J 20*, 2435–2442.

105 Sparrer, H. E., Santoso, A., Szoka, F. C., Jr., and Weissman, J. S. (2000). Evidence for the prion hypothesis: induction of the yeast [PSI+] factor by in vitro-converted Sup35 protein. *Science 289*, 595–599.

106 Speransky, V. V., Taylor, K. L., Edskes, H. K., Wickner, R. B., and Steven, A. C. (2001). Prion filament networks in [URE3] cells of Saccharomyces cerevisiae. *J Cell Biol 153*, 1327–1336.

107 Tanaka, M., Chien, P., Naber, N., Cooke, R. and Weissman, J. S. (2004). Conformation Variations in Infectious Particles Determine Prion Differences. *Nature 428*, 323–328.

108 Tang, H. Y., Xu, J., and Cai, M. (2000). Pan1p, End3p, and S1a1p, three yeast proteins required for normal cortical actin cytoskeleton organization, associate with each other and play essential roles in cell wall morphogenesis. *Mol Cell Biol 20*, 12–25.

109 Taylor, J. P., Hardy, J., and Fischbeck, K. H. (2002). Toxic proteins in neurodegenerative disease. *Science 296*, 1991–1995.

110 Taylor, K. L., Cheng, N., Williams, R. W., Steven, A. C., and Wickner, R. B. (1999). Prion domain initiation of amyloid formation in vitro from native Ure2p. *Science 283*, 1339–1343.

111 Tuite, M. F., and Cox, B. S. (2003). Propagation of yeast prions. *Nat Rev Mol Cell Biol 4*, 878–890.

112 Tuite, M. F., Mundy, C. R., and Cox, B. S. (1981). Agents that cause a high frequency of genetic change from [psi+] to [psi−] in Saccharomyces cerevisiae. *Genetics 98*, 691–711.

113 Uptain, S. M., and Lindquist, S. (2002). Prions as protein-based genetic elements. *Annu Rev Microbiol 56*, 703–741.

114 Uptain, S. M., Sawicki, G. J., Caughey, B., and Lindquist, S. (2001). Strains of [PSI(+)] are

distinguished by their efficiencies of prion-mediated conformational conversion. *EMBO J 20*, 6236–6245.

115 Weissmann, C. (1991). Spongiform encephalopathies. The prion's progress. *Nature 349*, 569–571.

116 Weissmann, C. (1999). Molecular genetics of transmissible spongiform encephalopathies. *J Biol Chem 274*, 3–6.

117 Wickner, R. B. (1994). [URE3] as an altered URE2 protein: evidence for a prion analog in Saccharomyces cerevisiae. *Science 264*, 566–569.

118 Yang Fan, C., Lee, S., Ren, H. Y., and Cyr, D. M. (2003). Exchangeable chaperone modules contribute to specification of Type I and Type II Hsp40 cellular function. *Mol Biol Cell.*

119 Young, J. C., Barral, J. M., and Ulrich Hartl, F. (2003). More than folding: localized functions of cytosolic chaperones. *Trends Biochem Sci 28*, 541–547.

120 Zhou, P., Derkatch, I. L., and Liebman, S. W. (2001). The relationship between visible intracellular aggregates that appear after overexpression of Sup35 and the yeast prion-like elements [PSI(+)] and [PIN(+)]. *Mol Microbiol 39*, 37–46.

121 Zhou, P., Derkatch, I. L., Uptain, S. M., Patino, M. M., Lindquist, S., and Liebman, S. W. (1999). The yeast non-Mendelian factor [ETA+] is a variant of [PSI+], a prion-like form of release factor eRF3. *EMBO J 18*, 1182–1191.

36
Polyglutamine Aggregates as a Model for Protein-misfolding Diseases

Soojin Kim, James F. Morley, Anat Ben-Zvi, and Richard I. Morimoto

36.1
Introduction

Protein aggregation is common to a number of human neurodegenerative diseases including Alzheimer's disease, Parkinson's disease, prion disorders, amyotrophic lateral sclerosis, and polyglutamine diseases (see Chapters 33–35). The correlation between the appearance of inclusion bodies and neuropathology has led to a widely held hypothesis that aggregates are toxic. Recent studies, however, suggest that toxicity of misfolded proteins correlates imperfectly with aggregates, leading to the proposal that other intermediate species, distinct from the monomers and fibrils, may also contribute to neurotoxicity. While the identity of the "toxic species" remains an issue of debate, the common underlying features in these diseases are the self-association and accumulation of the misfolded proteins as visible subcellular aggregates.

In this chapter, we discuss various approaches to studying misfolded and aggregation-prone proteins using polyglutamine-containing proteins as a model system, and address the role of misfolded proteins in toxicity. Finally, we discuss various modulators of protein aggregation.

36.2
Polyglutamine Diseases

36.2.1
Genetics

Polyglutamine disorders refer to a group of nine neurodegenerative diseases including Huntington's disease (HD); spinal and bulbar muscular atrophy (SBMA); dentatorubral-pallidoluysian atrophy (DRPLA); and spinocerebellar ataxia (SCA)-1, -2, -3, -6, -7, and -17, in which expansion of glutamine repeats in the affected protein is directly linked to neuropathology [1–4]. For each of these disorders, the dis-

Protein Folding Handbook. Part II. Edited by J. Buchner and T. Kiefhaber

ISBN: 3-527-30784-2

ease is dominantly inherited, with the exception of SBMA. Genetic analysis of HD patient populations has revealed that the huntingtin gene contains more than 35–40 CAG repeats (encoding the amino acid glutamine, Q), whereas the huntingtin gene from normal individuals contains fewer than 30–34 CAG repeats [5–7]. Similar genetic features exist for the other polyglutamine diseases with regard to the relationship between normal and pathogenic repeat lengths and neurotoxicity, with the exception of SCA6, which contains 21–33 CAG repeats in the disease state and 4–18 CAG repeats in normal alleles [1]. These observations support the hypothesis of a polyglutamine-length threshold beyond which pathogenic mechanisms are triggered [2]. In addition, genotype-phenotype studies have established a strong inverse correlation between repeat length and the age of onset such that an increase in the number of CAG repeat leads to an earlier manifestation of the disease with more severe symptoms, indicating a direct involvement of polyglutamine length in the disease progression.

36.2.2
Polyglutamine Diseases Involve a Toxic Gain of Function

Polyglutamine diseases exhibit genetic dominance such that neuropathologies of heterozygous and homozygous individuals are nearly identical for Huntington's disease [8]. Huntingtin or ataxin-1 knockout mice or deletion in humans exhibited phenotypes that are distinct from the respective neuropathologies, suggesting that polyglutamine diseases are not the result of downregulation of the respective affected proteins but rather due to a gain-of-function toxicity associated with the polyglutamine expansion [9–14]. A gain-of-function hypothesis is further supported by observations that expression of either full-length or truncated mutant genes containing the expanded glutamine repeat or the expanded polyglutamine alone expressed in mice, *Drosophila*, and *C. elegans* is sufficient to cause neurological dysfunctions [15–19]. Taken together, these observations support the hypothesis that polyglutamine expansion is the principal cause of the toxic gain of function associated with expression of these mutant proteins.

36.3
Polyglutamine Aggregates

36.3.1
Presence of the Expanded Polyglutamine Is Sufficient to Induce Aggregation in Vivo

The presence of protein aggregates is a pathological hallmark of polyglutamine diseases. Nuclear inclusion bodies have been identified in postmortem brains from HD, SCA-1, SCA-3, SCA-7, DRPLA, and SBMA patients [2, 20–26]. In all cases, these inclusion bodies contain the mutant proteins with the expanded polyglutamines. For huntingtin, only the N-terminal fragment containing the polyglutamine tract was detected in the inclusions, suggesting a role for proteolytic cleavage in huntingtin aggregation [20]. These inclusion bodies also stained positive, by immunohistochemistry, for a common cohort of proteins including ubiquitin, molecular chaperones, and components of proteasomes [20, 21, 23, 27–31]. As these factors are typically associated with the cellular response to misfolded proteins, it has been proposed that the polyglutamine expansion leads to misfolding and altered conformational states and, subsequently, to aggregation of the affected protein.

Evidence that the expanded glutamine stretch is the source of protein aggregation comes from observations that expression of expanded glutamine repeats alone or in the context of the truncated mutant proteins induces formation of protein aggregates in a number of model systems. In all model organisms studied, including *E. coli*, *S. cerevisiae*, *Drosophila*, *C. elegans*, mice, and mammalian cell culture, expression of polyglutamine lengths above 40Q led to the formation of large visible aggregates [15–19, 21, 27, 28, 32–38]. Moreover, transgenic mice expressing expanded glutamine repeats within the context of a heterologous gene, hypoxanthine phosphoribosyltransferase (HPRT), exhibited multiple features of polyglutamine diseases, including formation of nuclear inclusion bodies and death of motor neurons, demonstrating that expanded polyglutamine sequence alone is sufficient to induce aggregation and toxicity [39].

36.3.2
Length of the Polyglutamine Dictates the Rate of Aggregate Formation

The rate of aggregate formation and the degree of toxicity are directly correlated to polyglutamine length. Expression of various lengths of glutamine repeats in model organisms, such as *E. coli*, *S. cerevisiae*, *Drosophila*, *C. elegans*, or cell culture, demonstrated that expression of sub-threshold glutamine length ($<$ 40Q) did not affect the viability of the host cells and that the polyglutamine proteins expressed showed a diffused localization [17–19, 21, 29–33, 36–38]. In contrast, cells expressing polyglutamine lengths above this threshold induced toxicity and readily formed punctate aggregated structures that were detergent-insoluble. Toxicity and aggregation frequency increased with polyglutamine length. In *C. elegans* expressing YFP-fused polyglutamines in muscle cells, diffuse fluorescence was detected in young adult *C. elegans* expressing 0Q, 19Q, 29Q, and 33Q, whereas animals expressing 44Q, 64Q, or 82Q exhibited only punctate fluorescent structures corresponding to protein aggregates. 35Q- and 40Q-expressing animals, however, displayed polymorphic distribution of the polyglutamine-YFP with diffuse fluorescence in some cells and punctate fluorescence in others (Figure 36.1A). A direct correlation was observed between toxicity and aggregation. Animals expressing 40Q or 82Q exhibited a rapid loss in motility in contrast to 19Q-, 29Q-expressing or wild-type animals, which were unaffected (Figure 36.1B, C). Similar toxic effects of increasing polyglutamine length have also been demonstrated in *E. coli*, in *Drosophila* eye development, and in cell culture models [19, 21, 29–32, 36–38, 40, 41]. These observations that aggregate formation and toxicity are polyglutamine length-dependent and that model organisms are also sensitive to similar polyglutamine-length thresholds

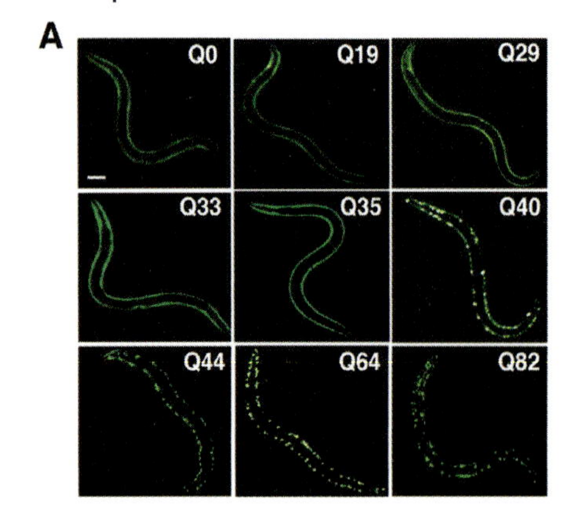

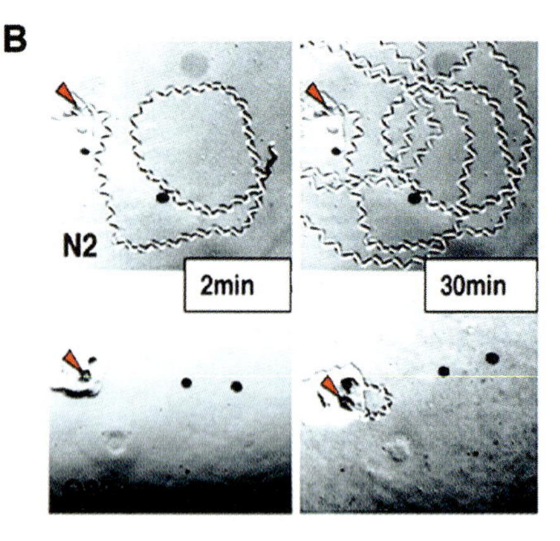

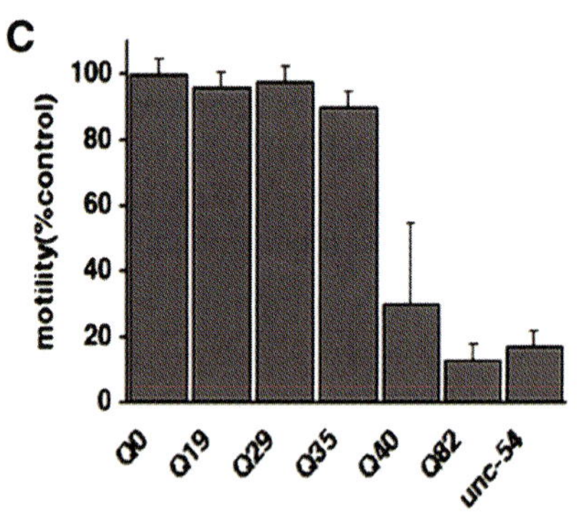

Fig. 36.1. Polyglutamine length-dependent aggregation and motility defect in *C. elegans*. (A) Epifluoresence micrographs of 3- to 4-day-old *C. elegans* expressing different lengths of polyQ-YFP (0Q, 19Q, 29Q, 33Q, 35Q, 40Q, 44Q, 64Q, 82Q). Bar = 0.1 mm. (B) Time lapse micrographs illustrating tracks left by 5-day-old wild-type (N2) and 82Q animals 2 min and 30 min after being placed at the position marked by the red arrow. (C) Quantitation of motility index for 4- to 5-day-old 0Q, 19Q, 29Q, 35Q, 40Q, 82Q, and motility defective *unc-54*(r293) animals. Data are mean ± SD for at least 50 animals of each type as a percentage of N2 (wild-type) motility.

(40Qs) as detected in human diseases indicate a common underlying biochemical principle for polyglutamine aggregation and toxicity.

The kinetics of polyglutamine aggregation has been studied intensively in vitro using synthetic polyglutamine peptides and recombinant proteins (see Chapters 6 and 37). Studies using purified polyglutamines were initially hampered by the strong propensity for even short polyglutamine peptides to precipitate in aqueous solution [42]. These difficulties have been overcome, to some extent, by flanking the polyglutamine peptides with basic residues to improve their solubility, by employing mixtures of aqueous and organic solvents (such as trifluoroacetic acid and hexafluoroisopropanol), or by generating chimeric proteins in which the polyglutamine is fused to a soluble protein such as GST and aggregate formation is initiated by proteolytic cleavage, liberating the polyglutamine fragment [43–45].

One striking observation from biochemical studies on polyglutamine aggregation was that the correlation between the repeat length and aggregation kinetics in vitro paralleled closely with that observed in vivo. An Htt exon 1-GST protein containing an expansion of 51Q formed insoluble aggregates as detected by filter-trap assay upon protease cleavage of the GST moiety, whereas peptides of 20Q or 30Q remained soluble [44]. The aggregation kinetics of peptides released from cleaved Htt exon 1-GST fusion proteins or synthetic polyglutamine peptides was positively correlated to the increasing glutamine repeat length, whereas the lag time for aggregation was inversely correlated with polyglutamine length [45–47]. These data suggest that the intrinsic biophysical parameters governing polyglutamine aggregation in vitro is analogous to the relationship between repeat length and aggregation observed in vivo.

36.3.3 Polyglutamine Aggregates Exhibit Features Characteristic of Amyloids

The structural determinant for polyglutamine aggregation is hypothesized to be the formation of stable β-sheet interactions. Perutz and co-workers proposed that repetitive sequences of polar residues are capable of linking β-strands together through hydrogen bonds between their main-chain amides and their polar side charges [43]. Consequently, formation of intramolecular hydrogen bonds of amide groups in an expanded glutamine stretch may provide the biochemical basis for polyglutamine self-association. Studies on a short polyglutamine peptide (D_2-Q_{15}-K_2) demonstrated that glutamine repeats form oligomers consisting of β-pleated sheets [43]. Additionally, circular dichroism (CD) analysis of recombinant or synthetic polyglutamine peptides demonstrated that while a monomeric polyglutamine exists as a random coil, oligomers of the expanded glutamines have substantial β-structures, suggesting that a conformational switch to a β-sheet-rich state is key to polyglutamine aggregation [42, 43, 45, 47, 48]. An in vitro analysis using two stretches of 9Qs linked together by Gly-Pro pairs (predicted to induce β-turns) revealed that the minimal unit required for stable β-hairpin structures is 40Qs [49]. These results are intriguing since 40Q is the threshold polyglutamine length for the aggregation phenotype in human disease. These observations have contributed to the hypothesis that stabilization of expanded polyglutamines in β-sheet conformation promotes self-association of the mutant proteins.

Electron microscopic (EM) studies on purified proteins have revealed that polyglutamine expansions above the threshold length form fibrous, ribbon-like amyloid structures, whereas sub-threshold-length glutamine repeats form oligomeric globular or amorphous structures [44, 46, 50–52]. Staining of these purified aggregates with amyloidophilic dyes, such as Thioflavin T or Congo red, has established that polyglutamine aggregates have structural features common to other amyloid-forming proteins [44, 45, 53]. Formation of fibrous amyloid structures consisting of β-sheets has also been observed for other amyloidogenic proteins including Aβ-peptide, prions, α-synuclein, and superoxide dismutase [54–57]. As the folding of β-sheets is driven predominantly by backbone interactions, it has been proposed that formation of amyloid structures is independent of side chain interactions [58]. If so, protein aggregation may be primary sequence–independent, as stabilization of β-sheet structures is the common mechanism behind aggregate formation.

36.3.4
Characterization of Protein Aggregates in Vivo Using Dynamic Imaging Methods

In vivo studies on polyglutamine aggregates using the live-cell imaging technique fluorescent recovery after photobleaching (FRAP) provided insights as to the organization, composition, and properties of self-associated polyglutamine structures without disrupting the cell [59–62]. Monitoring fluorescence recovery following photobleaching of fluorescent proteins containing polyglutamine expansions has revealed that the mobility of proteins within aggregates is very slow and essentially undetectable whether in cultured cells or in *C. elegans* (Figure 36.2). The nature of the molecular interactions of aggregate-associated proteins can be examined in cells by using fluorescence resonance energy transfer (FRET) analysis using pairs of differently fluorescently labeled proteins [62, 63] (see Chapter 17 in Part I). For example, co-expression of polyglutamine-expansion proteins fused to either YFP or CFP exhibited energy transfer within polyglutamine aggregates, indicating 40- to 100-Å distances between polyglutamine molecules in the aggregate. These results demonstrate that polyglutamine-expansion proteins form immobile aggregates comprised of self-associated interactions that are highly compacted and ordered. Taken together, the complement of in vitro and in vivo studies supports the hy-

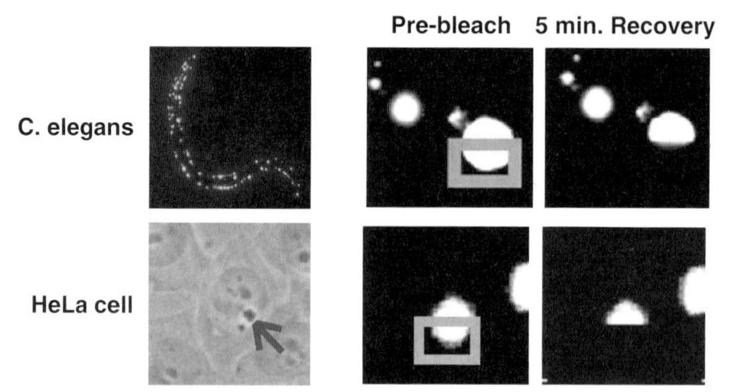

Fig. 36.2. Polyglutamine aggregates are immobile. FRAP analysis of *C. elegans* (top) and Hela cells (bottom). Light-colored foci in *C. elegans* and dark-colored foci in aHeLa cell (arrow) are Q82-GFP aggregates (left panels). Higher magnification fluorescence images of the aggregates before (pre-bleach) and 5 min after photobleaching (5-min recovery). Boxes indicate the area that was subjected to photobleaching. Fluorescence in the aggregate did not recover fully even after two hours of monitoring, indicating a stable structure of polyglutamine aggregate.

pothesis that misfolded proteins are stabilized by adopting β-sheet structures, which serve as the basis for the formation of ordered and immobile aggregates.

36.4 A Role for Oligomeric Intermediates in Toxicity

Despite the prevalence of observations that pathology is associated with the appearance of protein aggregates, a question that remains unresolved is the identity of the "toxic species" or whether there are multiple "toxic species." Demonstrations that the localization of mutant ataxin-1 to the nucleus or that the blocking of ubiquination of huntingtin accelerates toxicity in the absence of visible aggregates have led to the proposal that aggregates could be protective or that toxicity may be induced by microscopically undetectable species [42, 64–67]. These visually undetectable species have become of particular interest since non-fibrous oligomeric intermediates have been detected at the earlier phases of misfolding and aggregation as well as at a very low temperature during aggregation of the purified huntingtin fragments in vitro [45, 51, 52]. The observation that the formation of mature fibers, but not early globular or protofibrillar species, was inhibited by the amyloid-binding dye Congo red suggests that these intermediates may be structurally distinct from β-hairpin-containing structures [52]. Although the role of oligomeric intermediates as toxic species in polyglutamine diseases is not established, cell death caused by addition of protofilaments (Aβ-peptide) to the media of cultured neurons suggests that polyglutamine toxicity may be induced in a similar fashion [68–71]. A common conformational state shared among intermediates species, rather than the presence of a specific sequence, is responsible for toxicity, since addition of other non-disease-related, small globular proteins such as PI3-SH3 (SH3 domain from bovine phosphatidyl-inositol-3′-kinase) or HypF-N (the N-terminal "acylphosphatase-like" domain of the *E. coli* HypF protein) to neuronal cultures also induces cell death [72]. Additional support that aggregation-prone proteins can adopt a common conformational state comes from immunological studies in which an antibody, derived against a small soluble oligomeric state of Aβ-peptide, detects protofilament structures but not soluble monomers or visible aggregates [73]. A striking property of this antibody is the ability to detect not only the protofilaments of Aβ but also intermediate species generated by α-synuclein, prion, and polyglutamine-expansion proteins. Taken together, these studies support the presence of a common conformational state associated with intermediate species in otherwise distinct proteins with divergent sequences.

36.5 Consequences of Misfolded Proteins and Aggregates on Protein Homeostasis

Immunolocalization studies of polyglutamine protein aggregates have established that a number of cellular proteins, including molecular chaperones (Hsp70, Hdj1,

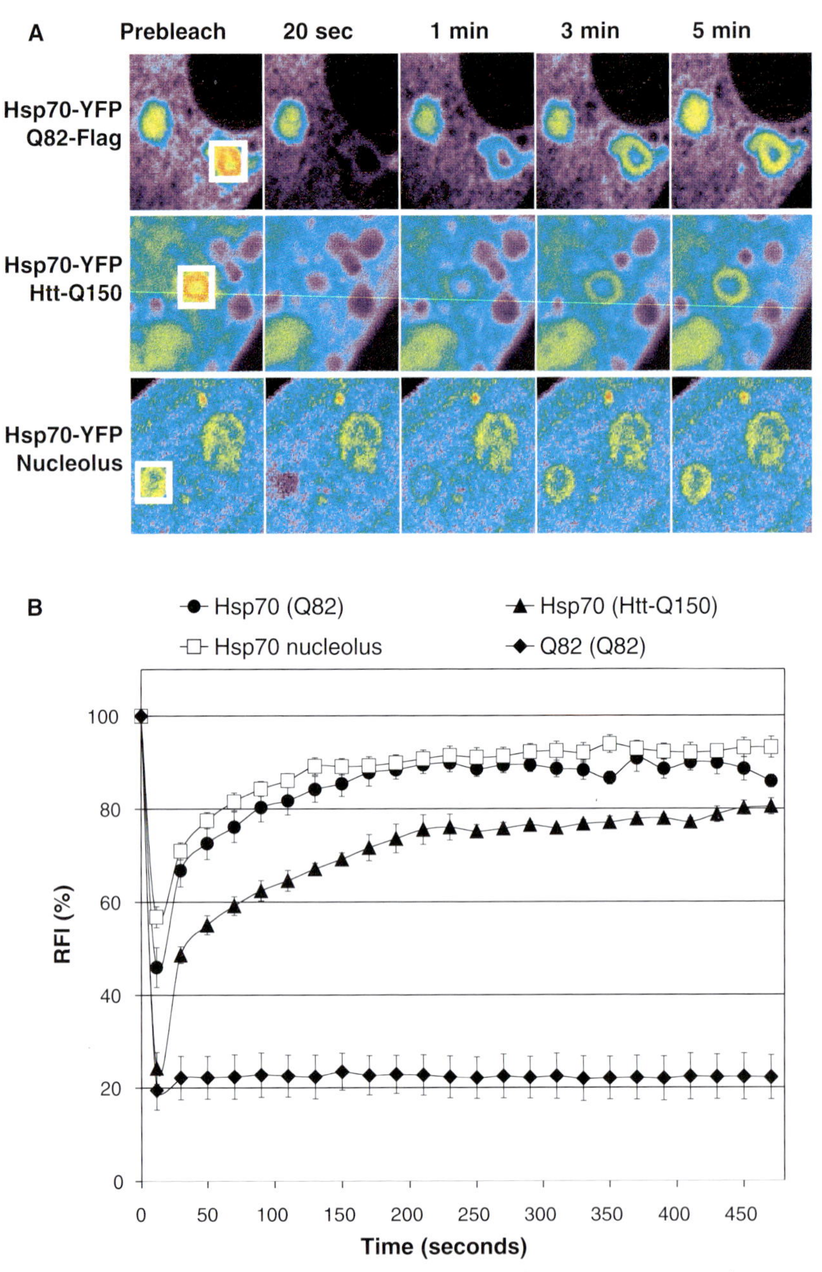

Fig. 36.3. Hsp70 associates transiently with polyglutamine aggregates via chaperon-substrate interactions. (A) FRAP analysis of Hsp70-YFP associated with polyglutamine aggregates. Pseudo-color images represent fluorescence intensity. Hsp70-YFP associated with Q82-Flag aggregate (top panels), Hsp70-YFP associated with Htt-Q150 aggregate (middle panels), and Hsp70-YFP in nucleolus following heat shock at 45 °C for 30 min and recovery at 37 °C for 1–2 h (bottom panels) were photobleached in the boxed area. Single

and Hdj2), proteasome subunits (20S), structural proteins (neurofilaments), and short glutamine repeat–containing transcription factors (TBP and CBP), colocalize with polyglutamine aggregates [27, 28, 74–77]. A question that follows from studies on fixed cells or biochemical analyses of isolated aggregates is whether these aggregate-associated proteins are stably or transiently bound in vivo. To address this question, FRAP analysis was used to show that the basal transcription factors TBP (TATA-binding protein) and CBP (CREB-binding protein) bind stably and do not diffuse away from huntingtin or ataxin-3 aggregates at a detectable rate [59, 62]. These results suggest that polyglutamine toxicity stems from irreversible sequestration of essential transcription factors.

Whereas transient association of molecular chaperones (Figure 36.3A, B) with polyglutamine aggregates likely represents a protective cellular response to a disruption of protein homeostasis, irreversible sequestration of transcription factors into polyglutamine aggregates could lead to the loss of their normal cellular functions. RT-PCR, differential display, and DNA microarray approaches have demonstrated altered expression of numerous genes in transgenic models of polyglutamine diseases [78–80]. Many of these changes can be observed before the onset of symptoms, arguing for a causal role of transcriptional dysregulation in toxicity.

CBP has been studied as a sequestration model of toxicity; CBP contains 19 glutamine repeats at its C-terminus, and deletion of this region abolishes the interaction with polyglutamine aggregates [81]. Expression of expanded polyglutamine downregulates soluble CBP and CBP-dependent gene expressions, whereas overexpression of CBP rescues transcriptional dysregulation and cellular toxicity caused by expanded polyglutamines. Depletion of CBP activity has a role in toxicity, as brain-specific chromosomal deletion of CREB in mice leads to neurodegeneration, suggesting that normal regulation of CBP target genes is required for neuronal function and survival [82]. Taken together, these observations suggest that sequestration of essential cellular factors by polyglutamine aggregates may contribute to cellular toxicity.

The range of cellular proteins that polyglutamine aggregates recruit may be large given that short glutamine repeats or polar-rich motifs can interact with polyglutamine-expansion proteins. Recruitment of normally soluble polyglutamine lengths (~19Q) into aggregates composed of long stretches (~82Q) has been observed both in vitro and in vivo [35, 45]. In addition, the interaction between yeast prion determinants consisting of polar (glutamine/asparagine)-rich motifs and

scanned images were taken before photobleaching (pre-bleach) and at the indicated time points following photobleaching. Hsp70 association with polyglutamine or huntingtin aggregates and heat-shocked nucleolus exhibited fast recovery indicative of transient interactions. (B) Quantitative FRAP analysis of the Hsp70-YFP associated with polyglutamine aggregates or localized to the nucleolus of a heat-shocked cell as in (A). Q82-Flag or Htt-Q150 was used to seed aggregates as indicated. The relative fluorescence intensity (RFI) was determined for each time point and was represented as the average of analysis of five cells. Error bars indicate the standard error of the mean (SEM). Data from Q82-GFP analysis was included as a reference for an immobile protein.

polyglutamine stretches has also been detected [83]. These results together with bioinformatics studies on eukaryotic genomes, which propose that 1–5% of proteins have polar-rich motifs, suggest that a substantial complement of proteins could be sequestered into polyglutamine aggregates [84].

However, it is also clear that some of the proteins that associate with aggregates do not have glutamine- or polar-rich motifs. For these proteins, it has been hypothesized that localization to aggregates occurs by virtue of their interactions with sequences other than the polyglutamine repeats or indirectly by association with other recruited proteins [85, 86]. Yeast two-hybrid screens and biochemical methods have identified a number of cellular factors that associate with huntingtin or ataxia-1, but the role of these interacting cellular proteins in polyglutamine toxicity remains to be elucidated [85, 87–89].

Recruitment of cellular proteins by polyglutamine aggregates reveals that an underlying mechanism for polyglutamine toxicity may be due to sequestration of essential cellular factors. It may be that expansion of polyglutamine in the affected proteins causes misfolding, which abolishes or enhances interactions with its normal associating partners or creates new interactions. Whether such interactions require the presence of large aggregates is uncertain, since smaller oligomeric species should be as effective in associating with other cellular proteins. A study using purified polyglutamine peptides demonstrated that oligomers are more efficient in propagating aggregation than are large aggregates, which supports the hypothesis that oligomers are more toxic than large aggregates [45, 72]. It may be that large aggregates are not as toxic, since most of the interactive surfaces are occupied, ordered amyloidal structures, although this hypothesis remains untested.

36.6 Modulators of Polyglutamine Aggregation and Toxicity

Whereas it can take decades for aggregates to form in human disease, the same process can occur in vitro in a matter of hours. This reveals the importance of in vivo genetic modifiers that regulate protein-folding homeostasis even when challenged by expression of highly aggregation-prone proteins. Identification of the factors that modulate the aggregation process and toxicity will be important in understanding disease pathology and invaluable for the development of therapeutic strategies.

36.6.1 Protein Context

The influence of the protein context in aggregation was first demonstrated in studies on ataxin-1 where deletion of the self-association domain, distinct from the polyglutamine-expansion region, prevented the formation of visible aggregates [64]. Similarly, mutation of an Akt phosphorylation site (S776A) decreased aggregation and neuropathology induced by ataxin-1 despite the presence of 82Qs [89].

Phosphorylation of serine 776 has been suggested to stabilize ataxin-1 through interactions with 14-3-3 (a multifunctional regulatory protein), thus exacerbating aggregate formation and neurodegeneration [88]. These results offer the insight that aggregation and neurotoxicity, even in the presence of the polyglutamine expansion, can be modulated by other sequences within the disease-causing protein.

36.6.2 Molecular Chaperones

The role of molecular chaperones as modifiers of polyglutamine-induced toxicity and aggregation has been well established; however, there is no mechanistic understanding of how chaperones suppress or modulate such phenotypes [90, 91]. In *Drosophila*, p-element insertion into the dHdj1 gene or knocking down HSPA1L expression enhanced polyglutamine-mediated toxicity [40, 41, 92]. Moreover, overexpression of Hsp70, Hdj1, and/or Hdj2 in *Drosophila*, mice, and cell culture models reduced polyglutamine aggregate toxicity (Table 36.1). Similar modulatory effects on polyglutamine proteins were also observed with elevated expression of Hsp104 in *S. cerevisiae*, *C. elegans*, and cell culture [34, 35, 93]. In contrast, overexpression of Hsp70 alone had inconsistent results on suppression of aggregate formation, whereas overexpression of co-chaperones, Hdj1 and Hdj2, yielded more-consistent results on reduction of both aggregation and toxicity (Table 36.1). Co-expression of Hsp70 and co-chaperones consistently exhibited synergistic effects on aggregation and toxicity [31, 94]. Taken together, these observations strongly support the ability of molecular chaperones to have potent effects on protein aggregation and toxicity while leaving unanswered questions on the mechanism and chaperone specificity.

There are several possible mechanisms by which molecular chaperones can modulate protein aggregation. The property of molecular chaperones to prevent aggregate formation is well established from a large number of in vitro studies in which the addition of molecular chaperones to heat- or chemically denatured substrates reduced aggregation and subsequently enhanced the refolding to the native state in an ATP-dependent manner [95, 96]. Similarly, in vitro studies using purified huntingtin exon-1 demonstrated that Hsc70 with Hdj1 prevented polyglutamine aggregate formation, thus supporting the ability of chaperones to prevent amyloidogenesis [94]. From these studies, it seems self-evident that the major effect of chaperone overexpression is to prevent aggregation. However, aggregation and toxicity were reversed by preventing further huntingtin expression, suggesting that chaperones can act on preformed aggregates to dissociate and presumably degrade them [97]. Are molecular chaperones responsible for dissociating preformed aggregates? A number of in vitro studies demonstrated that DnaK, the Hsp70 bacterial homologue, promotes disaggregation and reactivation of stable preformed protein aggregates both in vitro and in vivo [95, 98–100]. In addition, yeast Hsp104 and Hsp70 have also been shown to have similar effects on preformed aggregates [101, 102]. Whether mammalian molecular chaperones also have this capacity to dissociate preformed amyloids, however, remains to be demonstrated. There are

Tab. 36.1. Effects of molecular chaperones on polyglutamine-mediated aggregation and toxicity. Summary of the overexpressed molecular chaperones (as indicated) and their effects on different types of polyglutamine aggregates as determined by reduction of aggregation and/or toxicity. The list does not include studies from *S. cerevisiae* or *E. coli* models (n.a.: information was not available).

Family	*Overexpressed chaperones*	*Polyglutamine protein*	*Model system*	*Reduced aggregates*	*Reduced cell death*	*Reference*
Hsp60	apoGroEL	EGFP-HDQ74	COS7	+	+	93
	apoGroEL	EGFP-HDQ74	PC12	+	+	93
Hsp40	hdj1	HA-Q78	COS7	+	n.a.	29
	hdj1	NLS-ataxin3 (Q78)	COS7	+	n.a.	29
	hdj1	httEx1-Q74 (GFP)	Sk-N-SH	+	+	113
	hdj1	GFP-Q80	PC12	+	+	29
	hdj1	NLS-ataxin3 (Q78)	PC12	+	n.a.	29
	Hsp40 (hdj1)	120Q (HD exon1)	HEK293	+	+	111
	hdj1	tNhtt-150Q (GFP)	Neuro2a-ecdy	+	+	30
	dhdj1	127Q	*Drosophila* eye	–	+	92
	dhdj1	MJDtr-Q78	*Drosophila* eye	–	+	112
	hdj2	ataxin-1-GFP (Q92)	HeLa	+	n.a.	27
	hdj2	ARQ48 (GFP)	HeLa	+	n.a.	28
	hdj2	HA-Q78	COS7	+	n.a.	29
	hdj2	NLS-ataxin3 (Q78)	COS7	+	n.a.	29
	hdj2	GFP-Q80	PC12	–	n.a.	29
	hdj2	NLS-ataxin3 (Q78)	PC12	–	n.a.	29
	hdj2	tAR97-GFP	Neuro2a	–	–	31
Hsp70	Hsp70	HA-Q78	COS7	–	n.a.	29
	Hsp70	NLS-ataxin3 (Q78)	COS7	–	n.a.	29
	Hsp70	httEx1-Q74 (GFP)	COS7	–	–	113
	Hsp70	120Q (HD exon1)	HEK293	–	+	111
	Hsp70	tAR97-GFP	Neuro2a	+	+	31
	Hsp70	tNhtt-150Q (GFP)	Neuro2a-ecdy	–	n.a.	30
	Hsc70	tNhtt-150Q (GFP)	Neuro2a-ecdy	+	–	30
	Hsp70	httEx1-Q103 (GFP)	SK-N-SH	+	+	113
	HSPA1L (Hsp70)	MJDtr-Q78	*Drosophila* eye	–	+	40
	Hsp70	MJDtr-Q78	*Drosophila* eye	–	–	112
	Hsp70	SCA1	SCA1 mouse	–	+	114
Hsp70 & Hsp40	Hsp70 + Hsp40	tAR97-GFP	Neuro2a	+	+	31
	Hsc70 + hdj1	tNhtt-150Q (GFP)	Neuro2a-ecdy	–	+	30
	Hsp70 + hdj2	tAR97-GFP	Neuro2a	–	–	31
	Hsp70 + dhdj1	MJDtr-Q78	*Drosophila* eye	–	+	112
Hsp104	Hsp104	EGFP-HDQ74	COS7	+	+	93
	Hsp104	EGFP-HDQ74	PC12	+	+	93
	Hsp104	Q82-GFP	*C. elegans*	+	n.a.	35

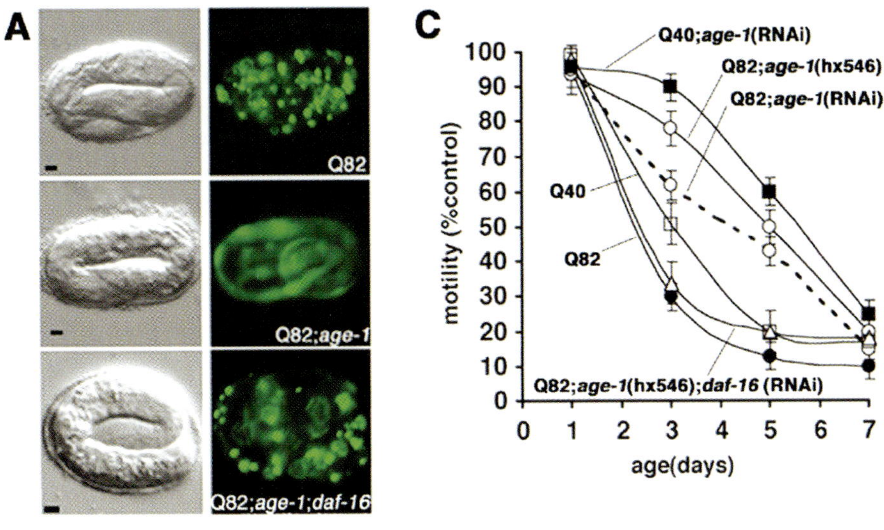

B

% of aggregates versus wild type background

	background	day 1	day 2	N
Q82 line 1	*age-1* (hx546)	55 ± 5	57 ± 4	45
	age-1 (RNAi)	69 ± 5	71 ± 3	60
Q82 line 2	*age-1* (hx546)	52 ± 3	48 ± 4	108
	daf-16 (RNAi)	95 ± 6	98 ± 8	90
	age-1 (hx546); *daf-16* (RNAi)	98 ± 4	98 ± 6	90
Q40	*age-1* (RNAi)	47 ± 7	38 ± 6	60

Fig. 36.4. An extended lifespan mutation delays polyglutamine aggregate accumulation and onset of toxicity. (A) DIC (left panels) and epifluorescence (right panels) micrographs showing embryos expressing Q82 in wild-type (top), *age-1* (hx546) (middle), or *age-1* (hx546); *daf-16* (RNAi) (bottom) genetic backgrounds. (B) Aggregate accumulation in larval animals expressing Q40 or Q82 in the indicated genetic backgrounds relative to aggregate accumulation in wild-type background (data are mean ± SEM). Aggregate formation in Q82-expressing embryo was inhibited in *age-1* background. The suppression of aggregation by *age-1* was dependent on *daf-16*. (C) Motility index for animals expressing Q40 or Q82 in the indicated genetic backgrounds (data are mean ± SEM for 30 animals of each type). Motility of non-transgenic wild-type and *age-1* animals was similar to that of wild-type animals (N2). The motility defects observed paralleled the degree of aggregate formation.

indications from an in vivo study on the association of human Hsp70 with polyglutamine aggregate that the chaperone may associate with nonnative proteins in the aggregate [62]. FRAP analysis of YFP-tagged Hsp70 showed transient interactions with aggregates and exhibited association kinetics identical to the interactions with unfolded substrates localized to the nucleolus during heat shock (Figure 36.3A, B). Furthermore, the association of Hsp70 with aggregates was dependent on chaperone activity, as mutations in Hsp70 that rendered the chaperone inactive prevented association with polyglutamine aggregates. This suggests that Hsp70 binding to misfolded polyglutamine proteins or other cellular proteins trapped in the polyglutamine aggregates reflects a productive intent to dislodge, refold, and/or target aggregate for degradation. Taken together, these observations suggest that aggregate formation reflects an imbalance in the kinetics among association, prevention, and dissociation by molecular chaperones.

36.6.3 Proteasomes

The proteasome is another central component in protein quality control that is essential for the clearance of damaged proteins and aggregates. Observations that proteasomes associate with polyglutamine aggregates suggest that proteasome activity could be involved in aggregate dissociation [27, 28, 103, 104]. In support of this hypothesis, inhibition of proteasome activity delayed aggregate clearance in tet-regulated huntingtin mice and in 293 tet-off cells [105, 106]. Conversely, the presence of aggregates results in the inhibition of proteasome activity [107, 108]. It is unclear whether there are distinct populations of soluble or aggregate-associated proteasomes and whether this would correspond to different proteasome activities. It seems likely that aggregate formation is a consequence of a competition between proteasome-dependent degradation of the misfolded proteins and the inhibitory effect of aggregates on proteasome activity.

36.6.4 The Protein-folding "Buffer" and Aging

The late age of onset observed in patients with neurodegenerative diseases seems to indicate a decline in the ability of a cell's "protein-folding buffer" to cope with persistent misfolded proteins. The observation of age-dependent aggregation is not limited to humans but can be recapitulated in a *C. elegans* polyglutamine model. The ability to visualize fluorescently tagged polyglutamine proteins in living, transparent animals has allowed monitoring of the age-dependent formation of polyglutamine aggregates in animals expressing a range of glutamines from 0Q to 82Q (Figure 36.1A) [61]. In this model, an age-dependent intensification of aggregate formation that paralleled a decline in motility was observed for all animals expressing polyglutamine above 29Qs.

What is the link between aging and the cell's protein-folding buffer? The relationship between the rate of aging and the progressive nature of polyglutamine-mediated phenotypes was observed in *C. elegans* expressing 82Q in the *age-1* (hx546) background, a mutation in the insulin-like signaling pathway that extends lifespan [61]. These mutant animals lived longer and exhibited delayed aggregate formation and motility defects relative to an 82Q-expressing animal in the wild-type background (Figure 36.4, see p. 1191). Both the delay in aggregate formation and the loss of motility in the *age-1* background were dependent on the activity of Daf-16 (a forkhead transcription factor known to act downstream of Age-1). These

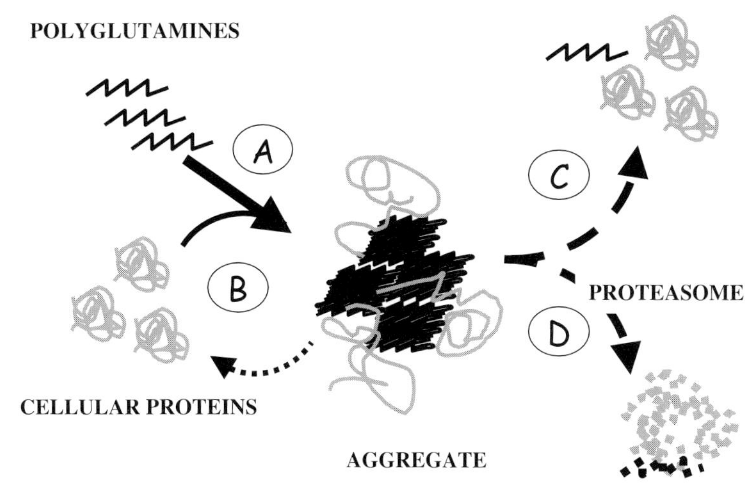

Fig. 36.5. Role of protein aggregation modifiers in reducing aggregate formation. Molecular modifiers can regulate different processes during protein aggregate formation and dissociation. Aggregate modifiers can prevent or delay oligomerization (A), prevent or delay interaction between the misfolded and aggregated proteins with other cellular proteins (B), alter the stability of the aggregate to facilitate dissociation by removing and refolding aggregating proteins along with the associated cellular proteins (C), or facilitate degradation of the components of aggregates (D).

observations indicate that the modulators of aging may also regulate factors involved in protein-folding homeostasis. Indeed, downregulation of Daf-16 leads to decreased expression of molecular chaperones [109]. Downregulation of Hsf-1, a master transcriptional regulator for protein-folding quality control, also causes premature aging and affects many of the same molecular chaperones that are involved in aging [110]. These findings suggest that the ability of cells to sense and respond adequately to the appearance of misfolded and aggregate-prone proteins diminishes with aging as a consequence of altered molecular chaperone levels [61].

36.6.5 Summary

A common denominator among the modulators of protein aggregation discussed in this review appears to be the ability to shield the interactive surfaces of the misfolding proteins from the rest of the cellular environment (Figure 36.5). Whether it is the protein context or protein homeostasis machinery, the prevention of intra- and inter-molecular interactions is a major requirement in the molecular events regulating protein aggregation. Molecular modifiers may also have the capacity to dissociate large aggregates. It may be that the same set of modulators is involved in both processes, as Hsp70 appears to function in both aggregate formation and dissolution. Almost certainly, there are other, still unidentified protein-folding modulators that can prevent aggregation by facilitating the refolding or degradation of misfolded proteins. In all cases, the ability to clear the cell of these noxious proteins may be the first line of defense against the appearance and persistence of such toxic entities. As we decipher how cells combat protein aggregates, the identification of other novel regulators of protein misfolding and aggregation that can be manipulated will be invaluable in understanding the course of protein-misfolding diseases.

36.7 Experimental Protocols

36.7.1 FRAP Analysis

Background

The ability to attach a fluorescent tag to a protein of interest, such as green fluorescent protein (GFP), allows researchers to study the protein in living cells without disrupting the cellular environment. A fluorescently tagged protein is useful in following the protein localization but is not limited to this, as mobility of the fluorescently tagged protein is also informative of its microenvironments. Protein mobility is affected by its size and shape, the viscosity of the environment, and interactions with organelles in the cell or with other proteins. Therefore, delay in mobility, compared to the predicted mobility, is a reflection of its biophysical state. For example, if a protein is bound to a cellular structure, its mobility is expected to be significantly slower than the soluble, unattached form.

One method that has become useful in deciphering the protein's mobility is fluorescence recovery after photobleaching (FRAP). In FRAP, a short pulse of intense laser light irreversibly destroys fluorescence of the molecules (photobleaching). The recovery of fluorescence in the photobleached area, therefore, is a result of diffusional exchange between the bleached and unbleached fluorescent molecules. By monitoring the time required for fluorescence recovery in the photobleached region, one can determine the speed at which the protein of interest travels from point A to point B. Therefore, we have used FRAP methodology to study mobility of the proteins associated with the polyglutamine aggregates to further understand the microenvironment of the co-associated protein as well as the polyglutamine aggregates themselves.

Microscope

FRAP analysis is usually done using a confocal microscope equipped with the ability to define a region of interest (ROI) for photobleaching and to acquire time series images at defined time intervals (preferably in milliseconds).

Software

Most commercially available confocal microscope operating software allows programming of the FRAP routines as macros, but some training is required. FRAP protocols using a specific operating software are usually available through the individual companies. Average fluorescent intensity can be determined using different imaging software, such as Metamorph (Universal Imaging Corp., Downingtown, PA) or Image J, which can be downloaded at no cost from http://rsb.info.nih.gov/ij/.

Sample Preparations

FRAP analysis was done with tissue cultured cells grown in a glass-bottom chamber (MatTreK Corp., Ashland, MA) or *C. elegans* mounted on a 2% agar pad glass slide and immobilized in 1 mM levamisole.

Controls for General Photobleaching

An important aspect in generating useful FRAP images is to include a control for general photobleaching. Although the general photobleaching is minimal with current confocal microscopes, especially since minimal laser power is used to scan the images, it is generally a good idea to have a non-photobleached sample. This control not only would determine the level of general photobleaching but also could be used to monitor lateral movements of the target cell during image acquisition. For this, the best control would be an adjacent cell expressing the fluorescent protein. However, depending on the transfection efficiency, it may be difficult to find an adjacent fluorescent cell. As a control, we tried an un-photobleached area of the aggregates as well as an adjacent cell (when possible) and found no significant differences. Therefore, we used the un-photobleached area next to the aggregates as control for our experiments.

Image Acquisition

The following protocol is outlined for FRAP analysis using a GFP-tagged sample; only a general description of the bleaching routine is outlined.

1. Scan a single image using a low laser power (at 488 nm) and save the image.
   The level of the laser power should be sufficiently low not to photobleach the sample and yet generate a good image. This condition will vary depending on the microscope. Scan the image near saturation but do not exceed it. (Prebleach)
2. Define the region of interest (ROI) for the photobleaching.
3. Set the laser power to maximum (100%) power. GFP (488 nm) and YFP (514 nm) lasers are better in photobleaching.
4. Scan the ROI (bleaching area) at maximum laser power until the fluorescence is no longer detectable.
   Optimal time for photobleaching can be determined using a fixed fluorescent sample and photobleaching at different durations (~3–10 s) until the fluorescence (only at ROI) is undetectable. Choose the minimal time for complete photobleaching since prolonged exposure to high-intensity laser can be harmful to the cell and also cause photobleaching outside of the ROI. (Bleach)
5. Reset the laser and the scan setting to the same condition as the pre-bleach condition.
6. Scan the image at desired intervals. This interval can range from microseconds to minutes depending on the mobility of the protein that is being observed.

Data Analysis

Determine the average fluorescence intensity of the area that was (1) subjected to photobleaching and (2) the adjacent control for each time point. Relative fluorescence intensity (RFI) is calculated using the following equation:

$$RFI = (Net/N1t)/(Ne0/N10)$$

where *Net* = average fluorescence intensity of the ROI (photobleached) area at time *x*;
N1t = average fluorescence intensity of the control (non-photobleached) area at time *x*;
Ne0 = average fluorescence intensity of the ROI at time zero (pre-bleaching);
N10 = average fluorescence intensity of the control area at time zero.

FRAP References

The following are excellent references for FRAP analysis:

Spector, D. L., Goldman, R. D., Leinwand, L. A. (1998) *Cells: a laboratory manual; Volume 2: Light Microscopy and Cell Structure.* pp. 79.1–79.23. Cold Spring Harbor Laboratory press.

Lippincott-Schwartz, J., Snapp, E., Kenworthy, A. (2001) Studying protein dynamics in living cells. *Nat. Rev. Mol. Cell. Biol*, 2,444–456.

References

1 Zoghbi, H. Y. & Orr, H. T. (2000). Glutamine repeats and neurodegeneration. *Annu Rev Neurosci* 23, 217–47.

2 Gusella, J. F. & MacDonald, M. E. (2000). Molecular genetics: unmasking polyglutamine triggers in neurodegenerative disease. *Nat Rev Neurosci* 1, 109–15.

3 Orr, H. T. (2001). Beyond the Qs in the polyglutamine diseases. *Genes Dev* 15, 925–32.

4 Nakamura, K., Jeong, S. Y., Uchihara, T., Anno, M., Nagashima, K., Nagashima, T., Ikeda, S., Tsuji, S. & Kanazawa, I. (2001). SCA17, a novel autosomal dominant cerebellar ataxia caused by an expanded polyglutamine in TATA-binding protein. *Hum Mol Genet* 10, 1441–8.

5 Andrew, S. E., Goldberg, Y. P., Kremer, B., Telenius, H., Theilmann, J., Adam, S., Starr, E., Squitieri, F., Lin, B. & Kalchman, M. A. (1993). The relationship between trinucleotide (CAG) repeat length and clinical features of Huntington's disease. *Nature genetics.* 4(4), 398–403.

6 Snell, R. G., MacMillan, J. C., Cheadle, J. P., Fenton, I., Lazarou, L. P., Davies, P., MacDonald, M. E., Gusella, J. F., Harper, P. S. & Shaw, D. J. (1993). Relationship between trinucleotide repeat expansion and phenotypic variation in Huntington's disease. *Nat Genet* 4, 393–7.

7 Duyao, M., Ambrose, C., Myers, R., Novelletto, A., Persichetti, F., Frontali, M., Folstein, S., Ross, C., Franz, M., Abbott, M. & et al. (1993). Trinucleotide repeat length instability and age of onset in Huntington's disease. *Nat Genet* 4, 387–92.

8 Wexler, N. S., Young, A. B., Tanzi, R. E., Travers, H., Starosta-

Rubinstein, S., Penney, J. B., Snodgrass, S. R., Shoulson, I., Gomez, F., Ramos Arroyo, M. A. & et al. (1987). Homozygotes for Huntington's disease. *Nature* 326, 194–7.

9 Duyao, M. P., Auerbach, A. B., Ryan, A., Persichetti, F., Barnes, G. T., McNeil, S. M., Ge, P., Vonsattel, J. P., Gusella, J. F., Joyner, A. L. & et al. (1995). Inactivation of the mouse Huntington's disease gene homolog Hdh. *Science* 269, 407–10.

10 Nasir, J., Floresco, S. B., O'Kusky, J. R., Diewert, V. M., Richman, J. M., Zeisler, J., Borowski, A., Marth, J. D., Phillips, A. G. & Hayden, M. R. (1995). Targeted disruption of the Huntington's disease gene results in embryonic lethality and behavioral and morphological changes in heterozygotes. *Cell* 81, 811–23.

11 Zeitlin, S., Liu, J. P., Chapman, D. L., Papaioannou, V. E. & Efstratiadis, A. (1995). Increased apoptosis and early embryonic lethality in mice nullizygous for the Huntington's disease gene homologue. *Nat Genet* 11, 155–63.

12 Matilla, A., Roberson, E. D., Banfi, S., Morales, J., Armstrong, D. L., Burright, E. N., Orr, H. T., Sweatt, J. D., Zoghbi, H. Y. & Matzuk, M. M. (1998). Mice lacking ataxin-1 display learning deficits and decreased hippocampal paired-pulse facilitation. *J Neurosci* 18, 5508–16.

13 Ambrose, H. J., Byrd, P. J., McConville, C. M., Cooper, P. R., Stankovic, T., Riley, J. H., Shiloh, Y., McNamara, J. O., Fukao, T. & Taylor, A. M. (1994). A physical map across chromosome 11q22-q23 containing the major locus for ataxia telangiectasia. *Genomics* 21, 612–9.

14 Davies, A. F., Mirza, G., Sekhon, G., Turnpenny, P., Leroy, F., Speleman, F., Law, C., van Regemorter, N., Vamos, E., Flinter, F. & Ragoussis, J. (1999). Delineation of two distinct 6p deletion syndromes. *Hum Genet* 104, 64–72.

15 Bates, G. P., Mangiarini, L., Mahal, A. & Davies, S. W. (1997). Transgenic models of Huntington's disease. *Hum Mol Genet* 6, 1633–7.

16 Rubinsztein, D. C. (2002). Lessons from animal models of Huntington's disease. *Trends Genet* 18, 202–9.

17 Marsh, J. L., Pallos, J. & Thompson, L. M. (2003). Fly models of Huntington's disease. *Hum Mol Genet* 12 Spec No 2, R187–93.

18 Faber, P. W., Alter, J. R., MacDonald, M. E. & Hart, A. C. (1999). Polyglutamine-mediated dysfunction and apoptotic death of a Caenorhabditis elegans sensory neuron. *Proc Natl Acad Sci USA* 96, 179–84.

19 Parker, J. A., Connolly, J. B., Wellington, C., Hayden, M., Dausset, J. & Neri, C. (2001). Expanded polyglutamines in Caenorhabditis elegans cause axonal abnormalities and severe dysfunction of PLM mechanosensory neurons without cell death. *Proc Natl Acad Sci USA* 98, 13318–23.

20 DiFiglia, M., Sapp, E., Chase, K. O., Davies, S. W., Bates, G. P., Vonsattel, J. P. & Aronin, N. (1997). Aggregation of huntingtin in neuronal intranuclear inclusions and dystrophic neurites in brain. *Science* 277, 1990–3.

21 Paulson, H. L., Perez, M. K., Trottier, Y., Trojanowski, J. Q., Subramony, S. H., Das, S. S., Vig, P., Mandel, J. L., Fischbeck, K. H. & Pittman, R. N. (1997). Intranuclear inclusions of expanded polyglutamine protein in spinocerebellar ataxia type 3. *Neuron* 19, 333–44.

22 Skinner, P. J., Koshy, B. T., Cummings, C. J., Klement, I. A., Helin, K., Servadio, A., Zoghbi, H. Y. & Orr, H. T. (1997). Ataxin-1 with an expanded glutamine tract alters nuclear matrix-associated structures. *Nature* 389, 971–4.

23 Holmberg, M., Duyckaerts, C., Durr, A., Cancel, G., Gourfinkel-An, I., Damier, P., Faucheux, B., Trottier, Y., Hirsch, E. C., Agid, Y. & Brice, A. (1998). Spinocerebellar ataxia type 7 (SCA7): a neurodegenerative disorder with neuronal intra-

nuclear inclusions. *Hum Mol Genet* 7, 913–8.

24 Hayashi, Y., Kakita, A., Yamada, M., Egawa, S., Oyanagi, S., Naito, H., Tsuji, S. & Takahashi, H. (1998). Hereditary dentatorubral-pallidoluysian atrophy: ubiquitinated filamentous inclusions in the cerebellar dentate nucleus neurons. *Acta Neuropathol (Berl)* 95, 479–82.

25 Li, M., Miwa, S., Kobayashi, Y., Merry, D. E., Yamamoto, M., Tanaka, F., Doyu, M., Hashizume, Y., Fischbeck, K. H. & Sobue, G. (1998). Nuclear inclusions of the androgen receptor protein in spinal and bulbar muscular atrophy. *Ann Neurol* 44, 249–54.

26 Ross, C. A. (1997). Intranuclear neuronal inclusions: a common pathogenic mechanism for glutamine-repeat neurodegenerative diseases? *Neuron* 19, 1147–50.

27 Cummings, C. J., Mancini, M. A., Antalffy, B., DeFranco, D. B., Orr, H. T. & Zoghbi, H. Y. (1998). Chaperone suppression of aggregation and altered subcellular proteasome localization imply protein misfolding in SCA1. *Nat Genet* 19, 148–54.

28 Stenoien, D. L., Cummings, C. J., Adams, H. P., Mancini, M. G., Patel, K., DeMartino, G. N., Marcelli, M., Weigel, N. L. & Mancini, M. A. (1999). Polyglutamine-expanded androgen receptors form aggregates that sequester heat shock proteins, proteasome components and SRC-1, and are suppressed by the HDJ-2 chaperone. *Hum Mol Genet* 8, 731–41.

29 Chai, Y., Koppenhafer, S. L., Bonini, N. M. & Paulson, H. L. (1999). Analysis of the role of heat shock protein (Hsp) molecular chaperones in polyglutamine disease. *J Neurosci* 19, 10338–47.

30 Jana, N. R., Tanaka, M., Wang, G. & Nukina, N. (2000). Polyglutamine length-dependent interaction of Hsp40 and Hsp70 family chaperones with truncated N-terminal huntingtin: their role in suppression of aggregation and cellular toxicity. *Hum Mol Genet* 9, 2009–18.

31 Kobayashi, Y., Kume, A., Li, M., Doyu, M., Hata, M., Ohtsuka, K. & Sobue, G. (2000). Chaperones Hsp70 and Hsp40 suppress aggregate formation and apoptosis in cultured neuronal cells expressing truncated androgen receptor protein with expanded polyglutamine tract. *J Biol Chem* 275, 8772–8.

32 Onodera, O., Roses, A. D., Tsuji, S., Vance, J. M., Strittmatter, W. J. & Burke, J. R. (1996). Toxicity of expanded polyglutamine-domain proteins in Escherichia coli. *FEBS Lett* 399, 135–9.

33 Meriin, A. B., Zhang, X., He, X., Newnam, G. P., Chernoff, Y. O. & Sherman, M. Y. (2002). Huntington toxicity in yeast model depends on polyglutamine aggregation mediated by a prion-like protein Rnq1. *J Cell Biol* 157, 997–1004.

34 Krobitsch, S. & Lindquist, S. (2000). Aggregation of huntingtin in yeast varies with the length of the polyglutamine expansion and the expression of chaperone proteins. *Proc Natl Acad Sci USA* 97, 1589–94.

35 Satyal, S. H., Schmidt, E., Kitagawa, K., Sondheimer, N., Lindquist, S., Kramer, J. M. & Morimoto, R. I. (2000). Polyglutamine aggregates alter protein folding homeostasis in Caenorhabditis elegans. *Proc Natl Acad Sci USA* 97, 5750–5755.

36 Cooper, J. K., Schilling, G., Peters, M. F., Herring, W. J., Sharp, A. H., Kaminsky, Z., Masone, J., Khan, F. A., Delanoy, M., Borchelt, D. R., Dawson, V. L., Dawson, T. M. & Ross, C. A. (1998). Truncated N-terminal fragments of huntingtin with expanded glutamine repeats form nuclear and cytoplasmic aggregates in cell culture. *Hum Mol Genet* 7, 783–90.

37 Moulder, K. L., Onodera, O., Burke, J. R., Strittmatter, W. J. & Johnson, E. M., Jr. (1999). Generation of neuronal intranuclear inclusions by polyglutamine-GFP: analysis of inclusion clearance and toxicity as a function of polyglutamine length. *J Neurosci* 19, 705–15.

38 Sato, A., Shimohata, T., Koide, R., Takano, H., Sato, T., Oyake, M., Igarashi, S., Tanaka, K., Inuzuka, T., Nawa, H. & Tsuji, S. (1999). Adenovirus-mediated expression of mutant DRPLA proteins with expanded polyglutamine stretches in neuronally differentiated PC12 cells. Preferential intranuclear aggregate formation and apoptosis. *Hum Mol Genet* 8, 997–1006.

39 Ordway, J. M., Tallaksen-Greene, S., Gutekunst, C. A., Bernstein, E. M., Cearley, J. A., Wiener, H. W., Dure, L. S. t., Lindsey, R., Hersch, S. M., Jope, R. S., Albin, R. L. & Detloff, P. J. (1997). Ectopically expressed CAG repeats cause intranuclear inclusions and a progressive late onset neurological phenotype in the mouse. *Cell* 91, 753–63.

40 Warrick, J. M., Chan, H. Y., Gray-Board, G. L., Chai, Y., Paulson, H. L. & Bonini, N. M. (1999). Suppression of polyglutamine-mediated neurodegeneration in Drosophila by the molecular chaperone HSP70. *Nat Genet* 23, 425–8.

41 Fernandez-Funez, P., Nino-Rosales, M. L., de Gouyon, B., She, W. C., Luchak, J. M., Martinez, P., Turiegano, E., Benito, J., Capovilla, M., Skinner, P. J., McCall, A., Canal, I., Orr, H. T., Zoghbi, H. Y. & Botas, J. (2000). Identification of genes that modify ataxin-1-induced neurodegeneration. *Nature* 408, 101–6.

42 Perutz, M. F. (1999). Glutamine repeats and neurodegenerative diseases: molecular aspects. *Trends Biochem Sci* 24, 58–63.

43 Perutz, M. F., Johnson, T., Suzuki, M. & Finch, J. T. (1994). Glutamine repeats as polar zippers: their possible role in inherited neurodegenerative diseases. *Proc Natl Acad Sci USA* 91, 5355–8.

44 Scherzinger, E., Lurz, R., Turmaine, M., Mangiarini, L., Hollenbach, B., Hasenbank, R., Bates, G. P., Davies, S. W., Lehrach, H. & Wanker, E. E. (1997). Huntingtin-encoded polyglutamine expansions form amyloid-like protein aggregates in vitro and in vivo. *Cell* 90, 549–58.

45 Chen, S., Berthelier, V., Yang, W. & Wetzel, R. (2001). Polyglutamine aggregation behavior in vitro supports a recruitment mechanism of cytotoxicity. *J Mol Biol* 311, 173–82.

46 Scherzinger, E., Sittler, A., Schweiger, K., Heiser, V., Lurz, R., Hasenbank, R., Bates, G. P., Lehrach, H. & Wanker, E. E. (1999). Self-assembly of polyglutamine-containing huntingtin fragments into amyloid-like fibrils: implications for Huntington's disease pathology. *Proc Natl Acad Sci USA* 96, 4604–9.

47 Chen, S., Berthelier, V., Hamilton, J. B., O'Nuallain, B. & Wetzel, R. (2002). Amyloid-like features of polyglutamine aggregates and their assembly kinetics. *Biochemistry* 41, 7391–9.

48 Tanaka, M., Machida, Y., Nishikawa, Y., Akagi, T., Hashikawa, T., Fujisawa, T. & Nukina, N. (2003). Expansion of polyglutamine induces the formation of quasi-aggregate in the early stage of protein fibrillization. *J Biol Chem* 278, 34717–24.

49 Thakur, A. K. & Wetzel, R. (2002). Mutational analysis of the structural organization of polyglutamine aggregates. *Proc Natl Acad Sci USA* 99, 17014–9.

50 Georgalis, Y., Starikov, E. B., Hollenbach, B., Lurz, R., Scherzinger, E., Saenger, W., Lehrach, H. & Wanker, E. E. (1998). Huntingtin aggregation monitored by dynamic light scattering. *Proc Natl Acad Sci USA* 95, 6118–21.

51 Tanaka, M., Morishima, I., Akagi, T., Hashikawa, T. & Nukina, N. (2001). Intra- and intermolecular beta-pleated sheet formation in glutamine-repeat inserted myoglobin as a model for polyglutamine diseases. *J Biol Chem* 276, 45470–5.

52 Poirier, M. A., Li, H., Macosko, J., Cai, S., Amzel, M. & Ross, C. A. (2002). Huntingtin spheroids and protofibrils as precursors in polyglutamine fibrilization. *J Biol Chem* 277, 41032–7.

53 Sanchez, I., Mahlke, C. & Yuan, J. (2003). Pivotal role of oligomerization in expanded polyglutamine neurodegenerative disorders. *Nature* 421, 373–9.

54 Sipe, J. D. & Cohen, A. S. (2000). Review: history of the amyloid fibril. *J Struct Biol* 130, 88–98.

55 Rochet, J. C. & Lansbury, P. T., Jr. (2000). Amyloid fibrillogenesis: themes and variations. *Curr Opin Struct Biol* 10, 60–8.

56 Taylor, J. P., Hardy, J. & Fischbeck, K. H. (2002). Toxic proteins in neurodegenerative disease. *Science* 296, 1991–5.

57 Temussi, P. A., Masino, L. & Pastore, A. (2003). From Alzheimer to Huntington: why is a structural understanding so difficult? *EMBO J* 22, 355–61.

58 Fandrich, M. & Dobson, C. M. (2002). The behaviour of polyamino acids reveals an inverse side chain effect in amyloid structure formation. *EMBO J* 21, 5682–90.

59 Chai, Y., Shao, J., Miller, V. M., Williams, A. & Paulson, H. L. (2002). Live-cell imaging reveals divergent intracellular dynamics of polyglutamine disease proteins and supports a sequestration model of pathogenesis. *Proc Natl Acad Sci USA*.

60 Stenoien, D. L., Mielke, M. & Mancini, M. A. (2002). Intranuclear ataxin1 inclusions contain both fast- and slow-exchanging components. *Nat Cell Biol* 4, 806–10.

61 Morley, J. F., Brignull, H. R., Weyers, J. J. & Morimoto, R. I. (2002). The threshold for polyglutamine-expansion protein aggregation and cellular toxicity is dynamic and influenced by aging in Caenorhabditis elegans. *Proc Natl Acad Sci USA* 99, 10417–22.

62 Kim, S., Nollen, E. A., Kitagawa, K., Bindokas, V. P. & Morimoto, R. I. (2002). Polyglutamine protein aggregates are dynamic. *Nat Cell Biol* 4, 826–31.

63 Rajan, R. S., Illing, M. E., Bence, N. F. & Kopito, R. R. (2001). Specificity in intracellular protein aggregation and inclusion body formation. *Proc Natl Acad Sci USA* 98, 13060–5.

64 Klement, I. A., Skinner, P. J., Kaytor, M. D., Yi, H., Hersch, S. M., Clark, H. B., Zoghbi, H. Y. & Orr, H. T. (1998). Ataxin-1 nuclear localization and aggregation: role in polyglutamine-induced disease in SCA1 transgenic mice. *Cell* 95, 41–53.

65 Saudou, F., Finkbeiner, S., Devys, D. & Greenberg, M. E. (1998). Huntingtin acts in the nucleus to induce apoptosis but death does not correlate with the formation of intranuclear inclusions. *Cell* 95, 55–66.

66 Sisodia, S. S. (1998). Nuclear inclusions in glutamine repeat disorders: are they pernicious, coincidental, or beneficial? *Cell* 95, 1–4.

67 Soto, C. (2003). Unfolding the role of protein misfolding in neurodegenerative diseases. *Nat Rev Neurosci* 4, 49–60.

68 Lambert, M. P., Barlow, A. K., Chromy, B. A., Edwards, C., Freed, R., Liosatos, M., Morgan, T. E., Rozovsky, I., Trommer, B., Viola, K. L., Wals, P., Zhang, C., Finch, C. E., Krafft, G. A. & Klein, W. L. (1998). Diffusible, nonfibrillar ligands derived from Abeta1-42 are potent central nervous system neurotoxins. *Proc Natl Acad Sci USA* 95, 6448–53.

69 Hartley, D. M., Walsh, D. M., Ye, C. P., Diehl, T., Vasquez, S., Vassilev, P. M., Teplow, D. B. & Selkoe, D. J. (1999). Protofibrillar intermediates of amyloid beta-protein induce acute electrophysiological changes and progressive neurotoxicity in cortical neurons. *J Neurosci* 19, 8876–84.

70 Pillot, T., Drouet, B., Queille, S., Labeur, C., Vandekerchkhove, J., Rosseneu, M., Pincon-Raymond, M. & Chambaz, J. (1999). The nonfibrillar amyloid beta-peptide induces apoptotic neuronal cell death: involvement of its C-terminal fusogenic domain. *J Neurochem* 73, 1626–34.

71 Zhu, Y. J., Lin, H. & Lal, R. (2000). Fresh and nonfibrillar amyloid beta protein(1-40) induces rapid cellular degeneration in aged human

fibroblasts: evidence for AbetaP-channel-mediated cellular toxicity. *Faseb J* 14, 1244–54.

72 Bucciantini, M., Giannoni, E., Chiti, F., Baroni, F., Formigli, L., Zurdo, J., Taddei, N., Ramponi, G., Dobson, C. M. & Stefani, M. (2002). Inherent toxicity of aggregates implies a common mechanism for protein misfolding diseases. *Nature* 416, 507–11.

73 Kayed, R., Head, E., Thompson, J. L., McIntire, T. M., Milton, S. C., Cotman, C. W. & Glabe, C. G. (2003). Common structure of soluble amyloid oligomers implies common mechanism of pathogenesis. *Science* 300, 486–9.

74 Perez, M. K., Paulson, H. L., Pendse, S. J., Saionz, S. J., Bonini, N. M. & Pittman, R. N. (1998). Recruitment and the role of nuclear localization in polyglutamine-mediated aggregation. *J Cell Biol* 143, 1457–70.

75 Kazantsev, A., Preisinger, E., Dranovsky, A., Goldgaber, D. & Housman, D. (1999). Insoluble detergent-resistant aggregates form between pathological and nonpathological lengths of polyglutamine in mammalian cells. *Proc Natl Acad Sci USA* 96, 11404–9.

76 Nagai, Y., Onodera, O., Chun, J., Strittmatter, W. J. & Burke, J. R. (1999). Expanded polyglutamine domain proteins bind neurofilament and alter the neurofilament network. *Exp Neurol* 155, 195–203.

77 Schmidt, T., Lindenberg, K. S., Krebs, A., Schols, L., Laccone, F., Herms, J., Rechsteiner, M., Riess, O. & Landwehrmeyer, G. B. (2002). Protein surveillance machinery in brains with spinocerebellar ataxia type 3: redistribution and differential recruitment of 26S proteasome subunits and chaperones to neuronal intranuclear inclusions. *Ann Neurol* 51, 302–10.

78 Li, S. H., Cheng, A. L., Li, H. & Li, X. J. (1999). Cellular defects and altered gene expression in PC12 cells stably expressing mutant huntingtin. *J Neurosci* 19, 5159–72.

79 Lin, X., Antalffy, B., Kang, D., Orr, H. T. & Zoghbi, H. Y. (2000). Polyglutamine expansion down-regulates specific neuronal genes before pathologic changes in SCA1 [see comments]. *Nat Neurosci* 3, 157–63.

80 Sugars, K. L. & Rubinsztein, D. C. (2003). Transcriptional abnormalities in Huntington disease. *Trends Genet* 19, 233–8.

81 Nucifora, F. C., Jr., Sasaki, M., Peters, M. F., Huang, H., Cooper, J. K., Yamada, M., Takahashi, H., Tsuji, S., Troncoso, J., Dawson, V. L., Dawson, T. M. & Ross, C. A. (2001). Interference by huntingtin and atrophin-1 with cbp-mediated transcription leading to cellular toxicity. *Science* 291, 2423–8.

82 Mantamadiotis, T., Lemberger, T., Bleckmann, S. C., Kern, H., Kretz, O., Martin Villalba, A., Tronche, F., Kellendonk, C., Gau, D., Kapfhammer, J., Otto, C., Schmid, W. & Schutz, G. (2002). Disruption of CREB function in brain leads to neurodegeneration. *Nat Genet* 31, 47–54.

83 DePace, A. H., Santoso, A., Hillner, P. & Weissman, J. S. (1998). A critical role for amino-terminal glutamine/asparagine repeats in the formation and propagation of a yeast prion. *Cell* 93, 1241–52.

84 Michelitsch, M. D. & Weissman, J. S. (2000). A census of glutamine/asparagine-rich regions: implications for their conserved function and the prediction of novel prions. *Proc Natl Acad Sci USA* 97, 11910–5.

85 Suhr, S. T., Senut, M. C., Whitelegge, J. P., Faull, K. F., Cuizon, D. B. & Gage, F. H. (2001). Identities of sequestered proteins in aggregates from cells with induced polyglutamine expression. *J Cell Biol* 153, 283–94.

86 Harjes, P. & Wanker, E. E. (2003). The hunt for huntingtin function: interaction partners tell many different stories. *Trends Biochem Sci* 28, 425–33.

87 Koshy, B., Matilla, T., Burright,

E. N., Merry, D. E., Fischbeck, K. H., Orr, H. T. & Zoghbi, H. Y. (1996). Spinocerebellar ataxia type-1 and spinobulbar muscular atrophy gene products interact with glyceraldehyde-3-phosphate dehydrogenase. *Hum Mol Genet* 5, 1311–8.

88 Chen, H. K., Fernandez-Funez, P., Acevedo, S. F., Lam, Y. C., Kaytor, M. D., Fernandez, M. H., Aitken, A., Skoulakis, E. M., Orr, H. T., Botas, J. & Zoghbi, H. Y. (2003). Interaction of Akt-phosphorylated ataxin-1 with 14-3-3 mediates neurodegeneration in spinocerebellar ataxia type 1. *Cell* 113, 457–68.

89 Emamian, E. S., Kaytor, M. D., Duvick, L. A., Zu, T., Tousey, S. K., Zoghbi, H. Y., Clark, H. B. & Orr, H. T. (2003). Serine 776 of ataxin-1 is critical for polyglutamine-induced disease in SCA1 transgenic mice. *Neuron* 38, 375–87.

90 Opal, P. & Zoghbi, H. Y. (2002). The role of chaperones in polyglutamine disease. *Trends Mol Med* 8, 232–6.

91 Muchowski, P. J. (2002). Protein misfolding, amyloid formation, and neurodegeneration: a critical role for molecular chaperones? *Neuron* 35, 9–12.

92 Kazemi-Esfarjani, P. & Benzer, S. (2000). Genetic suppression of polyglutamine toxicity in Drosophila. *Science* 287, 1837–40.

93 Carmichael, J., Chatellier, J., Woolfson, A., Milstein, C., Fersht, A. R. & Rubinsztein, D. C. (2000). Bacterial and yeast chaperones reduce both aggregate formation and cell death in mammalian cell models of Huntington's disease. *Proc Natl Acad Sci USA* 97, 9701–5.

94 Muchowski, P. J., Schaffar, G., Sittler, A., Wanker, E. E., Hayer-Hartl, M. K. & Hartl, F. U. (2000). Hsp70 and hsp40 chaperones can inhibit self-assembly of polyglutamine proteins into amyloid-like fibrils. *Proc Natl Acad Sci USA* 97, 7841–6.

95 Ben-Zvi, A. P. & Goloubinoff, P. (2001). Review: mechanisms of disaggregation and refolding of stable protein aggregates by molecular chaperones. *J Struct Biol* 135, 84–93.

96 Hartl, F. U. & Hayer-Hartl, M. (2002). Molecular chaperones in the cytosol: from nascent chain to folded protein. *Science* 295, 1852–8.

97 Yamamoto, A., Lucas, J. J. & Hen, R. (2000). Reversal of neuropathology and motor dysfunction in a conditional model of Huntington's disease. *Cell* 101, 57–66.

98 Skowyra, D., Georgopoulos, C. & Zylicz, M. (1990). The E. coli dnaK gene product, the hsp70 homolog, can reactivate heat-inactivated RNA polymerase in an ATP hydrolysis-dependent manner. *Cell* 62, 939–44.

99 Diamant, S., Ben-Zvi, A. P., Bukau, B. & Goloubinoff, P. (2000). Size-dependent disaggregation of stable protein aggregates by the DnaK chaperone machinery. *J Biol Chem* 275, 21107–13.

100 Mogk, A., Tomoyasu, T., Goloubinoff, P., Rudiger, S., Roder, D., Langen, H. & Bukau, B. (1999). Identification of thermolabile *Escherichia coli* proteins: prevention and reversion of aggregation by DnaK and ClpB. *EMBO J* 18, 6934–49.

101 Parsell, D. A., Kowal, A. S., Singer, M. A. & Lindquist, S. (1994). Protein disaggregation mediated by heat-shock protein Hsp104. *Nature* 372, 475–8.

102 Glover, J. R. & Lindquist, S. (1998). Hsp104, Hsp70, and Hsp40: a novel chaperone system that rescues previously aggregated proteins. *Cell* 94, 73–82.

103 Chai, Y., Koppenhafer, S. L., Shoesmith, S. J., Perez, M. K. & Paulson, H. L. (1999). Evidence for proteasome involvement in polyglutamine disease: localization to nuclear inclusions in SCA3/MJD and suppression of polyglutamine aggregation in vitro. *Hum Mol Genet* 8, 673–82.

104 Wyttenbach, A., Carmichael, J., Swartz, J., Furlong, R. A., Narain, Y., Rankin, J. & Rubinsztein, D. C. (2000). Effects of heat shock, heat shock protein 40 (HDJ-2), and proteasome inhibition on protein aggregation in cellular models of Huntington's disease. *Proc Natl Acad Sci USA* 97, 2898–903.

105 Martin-Aparicio, E., Yamamoto, A., Hernandez, F., Hen, R., Avila, J. & Lucas, J. J. (2001). Proteasomal-dependent aggregate reversal and absence of cell death in a conditional mouse model of Huntington's disease. *J Neurosci* 21, 8772–81.

106 Waelter, S., Boeddrich, A., Lurz, R., Scherzinger, E., Lueder, G., Lehrach, H. & Wanker, E. E. (2001). Accumulation of mutant huntingtin fragments in aggresome-like inclusion bodies as a result of insufficient protein degradation. *Mol Biol Cell* 12, 1393–407.

107 Bence, N. F., Sampat, R. M. & Kopito, R. R. (2001). Impairment of the ubiquitin-proteasome system by protein aggregation. *Science* 292, 1552–5.

108 Verhoef, L. G., Lindsten, K., Masucci, M. G. & Dantuma, N. P. (2002). Aggregate formation inhibits proteasomal degradation of polyglutamine proteins. *Hum Mol Genet* 11, 2689–700.

109 Hsu, A. L., Murphy, C. T. & Kenyon, C. (2003). Regulation of aging and age-related disease by DAF-16 and heat-shock factor. *Science* 300, 1142–5.

110 Morley, J. F. & Morimoto, R. I. (2004). Regulation of longevity in Caenorhabditis elegans by heat shock factor and molecular chaperones. *Mol Biol Cell* 15, 657–64.

111 Zhou, H., Li, S. H. & Li, X. J. (2001). Chaperone suppression of cellular toxicity of huntingtin is independent of polyglutamine aggregation. *J Biol Chem* 276, 48417–24.

112 Chan, H. Y., Warrick, J. M., Gray-Board, G. L., Paulson, H. L. & Bonini, N. M. (2000). Mechanisms of chaperone suppression of polyglutamine disease: selectivity, synergy and modulation of protein solubility in Drosophila. *Hum Mol Genet* 9, 2811–20.

113 Wyttenbach, A., Sauvageot, O., Carmichael, J., Diaz-Latoud, C., Arrigo, A. P. & Rubinsztein, D. C. (2002). Heat shock protein 27 prevents cellular polyglutamine toxicity and suppresses the increase of reactive oxygen species caused by huntingtin. *Hum Mol Genet* 11, 1137–51.

114 Cummings, C. J., Sun, Y., Opal, P., Antalffy, B., Mestril, R., Orr, H. T., Dillmann, W. H. & Zoghbi, H. Y. (2001). Over-expression of inducible HSP70 chaperone suppresses neuropathology and improves motor function in SCA1 mice. *Hum Mol Genet* 10, 1511–8.

37
Protein Folding and Aggregation in the Expanded Polyglutamine Repeat Diseases

Ronald Wetzel

37.1
Introduction

There are at least 15 human diseases associated with the genetic expansion of a chromosomal trinucleotide repeat sequence [1–3]. Some of these occur in noncoding regions, while others are located in open reading frames coding for a variety of proteins. Most of the diseases associated with expressed protein repeats involve CAG repeats coding for polyglutamine (polyGln). The common feature of DNA that underlies the genetic nature of these diseases is the genetic instability of triplet repeats of some of the 64 possible DNA nucleotide triplets [1].

Currently there are nine known diseases involving expansion of a polyGln sequence [3]. While these diseases on average vary one from the other in details of brain pathology and symptoms [3], there are also significant similarities. The polyGln repeat length threshold separating the normal population from at-risk individuals is strikingly similar for eight of the nine diseases [3]. All nine diseases are progressive and uniformly lethal and tend to present in midlife [3]. Of particular significance to this chapter, polyGln aggregates have been demonstrated in each of these diseases [3].

The demonstration of polyGln-containing inclusions in patient material [4] and in animal and cellular models [5–7] also suggested their resemblance to other neurodegenerative diseases [8] involving protein deposition into amyloid or other aggregated deposits. As a subject for studying protein aggregation diseases at the molecular level, however, the polyGln diseases have pluses and minuses. On the one hand, the uniform monotony of glutamine that constitutes the pathogenic core of the disease proteins does not inspire immediate insights into disease mechanisms or protein-folding mechanisms. Indeed, standard site-directed mutagenesis approaches are technically difficult at the DNA level, due to the lack of unique internal restriction sites [9], and in any case may be of questionable value at the protein level, due to the likely ability of the sequence to shift and slide while aggregating so as to accommodate, and negate the effect of, potential modifying mutations [10]. On the other hand, there is something attractive about the simplicity of a homopolymer as a subject for addressing fundamental problems in protein folding and

Protein Folding Handbook. Part II. Edited by J. Buchner and T. Kiefhaber

ISBN: 3-527-30784-2

aggregation; for example, studying a real-world homopolymer should provide data more easily related to theoretical protein-folding studies that are often based on highly simplified model structures. A significant additional attraction to studying these diseases is the prospect of contributing to the development of therapies. As summarized below, protein-folding and aggregation studies of these sequences in vitro are making significant contributions to a better understanding of the molecular basis of these diseases and along the way are providing unique systems for better characterizing the role of protein structure in ordered aggregation processes.

37.2
Key Features of the Polyglutamine Diseases

The literature on the clinical, genetic, cellular, and biochemical aspects of the expanded polyGln diseases has grown enormously since the discovery of the disease genes over the past decade. There are excellent books reviewing the various aspects of each disease [11, 12], and a steady stream of review articles and essays discuss possible disease mechanisms from various points of view [2, 13–31]. These reviews reflect the lively, ongoing debate on the importance of polyGln oligomers and aggregates in the disease mechanism. The discussion here is very brief and necessarily centered on the toxicity of polyGln peptides, especially the possibility that polyGln aggregates play a toxic role.

37.2.1
The Variety of Expanded PolyGln Diseases

There are currently nine human diseases linked to a genetic expansion in an in-frame, translated CAG repeat [3], of which the most recently identified is spinocerebellar ataxia 17, involving expansions in the transcription factor TBP, or TATA-binding protein [32]. Table 37.1 lists the known diseases, the disease protein involved, its molecular weight and subcellular localization, and the CAG repeat lengths for the normal and pathological ranges. The most well known and most common of these diseases is Huntington's disease (HD). There are a number of genetic ataxias whose disease gene has not yet been identified [33, 34], and at least some of these may eventually be found to also involve CAG repeat expansion.

37.2.2
Clinical Features

Except for the X-linked spinal and bulbar muscular atrophy (SBMA), involving CAG repeat expansions in the gene for the androgen receptor, the other eight known polyGln diseases exhibit autosomal dominant genetics [3]. Huntington's disease normally presents as a motor disorder, although a significant minority of patients present with psychiatric symptoms [12]. The expanded CAG repeat ataxias involve a wide range of symptoms and significant variability [3]. Early-onset forms

Tab. 37.1. Overview of expanded CAG repeat diseases.[a]

Disease	*Gene product*	*Cellular localization*	*MW (kDa)*	*Normal CAG repeat range*	*Mutant CAG repeat range*
Huntington's	huntingtin	Cytoplasm/nucleus	348	6–39	36–>200
DRPLA[b]	atrophin-1	Cytoplasm/nucleus	190	3–35	49–88
SBMA[c]	androgen rec.	Cytoplasm/nucleus	104	9–33	38–65
SCA1[d]	ataxin-1	Cytoplasm/nucleus	87	6–44	39–83
SCA2[d]	ataxin-2	Cytoplasm	145	13–33	32–>200
SCA3[d]/MJD[e]	ataxin-3	Cytoplasm/nucleus	46	3–40	54–89
SCA6[d]	CACNA1A[f]	Cytoplasm	280	4–19	20–33
SCA7[d]	ataxin-7	Cytoplasm/nucleus	95	4–35	37–306
SCA17[d]	TBP[g]	Nucleus	38	24–44	46–63

[a] Data from Ref. [12].
[b] Dentatorubral-pallidoluysian atrophy
[c] Spinal and bulbar muscular atrophy
[d] Spinocerebellar ataxia
[e] Machado-Joseph Disease
[f] α_{1A} voltage-dependent calcium channel
[g] TATA box-binding protein

tend to present with distinct, more-aggressive symptoms that set them apart from the later-onset forms. The large variations in the course of disease for patients with the same CAG repeat length suggest important roles for secondary factors, including environment and modifying genes, and these are currently under investigation in those diseases with statistically significant amounts of patient data.

37.2.2.1 **Repeat Expansions and Repeat Length**

DNA triplet expansions leading to genetic disease can occur in expressed open reading frames or in non-coding regions [1]. Of the possible triplet repeat sequences, only a few are known to be involved in triplet expansion genetic disorders [1]. Although it is presumed that there are unique structural aspects to the DNA of those triplets whose instability during one or more enzymatic process in the cell leads to expansion-based diseases, the structural basis of repeat instability is unclear and continues to be evaluated [35]. The stability of the CAG repeat, in particular, appears to depend on a number of factors, including the length of the repeat before expansion, its sequence context, and whether the donor of that gene is the mother or the father [3, 36]. The normal role of benign repeat lengths of the polyGln sequence in the various proteins in which it is found remains a mystery. In fact, since homologues of the same protein often vary considerably in their polyGln repeat length, while other parts of the protein may show very little variation, it is possible that polyGln in most cases is not functionally important. The hypothesis that normal-length polyGln sequences serve no important function that might be compromised upon sequence expansion is consistent with the widely accepted view that expanded CAG repeat diseases are predominantly gain-of-function disorders [3].

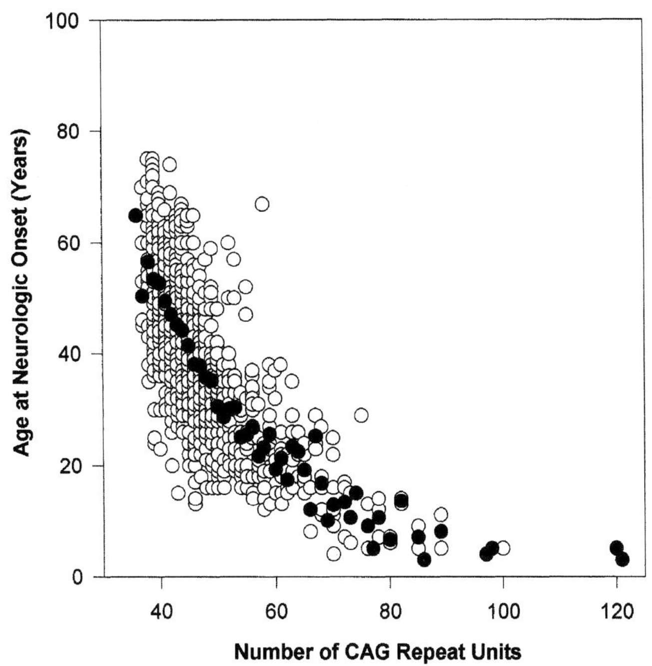

Fig. 37.1. *HD* CAG repeat length is correlated with the age at onset of neurologic symptoms. The relationship between the expanded *HD* CAG repeat length ("Number of CAG Repeat Units") and the age at neurologic onset is given for 1070 HD patients reported in the literature. The mean age at onset for any given *HD* CAG repeat length is depicted by filled symbol. Power regression analysis reveals a significant inverse correlation between expanded CAG repeat size and age at onset ($r = -0.87$, $P < 0.0001$), although individuals with the same expanded repeat exhibit widely different ages at onset, suggesting the existence of modifiers of the disease process. Figure and analysis courtesy of Marcy MacDonald.

Table 37.1 shows the clear separation of benign repeat lengths from pathological repeat lengths in the expanded CAG repeat diseases. In all cases but one (SCA 6, which uniquely involves a resident transmembrane channel protein), the pathological cutoff lies in the mid- to upper-30s. In HD, within this critical region, repeat length also influences penetrance, the percentage of individuals with the mutation that eventually develop disease symptoms [37]. Within the pathological repeat length range, age of onset correlates strongly with repeat length, as shown for HD in Figure 37.1. One of the clinical characteristics of the disease, recognized prior to the discovery of the disease genes, is that of anticipation, in which a parent presents with a disease only after it is already recognized in a child. With the discovery of the dynamic genetic nature of the triplet repeat diseases, anticipation is now understood by the fact that repeat expansions, especially when passed on by the father, can be large, and large repeat expansions tend to dramatically decrease age of onset (Figure 37.1) [1].

37.2.3
The Role of PolyGln and PolyGln Aggregates

Although a few genetic ataxias involve CAG expansions in non-coding regions [33, 34], the bulk of the characterized disorders are associated with expression of the polyGln sequence at the protein level. Loss of function may play a role in some mutations in the gene associated with SCA 6 [3]. Further, a decline in normal levels of active protein, perhaps mediated by aggregation, may also contribute to symptoms in some expanded CAG repeat diseases [38, 39]. On the whole, however, the dominant effect of polyGln expansion appears to be a gain of toxic function. It seems very unlikely that the expanded polyGln disorders, with their often overlapping clinical features and very similar genetics, do not share intimate details of their molecular mechanisms, and it is further unlikely that the molecular mechanisms involving these otherwise unrelated protein sequences are not centered in the behavior of the polyGln sequence. This is further supported by the generation of animal models based on expression of artificial proteins containing in-frame CAG repeats spliced to proteins not known to be associated with disease [5, 40].

Opinion in the field has been divided as to whether expanded polyGln is cytotoxic due to the induction, within the soluble monomer, of a toxic conformation or, alternatively, to the accelerated aggregation of the monomer into a toxic, oligomeric state. Although the ability of a series of antibodies to bind preferentially to expanded polyGln has been interpreted as evidence for the existence of a novel conformation within the monomer, recent studies suggest that the increased avidity observed for such antibodies can be explained by the multiplicity of independent, oligoGln-binding sites in the expanded sequence (a linear lattice effect) [41]. Most other data (reviewed in Chapter 3) is consistent with no significant change in the conformation of bulk-phase polyGln upon its sequence expansion.

In contrast, substantial evidence points to a central role for polyGln aggregation in the disease process [25, 27–29], beginning with the observation of intraneuronal, nuclear inclusions (NII) in cell and animal models of polyGln disease [5–7]. Similar aggregates are found in brain tissue from HD and other diseases listed in Table 37.1, but not in normal, aged control brain [3, 4]. Genetically identified suppressers of polyGln toxicity, such as PQE1 [42] and dHDJ1 [40], have been shown to be aggregation suppressors in vitro (see Section 37.8). Further, designed polypeptide-based inhibitors of polyGln aggregation also protect cells from the toxic effects of expanded polyGln sequences [43] and polyGln aggregates [44] (see Section 37.8.1). As discussed in detail below, the repeat length–dependent aggregation properties of polyGln closely mirror the repeat-length dependence of disease risk shown in Table 37.1. Also as discussed below, polyGln aggregates made in vitro and delivered to the nuclei of cells in culture are very cytotoxic [45]. Although the appearance of visible aggregates does not always closely correlate with cytotoxicity in cell and animal models [2, 17, 21], there are reasonable explanations for how such observations might still be consistent with cytotoxic aggregates. Some of these rationales will be alluded to in this chapter.

37.3 PolyGln Peptides in Studies of the Molecular Basis of Expanded Polyglutamine Diseases

If the effect of sequence expansion on the physicochemical properties of the polyGln sequence plays a central role in the expanded CAG repeat diseases, then it is of particular importance to study polyGln behavior with respect to repeat length in well-defined biophysical experiments. Chapter 3 consists of an overview of studies using either chemically synthesized peptides or recombinant proteins to conduct studies on the conformational preferences of soluble polyGln and on the aggregation efficiency of these sequences. Later chapters expand into details of certain aspects of their biophysical behavior, concentrating on aggregation. The poor solution behavior of the polyGln sequence has provided substantial barriers to conducting such studies, and much has been learned simply in the effort to produce molecules amenable to analysis.

37.3.1 Conformational Studies

Initial studies on the favored conformation of the polyGln sequence gave conflicting results. Working with chemically synthesized polyGln sequences well below the pathological threshold repeat length, Altschuler et al. found the sequence to exist in random coil [46], while Sharma et al. found peptides of similar length to be dominated by β-sheet [47]. This discrepancy was resolved with the later realization, as discussed in Chapter 4, of the importance of applying rigorous methods of peptide disaggregation as part of the solubilization protocol [48]. Using such protocols, Chen et al. showed that polyGln sequences in solution in PBS exhibit the signature CD spectrum of a statistical coil [49]. More importantly, they also showed that the CD spectra of polyGln peptides both below and above the pathological threshold repeat length give essentially identical CD spectra [49], providing the first experimental indication that a pure conformational change within a monomeric polyGln protein is not likely to underlie the repeat-length dependence of clinical features. Similar results were later obtained with various polyGln repeat lengths within the context of *E. coli*-expressed fusion proteins containing either naked polyGln sequences [50] or polyGln sequences embedded in a fragment of *huntingtin* [41]. Importantly, these studies also show that results obtained with synthetic peptides, such as those described here, are not compromised by the disaggregation protocol.

The above discussion notwithstanding, the CD spectrum generally assigned to the random coil state is not unambiguous. In fact, a recent study has confirmed earlier suggestions that polypeptides exhibiting the random coil CD spectrum might actually exist in the ordered conformation known as polyproline type II helix [51]. Indeed, recent studies suggest that the amino acid glutamine is particularly comfortable in this form of secondary structure [52]. While the exact nature of the conformation of monomeric polyGln may ultimately be of significant importance to understanding disease mechanisms, the main important lesson to date from solution studies is that there appears to be no difference between the solution structures of polyGln sequences below and above the pathological repeat length range. Absent some more subtle conformational difference between short and long polyGln, some other aspect of the polyGln sequence must underlie the repeat-length dependence of disease risk.

37.3.2
Preliminary in Vitro Aggregation Studies

With the revelation that expanded polyGln sequences are responsible for the diseases listed in Table 37.1, and in light of the importance of protein aggregation in other neurodegenerative diseases such as Alzheimer's disease, Perutz speculated that polyGln expansion might lead to an increased propensity of proteins containing these sequences to aggregate and thereby cause disease [53]. Working with a Q_{15} sequence flanked by pairs of charged residues, the Perutz lab reported facile aggregation of the peptide in aqueous solution to yield aggregates exhibiting the cross-β structure characteristic of amyloid [53]. In analogy to the leucine zippers designed by nature to promiscuously self-associate in the creation of dimeric transcription factors, Perutz coined the term "polar zipper" for the tendency of polyGln sequences to aggregate, presenting a model in which hydrogen bonding by side chain amide groups contributes to the stability of the aggregate [53].

Scherzinger et al. were able to overcome many of the limitations encountered in the above preliminary studies by using a recombinantly expressed form of glutathione S-transferase (GST) fused with the exon 1 fragment of *huntingtin* containing the polyGln sequence [54, 55]. Fusion of GST with *htt* exon 1 renders the protein reasonably well behaved so that it can be extracted and partially purified while maintaining it in a soluble form. By the ingenious use of an installed protease site, Scherzinger et al. were able to cleave this soluble protein between its GST and *htt* components in a controlled reaction, releasing the polyGln portion in situ so that its aggregation could be monitored in a synchronized reaction [54]. By doing so, and by following the generation of aggregate using a filter trap assay, they were able to observe a clear increase in the aggressiveness of spontaneous aggregation as polyGln repeat length increased, with a significant increase in aggregation rate as repeat length increased through the pathological cutoff range [54, 55]. This correspondence between the repeat length threshold for disease risk and the threshold for rapid aggregation continues to be one of the most striking features of polyGln behavior in solution, and it has been invoked as supporting evidence for hypotheses involving aggregation in the disease mechanism.

37.3.3
In Vivo Aggregation Studies

In addition to contributing to our understanding of the disease mechanism, cell and animal models involving recombinant expression of polyGln-containing disease proteins have contributed greatly to our knowledge of the aggregation process itself. The dependence of aggregation efficiency on polyGln repeat length was characterized qualitatively in a number of in vivo experiments comparing a limited variety of repeat length constructs [6, 7, 54, 56]. At the same time, it is also possible for non-pathological repeat length polyGlns to form aggregates under certain conditions in the cell [57], emphasizing that even sequences below the pathological threshold have some potential to aggregate. Transgene experiments have implicated a modulating role for flanking sequences and/or flanking domains in the aggregation of expanded polyGln, consistent with the possibility that proteolytic processing may play an important role in controlling aggregation and toxicity in the cell. For example, mice expressing exon 1–like fragments of *htt* tend to develop aggregates (and pathology) more rapidly and aggressively than do mice expressing full-length *htt* [26]. Besides the mammalian systems mentioned here, a number of other organisms, such as yeast [58], *D. melanogaster* and *C. elegans* [59], have been used to learn details about how polyglutamine aggregates in the cellular environment.

The promiscuity of polyGln aggregation was noted early on in the observation that an aggregate initiated by an expanded polyGln sequence is capable of recruiting the normal-length polyGln protein into the deposit [57]. More recently, recruitment of the normal allele of *htt* by expanded polyGln htt aggregates has also been observed, suggesting that expanded CAG repeat diseases may include an overlay of loss of function in the disease mechanism [39]. The possible role of generalized polyGln aggregation in the cell was also suggested by the observation that expression of expanded polyGln containing a nuclear localization signal (NLS) could mediate nuclear localization of green fluorescent protein (GFP) linked to a normal-length polyGln [60]. One of the intriguing possible mechanisms of polyGln cytotoxicity is that cellular polyGln aggregates might deplete the soluble cellular pools of other polyGln-containing factors by recruiting them into the aggregate in an ongoing elongation reaction [57, 61, 62]. Such mechanisms are supported by the observation of co-localization of TBP [57, 61] and CBP (CREB-binding protein) [60, 63] to cellular aggregates in tissue from patients and/or from cell and animal models.

37.4
Analyzing Polyglutamine Behavior With Synthetic Peptides: Practical Aspects

Given the limited success of previous studies of polyGln behavior using synthetic peptides and the comparative success of studies using recombinant proteins in vitro and in vivo (see Chapter 3), one might legitimately question the value of putting further efforts into the chemical approach. There are a number of reasons for doing so, however.

While polypeptides, even unnatural fusions, expressed in a cellular environment might be considered to be "more native" than chemically synthesized peptides exposed, even transiently, to organic solvents, the situation is complicated by the central role of the prior aggregation state of a protein in biasing its aggregation behavior in vitro. This is illustrated by early experiments in the Alzheimer's disease field, in which studies of the amyloid peptide Aβ were confused by investigators' ignorance of the aggregation state of the peptide and the importance it plays in cytotoxicity and aggregation. This led to an odd lore concerning certain manufacturers' batches of synthetic peptide as being "good" (cytotoxic right out of the vial) or "bad" (nontoxic) batches. Eventually, both cytotoxicity [64] and aggressive aggregation ability [65] were clearly shown to be attributable to traces of aggregates resident in the vialed material. As shown below, methods now exist for rigorously removing preexisting aggregates from synthetic peptides. Such methods are chemically benign, allow complete removal of the organic solvent, and are compatible with peptides whose natural state as a soluble monomer in native buffer is relatively unstructured. In contrast, recombinant proteins are potentially much more complex and, once nonnative assembly occurs, the unnatural oligomeric state may be much more difficult to recognize and reverse. For example, expression of a GST-AT-3 fusion containing a Q_{27} sequence led to an enriched, soluble protein that, however, migrated as an oligomer on native gel filtration [66]; similar soluble oligomers can be observed in isolation of *htt* exon-1 fusion proteins (G. Thiagaragan and R. Wetzel, unpublished). Clearly, it is possible, at least with relatively short polyGln repeats, to prepare monomeric, folded proteins by recombinant expression [41], but a great level of care and analytical scrutiny is required to insure that the preparation is aggregate-free, and the challenges increase dramatically with increasing repeat length. Thus, there may be situations in which chemically synthesized peptides, exposed transiently to non-aqueous solvents to effect efficient disaggregation, are "more native" than a recombinant protein prepared and maintained in aqueous buffers.

An additional attractive feature of focusing on chemically synthesized polyGln peptides is that optical studies are not complicated by the presence of other segments of polypeptide. For example, this allows relatively clean interpretation of the CD spectra of subject polypeptides without having to be concerned about the contributions of other domains and how those other domains may or may not also be changing during the experiment. Work with synthetic peptides also allows facile use of chemical tags such as biotin [67] and fluorescein [45], as well as nonnatural amino acids such as D-amino acids [10]. Synthetic peptides often make it more possible to conduct studies requiring relatively large amounts of material and to accomplish the studies in a shorter time.

Finally, while it is clear that the most appropriate studies, if they can be done on pristine protein samples, are those that are chemically and physically most similar to the toxic fragments generated in vivo, it is also true that studies on simple polyGln peptides will allow baselines to be established for their biophysical behavior, so that further studies on the sequence in the natural settings of the disease proteins can be placed into context and the specific effects of sequence context be cleanly appreciated.

37.4.1 Disaggregation of Synthetic Polyglutamine Peptides

In the review of the aggregation literature in Section 37.3.2, one troubling inconsistency stands out. Work with recombinant protein fragments yields results on the repeat-length dependence of aggregation that mirror the repeat-length dependence of disease risk in HD and other expanded CAG diseases, with sequences up to about Q_{35} being relatively resistant to aggregation. In contrast, however, previous work with chemically synthesized peptides established an apparent solubility ceiling of Q_{15}–Q_{22}, above which it appeared that peptides could not be studied in solution [46, 47, 53]. This appeared to be so even when charged flanking residues were included in an attempt to improve peptide solubility. However, we now understand that this apparent discrepancy between the transient solubilities of recombinant and chemically synthesized peptides is due to the presence, in the latter, of residual polyGln aggregates, leading to a situation not unlike that described above for $A\beta$ peptides.

This is dramatically demonstrated by two examples. Although attempts to directly dissolve a $K_2Q_{44}K_2$ sequence in aqueous buffer failed to give significant soluble peptide [48], pretreatment of a $K_2Q_{44}K_2$ peptide (or a $K_2Q_{20}K_2$ peptide) with a trifluoroacetic acid/hexafluoroisopropanol disaggregation protocol adapted from the $A\beta$ literature [68] yields peptides that display excellent initial solubility in aqueous buffers [48]. Even more dramatic is the effect this protocol has on the peptide $K_2Q_{15}K_2$. This peptide readily dissolves directly from the synthetic lyophilized powder in low salt, pH 3 buffer to yield a transparent, low light-scattering solution. When this solution is adjusted to phosphate-buffered saline conditions, however, aggregation occurs very rapidly to form a protofibril-like product [48]. In contrast, when this same peptide is exposed to the TFA/HFIP disaggregation protocol prior to solubilization in water and then is adjusted to PBS conditions, the peptide remains in solution for months at 37 °C, finally reaching an equilibrium position of aggregation after about six months [48]. Although the composition of $K_2Q_{15}K_2$ directly dissolved in water has not been further investigated, it is most likely that the suspended peptide exists as a mixture of dissolved monomer and microaggregates, the latter of which seed aggregation by the former when the solution is adjusted to PBS conditions. This tediously slow aggregation kinetics of the $K_2Q_{15}K_2$ peptide after disaggregation now qualitatively matches the behavior of short polyGln sequences in recombinant proteins and is also consistent with the lack of disease risk in humans with short polyGln sequences in *htt* and other disease proteins.

This change in the character of the $K_2Q_{15}K_2$ peptide is not a consequence of any covalent chemical change in the peptide [48], and no significant amounts of organic solvent remain after the treatment. To ensure that all traces of aggregates are removed, a preparative ultracentrifuge step is included. The sequential treatment of the lyophilized, synthetic peptide with TFA, followed by HFIP, as described for treatment of $A\beta$ peptides [68], is adequate for disaggregating polyGln peptides up to repeat length 40. Above this, the synthetic peptides do not dissolve well in TFA. However, it was found that peptides of higher repeat length dissolve well in a 1:1 mixture of TFA and HFIP and that, with sufficient incubation time and care, they can be completely disaggregated in this solvent mixture [48]. This treatment also appears to dissolve other aggregation-prone peptides that are resistant to the sequential method. The method is described in detail in the experimental Section 37.10.

37.4.2 Growing and Manipulating Aggregates

Although many studies on disease mechanism in HD and other protein aggregation diseases involve analysis of aggregation behavior, the study of the aggregation products is also important, since it is likely that it is the aggregates themselves, or a continuing aggregation process supported by these aggregates, that are responsible for cytotoxicity. In analogy to the above discussion on working with the peptides, therefore, a few words must also be said about the growth and handling of aggregates. Simple polyGln peptides incubated at 37 °C grow into a variety of morphologically related structures, as reviewed in Section 37.6.1. In general, these aggregates appear to be variations on the assembly of narrow protofilaments, in analogy to the substructures of amyloid fibrils. The aggregate formation reaction is quite reproducible, so long as the peptides are processed as described above and growth conditions are held constant. The aggregates are quite stable, with critical concentrations well below micromolar for polyGln peptides of repeat length 30 or more [49]. Aggregate suspensions in PBS can be snap-frozen in liquid nitrogen and stored at −80 °C for extended periods without noticeable change in properties.

37.4.2.1 Polyglutamine Aggregation by Freeze Concentration

Simple polyGln peptides dissolved in PBS, when incubated for one day or more in the frozen state at −5 to −20 °C, are highly aggregated upon thawing [66]. This is true even for Q_{15} peptides, which, when incubated at 37 °C in the completely disaggregated state, require months to achieve significant levels of aggregation. Studies on the mechanism of this effect showed that aggregate formation under these conditions is due to the process of freeze-concentration [69, 70]. Peptide solutions in buffer stored frozen at temperatures above the eutectic points of the buffer components consist of a solid ice phase and fissures of liquid phase containing most of the solutes, including the peptide [69, 70]. Since the liquid phase is only a small volume of the total, the solute concentrations in the liquid phase are enormous, with, for example, NaCl concentrations of several molar resulting from the freezing of a simple PBS ([NaCl] $\sim$ 0.15 M) solution. These conditions often result in the formation of protein aggregates, which is why it is often advisable to store protein solutions either above 0 °C or below 0 °C with freezing-point depression agents such as glycerol. Interestingly, because ice structure at modest freezing temperatures is rather unstable, solutions of polyGln peptides may be snap-frozen in liquid nitrogen, but if they are then stored at −20 °C, they will still develop ag-

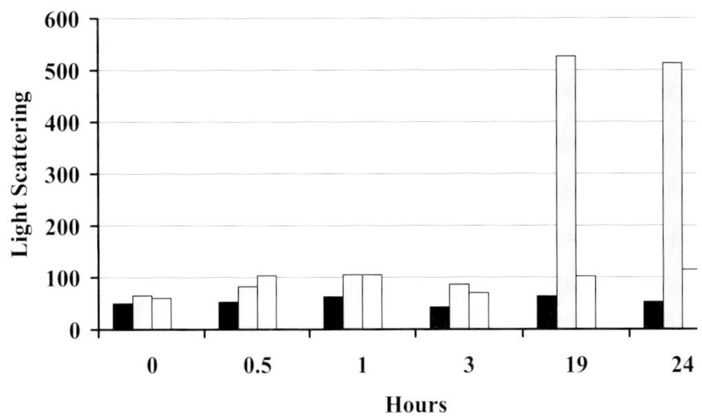

Fig. 37.2. Time course of freeze-concentration aggregation monitored by light scattering. Samples of 10 μM disaggregated $K_2Q_{37}K_2$ in PBS were incubated at 37 °C (solid black bars), at −10 °C in the liquid state (open bars), or at −10 °C in the frozen state (gray bars). The latter sample was snap-frozen in liquid nitrogen then transferred to the −10 °C bath. From Ref. [66] with permission from the American Chemical Society.

gregates as the ice structure equilibrates to that normally formed at the −20 °C storage temperature. Figure 37.2 shows the time course of aggregate formation in the ice state versus liquid state in the −5 to −10 °C range [66].

The freeze-concentration-induced aggregation of polyGln has several implications. First, there is no reason to suspect that any polyGln sequence, even in the context of other protein sequences, will not be susceptible to this effect. We have found that polyGly peptides can be stored for months at −80 °C if they are first disaggregated, adjusted to low concentrations, and snap-frozen in liquid nitrogen before storage. It is recommended that all polyGln-containing proteins be stored with similar care or not frozen at all. Interestingly, we found that the Aβ(1-40) peptide does not form amyloid by freeze-concentration. Thus, the same level of care required for working with polyGln sequences may not be required when working with other amyloid systems. It is important to be aware, however, of the potential for the seemingly benign freezing process to wreck havoc with samples of any protein, especially aggregation-prone proteins.

37.4.2.2 **Preparing Small Aggregates**

In addition to this potential for undesired aggregation, the freeze-concentration-induced aggregation of polyGln sequences is important because it generates aggregates with somewhat different properties compared to those grown from the same peptide at 37 °C [66]. As discussed in Section 37.6, these aggregates appear to be more like the protofibril assembly intermediates of other amyloid growth reactions than like the mature aggregates grown at 37 °C. These aggregates, once formed, are stable and do not assemble further on incubation at 37 °C. On a weight basis, aggregates prepared by freeze-concentration are much better seeds for fibril elongation than are aggregates grown at 37 °C [49]. At the same time,

37 °C aggregation reactions seeded (and hence greatly accelerated) with aggregates prepared by freeze-concentration produce aggregates indistinguishable from those grown by spontaneous aggregation at 37 °C, indicating that the substructures of these aggregates must be quite similar.

We have used freeze-concentration to produce aggregates of small sizes to facilitate their uptake into mammalian cells [45]. After aggregates are grown in frozen buffer, reactions are thawed and the aggregates collected by centrifugation. Aggregates are sonicated with a probe sonicator and then pushed through a series of molecular weight cutoff membrane filters to produce the finest possible aggregate size. This procedure is critical for the ability of mammalian cells to take up naked polyGln aggregates [45] (see Section 37.7.1.1).

37.5
In vitro Studies of PolyGln Aggregation

The central importance of the expansion of the polyGln repeat in this family of neurodegenerative diseases logically leads to the study of the biophysical properties of various repeat lengths of polyGln in search of clues to molecular mechanisms. Coincidentally, focusing on the polyGln sequence in isolation also makes accessible certain experimental approaches not possible with more complex proteins. For example, circular dichroism (CD) spectra can be cleanly interpreted in terms of what the polyGln sequence is doing, without being concerned about sequences and other elements of secondary structure. Solution concentrations of non-aggregated peptide can be quantified with great precision and accuracy using reverse-phase HPLC, as described in the experimental Section. Insolubility can be cleanly interpreted in terms of polyGln aggregation, without worrying about the aggregation tendency of other protein sequences in unnatural fusion proteins or truncated native proteins. The data from studies on isolated polyGln sequences establish a baseline for the fundamental behavioral trends of the polyGln sequence, so that further studies with sequences more closely resembling the complex structures of disease proteins can be put into context.

37.5.1
The Universe of Protein Aggregation Mechanisms

The broad phenomenon of particulate aggregation is systematized in the field of colloidal assembly and coagulation. The aggregation of proteins, especially in response to heat and as a side reaction in protein folding, has been viewed historically as a problem in colloidal assembly [71]. Early generalized models of the kinetics of coagulation involved models of particles undergoing hierarchical assembly into larger and larger clusters. Cluster formation in these models was considered to be controlled either by simple diffusion limits or by an energy barrier that determined the productiveness of each collision, but the assumption in either case was that each productive collision was irreversible [72]. However, since many protein aggregation reactions such as crystallization and amyloid fibril formation reach equilibrium positions in which both aggregates and monomers are populated, the possibility of productive collisions being reversible must be considered, which complicates the modeling [72]. Protein crystallization and amyloid formation may differ from classical colloidal assembly in another respect; rather than a hierarchical assembly mechanism involving the interactions of intermediates of ever-increasing size, protein crystals and amyloids, once formed, might be capable of growing by the addition of monomeric or oligomeric protein units [72]. This could occur, for example, if the protein in the aggregated state projects a docking surface that is capable of recruiting bulk-phase monomer very efficiently, compared to the relatively inefficient process of nucleation.

Such a nucleation-dependent polymerization model has been developed [73, 74] and successfully applied to hemoglobin aggregation [75]. The model also takes into account the potential reversibility of steps in the aggregation process and thus treats the nucleation process as an unfavorable pre-equilibrium before a series of increasingly favorable elongation steps; that is, the kinetic nucleus is viewed as the least stable species on the aggregation pathway. The unfavorability of the nucleation step gives rise to a "lag phase," during which no mature aggregates are formed and negligible amounts of monomer are consumed, followed by a rapid growth phase. Since this is a relatively simple model that has been successfully applied to protein systems, and since there is no evidence from in vitro studies of a more coagulation-based assembly mechanism for polyGln aggregation, this model has been applied to polyGln aggregation data, as described in the following section.

37.5.2
Basic Studies on Spontaneous Aggregation

Rigorously disaggregated polyglutamine peptides (Section 37.4.1) incubated in PBS at 37 °C undergo a spontaneous aggregation process that generally exhibits a lag phase typical of nucleated growth polymerization. These lag times become shorter as the repeat length gets longer, with aggregation becoming much more aggressive at repeat lengths about 35 [49]. These results agree with previous observations based on studies of exon I fragments of *huntingtin* [55]. Following the kinetics by a number of different windows on the aggregation process – light scattering, ThT fluorescence, CD, and solubility – reveals overlapping growth curves (Figure 37.3), suggesting that there are no major hidden species in the process [76]. The fact that the CD transition, from a spectrum consistent with random coil to one consistent with β-sheet, coincident with the reaction as monitored by aggregate formation and peptide solubility suggests that there is no pre-assembly in the bulk-phase monomer to a β-sheet containing monomeric species before aggregation occurs [76]. This interpretation is reinforced by an experiment in which the aggregation reaction is stopped midway and centrifuged [76]. The resuspended centrifugation pellet gives a typical β-sheet CD spectrum, while the supernatant gives the same random coil signature spectrum as seen in the starting reaction mixture (Figure 37.4). The implication is that the mechanism by which

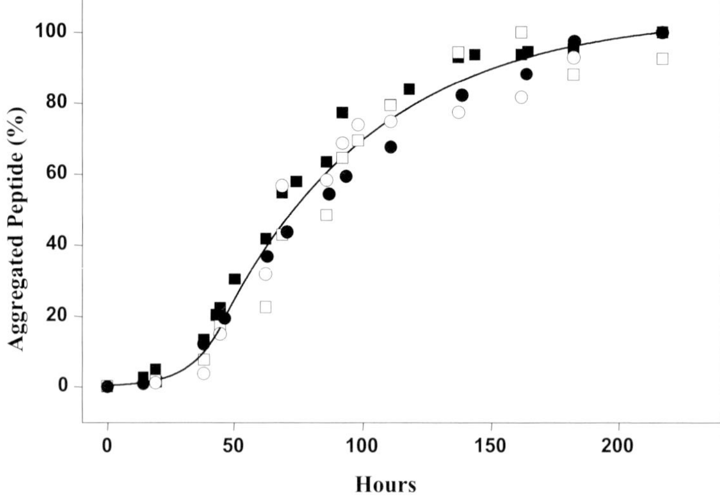

Fig. 37.3. Spontaneous aggregation of a polyGln peptide monitored by four independent measures. A sample of 66 μM freshly disaggregated $K_2Q_{42}K_2$ in 10 mM Tris-TFA, pH 7.0, was incubated at 37 °C and aggregation was monitored by circular dichroism signal at 200 nm (■), determining remaining soluble monomer by HPLC (●), light scattering (□), and thioflavin T fluorescence (○). From Ref. [76] with permission from the National Academy of Sciences (USA).

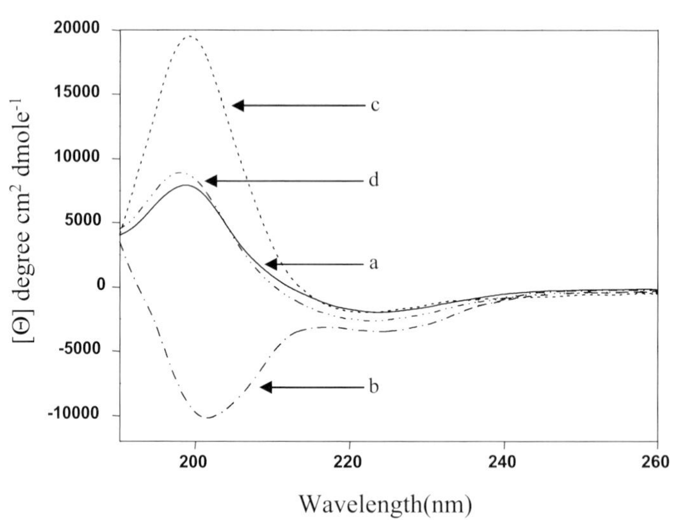

Fig. 37.4. Circular dichroism spectra of various fractions of a polyGln aggregation reaction midpoint. The reaction described in the legend of Figure 37.3 was sampled after 90-h incubation and CD spectra were collected after various treatments: no treatment (a); ultracentrifugation supernatant (b); resuspended ultracentrifugation pellet (c); sum of the supernatant and pellet spectra (d). Adapted from Ref. [76] with permission from the National Academy of Sciences (USA).

the bulk-phase polyGln peptides take on β-sheet characteristics occurs either through a concerted mechanism, in which molecules take on β-structure as they become ensconced onto the end of the growing aggregate, or through a mechanism in which isolated monomers rapidly add to aggregates as soon as they acquire β-structure.

37.5.3
Nucleation Kinetics of PolyGln

The increasing efficiency with which peptides undergo spontaneous aggregation (see previous section) suggests that it may be this property of the polyGln sequence that is responsible for the critical importance of repeat length in the expanded CAG repeat diseases. The source of the repeat-length dependence of aggregation is not immediately obvious, however. To investigate whether the repeat-length effect is played out at the level of nucleation and to learn more about the nucleation process, we applied a successful model for the nucleation-dependent aggregation of sickle hemoglobin [74] to the polyGln aggregation process [76]. In this model, the formation of an oligomeric kinetic nucleus from a bulk phase of monomeric proteins is treated as a pre-equilibrium, the favorability of which is dependent on the equilibrium constant and the number of monomers that cluster together to form the nucleus. This latter number, which is of particular importance because it becomes an exponential factor in the equation describing nucleation kinetics, is called the "critical nucleus," n^*. Whether any particular nucleus collapses back to n^* monomers, or progresses on to committed aggregate growth, depends on the efficiency with which it is stabilized through the elongation of the nucleus into a growing aggregate, as determined by the rate constant describing this elongation step. Since the number of kinetic nuclei formed during a nucleation-dependent polymerization reaction is vanishingly small and cannot be directly observed, the concentration of nuclei present at any one time is inferred by the kinetics with which observable aggregates develop, as described by the rate constant for elongation of aggregates. The overall nucleation kinetics equation (Eq. (1)) emerging from this model describes Δ, the increase in the concentration of monomers that have been converted into aggregates at time t, as it depends on the concentration of monomers in the bulk phase, c, the nucleation (pre-)equilibrium constant K_{n^*}, and the rate constants for elongation of the nucleus and aggregate, k_+, which, for simplicity, are assumed to be identical.

$$\Delta = 1/2k_+{}^2 K_{n^*} c^{(n^*+2)} t^2 \qquad (1)$$

Eq. (1) accounts for the time lag observed for nucleation-dependent growth kinetics and for the concentration dependence of the lag phase and the underlying nucleation kinetics. It also predicts two other features of nucleation kinetics: (1) that the kinetics should be dependent on *time*^{2}, and (2) that a log-log plot of the term emerging from the t^2 plot, with respect to monomer concentration, will yield a line of slope $n^* + 2$, from which n^*, the number of monomers in the kinetic

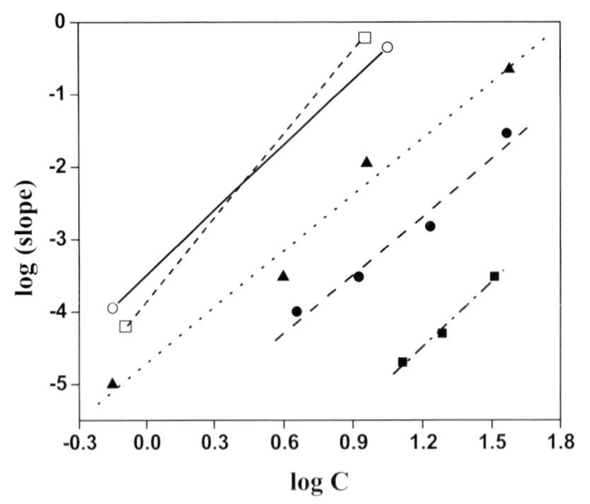

Fig. 37.5. Critical nucleus analysis for polyGln peptides of various lengths. Nucleation kinetics for each peptide at several concentrations was determined as described in the text and analyzed by t^2 plots. The log of the slope of the t^2 plot is plotted against the log of peptide concentration. The slope of the straight line fitting this data is $n^* + 2$, where n^* is the critical nucleus. Shown are data for $K_2Q_{28}K_2$ (■), $K_2Q_{36}K_2$ (●), and $K_2Q_{47}K_2$ (▲). Also shown are theoretical lines showing slopes of 3 (○) and 4 (□). From Ref. [76] with permission from the National Academy of Sciences (USA).

nucleus, can be derived. In fact, analysis of polyGln peptides with repeat lengths below and above the pathological threshold region shows linear t^2 plots at all peptide concentrations examined [76]. Figure 37.5 shows that plotting the log of the slopes of the t^2 plots at each concentration, against the log of the concentration, gives a straight line for each repeat length. The equivalence of the slopes for peptides above, at, and below the pathological threshold, as shown in Figure 37.5, shows that, within this range, repeat length does not alter the size of the critical nucleus. Table 37.2 shows the surprising result that the kinetics analyses for these peptides yield slopes of 3, reducing, therefore, to $n^* = 1$ in each case. Table 37.2 also shows that the source of the more aggressive nucleation kinetics, as polyGln repeat length increases, is therefore the larger value of the $k_+{}^2K_{n^*}$ term, rather than a change in the value of n^*. Since it was previously shown that elongation rates do not change dramatically as polyGln repeat lengths change [49], it can

Tab. 37.2. Kinetic parameters for polyGln nucleation of aggregation.[a]

	Q_{28}	Q_{36}	Q_{47}
Slope of the log-log plot	2.98	2.68	2.59
Calculated critical nucleus (n^*) from slope	0.98	0.68	0.59
$k_+{}^2K_{n^*}$ ($M^{-2}s^{-2}$)	0.001128	0.09193	1.8304

[a] Data from Ref. [76].

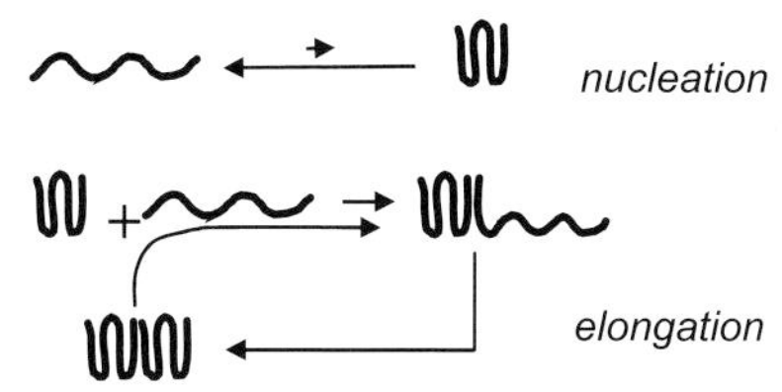

Fig. 37.6. A model for the nucleation-dependent polymerization of polyGln peptides. Reversible formation of the monomeric nucleus, the least stable species on the aggregation pathway, is visualized as an unfavorable folding reaction. Elongation is visualized as a cyclic two-step process involving addition of bulk-phase polyGln monomer to the nucleus or growing aggregate, followed by consolidation of the structure of the newly added monomer to extend the structure of the aggregate. From Ref. [76] with permission from the National Academy of Sciences (USA).

be concluded that the major source of the increased aggressiveness of polyGln aggregation above the pathological threshold is an increase in the nucleation (pre-)equilibrium constant K_{n^*}.

The result of this analysis, i.e., that the kinetic nucleus for nucleation-dependent polymerization of polyGln sequences contains a single molecule of polyGln, is a surprising, if not shocking, result. The whole theoretical basis for nucleation in nucleation-dependent polymerization and colloidal assembly theories is a model in which the nucleus is itself an oligomer on the assembly pathway. Based on general colloidal assembly models [72] and on water condensation analysis [77], the stability derived from the growth of the nucleus is basically attributed to an improved surface-to-volume quotient for this particulate cluster [78]. The recognition, however, that monomeric polyGln peptides, even long ones, assume a non-β-sheet conformation in solution makes conceptually possible a model in which the nucleation event is an unfavorable folding reaction within the monomer population. Figure 37.6 illustrates this model schematically, showing bulk-phase monomer as random coil and the nucleus as a monomer collapsed into an antiparallel β-sheet conformation. Interestingly, despite the fact that the model for the nucleation process was developed based on the assumption of an oligomeric critical nucleus, the mathematical analysis growing out of this model is sufficiently robust that it also describes quite well the situation in which $n^* = 1$. The reason for this is that the most fundamental definition of the critical nucleus is more mathematical than physical, stating only that the nucleus is the least stable species on the assembly pathway. The critical nucleus for polyGln assembly into amyloid-like aggregates achieves this distinction by being the product of a highly unfavorable folding reaction. How unfavorable these reactions are in the polyGln series is presently unknown, since the $k_+{}^2 K_{n^*}$ term cannot be deconvoluted due to (1) a present inability to independently determine the value of k_+ and (2) the inability to independently observe the nucleation equilibrium without its leading to aggregation. Direct observation of the nucleation process, determination of K_{n^*}, and derivation of molar elongation rate constants remain challenges for future research.

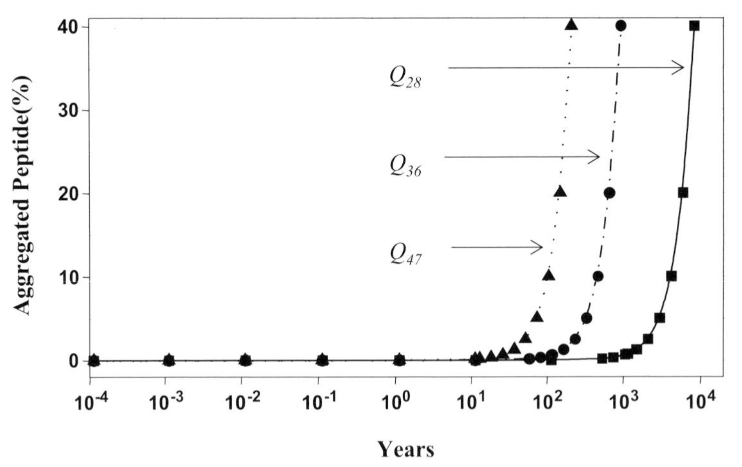

Fig. 37.7. Calculated nucleation kinetics curves for 100-pM polyGln peptides in PBS at 37 °C, based on kinetic parameters derived from in vitro studies at higher concentrations. From Ref. [76] with permission from the National Academy of Sciences (USA).

The ability of the nucleation kinetics model to accommodate polyGln aggregation made possible the calculated estimation of the expected kinetics curves for nucleation-dependent polymerization of polyGln at relatively low steady-state concentrations, i.e., of magnitudes similar to those expected in vivo. Figure 37.7 shows the resulting curves for assumed polyGln concentrations of 100 pM, which show predicted ages of onset for the three polyGln repeat lengths studied that differ by orders of magnitude and in a way that is consistent with disease statistics (Figure 37.1). The actual steady-state concentrations of the aggregation-prone proteolytic fragments of *huntingtin* or other expanded CAG repeat disease proteins are not known, and it is clear that the molecular complexity of cells and organisms is likely to foil attempts at adequately describing disease through test tube experiments in simple buffers alone. However, the simulation in Figure 37.7 illustrates how nucleation propensity, as controlled simply by repeat-length variation, is capable of contributing significantly to disease risk and age of onset. Ongoing studies of simple peptides in defined buffers are also providing quantitative descriptions of the roles of molecular crowding, molecular chaperones, downstream amino acid sequences, and other polyGln-containing molecules on impacting aggregation nucleation (A. Thakur, A. Bhattacharyya, G. Thiagaragan, and R. Wetzel, unpublished results).

37.5.4
Elongation Kinetics

The model for polyGln aggregate growth shown in Figure 37.6 shows that the nucleus elongates by successive productive encounters with bulk-phase monomer.

Although other models have been proposed for fibril elongation that may be followed by some proteins under some solution conditions [79], growth by monomer additions is likely to be the major pathway by which amyloid fibrils grow in vivo, especially in most systems where precursor protein concentrations are quite low. The kinetics of this process have been successfully modeled [80] as a simple bimolecular reaction between aggregate and monomer, such that the observed rate depends on monomer concentration, aggregate (molar) concentration, and the second-order rate constant. However, since in a heavily seeded reaction, aggregation mass builds primarily by the lengthening of the seeds, the molar concentration of aggregate (or the molar concentration of growth sites) does not change appreciably over the course of a simple elongation reaction, and the rate expression reduces to a pseudo-first-order kinetics equation involving monomer concentration and the first-order rate constant. (This rate constant is not a universal constant for this peptide, however, since it incorporates the molar concentration of fibrils used as seeds, which is not trivial to determine.)

Understanding more about the polyGln aggregate elongation process is important not only because of how this process contributes to the deposition of the disease protein in neurons but also because of the possible importance of the recruitment of other polyGln-containing proteins in the disease mechanism (Section 37.3.3). If the polyGln aggregate, or the aggregation process itself, is responsible for the neurotoxicity observed in these diseases, it is also important to learn how we might interfere with this process therapeutically. The need to discover and characterize aggregation inhibitors adds to the importance of having available one or more ways of quantifying elongation in vitro.

37.5.4.1 Microtiter Plate Assay for Elongation Kinetics

One convenient way of studying the elongation reaction is by immobilizing preformed aggregate on a surface, incubating the surface in a solution of monomer, and monitoring the growth of the aggregate seed. Aβ elongation has been studied by this approach in two ways, by using either fibrils fixed to microtiter plate wells and following elongation with radio-labeled monomer [81] or fibrils fixed to a dextran surface and following elongation by an increase in fibril mass as detected by surface plasmon resonance [82]. A modification of the microplate approach has been used to develop a polyGln elongation assay, which has been described in detail [67, 83]. In this assay, very small masses (typically 20–100 ng per well) of polyGln aggregates are immobilized onto microtiter plate wells. An aqueous solution of a disaggregated polyGln, with a biotin covalently attached to the N-terminus, is supplied at low concentration (typically 10 nM) and incubated. After incubation, excess biotinylated peptide is washed out and the amount of biotin remaining on the solid phase by virtue of a continuation of the polyGln aggregation process is measured with a streptavidin-linked reagent. High sensitivity is achieved through using streptavidin complexed with europium, which in turn can be measured with great sensitivity using time-resolved fluorescence. Since the microplate must be processed intact, kinetics are accomplished by phasing the start times of multiple reactions so that their common termination will deliver a time course. A

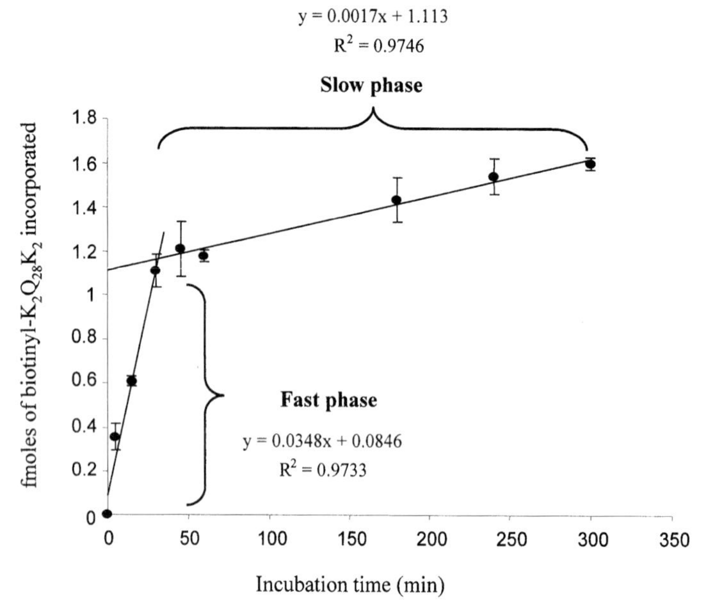

Fig. 37.8. Kinetics of polyGln aggregate elongation monitored by a microtiter plate–based assay. Twenty nanograms of $K_2Q_{28}K_2$ aggregates were immobilized by adsorption to each microtiter plate well, and the wells were incubated for various times with 10 nM biotinyl-$K_2Q_{28}K_2$. Signal was developed using europium-tagged streptavidin and time-resolved fluorescence. From Ref. [67] with permission from Elsevier.

typical kinetics progress curve for this polyGln elongation reaction is pictured in Figure 37.8. The division of the kinetics into fast and slow phases, as seen in Figure 37.8, has also been observed in the Aβ microplate assay [84]. As predicted by the kinetic model described above, the magnitude of the kinetic phases increases both with the increase of the amount of deposited polyGln aggregate and with an increase in monomer concentration. This assay is proving useful in working out some fundamental aspects of the polyGln elongation reaction [49], in screening for inhibitors of polyGln aggregate elongation (see following section), and in characterizing HD brain material.

37.5.4.2 Repeat-length and Aggregate-size Dependence of Elongation Rates

The repeat-length dependence of spontaneous polyGln aggregation is grounded in the repeat-length dependence of the unfavorable folding of the monomeric kinetic nucleus (Section 37.5.3) and is the simplest available explanation for how the gross features of the repeat-length effect on disease risk and age of onset are established. The repeat-length effect of the elongation phase of aggregate growth is also important, however. If the ability of polyGln aggregates to cause cell death and/or dysfunction is due to their ability to recruit other polyGln-containing proteins into inactivating aggregates, then the dependence of the efficiency of recruitment on polyGln repeat length will determine the range of polyGln-containing proteins that are at risk of being recruited. Experiments using the microtiter plate elongation assay show that there is, indeed, a length dependence of recruitment kinetics, but it is not nearly as dramatic as that for nucleation-dependent aggregation [49]. PolyGln peptides in the 5–10 repeat length range are not strongly recruited into preexisting aggregates. At repeat lengths in the Q_{15} to Q_{20} range, however, elongation rates become more substantial and continue to increase with increasing repeat length up to about Q_{40}, above which there appears to be little effect on rate of increasing repeat length. This suggests that polyGln proteins with repeat lengths in the 5–10 range are not likely to impact polyGln aggregation processes very much. However, at repeats of Q_{15} and above, the existence of polyGln proteins, and their relative concentrations and functions in the cell, may well impact expanded CAG repeat diseases. These results are consistent with observations that neuronal polyGln aggregates often contain CREB-binding protein (which has a repeat length of 18 Gln residues), but that the homologous protein P300, which has a polyGln repeat of only 6, is not recruited into cellular aggregates [85]. The rate of aggregate elongation, as determined by polyGln repeat length and concentration, may also significantly contribute to other aspects of polyGln aggregation and pathology (A. M. Bhattacharyya and R. Wetzel, manuscript in preparation).

Size also matters in considering the mass-normalized recruitment activity of polyGln aggregates. Microtiter plate experiments show that, on an equal mass basis, the ability of a preformed aggregate to serve as a template for polyGln elongation improves as the average particle size of the aggregate decreases [49]. The data support the idea that not all polyGln aggregates are functionally equivalent and that small aggregates are substantially better than the larger ones at supporting elongation. If polyGln aggregates are, in fact, toxic by virtue of their ability to recruit polyGln-containing proteins and thus remove them from their normal sites of action, it stands to reason that more recruitment-active aggregates would be more toxic. That is, aggregates too small to be easily detected by light/fluorescence microscopy may be far more cytotoxic than the relatively recruitment-inert aggregates that can be so visualized. These conclusions are further supported by staining cellular aggregates for recruitment activity, which will be discussed in Section 37.6.2.

37.6 The Structure of PolyGln Aggregates

Like other amyloid-like fibrils, the interior, atomic-level structure of the polyGln aggregate remains opaque, due to its inaccessibility to standard high-resolution methods and the limited resolution of the available methods. The following sections catalog what we do know about polyGln aggregate structure, with an emphasis on the extent to which these aggregates resemble other amyloids.

37.6.1
Electron Microscopy Analysis

Perutz and coworkers first showed that the polyGln aggregate has a fibrillar construction. The aggregates obtained from a Q_{15} peptide immediately upon adjusting a pH 3 solution of the synthetic product to pH 7 [53] resemble in the electron microscope the protofibrils observed as assembly intermediates in amyloid formation by other peptides, such as Aβ [86]. Similar structures are observed from Q_{15} or Q_{20} peptides after they are rigorously disaggregated and then incubated in PBS (Figure 37.9). In addition, aggregates of longer polyGln sequences take on this same protofibril-like appearance when they are grown in PBS at −20 °C by freeze-concentration (Section 37.4.4.1) (Figure 37.9). It is possible that aggregation stops at this stage at −20 °C because the diffusion of the protofibrils is limited by the ice lattice. The generation of protofibril-like structures from simple polyGln peptides in vitro has also been accomplished through the use of nonnative buffer conditions [87]; the relationship of these aggregates to those formed, for example, by freeze-concentration has not been explored.

When simple polyGln peptides longer than Q_{20} are incubated at 37 °C in PBS, they grow into more complex aggregated structures that appear in the EM as ribbons of about 20 nm in diameter and with substructures suggesting assembly from parallel-aligned protofibrils or protofilaments (Figure 37.9). Attempts to detect protofibril intermediates in these assembly reactions have not been successful, however (unpublished results), suggesting that under native conditions in simple PBS buffers protofibrils either do not exist, or have a very short lifetime. The observation of polyGln-associated protofibril-like structures in cells [88] may be due to the presence of flanking protein sequences and/or other molecules in the milieu altering the assembly pathway.

When longer polyGln peptides (Q_{45} or higher) are incubated in PBS at 37 °C, especially in low salt buffer, they grow into long filamentous aggregates that match the structures of typical amyloid fibrils (Figure 37.9). Incubation in vitro of fusion proteins containing *huntingtin* exon 1 fragments with expanded polyGln repeats also generates amyloid fibrils [54], and some EM studies suggest a fibrillar appearance of the substructures of polyGln inclusions in cells [4]. Thus, there appears to be a hierarchy of form in the world of polyGln aggregates, all of which exhibit a similar protofilament substructure that is capable of assembling in higher-ordered structures depending on polyGln repeat length, protein and cellular context, and growth conditions.

37.6.2
Analysis with Amyloid Dyes Thioflavin T and Congo Red

The early history of amyloids is associated with the discovery of the ability of certain heteroaromatic dyes to bind selectively to fibrils and, through binding, take on altered spectroscopic properties that can be monitored. In 1927, Divry reported that amyloid deposits in tissue, previously detected by a starch-like reaction to iodide

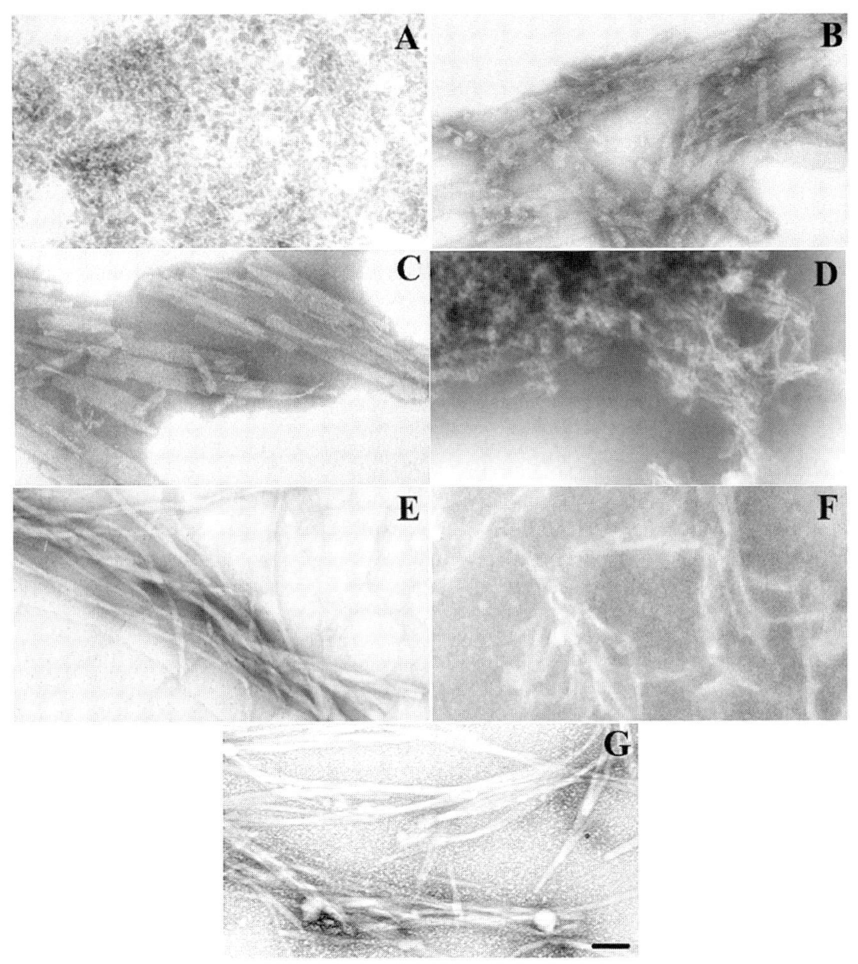

Fig. 37.9. Electron micrographs of various protein aggregates. (A) $D_2Q_{15}K_2$ aggregates grown in PBS at 37 °C. (B) $K_2Q_{20}K_2$ aggregates grown in PBS at 37 °C. (C) $K_2Q_{37}K_2$ aggregates grown in PBS at 37 °C. (D) $K_2Q_{37}K_2$ aggregates grown in PBS in the frozen state at −20 °C. (E) Fluorescein-tagged $K_2Q_{42}K_2$ aggregates grown in PBS at 37 °C. (F) $K_2Q_{42}K_2$ aggregates grown in Tris-HCl, pH 7.2 at 37 °C. (G) Aβ(1-40) fibrils grown in PBS at 37 °C. The bar represents 50 nm. From Ref. [66] with permission from the American Chemical Society.

staining, could be easily detected by staining the tissue with the dye Congo red (CR) followed by polarized light microscopy [89]. The success of the method, which remains the primary pathological stain for amyloid [90], depends not only on the ability of CR to bind to amyloid fibrils but also on the ability of the fibrils to confer onto the bound CR a macroscopic order that can be detected as birefringence in light microscopy. Since the resolution of light microscopy is limited by the wavelength range of visible light, birefringence is observed only if the ordered array of CR molecules that produces it extends into that size range. Since amyloid fibril diameters are in the 10 nm range, it is clear that for fibrils to be detected by CR birefringence, the fibrils have to be arranged in a higher level of order, that is, oriented non-randomly in the field. In fact, EM of fibrils often shows them to be loosely bundled together in an approximate parallel arrangement. This type of order is required in tissue for amyloid to be detected by CR birefringence. Perhaps this explains the lack of success in detecting polyGln aggregates in tissue by amyloid reagents. CR staining of polyGln aggregates grown in vitro produces a red cast, suggesting that the aggregates bind CR, but does not yield significant birefringence [66], suggesting that a lack of macroscopic order, rather than an absence of the fundamental amyloid fold, may be the explanation for the lack of signal. CR binding to amyloid in vitro can be detected by virtue of a spectral shift conferred onto CR upon binding [91], but there appear to be no data on whether polyGln aggregates produce this effect.

Another important dye for amyloid analysis is thioflavin T (ThT). ThT has the ability to bind to many different amyloid fibrils [92, 93], and when it does, the binding produces a characteristic change in ThT's fluorescent properties. The fluorescence yield upon ThT binding is linear with the number of ThT molecules bound, and therefore (under saturating conditions) also with the mass of amyloid, which allows the dye to be used to quantify amyloid in vitro [93]. PolyGln aggregates produce a typical amyloid-like ThT response [49]. PolyGln aggregation kinetics monitored by ThT track exactly, within experimental error, with other measures of aggregate formation (Figure 37.3).

The molecular mechanisms by which CR and ThT bind to polyGln and other amyloids are not known. CR, as a sulfonate, carries a strong negative charge, while ThT carries a positive charge, at neutral pH. The charged nature of these dyes has led to speculation that electrostatics play an important role in their binding to amyloids, most of which are derived from sequences possessing charged amino acids. PolyGln sequences carry only the neutral glutamine side chain, however, casting doubt on any overriding role for electrostatics in the binding of ThT. Although the synthetic polyGln molecules summarized here contain flanking Lys residues for peptide solubility, their aggregates, displaying only positively charged Lys side chains, are still capable of binding the positively charged ThT.

37.6.3
Circular Dichroism Analysis

As discussed in Section 37.5.2, the CD spectra of aggregated polyGln peptides are consistent with high levels of β-sheet (Figure 37.4), and the progress of the CD signal change at 200 nm that marks the progress of the two-state transition from soluble peptide to aggregate overlaps the aggregation kinetic curves determined by other means (Figure 37.3). The ease of obtaining an interpretable CD spectrum of the polyGln aggregate is somewhat unusual in the amyloid field, where attempts to collect CD spectra of amyloid fibrils in suspension are marred by high levels of light scattering. That polyGln aggregates are better behaved in

CD must be due to lower light scattering, presumably due to smaller average particle sizes. In fact, while Rayleigh light scattering in a fluorometer is a very sensitive means of following amyloid fibril formation by Aβ (B. O'Nuallain and R. Wetzel, unpublished observations), equivalent weight concentrations of polyGln aggregates scatter light much less well, although still to a useful extent [48, 76]. When polyGln aggregation reactions are carried beyond the point when the ThT reaction has reached plateau and the solution-phase peptide is essentially depleted, further, relatively small, changes occur in the CD spectra (S. Chen and R. Wetzel, unpublished). Similar changes have been seen in other β-sheet proteins, where they have been interpreted as reflecting increasing degrees of twist in the sheet [94].

37.6.4
Presence of a Generic Amyloid Epitope in PolyGln Aggregates

Recently, a class of antibody has been described that has the unusual, unanticipated ability to recognize a conformational epitope that seems to be present to some extent on all amyloid fibrils while not being present in the native state of the monomer [95, 96]. For example, the mouse MAb WO1 was generated by challenging mice with an Aβ(1-40) amyloid fibril immunogen, but the resulting antibody binds not only to Aβ fibrils but also to amyloid fibrils derived from IAPP, β2-microglobulin, transthyretin, immunoglobulin light chain, etc. [96]. This antibody also binds to the amyloid-like aggregates of synthetic polyGln peptides [96], offering another indication of the relatedness of the polyGln aggregate to the generic amyloid motif [66]. Conversely, some MAbs generated from mice immunized with polyGln aggregates are able to bind well to other amyloid fibrils such as Aβ fibrils (B. O'Nuallain, R. Wetzel, S. Ou, J. Ko, P. Patterson, unpublished), further indicating the presence of an amyloid epitope on the polyGln aggregate. Such pan-amyloid antibodies tend to be IgMs, limiting their use as tissue-staining reagents. These antibodies have a number of other uses, however, as diagnostics for the amyloid motif and as a means of quantifying certain features of fibrils [10].

37.6.5
Proline Mutagenesis to Dissect the Polyglutamine Fold Within the Aggregate

One of the central questions in amyloid research is how the constituent polypeptide folds upon itself to engage the amyloid structural motif. Amyloid fibrils clearly possess a cross-β structure, in which the peptide chains are perpendicular to the fibril axis, while the H-bonds between strands in the β-sheets are parallel to the axis [97]. The details of how the peptide chains are arranged is space within these simple, minimal restraints have not, however, been established for any amyloid. Mutagenesis, especially with proline residues [98], has been used in other amyloid systems to garner details about how the sequence of the amyloidogenic peptide folds within the fibril. In Aβ, for example, it is clear from a variety of techniques that the N-terminal 10–15 amino acids, as well as the C-terminal 3–4 amino acids, are not involved in tight β-sheet H-bonding [99]. Proline analysis, which is based on the observation that loops, floppy ends, and some turn positions can comfortably accept Pro residues, while β-sheets are destabilized by them, also suggests the location of turns in the Aβ fibril [99].

A major problem confronts the use of mutagenesis to dissect polyGln structure in the aggregate, however. The problem is the prospect that the polyGln sequence will be promiscuous in how it adopts its fold within the aggregate. This means that any particular sequence position of a Q_{40} sequence, let's say position 10, might be found in a number of environments when a series of Q_{40} molecules folds into the aggregate. Further, should position 10 be favored in a certain kind of structure in the aggregate, a mutation at position 10 that is unfavorable in that type of structure will likely simply lead to adjustments in the polyGln chain as it engages the aggregate structure, to place residue 10 in the least damaging – that is, least destabilizing – place in the aggregate.

A modified mutagenesis strategy, taking the above complication into account, has led to some insights into polyGln structure within the aggregate. It is based on the hypothesis that polyGln folds in the aggregate in a series of extended chain segments, of some unknown optimal length, alternating with turn segments [10]. This hypothesis was then tested by analysis of the aggregation kinetics of a series of mutant polyGln peptides composed of alternating segments of oligoGln and Pro-Gly sequences, the latter being especially comfortable in turn elements. It was found that the peptide PGQ_9, consisting of four Q_9 segments interspersed with three PG pairs, undergoes spontaneous aggregation essentially as fast as a Q_{42} peptide of the same length. Since peptides with shorter Q_N elements aggregate less rapidly, while a PGQ_{10} peptide aggregates only marginally more rapidly, it was concluded that the best model for how an unbroken polyGln sequence aggregates is PGQ_9. Since each turn may involve one or two Gln residues bordering the PG pair, this corresponds to extended chain segments of 7–8 residues interspersed with turns of 3–4 residues.

Two further mutagenesis tests on PGQ_9 support such models. To test the interpretation that the PG pairs are located in turns in the aggregate, a PGQ_9 peptide was synthesized containing D-prolines in place of L-prolines; this peptide aggregates more rapidly than the L-Pro form, consistent with data showing D-Pro-Gly to be more compatible in β-turns than L-Pro-Gly [100]. To test the interpretation that the oligoGln sequences between the PG pairs are in extended chain in the aggregate, mutant peptides were synthesized in which additional proline residues were inserted in the middle of one or more of these oligoGln sequences. When a single Pro residue is placed in the middle of either the second (PGQ_9P^2) or third (PGQ_9P^3) of the four Q_9 segments, these peptides are completely incapable of forming aggregates under conditions where PGQ_9 aggregates readily. This replicates the strong inhibition of amyloid formation when a single Pro residue is placed in the middle of a β-sheet region in Aβ peptides [98, 99] and strongly supports the model for the aggregate structure of PGQ_9. That the structure of the PGQ_9 aggregate is very similar to that of a Q_{42} aggregate was shown by a number of studies, including EM, WO1 binding (Section 37.6.4) and the abilities of each aggregate to seed elongation of the other peptide [10].

Interestingly, the aggregation of another Q_{42} analogue, containing only a single proline residue at the same residue position as the additional proline in PGQ_9P^2, is nearly as rapid as that of an unbroken Q_{42} peptide. This confirms the concerns about folding promiscuity in polyGln aggregation that guided the design of the PGQ_9 series. That is, simple mutagenesis experiments on unbroken polyGln will generally be very difficult to interpret with confidence, because of the ability of the polyGln sequence to accommodate to the mutation by slightly altering its fold in the aggregate.

The model of a repeat structure consisting of alternating extended chains and turns is compatible with either an antiparallel β-sheet motif or an irregular parallel β-helix motif of the kind seen in certain globular proteins [101]. Previously, Starikov et al. proposed an antiparallel β-sheet model, noting that the number of residues required to make a four-stranded version of such a model is very close to the pathological threshold for most expanded CAG repeat diseases [102]. Perutz and coworkers proposed a "water-filled nanotube" model for the polyGln aggregate [103], which is related to parallel β-helix models previously proposed for the amyloid structure [101, 104]. It is not clear whether the central core of water would be energetically favorable or whether a more stable variation might be to exclude the water to generate an irregular helix more like those seen in parallel β-helical proteins [101, 105]. However, the "ideal" water-filled nanotube model [103] might be expected to be destabilized by Pro or Pro-Gly insertions, given the uniform extended chain and continuous twisted β-sheet in the model. The ability of polyGln aggregates to tolerate certain periodicities of Pro-Gly insertion is also incompatible with aggregates consisting of very long strands of polyGln extended chain [27].

Interestingly, the mechanism of nucleation of polyGln aggregation does not seem to be altered in these Pro-Gly mutants. Thus, analysis of the concentration dependence of the early phases of spontaneous aggregation for the peptide PGQ_9 yields a value of $n^* = 1$ for the critical nucleus [10], the same number obtained for unbroken polyGln sequences.

37.7
Polyglutamine Aggregates and Cytotoxicity

There is little disagreement that the expansion of the polyGln sequence plays a major role in expanded CAG repeat diseases, but opinions differ considerably as to the mechanisms by which this effect plays out. Many reviews are available that summarize the data and argue for and against an active involvement of polyGln aggregates [2, 13–31]. One of the central barriers to reaching a consensus on the nature of the polyGln effect is the difficulty in characterizing and/or controlling the aggregation state of polyGln in cells. This problem has an impact in various ways. For example, in order to produce *htt* aggregates in a reasonable time frame in a cell or animal model, it is necessary to express unnaturally high concentrations of monomeric protein in the cell; if toxicity is observed, is it then due to the aggregates or to the high concentration of monomer required to make the aggregate? At the same time, genetic analyses for binding partners of expanded polyGln proteins may in some cases be mediated by an aggregated form of the bait molecule rather than the assumed monomeric form.

Semantics and sensitivity of detection may also play roles in the controversy over the toxic species in polyGln diseases. To a protein chemist, an aggregate is any oligomeric protein state held together by noncovalent interactions, even including native oligomers; thus, a nonnative dimer of huntingtin, for example, would be considered an aggregate. Some workers, however, interpret an absence of large, macroscopic cellular inclusions as definitively ruling out the involvement of aggregates in toxicity.

If the neuronal dysfunction and death characteristics of expanded CAG repeat diseases can in fact be attributed to polyGln aggregates, it remains to be determined how aggregates produce their effects at the molecular and cellular level. A number of theories have been put forward to explain the toxicity of protein aggregates in neurodegeneration. Aggregates might saturate and/or otherwise inactivate the cellular machinery, such as molecular chaperones, the ubiquitin-proteasome system [106], and the aggresome system [107], responsible for dealing with protein misfolding. Aggregates might insert into membranes and alter their properties [108]. One mechanism that is uniquely feasible for polyGln aggregation and that has received significant attention and confirmatory data is the recruitment-sequestration model [24, 57, 61, 62, 85]. Since it is well established that amyloid-like aggregates of one polyGln sequence are capable of being elongated by the addition of other polyGln sequences, it is possible that polyGln aggregates in the cell might recruit other polyGln-containing proteins. Many such proteins exist in the human genome [109–111], in particular among proteins involved in nucleic acid binding, such as transcription factors. In a pivotal paper, Ross and coworkers showed that expanded polyGln cytotoxicity in a cell model appears to be mediated by the recruitment of the transcription factor CREB-binding protein (CBP), which contains a Q_{18} repeat, into the polyGln aggregates [85]. Expression of high levels of a functional, polyGln-minus version of CBP protects cells from expanded polyGln toxicity [85].

As discussed below, synthetic polyGln peptides have contributed to the debate about the toxicity of expanded polyGln sequences by addressing some of these technical barriers.

37.7.1
Direct Cytotoxicity of PolyGln Aggregates

One way to more directly test the toxicity of aggregates would be to deliver polyGln aggregates produced in vitro into cells. This would eliminate the need for producing unnaturally high levels of monomeric polyGln, the toxicity of which then becomes an important question if the experimental result is to be properly interpreted. The ability to produce a variety of well-characterized polyGln aggregates in relative bulk, as described above, makes possible experiments involving delivery of aggregates to cells. In addition, the ability to deliver aggregates essentially composed of only the polyGln sequence addresses the question of the role of other sequence elements in cytotoxicity of expanded CAG repeat proteins.

37.7.1.1 Delivery of Aggregates into Cells and Cellular Compartments

The challenge in analyzing the cytotoxicity of exogenous aggregates is their delivery into the cell. In fact, liposome encapsulation does allow polyGln peptide aggregates to be taken up by cells in a relatively short time [45]. Interestingly, however, a control experiment of polyGln aggregate mixed directly with cells in culture also resulted in very good cellular uptake of the aggregate [45]. Sonication and membrane filtration (Section 37.4.2.2) to generate smaller average sizes of aggregates improves cellular uptake [45]. Uptake through the cellular membrane was confirmed by confocal microscopy on fluorescently tagged aggregates. The mechanism by which cells take up these aggregates is not clear, but it may be related to an electrostatic attraction between the negatively charged phospholipid membrane and the highly positively charged aggregates, which are composed of peptides consisting of polyGln flanked by pairs of Lys residues. The process does not require polyGln, since a control amyloid fibril is also readily taken up [45].

Since nuclear polyGln inclusions are often observed in cell and animal models, it is also of interest to deliver aggregates into the nucleus. In an attempt to do this, polyGln peptides were prepared that contain an N-terminal nuclear localization signal (NLS) peptide sequence as well as a fluorescein tag. In fact, aggregates of such peptides are not only taken up by the cell but also are transported into the nucleus, as confirmed by confocal microscopy of isolated nuclei [45]. A control amyloid fibril from a peptide containing a NLS is also readily taken up into the nucleus. The ability of the nuclear transporter to manage such aggregates is somewhat surprising, since the aggregates appear to exceed 100 nm in diameter (based on their ability to resist filtration through a 100-nm filter membrane), while the diameter of the nuclear pore is on the order of 26 nm [112]. It is possible that the multiple display of NLS signal peptides along the surface of the aggregate provides a repeat that allows it to "ratchet" through the pore [112]. It is also possible that the large aggregates mixed with and taken up by cells may transiently break apart to individual protofibrillar aggregates, as seen in EMs of these aggregates; these have diameters in the range of 5 nm.

It is very unlikely that aggregates get into cells or nuclei by way of full dissociation to monomers followed by reassembly on the other side of the membrane. These polyGln aggregates are very stable in vitro in physiological conditions [45]. Further, such a mechanism could not explain the ability of Q_{20} aggregates to invade the cell and nucleus, since the very slow aggregation kinetics of the monomer, especially at the concentrations likely to obtain in this scenario, rules out any efficient aggregation after dissociation.

37.7.1.2 Cell Killing by Nuclear-targeted PolyGln Aggregates

The ability to deliver various protein aggregates into cells and their nuclei makes it possible to study the toxicity of these aggregates. PolyGln, as well as control, aggregates delivered to the cytoplasm of either PC12 or Cos-7 cells in culture exhibit little or no toxicity to these cells when examined by a number of standard methods for determining cell death: propidium iodide exclusion, lactate dehydrogenase release, and MTT reduction [45]. In contrast, polyGln aggregates delivered to cell nuclei rapidly kill cells, with very similar results from all three measures of cell killing. Although amyloid fibrils of cold shock protein peptide B-1 (CspB1) are efficiently delivered to cell nuclei, they do not kill cells, indicating the importance of the polyGln sequence. Interestingly, nuclear-targeted aggregates of Q_{20} and Q_{42} peptides are equally effective at killing cells. This strongly suggests that any aggregated polyGln peptide can be toxic when delivered to the nucleus and reinforces the conclusion derived from the aggregation kinetics (Section 37.5.2) that the distinction between a toxic (Q_{42}) and benign (Q_{20}) repeat-length sequence is made at the level of aggregation efficiency (Section 37.5).

Although the ability of nuclear-localized polyGln aggregates to rapidly kill cells may thus be responsible for the neuronal loss observed in end-stage expanded CAG repeat brain stem, it remains possible that cytoplasmic polyGln aggregates might also have harmful effects that may be less dramatic and difficult to detect in a one-day cell culture. The cell death induced in cultured cells by nuclear uptake of polyGln aggregates appears to be related to an apoptotic process (W. Yang and R. Wetzel, manuscript in preparation).

37.7.2 Visualization of Functional, Recruitment-positive Aggregation Foci

One possible explanation for why easily detected polyGln aggregates do not correlate with cytotoxicity in all cases is that the wrong aggregates are being monitored. If the cell killing by expanded polyGln sequences is due to recruitment of cellular polyGln proteins by aggregates of the polyGln disease protein, then the ability of these aggregates to efficiently recruit other polyGln sequences – their "recruitment activity" – is of paramount importance to their cytotoxicity. Recruitment activity can be defined as a measure of the number of monomeric polyGln peptides that can be recruited into a polyGln aggregate of a certain mass in a certain time. Interestingly, even in vitro aggregates of simple polyGln peptides can exhibit different recruitment activities depending on the way they are prepared and on their sizes [49]. Larger aggregates exhibit lower recruitment activity than smaller aggregates, presumably because of the greater surface area of the latter [49] (Section 37.5.4).

Thus, if the recruitment hypothesis is valid, then the best way to gauge the polyGln aggregate burden of a cell or tissue would be to assess the recruitment activity in the tissue, either as a tissue stain or as an activity measurement akin to an enzymatic assay. Both of these approaches are feasible, because of the robustness and specificity of the polyGln elongation reaction, and both have been shown to work when using human brain material. Aggregates in an enriched fraction isolated from HD brain tissue, when fixed to microplate wells, exhibit the same ability to recruit biotinylated polyGln peptides as synthetic aggregates exhibit (V. Berthelier and R. Wetzel, unpublished). HD brain slices incubated with biotinylated polyGln pick up the peptide in small, punctate, intracellular centers called aggregation foci [113]. The ability to characterize cells and tissue for levels of functional polyGln aggregates should provide sharper tools for addressing the question of the role of polyGln aggregates in cell dystrophy and death in expanded CAG repeat diseases.

37.8 Inhibitors of polyGln Aggregation

If polyGln aggregates are toxic to cells, a viable therapeutic approach for expanded CAG repeat diseases would be the use of specific inhibitors of spontaneous polyGln aggregation and/or elongation. Such inhibitors clearly exist in nature. For example, several genetic screens in HD disease models reveal that increased expression of DnaJ class chaperones can protect cells and animals from polyGln toxicity. Investigations of the molecular basis of this effect reveal that recombinant HDJ1, a human DnaJ homologue, is very effective at inhibiting both spontaneous polyGln aggregation and seeded elongation, presumably by binding to nuclei or small aggregates and inhibiting their elongation with additional polyGln monomers (A. Bhattacharyya and R. Wetzel, unpublished). This suggests that inhibitors of polyGln aggregation can have a protective effect.

37.8.1 Designed Peptide Inhibitors

A similar result has been obtained with a structure-based elongation inhibitor. Not only is the peptide $PGQ_9\text{-}P^2$ incapable of spontaneously aggregating (see Section 37.6.5), it can also inhibit the aggregation of other polyGln peptides. In the microplate assay described earlier, $PGQ_9\text{-}P^2$ inhibits Q_{42} elongation with an IC_{50} in the low micromolar range (Figure 37.10) [44]. Furthermore, cells pre-incubated with $PGQ_9\text{-}P^2$ are protected, in a dose-dependent manner, from the cytotoxicity of polyGln aggregates in the cell model described in Section 37.7.1.2 [44]. This implies not only that aggregates are toxic but also that the basis of their toxicity is their ability to recruit other polyGln peptides in the cell. This in turn suggests that even cells that have already entered the aggregation pathway might be protected from aggregate toxicity by appropriate inhibitors.

37.8.2 Screening for Inhibitors of PolyGln Elongation

As discussed above, polypeptide-based inhibitors of polyGln aggregation exist. However, small molecules with comparable activities would presumably be better drug candidates. In order to discover and refine such inhibitors, viable assays are needed. A filter blot assay that allows screening of small compound libraries for inhibitors of the aggregation of the polyGln-containing *htt* exon 1 domain has iden-

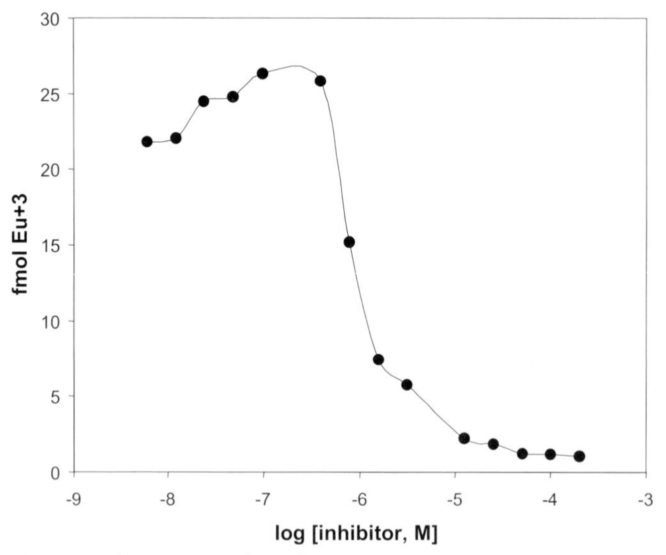

Fig. 37.10. Concentration-dependent inhibition of polyGln aggregate elongation by PGQ_9-based peptide inhibitors. Microtiter plate wells containing adsorbed Q_{45} polyGln aggregates (100 ng per well) were incubated with 10 nM biotinyl-Q_{28} and various concentrations of PGQ_9P^2 (●) for 45 min at 37 °C. The derived EC_{50} value is 1.1 μM.

tified a number of small molecule inhibitors [114]. The microtiter plate elongation assay described in Section 37.5.4.1 was initially designed for the same purpose [67, 83]. The assay has several important features, including reproducibility, insensitivity to small concentrations of DMSO and small pH changes, good throughput, and the ability to operate at low polyGln concentrations approaching physiological. This assay has been used to screen a number of small libraries of drug-like compounds as well as to determine dose-response curves for hits and analogues (V. Berthelier and R. Wetzel, unpublished). A typical screening result is shown in Figure 37.11.

37.9 Concluding Remarks

Historically, protein aggregation has been viewed as a nonspecific, amorphous process that is very difficult to study and unlikely to be relevant to events in the cell. Recent studies of amyloid-like phenomena, however, are showing that at least some aggregation processes are highly specific, more closely resembling crystallization phenomena. As described in this article, a logical approach that assumes such crystal-like packing and growth as a starting point for focused analytical experiments can make great progress in understanding the aggregation process both in vitro and in vivo. Far from being alien to the cell environment, it is now

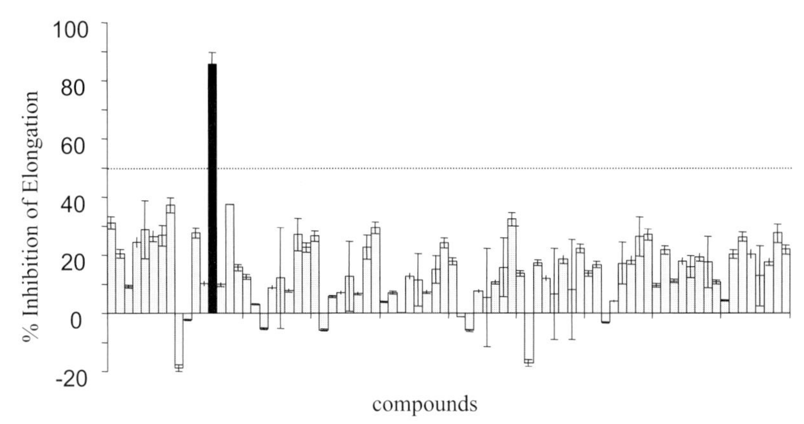

Fig. 37.11. Microtiter plate screen of a small compound library for inhibitors of biotinyl-polyGln elongation on adsorbed polyGln aggregates. Amount of elongation for each compound was compared to elongation without added compound, and the percent inhibition was calculated. In this plate, 1 out of 80 compounds tested gave >50% inhibition. Error bars from triplicate analysis.

known that aggregation is a continuing background problem in the life of the cell. Although cells can normally manage the process pretty well, in some cases, as in the amyloid-related neurodegenerative diseases, the normal cellular safeguards break down, are overwhelmed, or are isolated by cellular compartmentalization, and protein aggregation proceeds unabated, with devastating consequences. Protein aggregation, long swept under the rug in the protein-folding field, is now very much a viable subject for biophysical studies. While there has long been a perception that protein misfolding and aggregation are shadowy events occurring behind a veil impenetrable to the experimentalist, a variety of analytical methods exist [115], and improved methods are being developed [99, 116–120], that are capable of revealing these processes with analytical clarity.

Many techniques have been discussed or alluded to in this article. The protocols described in detail below represent central technologies necessary to getting started in polyGln studies, and they may be transportable, with modification, to studies of other protein aggregation phenomena as well.

37.10 Experimental Protocols

37.10.1 Disaggregation of Synthetic PolyGln Peptides

The procedures described for working with synthetic polyGln peptides have been developed through the almost exclusive use of polyGln peptides flanked by pairs of Lys residues, for better solution behavior of the peptides prior to their aggregation. Other flanking-charged residues work to varying extents. Peptides without flanking-charged residues are quite poorly behaved and aggregate very rapidly after being adjusted to PBS conditions [66].

Purification

Lyophilized, crude synthetic product is purified by reverse-phase HPLC as follows. Working in a chemical fume hood, 2 mg of the dry product is dissolved in 4 mL of a 1:1 mixture of trifluoroacetic acid (TFA) and hexafluoroisopropanol (HFIP) in a 20-mL Erlenmeyer flask and allowed to stand for up to 5 h at RT. Normally this should completely dissolve the sample; if there is any visible undissolved material after 5 h, the supernatant should be removed to a fresh flask. The volatile solvents are removed, in the fume hood, under a stream of argon gas, being careful to avoid splashing of the solution. The residue is dissolved in 5 mL of distilled water adjusted to pH 3 with TFA and filtered through a 0.45-μm membrane filter. The filtrate is loaded onto a 9.4×250-mm C3 HPLC column equilibrated to 0.05% TFA, 1% acetonitrile in water. The column is developed with a linear gradient of 1–21% acetonitrile in 0.05 % TFA at 4 mL min^{-1}. Column fractions are checked for purity by analytical HPLC under similar conditions and by mass spectrometry, and the fractions are pooled. Synthetic polyGln peptides longer than Q_{20} may have significant amounts of truncated side products, and these become more significant as repeat length increases; these truncated products are difficult to fully separate from the desired repeat length. Thus, most synthetic peptides longer than Q_{25} after purification will be mixtures of two to four different repeat lengths. A weighted average is calculated based on MS analysis of the purified pool, and this is the quoted repeat length. There is no indication that such mixtures behave substantially differently than homogeneous repeat length peptides. The pooled fractions are lyophilized and the powder is stored at −80 °C.

Disaggregation

Lyophilized purified peptide is dissolved in 1:1 TFA:HFIP to 0.5 mg mL^{-1} and incubated at RT as described above. For peptides of repeat length Q_{50} or above (and perhaps for shorter repeat lengths with alternative flanking sequences), it may be important to continue incubation in TFA/HFIP until completion is confirmed by a light-scattering test, as described in Ref. [48]. In this test, a small aliquot of the TFA/HFIP solution is removed and dried under argon, then further dried under vacuum, resuspended in pH 3 (TFA) distilled water, and brought to PBS conditions by addition of a one-ninth volume of 10× PBS buffer concentrate with stirring. The Rayleigh light scattering of this sample (in a fluorometer with both excitation and emission set to 450 nm, and with slit widths set such that aggregated versions of similar weight concentrations of polyGln give significant signals) should be essentially identical to that of buffer alone. If that is not the case, the incubation in TFA/HFIP should be continued, at least until the light scattering of the test reaches a low plateau.

When this point is reached, the bulk of the reaction is dried under an argon stream and then under vacuum. The residue is dissolved in pH 3 (TFA) distilled water to a concentration in the range of 100 μM or less, as required (higher concentrations may be possible, but such experiments should be carefully conducted and monitored to ensure that no continuing undesired aggregation process is initiated. This solution is centrifuged at 386 000 g for at least 3 h (or overnight) in a tabletop ultracentrifuge to remove any residual aggregates. The top two-thirds of the supernatant is carefully decanted and checked for concentration by HPLC as described in the next protocol. Appropriate buffer concentrate is then added to adjust the solution conditions, and additional buffer is added, as required, to bring the solution to the desired polyGln concentration. Ideally, this final solution should again be checked for polyGln concentration.

In general, it is advisable to carry out this entire procedure each time a new study is initiated, especially in the case of more aggregation-prone samples. As discussed in this chapter, freezing of polyGln solutions is generally not a good way to preserve their integrity, due to freeze-concentration-induced aggregation. Aggregates tend to build up slowly even in solutions stored at −80 °C. However, very low (low nanomolar) concentrations of polyGln peptides can be stored in PBS at −80 °C for one month, if 10% DMSO is included.

37.10.2
Determining the Concentration of Low-molecular-weight PolyGln Peptides by HPLC

As discussed in this chapter, monitoring the concentration of solution-phase, low-molecular-weight polyGln is probably the best way to conduct detailed aggregation kinetics studies. Obviously, while this technique leads to highly reproducible solubility values, other methods must be used for any given peptide and reaction conditions to confirm the nature of the aggregated product. The fundamental assumption of the approach described here is that the A_{220} weight-concentration extinction coefficient, based primarily on the absorbance of the amide bonds in the peptide, will be essentially identical for all polyGln peptides. While the same assumption is not valid for other peptides, since some amino acids in peptides make different contributions to the A_{220}, it is a reasonable assumption for polyGln peptides. By making this assumption, we can use the standard curve generated from a standard of a given repeat length to analyze the unknown concentrations of solutions of other repeat length polyGln peptides. This HPLC method works equally well for other peptides, such as the Alzheimer's disease amyloid plaque peptide Aβ [117], so long as a standard curve appropriate for that peptide is generated.

Preparation of the Standard

A synthetic polyGln peptide in the length range Q_{25} is purified and disaggregated as described above. A stock solution of an approximate concentration appropriate for use as a standard (~65 μM) is prepared in pH 3 water containing 20% formic acid and the solution is aliquoted into convenient volumes, snap-frozen in liquid

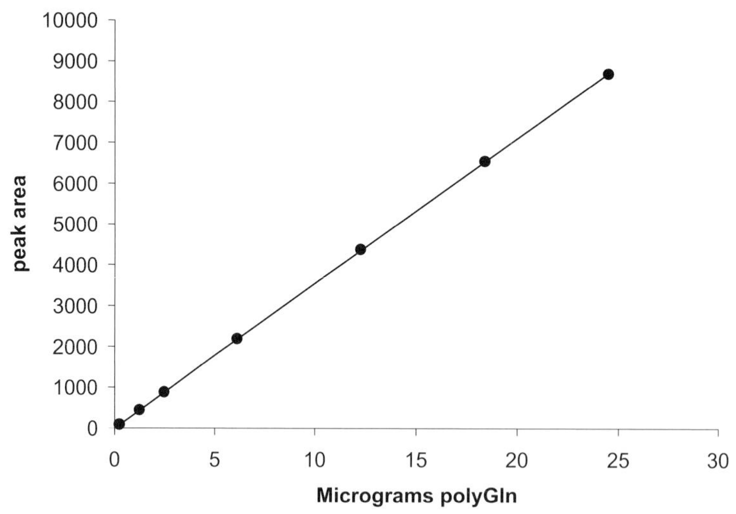

Fig. 37.12. HPLC standard curve relating peak area to mass of $K_2Q_{25}K_2$ injected into the system, as described in the experimental Section.

nitrogen, and stored at −80 °C. An aliquot of the stock is removed prior to freezing and diluted into pH 3 (TFA) water, the dilution factor is recorded, and the diluted solution is subjected to amino acid composition analysis to obtain a highly accurate concentration. The dilution factor is then applied to the concentration value obtained to generate a highly accurate concentration for the frozen standard. Peptides shorter than Q_{20}–Q_{25} may present problems when injected under certain solution conditions, such as eluting at somewhat earlier positions. Data from shorter peptides also are more dominated by the contribution of the Lys flanking residues and therefore may be less ideal standards for longer polyGln peptides. Peptides longer than Q_{25} may present problems because of a tendency to aggregate on long-term storage of the standard.

Generation of a Standard Curve

An HPLC system with a digital, computer-controlled variable wavelength detector is required. Aliquots of the polyGln standard generated above are injected onto an analytical reverse-phase HPLC column and developed with a 1–21% acetonitrile gradient in 0.05% TFA. The eluate is monitored at A_{220}. The polyGln peak is integrated and the area units from the integration are plotted with respect to peptide load. A typical standard curve for the standard $K_2Q_{25}K_2$ is shown in Figure 37.12. It must be emphasized that the standard curve is valid only for the HPLC system on which the data was collected. Different detectors will give substantially different curves. Standard curves from the same detector can also degrade over time, due, for example, to fouling of the detector cell windows altering the absorbance characteristics of the detector. Therefore, standard curves should be rerun every few months.

Use of the Standard Curve to Generate Solubility Data

Analysis of an aliquot of a centrifugation supernatant will yield a peak whose integrated area units will correspond to a certain weight of standard polyGln peptide. It is assumed that this value also corresponds to the same weight concentration of the test polyGln peptide, even though its repeat length is different. The molar concentration can then be calculated from the molecular weight of the test peptide. Extrapolation of the curve to very low, nonzero weight concentrations of polyGln should not be assumed to be linear, since a certain amount of low background irreversible absorption of peptide may occur. That is, the standard curve should be extended into the required analysis range. The standard curve shown in Figure 37.12 is linear to 0.25 μg of $K_2Q_{25}K_2$.

Detailed kinetics experiments involve many time points and multiple replicates and are most conveniently analyzed using an autosampler. Care must be taken, however, to ensure that there is no loss of signal, due to aggregation, in samples left standing in the autosampler for many hours. This is especially of concern for pathological length polyGln peptides. We have found that doping centrifugation supernatants with formic acid to a final concentration of 20% stabilizes polyGln peptide solutions for at least 20 h at RT in an autosampler tray. (It is important to emphasize that this method of stabilizing samples against aggregation while awaiting analysis may be effective only for centrifugation supernatants – i.e., for largely aggregate-free samples of polyGln; it appears not to be effective if there are significant aggregation seeds in the sample.) Whatever conditions are used for preserving sample integrity, it is a good idea to confirm that they are working by analyzing one or more samples at two times, spanning the range of times that samples are required to sit awaiting analysis. If the concentration of polyGln differs substantially in the repeated analyses, there may be a problem with time-dependent aggregation in the autosampler.

Acknowledgements

The author would like to thank Geetha Thiagarajan for help with the experimental Section of this chapter and Angela Williams, Valerie Berthelier, and Ashwani Thakur for permission to include unpublished data. I am also very grateful to Marcy MacDonald for providing Figure 37.1. I would like to acknowledge past and present members of my laboratory, in particular, Songming Chen, Valerie Berthelier, J. Bradley Hamilton, Wen Yang, Ashwani Thakur, Anusri Bhattacharyya, Brian O'Nuallain, Alex Osmand, Angela Williams, and Tina Richey, for their work and ideas in developing the body of work summarized here. I gratefully acknowledge the Hereditary Disease Foundation and the NIH (R01 AG19322) for sustained funding and Ethan Signer, Carl Johnson, and Nancy Wexler for their enthusiastic support.

References

1 Wilmot, G. R. & Warren, S. T. (1998). A new mutational basis for disease. In *Genetic Instabilities and Hereditary Neurological Diseases* (Wells, R. D. & Warren, S. T., eds.), pp. 3–12. Academic Press, San Diego.

2 Cummings, C. J. & Zoghbi, H. Y. (2000). Fourteen and counting: unraveling trinucleotide repeat diseases. *Hum Mol Genet* 9, 909–916.

3 Bates, G. P. & Benn, C. (2002). The polyglutamine diseases. In *Huntington's Disease* (Bates, G. P., Harper, P. S. & Jones, L., eds.), pp. 429–472. Oxford University Press, Oxford, U.K.

4 DiFiglia, M., Sapp, E., Chase, K. O., Davies, S. W., Bates, G. P., Vonsattel, J. P. & Aronin, N. (1997). Aggregation of huntingtin in neuronal intranuclear inclusions and dystrophic neurites in brain. *Science* 277, 1990–3.

5 Ordway, J. M., Tallaksen-Greene, S., Gutekunst, C. A., Bernstein, E. M., Cearley, J. A., Wiener, H. W., Dure, L. S. t., Lindsey, R., Hersch, S. M., Jope, R. S., Albin, R. L. & Detloff, P. J. (1997). Ectopically expressed CAG repeats cause intranuclear inclusions and a progressive late onset neurological phenotype in the mouse [In Process Citation]. *Cell* 91, 753–63.

6 Davies, S. W., Turmaine, M., Cozens, B. A., DiFiglia, M., Sharp, A. H., Ross, C. A., Scherzinger, E., Wanker, E. E., Mangiarini, L. & Bates, G. P. (1997). Formation of neuronal intranuclear inclusions underlies the neurological dysfunction in mice transgenic for the HD mutation. *Cell* 90, 537–48.

7 Paulson, H. L., Perez, M. K., Trottier, Y., Trojanowski, J. Q., Subramony, S. H., Das, S. S., Vig, P., Mandel, J. L., Fischbeck, K. H. & Pittman, R. N. (1997). Intranuclear inclusions of expanded polyglutamine protein in spinocerebellar ataxia type 3. *Neuron* 19, 333–44.

8 Martin, J. B. (1999). Molecular basis of the neurodegenerative disorders [published erratum appears in N Engl J Med 1999 Oct 28;341(18):1407]. *N Engl J Med* 340, 1970–80.

9 Chen, Y. W. (2003). Site-specific mutagenesis in a homogeneous polyglutamine tract: application to spinocerebellar ataxin-3. *Protein Eng* 16, 1–4.

10 Thakur, A. & Wetzel, R. (2002). Mutational analysis of the structural organization of polyglutamine aggregates. *Proc Natl Acad Sci USA* 99, 17014–17019.

11 Wells, R. D. & Warren, S. T. (1998). *Genetic Instabilities and Hereditary Neurological Diseases*, Academic Press, San Diego.

12 Bates, G. P., Harper, P. S. & Jones, L., Eds. (2002). Huntington's Disease. Oxford, U.K.: Oxford University Press.

13 Reddy, P. S. & Housman, D. E. (1997). The complex pathology of trinucleotide repeats. *Curr Opin Cell Biol* 9, 364–72.

14 Gusella, J. F. & MacDonald, M. E. (1998). Huntingtin: a single bait hooks many species. *Curr Opin Neurobiol* 8, 425–30.

15 Reddy, P. H., Williams, M. & Tagle, D. A. (1999). Recent advances in understanding the pathogenesis of Huntington's disease. *Trends Neurosci* 22, 248–55.

16 Lin, X., Cummings, C. J. & Zoghbi, H. Y. (1999). Expanding our understanding of polyglutamine diseases through mouse models. *Neuron* 24, 499–502.

17 Zoghbi, H. Y. & Orr, H. T. (1999). Polyglutamine diseases: protein cleavage and aggregation. *Curr Opin Neurobiol* 9, 566–70.

18 Paulson, H. L. (1999). Protein fate in neurodegenerative proteinopathies: polyglutamine diseases join the (mis)fold. *Am J Hum Genet* 64, 339–45.

19 Ferrigno, P. & Silver, P. A. (2000). Polyglutamine expansions: proteolysis, chaperones, and the dangers of promiscuity. *Neuron* 26, 9–12.

20 Tobin, A. J. & Signer, E. R. (2000). Huntington's disease: the challenge for cell biologists. *Trends Cell Biol* 10, 531–536.
21 Zoghbi, H. Y. & Orr, H. T. (2000). Glutamine repeats and neurodegeneration. *Annu Rev Neurosci* 23, 217–47.
22 Bates, G. (2000). Huntington's disease. In reverse gear. *Nature* 404, 944–5.
23 Cha, J. H. (2000). Transcriptional dysregulation in Huntington's disease. *Trends Neurosci* 23, 387–92.
24 McCampbell, A. & Fischbeck, K. H. (2001). Polyglutamine and CBP: fatal attraction? *Nat Med* 7, 528–30.
25 Ross, C. A. (2002). Polyglutamine pathogenesis: emergence of unifying mechanisms for Huntington's disease and related disorders. *Neuron* 35, 819–22.
26 Rubinsztein, D. C. (2002). Lessons from animal models of Huntington's disease. *Trends Genet* 18, 202–9.
27 Ross, C. A., Poirier, M. A., Wanker, E. E. & Amzel, M. (2003). Polyglutamine fibrillogenesis: the pathway unfolds. *Proc Natl Acad Sci USA* 100, 1–3.
28 Bates, G. (2003). Huntingtin aggregation and toxicity in Huntington's disease. *Lancet* 361, 1642–4.
29 Michalik, A. & Van Broeckhoven, C. (2003). Pathogenesis of polyglutamine disorders: aggregation revisited. *Hum Mol Genet* 12 Suppl 2, R173–86.
30 Paulson, H. (2003). Polyglutamine neurodegeneration: minding your Ps and Qs. *Nat Med* 9, 825–826.
31 Rudnicki, D. D. & Margolis, R. L. (2003). Repeat expansion and autosomal dominant neurodegenerative disorders: consensus and controversy. *Exp. Rev. Mol. Med.* 5, 1–24.
32 Koide, R., Kobayashi, S., Shimohata, T., Ikeuchi, T., Maruyama, M., Saito, M., Yamada, M., Takahashi, H. & Tsuji, S. (1999). A neurological disease caused by an expanded CAG trinucleotide repeat in the TATA-binding protein gene: a new polyglutamine disease? *Hum Mol Genet* 8, 2047–53.
33 Margolis, R. L. (2002). The spinocerebellar ataxias: order emerges from chaos. *Curr Neurol Neurosci Rep* 2, 447–56.
34 Albin, R. L. (2003). Dominant ataxias and Friedreich ataxia: an update. *Curr Opin Neurol* 16, 507–14.
35 Bowater, R. P. & Wells, R. D. (2001). The intrinsically unstable life of DNA triplet repeats associated with human hereditary disorders. *Prog Nucleic Acid Res Mol Biol* 66, 159–202.
36 Harper, P. S. & Jones, L. (2002). Huntington's disease: genetic and molecular studies. In *Huntington's Disease* (Bates, G. P., Harper, P. S. & Jones, L., eds.), pp. 113–158. Oxford University Press, Oxford, U.K.
37 Myers, R. H., Marans, K. S. & MacDonald, M. E. (1998). Huntington's Disease. In *Genetic Instabilities and Hereditary Neurological Diseases* (Wells, R. D. & Warren, S. T., eds.), pp. 301–323. Academic Press, San Diego.
38 Zajac, J. D. & MacLean, H. E. (1998). Kennedy's disease: clinical aspects. In *Genetic Instabilities and Hereditary Neurological Diseases* (Wells, R. D. & Warren, S. T., eds.), pp. 87–111. Academic Press, San Diego.
39 Cattaneo, E., Rigamonti, D., Goffredo, D., Zuccato, C., Squitieri, F. & Sipione, S. (2001). Loss of normal huntingtin function: new developments in Huntington's disease research. *Trends Neurosci* 24, 182–8.
40 Kazemi-Esfarjani, P. & Benzer, S. (2000). Genetic suppression of polyglutamine toxicity in Drosophila. *Science* 287, 1837–40.
41 Bennett, M. J., Huey-Tubman, K. E., Herr, A. B., West, A. P., Ross, S. A. & Bjorkman, P. J. (2002). A linear lattice model for polyglutamine in CAG expansion diseases. *Proc Natl Acad Sci USA* 99, 11634–11639.
42 Faber, P. W., Voisine, C., King, D. C., Bates, E. A. & Hart, A. C. (2002). Glutamine/proline-rich PQE-1 proteins protect Caenorhabditis elegans neurons from huntingtin

polyglutamine neurotoxicity. *Proc Natl Acad Sci USA* 99, 17131–6.

43 Kazantsev, A., Walker, H. A., Slepko, N., Bear, J. E., Preisinger, E., Steffan, J. S., Zhu, Y. Z., Gertler, F. B., Housman, D. E., Marsh, J. L. & Thompson, L. M. (2002). A bivalent Huntingtin binding peptide suppresses polyglutamine aggregation and pathogenesis in Drosophila. *Nat Genet* 25, 25.

44 Thakur, A. K., Yang, W., & Wetzel, R. (2004). Inhibition of polyglutamine aggregate cytotoxicity by a structure-based elongation inhibitor. *FASEB J.* 18, 923–925.

45 Yang, W., Dunlap, J. R., Andrews, R. B. & Wetzel, R. (2002). Aggregated polyglutamine peptides delivered to nuclei are toxic to mammalian cells. *Hum Mol Genet* 11, 2905–2917.

46 Altschuler, E. L., Hud, N. V., Mazrimas, J. A. & Rupp, B. (1997). Random coil conformation for extended polyglutamine stretches in aqueous soluble monomeric peptides. *J Pept Res* 50, 73–5.

47 Sharma, D., Sharma, S., Pasha, S. & Brahmachari, S. K. (1999). Peptide models for inherited neurodegenerative disorders: conformation and aggregation properties of long polyglutamine peptides with and without interruptions. *FEBS Lett* 456, 181–5.

48 Chen, S. & Wetzel, R. (2001). Solubilization and disaggregation of polyglutamine peptides. *Protein Science* 10, 887–891.

49 Chen, S., Berthelier, V., Yang, W. & Wetzel, R. (2001). Polyglutamine aggregation behavior in vitro supports a recruitment mechanism of cytotoxicity. *J. Mol. Biol.* 311, 173–182.

50 Masino, L., Kelly, G., Leonard, K., Trottier, Y. & Pastore, A. (2002). Solution structure of polyglutamine tracts in GST-polyglutamine fusion proteins. *FEBS Lett* 513, 267–72.

51 Shi, Z., Olson, C. A., Rose, G. D., Baldwin, R. L. & Kallenbach, N. R. (2002). Polyproline II structure in a sequence of seven alanine residues. *Proc Natl Acad Sci USA* 99, 9190–5.

52 Rucker, A. L., Pager, C. T., Campbell, M. N., Qualls, J. E. & Creamer, T. P. (2003). Host-guest scale of left-handed polyproline II helix formation. *Proteins* 53, 68–75.

53 Perutz, M. F., Johnson, T., Suzuki, M. & Finch, J. T. (1994). Glutamine repeats as polar zippers: their possible role in inherited neurodegenerative diseases. *Proc Natl Acad Sci USA* 91, 5355–8.

54 Scherzinger, E., Lurz, R., Turmaine, M., Mangiarini, L., Hollenbach, B., Hasenbank, R., Bates, G. P., Davies, S. W., Lehrach, H. & Wanker, E. E. (1997). Huntingtin-encoded polyglutamine expansions form amyloid-like protein aggregates in vitro and in vivo. *Cell* 90, 549–58.

55 Scherzinger, E., Sittler, A., Schweiger, K., Heiser, V., Lurz, R., Hasenbank, R., Bates, G. P., Lehrach, H. & Wanker, E. E. (1999). Self-assembly of polyglutamine-containing huntingtin fragments into amyloid-like fibrils: implications for Huntington's disease pathology. *Proc Natl Acad Sci USA* 96, 4604–9.

56 Onodera, O., Burke, J. R., Miller, S. E., Hester, S., Tsuji, S., Roses, A. D. & Strittmatter, W. J. (1997). Oligomerization of expanded-polyglutamine domain fluorescent fusion proteins in cultured mammalian cells. *Biochem Biophys Res Commun* 238, 599–605.

57 Perez, M. K., Paulson, H. L., Pendse, S. J., Saionz, S. J., Bonini, N. M. & Pittman, R. N. (1998). Recruitment and the role of nuclear localization in polyglutamine-mediated aggregation. *J Cell Biol* 143, 1457–70.

58 Sherman, M. Y. & Muchowski, P. J. (2003). Making yeast tremble: yeast models as tools to study neurodegenerative disorders. *Neuromolecular Med* 4, 133–46.

59 Driscoll, M. & Gerstbrein, B. (2003). Dying for a cause: invertebrate genetics takes on human neurodegeneration. *Nat Rev Genet* 4, 181–94.

60 Kazantsev, A., Preisinger, E.,

Dranovsky, A., Goldgaber, D. & Housman, D. (1999). Insoluble detergent-resistant aggregates form between pathological and nonpathological lengths of polyglutamine in mammalian cells. *Proc Natl Acad Sci USA* 96, 11404–9.

61 Huang, C. C., Faber, P. W., Persichetti, F., Mittal, V., Vonsattel, J. P., MacDonald, M. E. & Gusella, J. F. (1998). Amyloid formation by mutant huntingtin: threshold, progressivity and recruitment of normal polyglutamine proteins. *Somat Cell Mol Genet* 24, 217–33.

62 Preisinger, E., Jordan, B. M., Kazantsev, A. & Housman, D. (1999). Evidence for a recruitment and sequestration mechanism in Huntington's disease. *Philos Trans R Soc Lond B Biol Sci* 354, 1029–34.

63 McCampbell, A., Taylor, J. P., Taye, A. A., Robitschek, J., Li, M., Walcott, J., Merry, D., Chai, Y., Paulson, H., Sobue, G. & Fischbeck, K. H. (2000). CREB-binding protein sequestration by expanded polyglutamine. *Hum Mol Genet* 9, 2197–202.

64 Howlett, D. R., Jennings, K. H., Lee, D. C., Clark, M. S., Brown, F., Wetzel, R., Wood, S. J., Camilleri, P. & Roberts, G. W. (1995). Aggregation state and neurotoxic properties of Alzheimer beta-amyloid peptide. *Neurodegeneration* 4, 23–32.

65 Wood, S. J., Chan, W. & Wetzel, R. (1996). Seeding of Aβ fibril formation is inhibited by all three isotypes of apolipoprotein E. *Biochem.* 35, 12623–12628.

66 Chen, S., Berthelier, V., Hamilton, J. B., O'Nuallain, B. & Wetzel, R. (2002). Amyloid-like features of polyglutamine aggregates and their assembly kinetics. *Biochemistry* 41, 7391–7399.

67 Berthelier, V., Hamilton, J. B., Chen, S. & Wetzel, R. (2001). A microtiter plate assay for polyglutamine aggregate extension. *Anal Biochem* 295, 227–36.

68 Zagorski, M. G., Yang, J., Shao, H., Ma, K., Zeng, H. & Hong, A. (1999). Methodological and chemical factors affecting amyloid-β amyloidogenicity. *Meth Enzymol* 309, 189–204.

69 Franks, F. (1985). *Biophysics and biochemistry at low temperatures*, Cambridge University Press, Cambridge.

70 Franks, F. (1993). Storage stabilization of proteins. In *Protein Biotechnology: Isolation, Characterization and Stabilization* (Franks, F., ed.), pp. 489–531. Humana Press, New York.

71 Jaenicke, R. & Seckler, R. (1997). Protein Assembly In Vitro. In *Protein Misassembly* (Wetzel, R., ed.), Vol. 50, pp. 1–59. Academic Press, San Diego.

72 Anderson, V. J. & Lekkerkerker, H. N. (2002). Insights into phase transition kinetics from colloid science. *Nature* 416, 811–5.

73 Bishop, M. F. & Ferrone, F. A. (1984). Kinetics of nucleation-controlled polymerization. A perturbation treatment for use with a secondary pathway. *Biophys J* 46, 631–44.

74 Ferrone, F. (1999). Analysis of protein aggregation kinetics. *Meths. Enzymol.* 309, 256–274.

75 Cao, Z. & Ferrone, F. A. (1996). A 50th order reaction predicted and observed for sickle hemoglobin nucleation. *J Mol Biol* 256, 219–22.

76 Chen, S., Ferrone, F. & Wetzel, R. (2002). Huntington's Disease age-of-onset linked to polyglutamine aggregation nucleation. *Proc. Natl. Acad. Sci. USA* 99, 11884–11889.

77 Abraham, F. F. (1974). *Homogeneous Nucleation Theory*, Academic Press, New York.

78 Perutz, M. F. & Windle, A. H. (2001). Cause of neural death in neurodegenerative diseases attributable to expansion of glutamine repeats. *Nature* 412, 143–4.

79 Serio, T. R., Cashikar, A. G., Kowal, A. S., Sawicki, G. J., Moslehi, J. J., Serpell, L., Arnsdorf, M. F. & Lindquist, S. L. (2000). Nucleated conformational conversion and the replication of conformational information by a prion determinant. *Science* 289, 1317–21.

80 Naiki, H. & Gejyo, F. (1999). Kinetic analysis of amyloid fibril formation. *Meths Enzymol* 309, 305–318.
81 Esler, W. P., Stimson, E. R., Ghilardi, J. R., Felix, A. M., Lu, Y. A., Vinters, H. V., Mantyh, P. W. & Maggio, J. E. (1997). A beta deposition inhibitor screen using synthetic amyloid. *Nat Biotechnol* 15, 258–63.
82 Myszka, D. G., Wood, S. J. & Biere, A. L. (1999). Analysis of fibril elongation using surface plasmon resonance biosensors. In *Amyloid, Prions and other Protein Aggregates* (Wetzel, R., ed.), Vol. 309, pp. 386–402. Academic Press, San Diego.
83 Berthelier, V. & Wetzel, R. (2003). An assay for characterizing the in vitro kinetics of polyglutamine aggregation. In *Methods in Molecular Medicine: Neurogenetics* (N. T. Potter, e., ed.), pp. 295–303. Humana Press, New York.
84 Esler, W. P., Stimson, E. R., Jennings, J. M., Vinters, H. V., Ghilardi, J. R., Lee, J. P., Mantyh, P. W. & Maggio, J. E. (2000). Alzheimer's disease amyloid propagation by a template-dependent dock-lock mechanism. *Biochemistry* 39, 6288–95.
85 Nucifora, F. C., Jr., Sasaki, M., Peters, M. F., Huang, H., Cooper, J. K., Yamada, M., Takahashi, H., Tsuji, S., Troncoso, J., Dawson, V. L., Dawson, T. M. & Ross, C. A. (2001). Interference by huntingtin and atrophin-1 with cbp-mediated transcription leading to cellular toxicity. *Science* 291, 2423–8.
86 Harper, J. D., Wong, S. S., Lieber, C. M. & Lansbury, P. T. (1997). Observation of metastable Abeta amyloid protofibrils by atomic force microscopy. *Chem Biol* 4, 119–25.
87 Kayed, R., Head, E., Thompson, J. L., McIntire, T. M., Milton, S. C., Cotman, C. W. & Glabe, C. G. (2003). Common structure of soluble amyloid oligomers implies common mechanism of pathogenesis. *Science* 300, 486–9.
88 Poirier, M. A., Li, H., Macosko, J., Cai, S., Amzel, M. & Ross, C. A. (2002). Huntingtin spheroids and protofibrils as precursors in polyglutamine fibrilization. *J Biol Chem* 277, 41032–7.
89 Divry, P. & Florkin, M. (1927). Sur les proprietes optiques de l'amyloide. *C. R. Seances Soc Biol* 97, 1808–10.
90 Westermark, G. T., Johnson, K. H. & Westermark, P. (1999). Staining methods for identification of amyloid in tissue. *Methods Enzymol* 309, 3–25.
91 Klunk, W. E., Jacob, R. F. & Mason, R. P. (1999). Quantifying amyloid beta-peptide (Abeta) aggregation using the congo red-abeta (CR-abeta) spectrophotometric assay [In Process Citation]. *Anal Biochem* 266, 66–76.
92 Naiki, H., Higuchi, K., Hosokawa, M. & Takeda, T. (1989). Fluorometric determination of amyloid fibrils in vitro using the fluorscent dye, thioflavine T. *Anal. Biochem.* 177, 244–249.
93 LeVine, H. (1999). Quantification of β-sheet amyloid fibril structures with thioflavin T. *Meth Enzymol* 309, 274–84.
94 Woody, R. W. (1996). Theory of circular dichroism of proteins. In *Circular Dichroism and Conformational Analysis of Biomolecules* (Fasman, G. D., ed.), pp. 25–67. Plenum, New York.
95 Hrncic, R., Wall, J., Wolfenbarger, D. A., Murphy, C. L., Schell, M., Weiss, D. T. & Solomon, A. (2000). Antibody-Mediated Resolution of Light Chain-Associated Amyloid Deposits. *Am J Pathol* 157, 1239–1246.
96 O'Nuallain, B. & Wetzel, R. (2002). Conformational antibodies recognizing a generic amyloid fibril epitope. *Proc Natl Acad Sci USA* 99, 1485–1490.
97 Sunde, M. & Blake, C. (1997). The structure of amyloid fibrils by electron microscopy and X-ray diffraction. *Adv Protein Chem* 50, 123–159.
98 Wood, S. J., Wetzel, R., Martin, J. D. & Hurle, M. R. (1995). Prolines and amyloidogenicity in fragments of the Alzheimer's peptide β/A4. *Biochem.* 34, 724–730.
99 Williams, A., Portelius, E.,

Kheterpal, I., Guo, J., Cook, K., Xu, Y. & Wetzel, R. (2004). Mapping Aβ amyloid fibril secondary structure using scanning proline mutagenesis. *J. Mol. Biol.*, 335, 833–42.

100 Venkatraman, J., Shankaramma, S. C. & Balaram, P. (2001). Design of folded peptides. *Chem Rev* 101, 3131–52.

101 Wetzel, R. (2002). Ideas of order for amyloid fibril structure. *Structure* 10, 1031–1036.

102 Starikov, E. B., Lehrach, H. & Wanker, E. E. (1999). Folding of oligoglutamines: a theoretical approach based upon thermodynamics and molecular mechanics. *J Biomol Struct Dyn* 17, 409–27.

103 Perutz, M. F., Finch, J. T., Berriman, J. & Lesk, A. (2002). Amyloid fibers are water-filled nanotubes. *Proc Natl Acad Sci USA* 99, 5591–5.

104 Jenkins, J. & Pickersgill, R. (2001). The architecture of parallel beta-helices and related folds. *Prog Biophys Mol Biol* 77, 111–75.

105 Pickersgill, R. W. (2003). A primordial structure underlying amyloid. *Structure (Camb)* 11, 137–8.

106 Bence, N. F., Sampat, R. M. & Kopito, R. R. (2001). Impairment of the ubiquitin-proteasome system by protein aggregation. *Science* 292, 1552–5.

107 Kopito, R. R. (2000). Aggresomes, inclusion bodies and protein aggregation. *Trends Cell Biol* 10, 524–30.

108 Lashuel, H. A., Hartley, D., Petre, B. M., Walz, T. & Lansbury, P. T., Jr. (2002). Neurodegenerative disease: amyloid pores from pathogenic mutations. *Nature* 418, 291.

109 Margolis, R. L., Abraham, M. R., Gatchell, S. B., Li, S. H., Kidwai, A. S., Breschel, T. S., Stine, O. C., Callahan, C., McInnis, M. G. & Ross, C. A. (1997). cDNAs with long CAG trinucleotide repeats from human brain. *Hum Genet* 100, 114–22.

110 Hancock, J. M., Worthey, E. A. & Santibanez-Koref, M. F. (2001). A role for selection in regulating the evolutionary emergence of disease-causing and other coding CAG repeats in humans and mice. *Mol Biol Evol* 18, 1014–23.

111 Collins, J. R., Stephens, R. M., Gold, B., Long, B., Dean, M. & Burt, S. K. (2003). An exhaustive DNA micro-satellite map of the human genome using high performance computing. *Genomics* 82, 10–9.

112 Talcott, B. & Moore, M. S. (1999). Getting across the nuclear pore complex. *Trends Cell Biol* 9, 312–8.

113 Osmand, A. P., Berthelier, V. & Wetzel, R. (2002). Identification of aggregation foci, intracellular neuronal structures in the neocortex in Huntington's disease capable of recruiting polyglutamine. *Program 293.6, 2002 Abstract Viewer/Itinerary Planner.* Washington, DC: Society for Neuroscience, 2002. Online.

114 Heiser, V., Scherzinger, E., Boeddrich, A., Nordhoff, E., Lurz, R., Schugardt, N., Lehrach, H. & Wanker, E. E. (2000). Inhibition of huntingtin fibrillogenesis by specific antibodies and small molecules: Implications for Huntington's disease therapy. *Proc. Natl. Acad. Sci. (USA)* 97, 6739–6744.

115 Wetzel, R., Ed. (1999). Amyloid, Prions, and other Protein Aggregates. Methods in Enzymology. Edited by Abelson, J. N. & Simon, M. I. San Diego, CA: Academic Press.

116 Kheterpal, I., Zhou, S., Cook, K. D. & Wetzel, R. (2000). Abeta amyloid fibrils possess a core structure highly resistant to hydrogen exchange. *Proc Natl Acad Sci USA* 97, 13597–601.

117 Kheterpal, I., Williams, A., Murphy, C., Bledsoe, B. & Wetzel, R. (2001). Structural features of the Aβ amyloid fibril elucidated by limited proteolysis. *Biochem.* 40, 11757–11767.

118 Torok, M., Milton, S., Kayed, R., Wu, P., McIntire, T., Glabe, C. G. & Langen, R. (2002). Structural and Dynamic Features of Alzheimer's Abeta Peptide in Amyloid Fibrils Studied by Site-directed Spin Labeling. *J Biol Chem* 277, 40810–40815.

119 Petkova, A. T., Ishii, Y., Balbach, J. J., Antzutkin, O. N., Leapman, R. D., Delaglio, F. & Tycko, R. (2002). A structural model for Alzheimer's beta-amyloid fibrils based on experimental constraints from solid state NMR. *Proc Natl Acad Sci USA* 99, 16742–7.

120 Foguel, D., Suarez, M. C., Ferrao-Gonzales, A. D., Porto, T. C., Palmieri, L., Einsiedler, C. M., Andrade, L. R., Lashuel, H. A., Lansbury, P. T., Kelly, J. W. & Silva, J. L. (2003). Dissociation of amyloid fibrils of alpha-synuclein and transthyretin by pressure reveals their reversible nature and the formation of water-excluded cavities. *Proc Natl Acad Sci USA* 100, 9831–6.

38
Production of Recombinant Proteins for Therapy, Diagnostics, and Industrial Research by in Vitro Folding

Christian Lange and Rainer Rudolph

38.1
Introduction

Twenty years ago protein folding was regarded as a highly fascinating topic of pure basic research. Although tremendous progress has been made in our understanding of this enigmatic process, protein folding is still a very attractive field for academic research, as is documented in this handbook. However, at that time nobody would have anticipated that protein folding would today be a well-established downstream process in industrial production. Researchers in industry working on recombinant protein production are frequently confronted with misfolded, inactive material. They either can try to find a way to produce correctly folded native protein by using alternative expression systems or they can take up the challenge of in vitro protein folding. Since efficient processes are now available that allow medium- to large-scale production of recombinant proteins by in vitro folding, many recombinant proteins for therapy, diagnostics, and industrial applications are successfully and profitably produced via this process.

38.1.1
The Inclusion Body Problem

Before the advent of recombinant DNA technology, proteins for medical, diagnostic, or industrial applications were difficult to obtain from human body fluids, from animal or plant tissue, or from microorganisms. Potentially low availability and the risk of viral contamination or high antigenicity put severe limitations on the use of natural proteins in therapy. The extraction of proteins from natural sources was often very cost-intensive, requiring large amounts of raw material, water, and energy. Recombinant DNA technology has dramatically changed this situation. It has opened the way for the large-scale, low-cost production of almost any imaginable protein [1, 2]. Proteins that are available in only minute amounts from natural sources can now be produced in huge quantities in host cells such as *E. coli*. Expression levels up to 50% of the total cell protein can be obtained with standard expression systems (reviewed, e.g., in Ref. [3]). Using high-density fermentation

Protein Folding Handbook. Part II. Edited by J. Buchner and T. Kiefhaber

ISBN: 3-527-30784-2

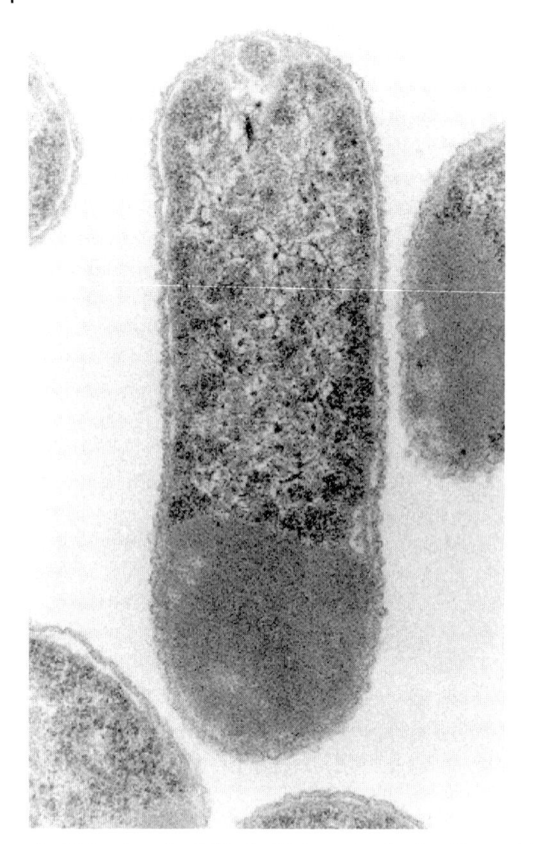

Fig. 38.1. *E. coli* cell filled with inclusion bodies (lower third of cell volume).

techniques, gram amounts of the desired proteins can be obtained per liter of fermentation broth [4–6].

There is, however, one major problem associated with overexpression in *E. coli*. Often the desired polypeptide is synthesized in these host cells, but not in the correctly folded, native form. Instead of the biologically active product, aggregates – inclusion bodies – are formed [7]. These dense particles, which may span the whole diameter of the host cell, are mostly biologically inactive and generally are insoluble in non-denaturing buffer systems (Figure 38.1). Inclusion bodies predominantly contain the overexpressed recombinant protein and only small amounts of host cell material such as inclusion body binding proteins (IBPs) [8], outer-membrane proteins, RNA polymerase, and ribosome components including ribosomal RNA as well as circular and nicked forms of plasmid DNA [9]. Some of these impurities may originate from co-precipitation of these cellular components together with the recombinant protein during inclusion body formation. However, most of the impurities found in inclusion body preparations may originate from incomplete removal of particulate host cell material during their isolation [10].

Inclusion body formation is not restricted to the overexpression of heterologous gene products in bacteria. Overexpression of autologous *E. coli* proteins in the cytosol of this organism has been observed to lead to inclusion body formation. For example, the overexpression of endogenous β-galactosidase in *E. coli* led to the formation of inclusion bodies that showed only partial enzymatic activity [11]. Inclusion body formation may also occur upon overexpression of proteins in eukaryotic host cells [12]. Apparently, high-level expression of any gene product beyond certain limits is sufficient to drive the recombinant protein into aggregation and inclusion body formation. Based on this observation, a simple kinetic model for inclusion body formation was proposed [13]. This model, which takes only the zero-order rate of protein synthesis and a subsequent competition between first-order folding and second-order aggregation into account, provides a fairly accurate description of inclusion body formation. In agreement with this model, inclusion body formation can be reduced or even prevented by decreasing the rate of protein synthesis. This can be achieved, for example, by cultivation at low temperature or by limited induction [14, 15].

Except for the presence of disulfide bonds, other molecular characteristics of proteins such as size, hydrophobicity, secondary structure content, etc., do not correlate well with the observed propensity for inclusion body formation. Small proteins containing few disulfide bonds are often expressed in soluble form, while inclusion body formation predominates upon cytosolic production of large proteins containing multiple disulfide bonds. Correct oxidative folding of more complex proteins is apparently not possible in the reducing cytosolic environment, in line with the finding that naturally secreted proteins mostly contain disulfide bonds, while the cysteine residues of intracellular proteins usually stay reduced [16]. The oxidative folding of secreted proteins occurs in the periplasm and the endoplasmic reticulum of prokaryotes and eukaryotes, respectively. These compartments offer optimum oxidizing conditions as well as thiol-disulfide isomerases for the formation of correct disulfide bonds [17–20]. Upon overexpression of disulfide-bonded proteins in the reducing bacterial cytosol, oxidative folding is massively impaired, and, as a consequence, inclusion body formation predominates even upon low-level expression.

By selecting for mutations in *E. coli*, which allow for disulfide bond formation in the cytoplasm, Beckwith and coworkers could isolate mutant strains with reduced or eliminated thioredoxin reductase activity [21–23]. These and other, further-improved, mutant strains showed increased cytosolic oxidative folding of disulfide-bonded proteins as compared to wild-type strains [24]. However, in spite of this recent progress in the soluble overexpression of heterologous proteins in *E. coli*, in many cases the formation of inclusion bodies still prevails.

The exact mechanism by which inclusion body formation is induced and the reason that the recombinant protein is deposited predominantly in the inclusion bodies are still unclear. Apart from the changes in the kinetic competition between folding and aggregation caused by changing the rate of protein synthesis, essential components of the chaperone machinery may be simply oversaturated upon massive overproduction of recombinant proteins. It is, however, not yet clear which chaperones are involved in this process. In some cases, inclusion body formation can be suppressed by co-overexpression of molecular chaperones such as GroE or DnaK and its cohort chaperones [25, 26]. Although this approach was quite efficient upon overexpression of some proteins, co-expression of chaperones had no effect whatsoever in other cases.

Because of the frequent, and sometimes frustrating, observation of inclusion bodies upon overexpression of recombinant proteins in *E. coli*, other host cell systems have been studied in detail for industrial protein production. Although very efficient in some cases, none of these expression systems offers the ultimate solution for recombinant protein production. Incorrect glycosylation in yeast and insect cells as well as long development times and high production costs in mammalian cell culture systems limit the general applicability of these alternatives. Therefore, enormous efforts have been made to develop efficient procedures for the reactivation of inactive bacterial inclusion body material by in vitro folding. These efforts relied heavily on the body of information available on in vitro protein folding that had been gathered over the years by basic research. Many protein folders from academia were approached by industry to help in setting up industrial folding processes. Since the main goal of academic research on protein folding was the identification of folding intermediates in order to understand the folding pathway of proteins in minute detail, the final goal of the industrial effort, which was to develop an efficient, reproducible, and cost-effective production process, sometimes came out of focus. In most previous academic in vitro folding studies, model proteins that showed more or less quantitative refolding under standard buffer conditions had been chosen. Many proteins of industrial relevance, on the other hand, form inactive aggregates under these conditions. In order to develop efficient processes, the "dark side" of in vitro folding, namely misfolding and aggregation, had to be taken into consideration. These unproductive side reactions had to be analyzed in detail in order to be overcome in maximizing the yield of the desired, correctly folded, native protein [10, 27–30].

Although subject to its specific problems and limitations, the industrial production of recombinant proteins for therapy, diagnostics, and research by overexpression into inclusion bodies and subsequent refolding has by now turned out to be an attractive and viable option. For example, Ernst et al. carefully compared the production cost for soluble expression of the recombinant heparinase in *E. coli* and for its production via inclusion bodies [31]. Mainly due to the low level of soluble expression, the isolation of the soluble protein was estimated to be 50% more expensive than the production of the enzyme by refolding of inclusion body material.

38.1.2
Cost and Scale Limitations in Industrial Protein Folding

Protein refolding is a relatively new unit operation in industrial processes for recombinant protein production. To date, there are no generally applicable protocols available for the optimization of a refolding process. There are, however, some guidelines. Protein-dependent variables make it difficult to estimate the final production cost before an extensive optimization of a given process has been carried out. It is impossible to predict the yield of refolding without analysis of a large number of parameters such as the maximum concentration at which refolding of a particular protein can be successfully carried out, optimum pH, temperature, time, and ionic strength, and the effect of one or more of many possible additives. While some proteins, like most of those described in this volume, refold quantitatively in simple buffer solutions, others may need sophisticated process design and cost-intensive additives to achieve the successful refolding of at least some of the starting material.

The protein concentration at which refolding can be performed best is a specific property of any particular protein. Because of the kinetic competition between first-order folding and higher-order unproductive aggregation, the yield of correct refolding decreases upon increasing the initial concentration of unfolded protein [10, 32]. While some proteins can be refolded to a high yield at high protein concentrations in the milligram per milliliter range, other proteins refold effectively only at concentrations on the order of micrograms per milliliter. For industrial processes, where comparatively huge amounts of proteins must be produced, the cost is extremely high if both the yield and the concentration of refolding are low. In an interesting case study, the process economics for the production of human tissue-type plasminogen activator (t-PA) by *E. coli* fermentation and in vitro folding were analyzed in detail. Human t-PA is a relatively complex protein, consisting of 527 amino acid residues that are organized in five structural domains. The native protein contains 17 disulfide bonds and one additional free cysteine and has a rather low solubility even in its native state. Because of the astronomical number of statistically possible disulfide bond pairings (2.2×10^{20}) and the low solubility of folding intermediates, t-PA poses an enormous problem for in vitro folding. This therapeutic protein activates plasmin at the surface of blood clots and is therefore ideally suited for treating patients suffering from acute myocardial infarction by inducing fibrinolysis. Its market potential justified the enormous efforts that were directed towards the optimization of its in vitro folding. Based on a low yield of refolding, to be obtained only at an extremely low protein concentration, Datar et al. calculated the process economics for the production of recombinant human t-PA from *E. coli* inclusion bodies [33]. The authors estimated that enormous refolding reactors (1500 m^3) would be necessary for production of the required amounts of protein. These reactors would account for 75% of the total equipment costs [34]. A simple comparison of the economics of the process involving the refolding of inclusion body protein from *E. coli* with an alternative process involving protein expression in Chinese hamster ovary cells led to the conclusion that the latter, despite the considerable costs of mammalian cell culture systems, was economically far more advantageous.

These studies seemed to demonstrate that it was clearly not feasible to set up an economic process for the production of recombinant human t-PA by bacterial overexpression and in vitro folding. However, a deletion variant of the plasminogen activator, Rapilysin, which was approved by the authorities in 1996 and has by now attained a considerable market share, is in fact produced by in vitro folding. A process was developed in which the protein is refolded in high yield at relatively high concentrations [35–37]. The clue to this success lay in careful process design and the inclusion of the low-molecular-weight additive L-arginine (L-Arg), which has proved effective for t-PA and many other proteins (cf. Section 38.3.4), in the refolding buffer. The presence of this additive led to a tremendous increase in the yield of refolding.

38.2 Treatment of Inclusion Bodies

38.2.1 Isolation of Inclusion Bodies

After cell lysis, inclusion bodies are usually harvested by solid-liquid separation. With this simple isolation procedure, the recombinant protein can be obtained in a relatively pure form, provided that all particulate matter of the *E. coli* host cells is disintegrated by a rigorous lysis procedure. Rather homogenous inclusion body isolates can be obtained by lysozyme treatment, subsequent high-pressure dispersion, and incubation with detergent at high salt concentrations [10]. This combination of cell disruption techniques provides for a nearly complete removal of cell wall debris and membrane fragments. As a consequence, highly enriched inclusion body material should be present in the pellet fraction of the subsequent sedimentation step.

After sedimentation, inclusion bodies form a very dense pellet. Impurities are difficult to extract from this glutinous material by later washing steps. Therefore, sedimentation should be performed after cell lysis and removal of membrane proteins by detergent and high salt concentrations (cf. Protocol 1). After the inclusion bodies have been collected by centrifugation, detergents and other buffer components that might interfere with the following solubilization and renaturation steps can be removed by repeated resuspension and sedimentation. This inclusion body isolation procedure, which combines different methods for cell lysis and disintegration, is rather robust and yields relatively pure inclusion body isolates (Figure 38.2), even in those cases where the lysis of *E. coli* is difficult due to changes in the cell wall morphology, which may occur as a response to the physiological stress caused by protein overexpression [38, 39].

38.2.2 Solubilization of Inclusion Bodies

Inclusion bodies are usually solubilized in high concentrations of denaturing agents such as 6 M guanidinium chloride (GdmCl) or 8 M urea. Guanidinium chloride, which is the stronger denaturant, may allow solubilization of inclusion body isolates that are resistant to urea. Since urea rapidly decomposes into cyanate

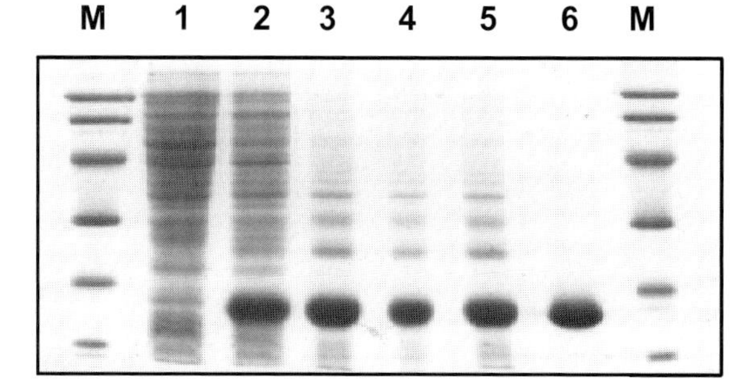

M:	Marker proteins
lane 1:	*E. coli* lysate from non-induced cells
lane 2:	*E. coli* lysate from induced cells
lane 3:	inclusion bodies
lane 4:	insoluble fraction after refolding
lane 5:	soluble fraction after refolding
lane 6:	purified nGLP1-R

Fig. 38.2. Expression in *E. coli*, inclusion body preparation, and purification of the N-terminal domain of human glucagon-like peptide 1 receptor. Reproduced from Ref. [54].

under certain solvent conditions (e.g., high pH values), care must be taken to avoid side chain modification of lysine and cysteine residues if this denaturant is used [40–42].

Although disulfide bonds are hardly formed in the reducing environment of the cytosol in vivo, random formation of these bonds, either within the same protein molecule or between different polypeptide chains, may occur by air oxidation during cell lysis and inclusion body isolation. This is of special importance in an industrial setting, where the incubation times may change upon scaling up the process. In order to guarantee the complete reduction of any accidentally formed disulfide bonds, reducing agents such as dithiothreitol (DTT), dithioerythritol (DTE), or β-mercaptoethanol should be present during solubilization. (cf. Protocol 2).

Provided that the level of overexpression is sufficiently high, the recombinant protein is usually found highly enriched in the inclusion body isolate (Figure 38.2). In this case, further purification of the recombinant protein in the denatured state prior to renaturation is not necessary. Contaminating proteins and other components of the host cells found in inclusion body isolates have no significant effect on protein refolding in most cases. Since the chromatographic resolution during the purification of unfolded proteins is sometimes poor and chromatographic purification represents a relatively cost-intensive unit operation, it should be avoided at this stage in industrial processes. In certain cases, however, contaminating proteins may co-precipitate during refolding of a given protein [43, 44]. The solubilized protein then has to be purified by, e.g., reversed-phase HPLC [45] or immobilized metal-affinity chromatography [46]. During research and process development, purification of the solubilized recombinant protein is essential if spectroscopic techniques such as fluorescence, circular dichroism, or light scattering are used to optimize refolding conditions.

38.3
Refolding in Solution

The following section aims to give an overview of the factors that have to be taken into account when dealing with the in vitro folding of a recombinant protein from inclusion bodies. Aggregation of unfolded protein and misfolded intermediates is in most cases the major complication during protein refolding. The ideas and methods that have been explored and used in order to overcome the problem posed by this unproductive side reaction will be discussed in some detail.

38.3.1
Protein Design Considerations

As already mentioned, the behavior of a protein upon refolding is a specific property of its given amino acid sequence. This property may be influenced by protein engineering, e.g., with respect to the pI value and the hydrophobicity. The refolding yield of a number of proteins could indeed be improved by protein design. The in vitro folding of human insulin-like growth factor was greatly enhanced by expressing it as a fusion protein with two Z domains derived from staphylococcal protein A [47]. This effect was mainly due to a more than 100-fold increase in the solubility of the reduced denatured polypeptide. In a similar fashion, the renaturation yield of granulocyte colony-stimulating factor could be systematically improved by fusion to various hydrophilic peptide sequences between 2 and 20 amino acids long [48].

Many important proteins naturally contain pro-sequences, which are later removed by specific proteolytic processing. These pro-sequences may also affect in vitro structure formation and act as natural folding helpers. The potentially beneficial "intramolecular chaperone" effect of pro-sequences was analyzed in detail for the folding of subtilisin [49] as well as for α-lytic protease [50]. Interestingly, the pro-region of α-lytic protease also promoted the correct folding of the enzyme when co-expressed in *trans*. In the case of human nerve growth factor, the pro-sequence was found to have a tremendously beneficial effect on both the rate and the yield of correct refolding and disulfide bond formation [51, 52].

Upon in vitro folding of proteins whose pro-sequences have been truncated or, more generally, of isolated protein domains, the refolding pathway may be corrupted by the absence of those parts of the protein that are essential for guiding structure formation towards the native state. This is especially true if any of the deleted parts strongly interact with the isolated domain in the native state. The energetic contribution of such an interaction may be essential for correct folding. In spite of these considerations, isolated domains have been successfully produced by in vitro folding. For example, the N-terminal extracellular domains of class B G protein-coupled receptors could be refolded in functional form [53, 54].

Finally, folding of polypeptides that naturally form part of a hetero-oligomeric protein may require the presence of all subunits of the complex for the formation of a thermodynamically stable, native state. The expression of fully functional tetrameric human hemoglobin in *E. coli*, for example, required the co-expression of the α and β globin genes [55]. Separate expression of the individual subunits resulted in a marked decrease in protein production as well as in inclusion body formation. In agreement with this observation, in vitro folding of β globin could proceed well only in the presence of α globin and the heme cofactor required for the formation of a functional tetramer [56].

38.3.2
Oxidative Refolding With Disulfide Bond Formation

Before designing a refolding strategy, it is essential to establish whether the target protein contains disulfide bonds in its native structure. While most cytosolic proteins contain only free cysteine residues, secreted proteins or those localized in cell organelles often form intramolecular cystine bridges [16]. Highly conserved cysteines in homologous sequences as well as the presence of export signals usually indicate the presence of disulfide bonds. In such a case, folding conditions must be chosen that facilitate their fast, efficient, and, above all, correct formation during oxidative refolding, since otherwise inactive, incorrectly disulfide-bonded products may accumulate, with potentially dramatic negative effects on the renaturation yield.

The number of possible combinations of disulfide bonds increases exponentially with the number of cysteines that are available for pairing (see Table 38.1). However, the problem is not as ill-behaved as it may look at first glance since pure statistics are usually overcome by effects that energetically favor the correct refolding pathway. Disulfide bond formation is directed towards the correct pairings by the conformational energy gained upon formation of the native structure [57–60]. Furthermore, during the in vivo folding of proteins, disulfide isomerases promote the reshuffling of disulfide bonds, avoiding kinetic trapping of misfolded conformations with incorrectly formed disulfide bridges (for recent reviews, see, e.g., Refs. [20, 61–63]).

The first experiments on in vitro folding with concomitant disulfide bond formation were performed using air as the oxidizing agent [64]. The rate and yield of reactivation that may be obtained by air oxidation have been found to be quite low. The reproducibility is usually low due to variations in surface-to-volume ratio, additional oxidation processes (e.g., methionine oxidation), and the presence of trace amounts of transition metal ions (especially Cu^{2+}) in variable concentrations in the buffer solutions. These metal ions act as efficient catalysts for air oxidation in a concentration range (low micromolar) that is experimentally difficult to control [65]. Because of these inherent drawbacks of air oxidation, more defined and efficient methods for disulfide bond formation in vitro have been developed.

Tab. 38.1. Number of possible combinations of $2n$ cysteines to form n disulfide bonds.

n	***Number of combinations***[1]
1	1
2	3
3	15
4	105
5	945
6	10 395
7	135 135
8	2 027 025
9	34 459 425
10	654 729 075
11	$1\,374\,931\,058 \times 10^{10}$
12	$3\,162\,341\,432 \times 10^{11}$
13	$7\,905\,853\,581 \times 10^{12}$
14	$2\,134\,580\,467 \times 10^{14}$
15	$6\,190\,283\,354 \times 10^{15}$
16	$1\,918\,987\,840 \times 10^{17}$
17	$6\,332\,659\,871 \times 10^{18}$
18	$2\,216\,430\,955 \times 10^{20}$
...	...

[1] $(2n-1) \cdot (2n-3) \cdot (2n-5) \cdot \ldots \cdot 3 \cdot 1$

As mentioned above, the reshuffling of incorrectly formed disulfide bridges is an essential step on the pathway towards the formation of native structure. In some cases, catalytic amounts of disulfide isomerase enzymes have been used for the efficient refolding of proteins (e.g., proinsulin) [66]. In general, however, thiol-disulfide exchange reactions between reduced polypeptide and low-molecular-weight thiols are used to facilitate disulfide bond shuffling [67, 68]. Reduced/oxidized glutathione (GSH/GSSG) are most commonly used as oxido-shuffling reagents, although other low-molecular-weight thiols such as cystine/cysteine, cystamine/cysteamine, or bis-β-hydroxyethyl disulfide/2-mercaptoethanol have been found to be equally effective in most cases. Since the thiolate anion is the active species in thiol-disulfide exchange, the pH of the refolding buffer must lie in the range of the pK_a of the thiol group (i.e., between pH 8.5 and pH 9.5).

Over the past few years, several low-molecular-weight thiol compounds have been designed with the aim of improving thiol-disulfide exchange during the oxidative refolding of proteins. For example, the synthetic dithiol $(\pm)$-*trans*-1,2-bis(2-mercaptoacetamido)cyclohexane (Vectrase P), which has a pK_a value of 8.3 and an $E^{0'}$ value of -0.24 V, was designed to mimic the properties of the active site of protein disulfide isomerases [69]. This compound improved the activation of scrambled RNAse A more efficiently than comparable monothiols. The synthetic dithiol compound also improved the refolding yield of human proinsulin [70]. Recently, various aromatic monothiols, characterized by low thiol pK_a values, were tested for their effect on rate and yield of the oxidative refolding of scrambled RNAse A [71, 72]. At slightly acidic pH values (pH 6.0), the aromatic thiols increased the refolding rate by a factor of 10 to 23 over that observed with glutathione. Since the nucleophilicity of the redox buffer thiolate determines the rate of thiol-disulfide exchange, the rate of RNAse A refolding was increased with decreasing thiol pK_a value.

Oxidative refolding is sometimes complicated by the low solubility of the aggregation-prone reduced-denatured polypeptide chains. This problem may be circumvented by the reversible redox modification of cysteine residues of a protein in its solubilized state prior to refolding. The formation of the mixed disulfide with glutathione introduces additional negatively charged residues on the polypeptide chain and may thereby reduce precipitation of unfolded protein during the initial states of oxidative refolding [73]. This idea has been demonstrated for the in vitro folding of human tissue-type plasminogen activator [37].

As an alternative to the complete removal of the reducing agent (DTT or DTE), which is present during solubilization (cf. Section 38.2.2), before oxidative refolding is carried out in a defined redox buffer, the solubilized protein solution (still containing DTT or DTE) may be directly diluted into a refolding buffer containing an oxidized monothiol (e.g., oxidized glutathione) [74]. Redox equilibration between the reduced dithiol and the oxidized monothiol will then result in predominantly oxidized dithiol and a redox buffer consisting of the reduced and oxidized monothiol. An elegant and simple approach towards oxidative refolding was demonstrated by Honda et al. [75]. They succeeded in refolding recombinant human growth/differentiation factor 5 (rhGDF-5) with a yield of 63% at high protein concentrations (2.4 mg mL^{-1}). The inclusion bodies were first solubilized in the presence of 32 mM cysteine/HCl for disulfide bond reduction. The solubilized protein solution was then diluted into a refolding buffer without any additional thiol reagent. Apparently, the reducing potential of the residual cysteine gradually decreased with time due to air oxidation, thus providing an oxidative redox-shuffling system composed of reduced and oxidized cysteine that was appropriate for rhGDF-5 refolding. It remains to be tested whether this simple and cost-effective protocol can be used for the refolding of other recombinant proteins at high concentrations.

38.3.3
Transfer of the Unfolded Proteins Into Refolding Buffer

The question of whether the buffer exchange from unfolding to folding conditions should be performed by dialysis or dilution is still controversial. On some occasions, higher folding yields might be obtained by dialysis as compared to rapid dilution of the denaturant. For a number of proteins, protocols comprising gradual removal of the denaturant have been reported. For example, mouse interleukin 4 [76], a recombinant fragment of bovine conglutinin [77], the fungal protease rhizopuspepsin [78], recombinant bovine guanylate cyclase-activating protein 1 [79], and human interleukin-21 [80] were obtained using stepwise dialysis for the removal of the denaturant. Recently, Umetsu et al. performed a detailed analysis of the refolding pathway of a single-chain antibody fragment during a complex dialysis protocol for the stepwise reduction of the GdmCl concentration [81].

For industrial processes, however, buffer exchange by dialysis presents considerable disadvantages with respect to reproducibility, handling, and ease of scale-up. Furthermore, many proteins form highly aggregation-prone folding intermediates at intermediate denaturant concentrations. Slow removal of the denaturant by dialysis then inevitably leads to quantitative and irreversible precipitation of these intermediates. In these cases, rapid removal of the denaturant by dilution reduces protein losses by aggregate formation, especially if carried out under properly controlled process conditions and in the presence of low-molecular-weight folding enhancers (cf. Section 38.3.4). Diluting the solubilized protein into refolding buffer is in most cases much more convenient and efficient than the removal of the denaturing agent by dialysis.

The rate of aggregation processes, which are of second or higher order, increases dramatically upon increasing the initial concentration of unfolded protein, while the first-order rate of productive folding processes does not change. Therefore, at high concentrations of denatured protein, aggregation processes predominate in the refolding reaction mixture, and this may represent a major difficulty for the design of an efficient refolding process. This problem can be at least partially overcome by process engineering (cf. Protocol 3). Since correctly folded protein does not participate in aggregation side reactions, high concentrations of native protein can be obtained by stepwise addition of the denatured protein to the refolding vessel [35]. By stepwise addition of the solubilized denatured protein at time intervals that are sufficiently large for the polypeptides to fold past the aggregation-prone early stages on the folding pathway, the concentration of unfolded protein and folding intermediates is kept sufficiently low at any given time during the refolding process (Figure 38.3). For example, this method has made possible the development of an economically feasible refolding process for Rapilysin, a variant of

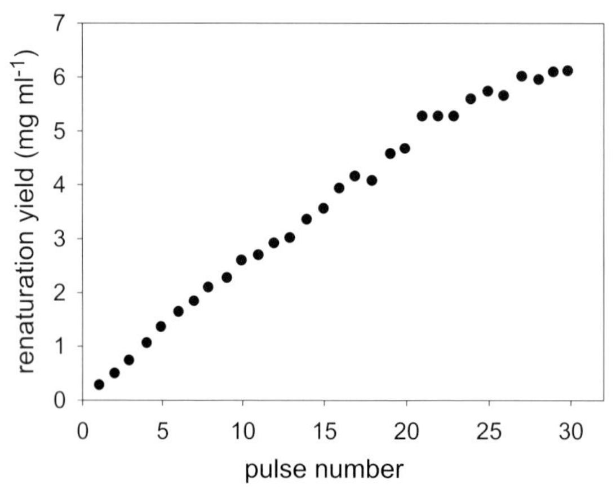

Fig. 38.3. Renaturation yield of proinsulin upon stepwise dilution of denatured protein into the refolding buffer. Reproduced from Ref. [82].

tissue-type plasminogen activator in refolding reactors that are about three orders of magnitude smaller than the huge and prohibitively expensive 1500-m^3 tanks that were considered necessary in the initial analysis [34].

38.3.4
Refolding Additives

One would be hard-pressed to give an exhaustive account of all low-molecular-weight compounds that have been found to improve in vitro folding of one or another model protein under some conditions. This section aims at giving a, necessarily incomplete, overview of additives that show no, in a biochemical sense, specific interaction with the refolded proteins but rather act as co-solvents that modify the solvation properties of the buffer for dissolved polypeptides and thereby influence the pathways of refolding and of its unproductive side reactions, potentially to great effect. Systematic studies by Timasheff and coworkers (see, e.g., Refs. [83, 84]) have greatly contributed to our understanding of the interplay among polypeptides, water, and co-solvents. Generally speaking, when using low-molecular-weight compounds as co-solvents in refolding reactions, the task is to strike a balance between, on the one hand, preserving the relative stability of the native state and, on the other hand, stabilizing denatured polypeptides and intermediates in solution in order to prevent them from following the path down to aggregation.

The earliest and most simple substances used to enhance refolding were the denaturing agents GdmCl and urea themselves, at non-denaturing concentrations [85]. Among other denaturing substances, alkyl ureas, carbonic acid amides [37], and alcohols [86] have been used to improve in vitro folding. The positive effect of theses substances on refolding is most probably caused by their very properties as denaturing agents, as they thermodynamically stabilize unfolded and misfolded forms of polypeptides in solution. As a consequence, the solubility of denatured protein and folding intermediates is increased and aggregation is reduced.

For a number of proteins, improved refolding has been observed upon addition of polyethylene glycol to the refolding buffer [87, 88]. Although polyethylene glycols do not act as denaturants, they have been found to decrease the thermostability of proteins with increasing average hydrophobicity of the amino acid sequence. An analysis of solvent-protein interactions indicated that the polyethylene glycols interact favorably with the hydrophobic side chains exposed upon unfolding at elevated temperatures [89]. Again, these hydrophobic interactions contribute to the stability of denatured polypeptides, folding intermediates, and misfolded forms in solution.

Detergents represent another class of substances that may be assumed to stabilize aggregation-prone states of polypeptides in solution by hydrophobic interactions. Indeed, detergents or phospholipid-detergent mixed micelles have been found to promote the correct refolding of various proteins [86, 90, 91]. Interestingly, the nonionic detergent Brij 58P was recently reported to be highly effective in preventing aggregation and promoting folding at very low, near-stoichiometric concentrations [92].

In other cases, osmolytes that generally stabilize the native states of proteins, such as sugars, polyalcohols, and trimethylamine oxide (TMAO), were found to be beneficial for the in vitro folding of various proteins [93, 94]. For example, a destabilized variant of hen egg white lysozyme, containing only three disulfide bonds, could be efficiently refolded in the presence of 20% glycerol [95]. In the case of human placental alkaline phosphatase, refolding could be achieved only in the presence of stabilizing sugars or Hofmeister salts [96]. In these cases efficient folding apparently depends on the relative stabilization of the native state and productive intermediates with respect to nonnative conformational states. However, the balance between the beneficial effect of stabilizing the desired native state of a protein and destabilizing the intermediates it has to pass on its folding pathway is apparently delicate. TMAO, for example, was recently found to hinder the productive refolding of lactate dehydrogenase [97]. Stabilizing osmolytes are therefore generally not the first choice as co-solvents for in vitro folding, unless the stability of the native state is an issue.

Several amino acids have been found to have a beneficial effect on the in vitro folding of certain proteins. For example, the presence of 5 M proline could prevent aggregation and, consequently, improve the refolding yield of bovine carbonic anhydrase II [98]. The refolding of lactate dehydrogenase, however, was negatively influenced by the presence of proline [97]. Recently, all naturally occurring amino acids (with the exception of cysteine, histidine, and the aromatic amino acids) were screened for their efficiency in suppressing the aggregation of lysozyme during thermal denaturation and during dilution of the reduced-carboxymethylated denatured protein from a urea solution into buffer [99]. In this study, the most basic amino acid, L-Arg, was found to be the most effective suppressor of aggregation. This is in line with the fact that L-Arg has probably been the most widely used low-molecular-weight refolding enhancer since it was first described in a patent application almost 20 years ago (cf. Ref. [35]). The presence of L-Arg as a co-solvent in concentrations above 0.4 M (up to 1.2 M) was found to promote the refolding of many different proteins. It was successfully used in the refolding of human tissue-type plasminogen activator [35], recombinant Fab antibody fragments [74], single-chain antibody fragments [100], engineered recombinant immunotoxins [101], human gamma interferon [102], human matrix metalloproteinase-7 [103], human interleukin-21 [80], and many others (Figure 38.4A).

Like GdmCl, L-Arg increases the solubility of unfolded polypeptides and of folding intermediates and prevents their aggregation. Although it contains a guanidino group, however, L-Arg, in contrast to GdmCl, was found to have no effect or only a minor effect on the thermodynamic stability of natively folded proteins [104, 105] (Figure 38.4B). Exactly how L-Arg exerts its effect as a suppressor of aggregation is still subject to discussion. As already mentioned, no dramatic relative stabilization of denatured polypeptides with respect to the native state by L-Arg could be observed. At molal concentrations above 0.1 mol kg^{-1}, L-Arg is preferentially excluded from the surface hydration sphere of proteins, which means it is expected to act as a weakly stabilizing osmolyte, while at lower concentrations it is preferentially bound and therefore should exert a slightly destabilizing effect [83, 106]. Be-

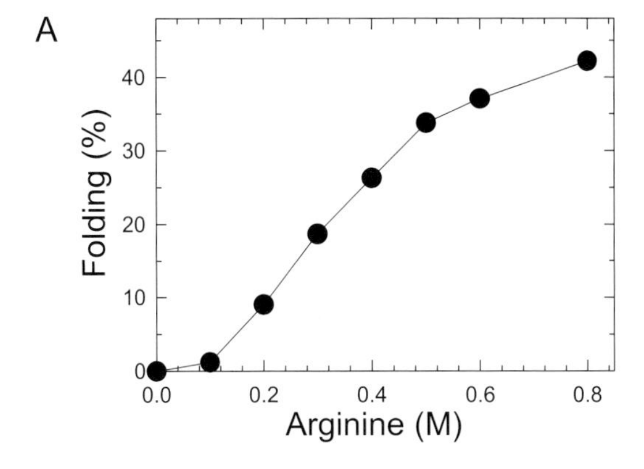

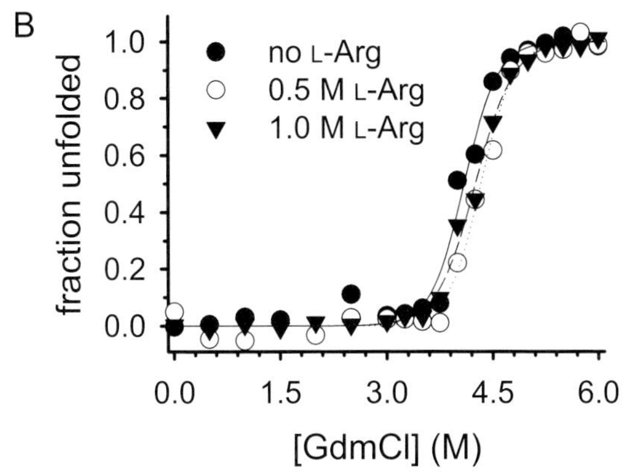

Fig. 38.4. (A) Refolding yield of human tissue-type plasminogen activator as a function of the concentration of L-Arg. Reproduced from Ref. [30]. (B) Influence of L-Arg on the GdmCl-induced folding/unfolding of hen egg white lysozyme. A final concentration of 20 μg mL^{-1} hen egg white lysozyme was incubated for 12 h at 20 °C in 50 mM HEPES/NaOH, pH 7, in the presence of increasing concentrations of GdmCl and 0 M, 0.5 M, and 1.0 M L-Arg, respectively, as indicated. The folding/unfolding transition was monitored by tryptophan fluorescence. Data are courtesy of Mr. Ravi Charan Reddy.

cause no straightforward classification as either a stabilizing or a destabilizing co-solvent is possible, more or less specific direct interactions of its functional groups with the functional groups of dissolved polypeptides will have to be involved to explain the effects of L-Arg as a refolding enhancer.

Recently, organic solvent systems with low water contents have received some attention as potential media for the in vitro folding of proteins. Rariy and Klibanov were able to show that it is possible to refold hen egg white lysozyme to its native state in 99.8% dry glycerol, albeit at low yields [107]. The same group later reported that in buffered organic media (1,ω-alkanediols, oligoethylene glycols, and *N*-methyl formamide), the presence of high concentrations of a stabilizing salt, namely, LiCl up to 2 M, resulted in markedly enhanced refolding yields as compared to the organic buffer system alone [108]. However, at water contents below 30% (w/w), the observed refolding yields for lysozyme were still extremely low. The renaturation of lysozyme was also reported in buffer systems containing the liquid organic salts ethyl ammonium nitrate and butyl ammonium nitrate [109]. These salts were found to have an effect similar to other denaturing agents (see above) by suppressing aggregation during refolding in concentrations up to approx. 0.5 M, while dissolving lysozyme in the pure liquid salts resulted in unfolded protein. With different liquid organic salts, however, efficient refolding of lysozyme and of human tissue-type plasminogen activator was observed up to salt concentrations above 90% (w/w) (Hauke Lilie, personal communication). Liquid organic salts represent a rather novel, and very versatile, class of polar organic solvents (for recent reviews, see, e.g., Refs. [110–112]). As the organic ions that form these salts may be easily derivatized, they can potentially be optimized as solvents for any given biochemical reaction. The potential of these "tailor-made" media for in vitro folding should be kept in view.

38.3.5
Cofactors in Protein Folding

It is a straightforward assumption that for proteins, whose structural integrity depends on the presence of strongly bound prosthetic groups or metal ions, these factors must be present during refolding in order to guide structure formation towards the stable native state. Cofactors that only weakly interact with the native state may also have a beneficial effect on in vitro folding by guiding folding intermediates towards native structure formation. Price et al. found that Ca^{2+}, although not strongly bound to bovine pancreatic DNAse, was necessary for correct disulfide bond formation [113]. Upon refolding of recombinant horseradish peroxidase C from *E. coli* inclusion bodies, the presence of Ca^{2+} and heme during the refolding process was essential for the formation of native, active enzyme [114]. In this case, protein folding was performed at a concentration of 220 μg mL^{-1} in redox buffer containing an additional 5 mM of $CaCl_2$ and 5 μM hemin. The refolding procedure was later successfully applied to other heme-containing proteins [115, 116]. Horse liver alcohol dehydrogenase requires Zn^{2+} for both enzymatic activity and stability. Upon refolding of this dimeric enzyme in the absence of Zn^{2+}, an inactive intermediate is formed. Addition of the metal ion initiates dimerization and recovery of native enzymatic activity [117]. Reactivation was observed only at moderate concentrations of Zn^{2+}, while higher concentrations of the metal ion seemed to sequestrate inactive intermediates in an inactive conformation.

Other cofactors such as NAD(P)H, ATP, or FADH can be essential for structure formation. Upon refolding of tetrameric yeast glyceraldehyde phosphate dehydrogenase, NAD^+ led to a significant increase in the final yield of reactivation [118]. In the case of proteins interacting with NAD(P)H, ATP, or DNA, the effect of these molecules during in vitro folding can be mimicked to a certain extent by including pyrophosphate in the refolding buffer. For example, human p53 tumor suppressor protein could be successfully refolded in the presence of 50 mM pyrophosphate [119].

38.3.6
Chaperones and Folding-helper Proteins

An elegant method to prevent aggregation during protein folding is to make use of the machinery designed by nature to promote correct folding in living cells, i.e., molecular chaperones and other folding-helper proteins such as protein disulfide isomerases and peptidyl-prolyl *cis/trans* isomerases (for recent reviews, see, e.g., Refs. [120, 121]). Both the yield and the rate of in vitro folding processes can be increased tremendously by chaperones and foldases [122]. Both classes of proteins are, however, not yet used in large-scale industrial production processes. The reason lies once again in cost considerations. Using molecular chaperones or foldases for in vitro protein folding would require the prior, cost-intensive production of these proteins. Since some chaperones and folding catalysts must be present in stoichiometric amounts to improve in vitro folding, large amounts of these compounds would be needed in production processes. ATP, which is essential for the function of many chaperones, would also add to the production costs.

Buchner et al. first reported the successful use of the chaperones GroEL/ES and DnaK, together with protein disulfide isomerase, as folding enhancers in the renaturation of a single-chain, antibody-based immunotoxin [123]. In this work, DnaK was found to retain its folding-helper activity when immobilized on a matrix. Immobilization of foldases allows for repeated use, thus reducing the necessary amount of chaperone and improving the process economy. This methodology was taken up by Fersht and coworkers, who co-immobilized a functional fragment of the *E. coli* chaperone GroEL, which was no longer ATP-dependent, together with DsbA, a bacterial disulfide isomerase, and peptidyl-prolyl isomerase [124, 125]. The disulfide-containing scorpion protein toxin Cn5 could be successfully refolded on the obtained multifunctionalized matrix. The use of chaperones and foldases will, however, need considerable further optimization before it can be implemented in industrial production processes.

38.3.7
An Artificial Chaperone System

It has been known for some time that cyclodextrins have a destabilizing effect on the native state of proteins [126]. This can be explained by a shift in the equilibrium in favor of the unfolded state, which is caused by the direct interaction of the cyclodextrins with buried hydrophobic side chains of the protein. During protein folding, this interaction may be exploited for transient shielding of hydrophobic regions of the unfolded polypeptide chain and thereby reducing unproductive aggregation [127].

This use of cyclodextrins in protein-folding processes was further improved in the artificial chaperone-assisted refolding process described by Gellman and co-workers [128, 129]. In this refolding protocol, chemically denatured protein (e.g., inclusion bodies solubilized by GdmCl) is first diluted in a "capture step" into a detergent solution. During this step, the denaturant is diluted to a non-denaturing concentration. However, protein folding, as well as aggregation, is still prevented by the formation of a protein-detergent complex. In a second step, a cyclodextrin is added to the solution to strip the protein of detergent, thereby inducing refolding. This procedure has by now been optimized with respect to the choice of detergent and the respective cyclodextrin. Efficient stripping of the detergent is achieved by its intercalation into the cyclodextrin ring [130, 131]. The artificial chaperone system was further improved by using β-cyclodextrin coupled to epichlorohydrin copolymers. This eliminated the unit operation of removal of the cyclodextrin-detergent complex in the liquid-phase artificial chaperone system [132]. Using the β-cyclodextrin beads in an expanded-bed mode allows in vitro folding on a large process scale. The only drawback of this elegant technique is the limited commercial availability of cyclodextrins coupled to hydrophilic resins.

38.3.8
Pressure-induced Folding

High hydrostatic pressure leads to reversible dissociation of oligomeric proteins [133], while pressures above 4 kbar lead to protein unfolding. Although folded proteins are highly incompressible, high hydrostatic pressures force conformational changes that lead to a reduction in the overall volume of the system. This effect can be described by taking into account the decrease in volume resulting from the exposure of hydrophobic groups from the interior of a protein, as well as the effect of pressure on the dynamic structure of water [134].

Lately, various attempts have been made to make use of high pressure for applied protein refolding processes. The basic idea is to shift the competition between correct folding and aggregate formation to the direction of the productive pathway. Since oligomeric proteins can be reversibly dissociated using high hydrostatic pressure, it could be assumed that the propensity for aggregate formation during in vitro folding should also be reduced upon increasing pressure. This hypothesis was confirmed by Gorovits and Horowitz [135]. They were able to show that at 2 kbar pressure the refolding yield of urea-denatured rhodanese increased to approx. 25% in contrast to a recovery of approx. 5% observed at ambient pressure. Upon further addition of 4 M glycerol during pressure treatment, the yield of refolding could be further increased up to 56%. Using P22 tail-spike protein as a model system, Foguel et al. showed that aggregates could be dissociated when subjected to hydrostatic pressures of 2.4 kbar [136]. Finally, St. John et al. demonstrated that native β-lactamase could be obtained by simply subjecting inclusion bodies to high pressure [137]. The effect of high pressure on the refolding of covalently cross-linked aggregates of lysozyme was analyzed concerning the effect of non-denaturing concentrations of GdmCl and the ratio of oxidized to reduced glutathione [138]. Under optimal conditions, refolding yields of up to 80% could be obtained.

The general applicability of high-pressure techniques for inclusion body solubilization and refolding remains to be demonstrated. Amorphous aggregates are formed upon decompression of RNAse A and of bovine pancreatic trypsin inhibitor after high-pressure treatment in the presence of reducing agents [139], indicating that, at least for some proteins, misfolded, nonnative states may be favored by high hydrostatic pressures.

38.3.9 Temperature-leap Techniques

The optimum temperature for in vitro folding is often significantly lower than the temperature at which a given protein functions under physiological conditions. For example, the optimum temperature for refolding of several mammalian proteins lies in the range of 5–15 °C. During in vivo folding, molecular chaperones that promote correct structure formation by preventing aggregation processes allow folding at higher temperatures than under in vitro conditions. Increasing temperatures during in vitro folding may lead to a stabilization of unproductive hydrophobic interactions and thus accelerate the rate of aggregate formation. Xie and Wetlaufer made use of the temperature dependency of the yield of refolding in designing an elegant and efficient folding protocol: the "temperature-leap tactic" [140]. Using bovine carbonic anhydrase II as a model system, refolding was optimal when first induced at 4 °C for 120 min, followed by a final folding phase after warming the solution to 36 °C. During the refolding of carbonic anhydrase II, an early intermediate is formed that is still susceptible to aggregation. If refolding is performed at elevated temperatures, this intermediate is largely converted to aggregates. If, however, the initial stage of refolding is allowed to proceed at lower temperatures, the early intermediate rearranges to a late intermediate that is more resistant to aggregation and can then be refolded at a higher rate at higher temperatures. The efficacy of this method, however, depends on the reaction pathway-dependent presence or absence of aggregation-prone refolding intermediates. Therefore, its applicability has to be checked and folding temperatures have to be optimized on a case-by-case basis. In another case, a short hydrophobic protein could be refolded after extensive purification of the solubilized inclusion bodies using a temperature-leap protocol [141]. In yet another example, oxidative renaturation of ovotransferrin from the fully reduced form was achieved by raising the temperature during folding [142]. Here the reduced protein was first diluted into refolding buffer containing 1 mM GSH and pre-incubated at 0 °C for varying times. To initiate the final oxidative folding of the preformed, still reduced, intermediates, 1 mM GSSG was added and the temperature was increased to 22 °C. The pre-incubation at 0 °C significantly decreased the aggregation, as measured by turbidity, during refolding.

38.3.10 Recycling of Aggregates

Upon in vitro folding, the complete conversion of the unfolded protein into its native biologically active form is rarely achieved. Instead, a significant amount of material is sequestered into inactive aggregates. These aggregates may comprise more than 90% of the initial solubilized protein material, depending on the particular protein and the refolding conditions. It should be possible to recover this material by recycling through a second round of solubilization and refolding. This was demonstrated in principle quite some time ago, using porcine lactate dehydrogenase as a model protein [143]. However, mistranslated forms or protein molecules containing nonnative covalent modifications may accumulate in the aggregated material and corrupt the correct refolding pathway during recycling. By using electron spray mass spectroscopy (ES-MS) for the analysis of the recycled aggregates after in vitro folding of the N-terminal domain of tissue inhibitor of metalloproteinase-2, it could be shown that even after partial purification the material was extremely heterogeneous in mass [144]. Thus, the need for the elaborate purification of the recycled protein, together with the relative ease of obtaining fresh inclusion bodies from the overexpression of a given target protein as starting material for in vitro folding, renders the recycling of aggregates unattractive.

38.4 Alternative Refolding Techniques

38.4.1 Matrix-assisted Refolding

The basic idea in matrix-assisted protein refolding is to make use of the specific interaction of the protein that is to be refolded with a solid support in order to prevent unspecific aggregation. To achieve this goal, three prerequisites must be fulfilled. First, protein binding to the matrix must be possible under denaturing conditions; second, the interaction between the protein and the matrix during the refolding step must allow for the flexibility of the polypeptide chain that is required for proper structure formation; and third, the interaction must be tight enough to prevent interchain interaction, which might result in aggregate formation.

The technology of matrix-assisted protein refolding was pioneered by Creighton [145, 146], using the transient interaction of the unfolded protein with an ion-exchange resin and the gradual removal of the denaturing agent to allow refolding. As an alternative, hydrophobic matrices were tested for the same purpose [147]. In another, slightly exotic, example, the MalK subunit of the *Salmonella typhimurium* maltoside transport complex was recovered in native form after solubilizing *E. coli* inclusion bodies in urea and subsequent binding of the material to a red agarose column, followed by refolding through buffer exchange [148].

Specific binding of solubilized protein to matrices can be achieved by using sequence tags, which must be chosen to allow for binding under denaturing conditions. For example, N- or C-terminal hexa-arginine tags were used to bind urea-unfolded α-glucosidase from *Saccharomyces cerevisiae* to polyanionic supports [149]. Careful optimization of the renaturation conditions was necessary, as ionic interactions of the refolding polypeptide with the matrix caused a drastic decrease in renaturation yields at low salt concentrations, while at high salt concentrations, renaturation was prevented by hydrophobic interactions with the matrix. Under optimum conditions, however, immobilized α-glucosidase could be renatured with a high yield at protein concentrations up to 5 mg mL^{-1}. Folding of the enzyme in solution had previously been feasible only at concentrations below 15 μg mL^{-1}.

Recently, binding of His-tagged proteins to Ni-chelate affinity material has been increasingly used for matrix-assisted refolding. In the case of cysteine-rich proteins or proteins containing multiple disulfide bonds, incorrect disulfide bond formation due to metal-catalyzed air oxidation may hamper proper renaturation. Bassuk et al. described a method for isolation and partial refolding of a reduced His-tagged protein on a Ni-nitrilotriacetic acid (NTA) column, followed by disulfide shuffling using a redox buffer system composed of reduced and oxidized glutathione [150]. In another case, an overnight linear gradient, starting with 8 M urea, 100 mM NaH_2PO_4, 10 mM Tris/HCl, pH 8.0, and 13 mM β-mercaptoethanol and finishing with refolding buffer (100 mM NaH_2PO_4, 10 mM Tris/HCl, pH 8.0, 2 mM GSH, and 0.2 mM GSSG), was successfully used for the refolding of the His-tagged recombinant human and mouse myelin oligodendrocyte glycoproteins on a Ni-NTA column [151]. Franken et al. recently reported a protocol for the refolding of matrix-bound His-tagged proteins, which includes a washing step with a buffer containing 60% isopropanol prior to refolding for the efficient removal of contaminants, namely detergents and bacterial endotoxins [152].

Apart from soluble proteins, membrane proteins also could be refolded after immobilizing the solubilized inclusion body material on a Ni-chelating matrix [153]. Here, the inclusion bodies were solubilized by urea in the presence of sodium *N*-lauryl sarcosinate and glycerol. After the binding to a Ni-chelate affinity matrix, folding was initiated by omitting urea from the elution buffer. With the protein still bound to the column, *N*-lauryl sarcosinate was exchanged against Triton X-100, and the refolded protein was finally eluted with imidazole.

Another interesting approach to matrix-assisted protein refolding was described by Berdichevsky et al. [154]. In this case, the folding of a single-chain Fv antibody fragment fused to a cellulose-binding domain from *Clostridium thermocellum* (CBD) was analyzed in the matrix-bound form and compared to standard refolding in solution. Interestingly, the CBD was found to retain its specific cellulose-binding capacity even in the presence of up to 6 M urea. Following solubilization of the inclusion bodies of the fusion protein in urea, the CBD recovered its native structure, while the Fv part was still unfolded. After binding to crystalline cellulose, the antibody fragment could then be refolded while bound to the matrix via its fusion partner, the already native CBD.

Protein refolding by transient interaction of folding intermediates with solid supports can also be achieved by using disulfide-carrying polymeric microspheres during oxidative refolding of disulfide bond-containing proteins [155]. Here, cystamine-functionalized microspheres were present during the oxidative refolding of reduced lysozyme. Due to transient binding of folding intermediates to the matrix by thiol-disulfide exchange, an increase in the yield of productive refolding could be achieved.

38.4.2
Folding by Gel Filtration

Surprisingly, unfolded proteins can in some cases be refolded simply by passing the solubilized, unfolded protein solution through a size-exclusion column equilibrated with refolding buffer [156–158]. Even more surprisingly, this refolding procedure often functions at relatively high protein concentrations. Several reasons may contribute to the feasibility of protein refolding by gel filtration. First, the unfolded polypeptide, characterized by a higher hydrodynamic radius than the chemical denaturant (GdmCl or urea) will run ahead of the denaturant and thus gradually move into conditions that promote correct refolding. Now either the protein can complete correct refolding and outrun the denaturant front, or higher aggregates may form, which then precipitate within the gel. The following denaturant front will re-dissolve these aggregates, giving them a renewed chance for correct folding. Furthermore, limited diffusion as well as physical separation of the refolding polypeptide chains within the pores of the gel may reduce the propensity for aggregate formation. Finally, transient interaction of the intermediates with the gel matrix may also contribute to a reduction in aggregate formation.

The use of size-exclusion chromatography for refolding was further improved by Gu et al., who used urea gradients to guarantee the gradual transfer of the unfolded protein to refolding conditions [159]. In a very thorough analysis, chromatographic refolding methods were compared to more traditional refolding techniques using *E. coli* inclusion bodies of recombinant single-chain antibody fragments [160]. Refolding by dilution and dialysis was compared to protein refolding using an immobilized metal affinity matrix and gel filtration. The highest yield was obtained using gel filtration with decreasing urea and, in parallel, increasing pH gradient. In the presence of the appropriate redox buffer, composed of reduced and oxidized glutathione, decreasing urea concentrations allowed structure formation, while increasing pH values promoted formation of disulfide bonds by thiol-disulfide exchange. This method allowed high-yield recovery of correctly refolded protein. Since the oxidative refolding of antibody fragments is a relatively slow process, low flow rates had to be used during gel filtration. In another study Gu et al. compared the efficacy of different gel-filtration media for size-exclusion chromatographic refolding of denatured bovine carbonic anhydrase B [161].

Although folding by gel filtration seems to be an elegant and promising approach to the renaturation of inclusion body material, several inherent difficulties may limit its practical applicability. Clogging of the gel-filtration column by aggregates can be a severe problem for many proteins, and the refolding conditions have to be carefully optimized with respect to buffer conditions, flow rate, and gradient volume in each case. Low maximum flow rates and the cost of large size-exclusion columns represent barriers to the scale-up of a process for industrial production.

38.4.3
Direct Refolding of Inclusion Body Material

As the properties of proteins vary so widely depending on their individual amino acid sequence, it is obviously to be expected that the inclusion bodies formed by different recombinant proteins also differ in their morphology. Using the same procedure for isolation may in one case yield inclusion body preparations that are extremely resistant to solubilization, while isolates of other proteins may be re-dissolved even under non-denaturing conditions. In general, inclusion body preparations of secreted proteins containing multiple disulfide bonds are quite stable, so that strong denaturants are a prerequisite for solubilization. Inclusion body isolates of intracellular proteins, on the other hand, are often found to be quite labile, so that the folded or partially folded protein contained in these inclusion bodies can be solubilized by a simple incubation in physiological buffer systems. For example, native protein could be extracted from inclusion body isolates obtained after massive overexpression of the endogenous *E. coli* β-galactosidase (R. Rudolph, unpublished observation). In another study, Tsumoto et al. reported the extraction of active green fluorescent protein (GFP) from inclusion bodies using non-denaturing solvent conditions [162]. In this case, the inclusion bodies already contained some native GFP, as indicated by their fluorescence. This native material, together with partially folded intermediates, could be solubilized using non-denaturing concentrations of GdmCl or, even more effectively, using L-Arg in concentrations ranging from 0.5 M to 2 M. In other cases, the mild detergent *N*-lauryl sarcosine was used to extract active protein from inclusion body material. Native *E. coli* RNA polymerase σ factor could be recovered by dissolving the inclusion bodies in buffer containing 0.4% (w/v) sarcosyl, followed by 10-fold dilution at 4 °C [163]. Actin could be recovered from inclusion body material by using 1.5% (w/v) sarcosyl in the presence of 1 mM ethylenediamine tetraacetic acid (EDTA) [164]. In the absence of divalent cations, only actin was extracted from the inclusion body material, while contaminating *E. coli* outer-membrane proteins were not solubilized.

Recombinant proteins that form more resistant inclusion bodies may still be directly solubilized and refolded by applying strongly alkaline buffer conditions. Although this procedure may appear relatively harsh, it has indeed been used in industrial processes. For example, prochymosin was extracted by suspending the inclusion body material in 0.05 M K_2HPO_4/K_2NaPO_4, 1 mM EDTA, 0.1 M NaCl, pH 10.7. After incubation for a least 1 h at 4 °C, the pH of the supernatant was adjusted to pH 8.0 by addition of concentrated HCl to induce final refolding [165]. In another case, bovine somatotropin was recovered by incubating inclusion bodies in dilute NaOH, pH 12.5. After 20 min at 25 °C, the inclusion bodies were completely dissolved. After adjusting the pH to 11.5, oxidative refolding proceeded, presumably by air oxidation, during a further incubation for 8 h [166].

38.5
Conclusions

Although almost all chapters in this handbook deal with fundamental aspects of both the in vivo and the in vitro structure formation of proteins, protein folding has lost its virgin status and become the subject of industrial exploitation for many years. By now a large number of proteins of industrial relevance are produced by processes involving refolding. Among the most prominent examples are recombinant human proteins of high market potential produced by in vitro folding from *E. coli* inclusion body material. These examples include human insulin, human granulocyte colony-stimulating factor, human growth hormone, and Rapilysin, an engineered variant of human tissue-type plasminogen activator. Other proteins used for drug discovery by screening or rational, structure-based design are also produced by in vitro folding.

In this chapter only a selection of the techniques that have been used for protein refolding could be mentioned. Unfortunately, many of the refolding protocols described in the literature work well for only one particular target protein. Folding protocols have to be optimized with respect to protein concentration, temperature, pH, redox conditions, additives, and specific methodology on a case-by-case basis. However, "standard procedures" have been developed over the last couple of years, which may serve as fairly general guidelines (cf. Protocol 3). Chances are high that native, biologically active protein can be obtained from inclusion body material by following these protocols. These procedures, however, may not straightforwardly lead to the most cost-efficient process for the large-scale production of a given protein. In designing an industrial process, other alternatives that are mentioned in this chapter should be kept in view as possibilities for the optimization of process economics.

38.6
Experimental Protocols

38.6.1
Protocol 1: Isolation of Inclusion Bodies

Equipment and Reagents

- Ultraturrax
- French press
- Centrifuge
- 0.1 M Tris/HCl pH 7, 1 mM EDTA
- Lysozyme
- $MgCl_2$
- DNAse

- 60 mM EDTA, 6% Triton X-100, 1.5 M NaCl, pH 7
- 0.1 M Tris/HCl pH 7, 20 mM EDTA

Method

1. Homogenize cells (5 g wet weight) in 25 mL 0.1 M Tris/HCl pH 7, 1 mM EDTA, at 4 °C, using an Ultraturrax (10 000 rpm).
2. Add 1.5 mg lysozyme per gram cells, mix briefly with the Ultraturrax, and incubate at 4 °C for 30 min.
3. Use sonication or high-pressure homogenization (French press) to disrupt the cells.
4. In order to digest DNA, add $MgCl_2$ to a final concentration of 3 mM and DNAse to a final concentration of 10 $\mu g\ mL^{-1}$, and incubate for 30 min at 4 °C.
5. Add 0.5 volume of 60 mM EDTA, 6% Triton X-100, 1.5 M NaCl, pH 7, and incubate for 30 min at 4 °C.
6. Pellet inclusion bodies (IB) by centrifugation at 31 000 *g* for 10 min at 4 °C.
7. Resuspend the pellet in 40 mL 0.1 M Tris/HCl pH 7, 20 mM EDTA, using the Ultraturrax.
8. Repeat centrifugation.
9. The IB pellet may be stored frozen at −20 °C for at least two weeks.

38.6.2
Protocol 2: Solubilization of Inclusion Bodies

Reagents

- 6 M GdmCl (or 8 M urea), 0.1 M Tris/HCl pH 8, 100 mM DTE (or DTT), 1 mM EDTA
- 1 M HCl
- 4 M GdmCl (or 6 M urea), 10 mM HCl

Method

1. Resuspend 50 mg IB pellet from Protocol 1 in 5 mL 6 M GdmCl (or 8 M urea), 0.1 M Tris/HCl pH 8, 100 mM DTE (or DTT), 1 mM EDTA, and incubate for 2 h at 25 °C.
2. Lower the pH of the solution to a value between pH 3 and pH 4 by drop-wise addition of HCl.
3. Remove debris by centrifugation at 10 000 g.
4. Completely remove the DTE (or DTT) by exhaustive dialysis (twice) against 500 mL of 4 M GdmCl (or 6 M urea), 10 mM HCl, for 2 h at 25 °C. Dialyze again at 4 °C overnight against 1 L of 4 M GdmCl (or 6 M urea). (In the case of proteins that do not contain disulfide bonds, the dialysis step may be omitted.)
5. Determine the protein concentration by a Bradford assay [167] using bovine serum albumin (BSA) in 4 M GdmCl (or 6 M urea), 10 mM HCl as standard protein.

Tab. 38.2. Buffer conditions for initial refolding screening.

Buffer	*pH*[1]	*Additive*	*GSH/GSSG*[2]	*EDTA*[3]
0.1 M Tris/HCl	8.5	–	5 mM/0.5 mM	1 mM
"	"	0.5 M L-Arg/HCl	"/"	"
"	"	1.0 M L-Arg/HCl	"/"	"
"	7.5	0.5 M L-Arg/HCl	"/"	"
"	9.5	"	"/"	"
"	8.5	"	0.5 mM/5 mM	"
"	"	"	–/–	"
"	"	0.5 M Na_2SO_4	5 mM/0.5 mM	"
"	"	20% (w/v) Sorbitol	"/"	"

[1] Care should be taken to adjust the pH value only after addition of all buffer components.
[2] Presence of the redox buffer system composed of oxidized and reduced glutathione is necessary only if the protein contains disulfide bonds; if the protein does **not** contain disulfide bonds, GSH/GSSG may be replaced by 5 mM DTT (or DTE). Stock solutions of GSH, DTT, and DTE should not be stored for longer than 24 h at 4 °C.
[3] EDTA should be added only if the native protein does not contain bound metal ions; if the protein **does** contain bound metal ions, low concentrations of the respective metal should be included (e.g., in fivefold molar excess over the protein).

6. Aliquots of the protein solution may be stored at –70 °C for at least two weeks.

38.6.3 Protocol 3: Refolding of Proteins

Refolding by Rapid Dilution of Denatured Protein

1. The optimal choice of the refolding buffer depends on the particular protein. Table 38.2 represents a series of buffer conditions that should be tested in an initial screening and may serve as starting points for further optimization.
2. Dilute the solubilized proteins obtained according to Protocol 2 into the pre-cooled refolding buffer (10 °C) under rapid mixing. The final protein concentration should be within the range of 10–30 μg mL^{-1}. This concentration should be obtained by diluting the solubilized inclusion bodies in a ratio of approx. 1:100 (v/v).
3. After 12 h refolding at 10 °C, the sample should be dialyzed against a dilute buffer system appropriate for the given protein, which must not contain L-Arg. Apart from improving refolding, L-Arg may also keep misfolded forms of the protein in solution and should therefore be removed. A precipitate may form during this dialysis step.
4. After removal of the precipitate by centrifugation, the biological activity of the refolded protein should be assessed using an appropriate assay. If no activity as-

say is available, solubility may be taken as a first indication for correct refolding. Upon quantification of the refolding yield, take into account volume changes that may have occurred during dialysis due to osmosis.

Refolding by Stepwise Addition of Denatured Protein

To achieve refolding at higher final concentrations of renatured protein, the dilution of the solubilized inclusion body material may be carried out in a stepwise manner, according to the following rules:

1. Determine the upper limit of the concentration of denatured protein (Δc) that can be refolded at high yield upon a single dilution into renaturation buffer.
2. Determine the time Δt that is required for renaturation to 90% of the final value.
3. Determine the maximum residual denaturant concentration that is compatible with native structure formation.
4. Refolding is performed by stepwise addition of concentration increments Δc of the denatured protein to the refolding buffer at time intervals Δt. The addition of denatured protein solution can be repeated until the maximum residual denaturant concentration is reached.
5. After refolding is complete, proceed with dialysis according to step 3 of the standard protocol (see above).

Acknowledgements

The group's ongoing work on the in vitro folding of proteins is supported by the German Federal Ministry of Education and Research (BMBF grant no. 03WKB01 A) and the Federal State (Land) of Saxony-Anhalt (grant no. 3537 A/0903 L).

References

1 Itakura, K., Hirose, T., Crea, R., Riggs, A. D., Heynecker, H. L., Bolivar, F., and Boyer, H. W. (1977). Expression in *Escherichia coli* of a chemically synthesized gene for the hormone somatostatin. *Science* **198**, 1056–1063.

2 Goeddel, D. V., Kleid, D. G., Bolivar, F., Heyneker, H. L., Yansura, D. G., Crea, R., Hirose, T., Kraszewski, A., Itakura, K., and Riggs, A. D. (1979). Expression in *Escherichia coli* of chemically synthesized genes for human insulin. *Proc. Natl. Acad. Sci. USA* **76**, 106–110.

3 Makrides, S. C. (1996). Strategies for Achieving High-Level Expression of Genes in *Escherichia coli. Microbiol. Rev.* **60**, 512–538.

4 Yee, L., and Blanch, H. W. (1992). Recombinant protein expression in high cell density fed-batch cultures of *Escherichia coli. Bio/Technology* **10**, 1550–1556.

5 Kleman, G. L., and Strohl, W. R.

Rao, Ch., and Ramakrishna, T. (1997). Co-refolding denatured-reduced hen egg white lysozyme with acidic and basic proteins. *FEBS Lett.* **418**, 363–366.

45 Ramage, P., Cheneval, D., Chvei, M., Graff, P., Hemmig, R., Heng, R., Kocher, H. P., Mackenzie, A., Memmert, K., Revesz, L., and Wishart, W. (1995). Expression, Refolding, and Autocatalytic Proteolytic Processing of the Interleukin-1β-converting Enzyme Precursor. *J. Biol. Chem.* **270**, 9378–9383.

46 Hochuli, E., Bannwarth, W., Döbeli, H., Gentz, R., and Stüber, D. (1988). Genetic Approach to Facilitate Purification of Recombinant Proteins with a Novel Metal Chelate Adsorbent. *Bio/Technology* **6**, 1321–1325.

47 Samuelsson, E., and Uhlen, M. (1996). Chaperone-like effect during in vitro refolding of insulin-like growth factor I using a solubilizing fusion partner. *Ann. N.Y. Acad. Sci.* **782**, 486–494.

48 Ambrosius, D., Dony, C., and Rudolph, R. (1996). Activation of recombinant proteins. *U.S. Patent* 5,578,710.

49 Zhu, X. L., Ohta, Y., Jordan, F., and Inouye, M. (1989). Pro-sequence of subtilisin can guide the refolding of denatured subtilisin in an intramolecular process. *Nature* **339**, 483–484.

50 Silen, J. L., and Agard, D. A. (1989). The α-lytic protease pro-region does not require a physical linkage to activate the protease domain in vivo. *Nature* **341**, 462–464.

51 Rattenholl, A., Ruoppolo, M., Flagiello, A., Monti, M., Vinci, F., Marino, G., Lilie, H., Schwarz, E., and Rudolph, R. (2001a). Pro-Sequence Assisted Folding and Disulfide Bond Formation of Human Nerve Growth Factor. *J. Mol. Biol.* **305**, 523–533.

52 Rattenholl, A., Lilie, H., Grossmann, A., Stern, A., Schwarz, E., and Rudolph, R. (2001b). The pro-sequence facilitates folding of human nerve growth factor from *Escherichia coli* inclusion bodies. *Eur. J. Biochem.* **268**, 3296–3303.

53 Grauschopf, U., Lilie, H., Honold, K., Wozny, M., Reusch, D., Esswein, A., Schaefer, W., Rücknagel, K. P., and Rudolph, R. (2000). The N-Terminal Fragment of Human Parathyroid Hormone Receptor 1 Constitutes a Hormone Binding Domain and Reveals a Distinct Disulfide Pattern. *Biochemistry* **30**, 8878–8887.

54 Bazarsuren, A., Grauschopf, U., Wozny, M., Reusch, D., Hoffmann, E., Schaefer, W., Panzner, S., and Rudolph, R. (2002). In vitro folding, functional characterization, and disulfide pattern of the extracellular domain of human GLP-1 receptor. *Biophys. Chem.* **96**, 305–318.

55 Hoffman, S. J., Looker, D. L., Roehrich, J. M., Cozart, P. E., Durfee, S. L., Tedesco, J. L., and Stetler, G. L. (1990). Expression of fully functional tetrameric human hemoglobin in *Escherichia coli. Proc. Natl. Acad. Sci. USA* **87**, 8521–8525.

56 Nagai, K., Perutz, M., and Poyart, C. (1985). Oxygen binding properties of human mutant hemoglobins synthesized in *Escherichia coli. Proc. Natl. Acad. Sci. USA* **82**, 7252–7255.

57 Creighton, T. E. (1978). Experimental Studies of Protein Folding and Unfolding. *Prog. Biophys. Molec. Biol.* **33**, 231–297.

58 Pace, C. N., and Creighton, T. E. (1986). The Disulphide Folding Pathway of Ribonuclease T_1. *J. Mol. Biol.* **188**, 477–486.

59 Weissman, J. S., and Kim, P. S. (1991). Reexamination of the folding of BPTI: predominance of native intermediates. *Science* **253**, 1386–1393.

60 Weissman, J. S., and Kim, P. S. (1992). Kinetic role of nonnative species in the folding of bovine pancreatic trypsin inhibitor. *Proc. Natl. Acad. Sci. USA* **89**, 9900–9904.

61 Frand, A. R., Cuozzo, J. W., and Kaiser, C. A. (2000). Pathways for protein disulphide bond formation. *Trends Cell Biol.* **10**, 203–210.

62 Woycechowsky, K. J., and Raines, R. T. (2000). Native disulfide bond formation in proteins. *Curr. Opin. Chem. Biol.* **4**, 533–539.

63 Hiniker, A., and Bardwell, J. C. A. (2003). Disulfide Bond Isomerization in Prokaryotes. *Biochemistry* **42**, 1179–1185.

64 Sela, M., White, F. H., and Anfinsen, C. B. (1957) Reductive cleavage of disulfide bridges in ribonuclease. *Science* **125**, 691–692.

65 Ahmed, A. K., Schaffer, S. W., and Wetlaufer, D. B. (1975). Nonenzymatic reactivation of reduced bovine pancreatic ribonuclease by air oxidation and by glutathione oxidoreduction buffers. *J. Biol. Chem.* **250**, 8477–8482.

66 Winter, J., Klappa, P., Freedman, R., Lilie, H., and Rudolph, R. (2002a). Catalytic Activity and Chaperone Function of Human Protein-disulfide Isomerase are Required for the Efficient Refolding of Proinsulin. *J. Biol. Chem.* **277**, 310–317.

67 Saxena, V. P., and Wetlaufer, D. B. (1970). Formation of three-dimensional structure in proteins. I. Rapid nonenzymic reactivation of reduced lysozyme. *Biochemistry* **9**, 5015–5023.

68 Wetlaufer, D. B. (1984). Nonenzymatic formation and isomerization of protein disulfides. *Meth. Enzymol.* **107**, 301–304.

69 Woycechowsky, K. J., Wittrup, K. D., and Raines, R. T. (1999). A small-molecule catalyst of protein folding in vitro and in vivo. *Chem. Biol.* **6**, 871–879.

70 Winter, J., Lilie, H., and Rudolph, R. (2002b). Recombinant expression and in vitro folding of proinsulin are stimulated by the synthetic dithiol Vectrase-P. *FEMS Microbiol. Lett.* **213**, 225–230.

71 Gough, J. D., Williams, R. H. Jr., Donofrio, A. E., and Lees, W. J. (2002). Folding Disulfide-Containing Proteins Faster with an Aromatic Thiol. *J. Am. Chem. Soc.* **124**, 3885–3892.

72 Gough, J. D., Gargano, J. M., Donofrio, A. E., and Lees, W. J. (2003) Aromatic Thiol pK_a Effects on the Folding Rate of a Disulfide Containing Protein. *Biochemistry* **42**, 11787–11797.

73 Light, A., and Higaki, J. N. (1987). Detection of Intermediate Species in the Refolding of Bovine Trypsinogen. *Biochemistry* **26**, 5556–5564.

74 Buchner, J., and Rudolph, R. (1991). Renaturation, Purification, and Characterization of Recombinant F_{ab}-Fragments Produced in *Escherichia coli*. *Bio/Technology* **9**, 157–162.

75 Honda, J., Andou, H., Mannen, T., and Sugimoto, S. (2000). Direct Refolding of Recombinant Human Growth Differentiation Factor 5 for Large-Scale Production Process. *J. Biosci. Bioeng.* **89**, 582–589.

76 Levine, A. D., Rangwala, S. H., Horn, N. A., Peel, M. A., Matthews, B. K., Leimgruber, R. M., Manning, J. A., Bishop, B. F., and Olins, P. O. (1995). High Level Expression and Refolding of Mouse Interleukin 4 Synthesized in *Escherichia coli*. *J. Biol. Chem.* **270**, 7445–7452.

77 Wang, J.-Y., Kishore, U., and Reid, K. B. M. (1995). A recombinant polypeptide, composed of the α-helical neck region and the carbohydrate recognition domain of conglutinin, self-associates to give a functionally intact homotrimer. *FEBS Lett.* **376**, 6–10.

78 Lowther, W. T., Majer, P., and Dunn, B. M. (1995). Engineering the substrate specificity of rhizopuspepsin: The role of Asp 77 of fungal aspartic proteinases in facilitating the cleavage of oligopeptide substrates with lysine in P_1. *Protein Sci.* **4**, 689–702.

79 Schrem, A., Lange, C., and Koch, K.-W. (1999). Identification of a Domain in Guanylyl Cyclase-activating Protein 1 that Interacts with a Complex of Guanylyl Cyclase and Tubulin in Photoreceptors. *J. Biol. Chem.* **274**, 6244–6249.

80 Asano, R., Kudo, T., Makabe, K., Tsumoto, K., and Kumagai, I. (2002). Antitumor activity of interleukin-21 prepared by novel refolding procedure

from inclusion bodies expressed in *Escherichia coli*. *FEBS Lett.* **528**, 70–76.

81 Umetsu, M., Tsumoto, K., Hara, M., Ashish, K., Goda, S., Adschiri, T., and Kumagai, I. (2003). How Additives Influence the Refolding of Immunoglobulin-folded Proteins in a Stepwise Dialysis System. *J. Biol. Chem.* **278**, 8979–8978.

82 Winter, J., Lilie, H., and Rudolph, R. (2002c). Renaturation of human proinsulin – a study on refolding and conversion to insulin. *Anal. Biochem.* **310**, 148–155.

83 Timasheff, S. N., and Arakawa, T. (1989). Stabilization of protein structure by solvents. in: Protein structure. A practical approach. Creighton, T. E. (ed.). IRL Press, Oxford, UK. pp. 331ff.

84 Timasheff, S. N. (2002). Protein Hydration, Thermodynamic Binding, and Preferential Hydration. *Biochemistry* **41**, 13473–13482.

85 Orsini, G., and Goldberg, M. E. (1978). The Renaturation of Reduced Chymotrypsinogen A in Guanidine HCl. Refolding versus Aggregation. *J. Biol. Chem.* 253, 3453–3458.

86 Wetlaufer, D. B., and Xie, Y. (1995) Control of aggregation in protein refolding: A variety of surfactants promote renaturation of carbonic anhydrase II. *Protein Sci.* **4**, 1536–1543.

87 Cleland, J. L., Hedgepeth, C., and Wang, D. I. C. (1992a). Polyethylene Glycol Enhanced Refolding of Bovine Carbonic Anhydrase B. *J. Biol. Chem.* **267**, 13327–13334.

88 Cleland, J. L., Builder, S. E., Swartz, J. R., Winkler, M., Chang, J. Y., and Wang, D. I. C. (1992b). Polyethylene Glycol Enhanced Protein Refolding. *Bio/Technology* **10**, 1013–1019.

89 Lee, L. L.-Y., and Lee, J. C. (1987). Thermal Stability of Proteins in the Presence of Poly(ethylene glycols). *Biochemistry* **26**, 7813–7819.

90 Tandon, S., and Horowitz, P. (1988). The effects of lauryl maltoside on the reactivation of several enzymes after treatment with guanidinium chloride. *Biochim. Biophys. Acta* **955**, 19–25.

91 Zardenata, G., Horowitz, P. M. (1994). Protein Refolding at High Concentrations Using Detergent/Phospholipid Mixtures. *Anal. Biochem.* **218**, 392–398.

92 Krause, M., Rudolph, R., and Schwarz, E. (2002). The non-ionic detergent Brij 58P mimics chaperone effects. *FEBS Lett.* **532**, 253–255.

93 Yancey, P. H., Clark, M. E., Hand, S. C., Bowlus, R. D., and Somero, G. N. (1982). Living with water stress: evolution of osmolyte systems. *Science* **24**, 1214–1222.

94 Bolen, D. W., and Baskakov, I. V. (2001). The osmophobic effect: natural selection of a thermodynamic force in protein folding. *J. Mol. Biol.* **310**, 955–963.

95 Sawano, H., Koumoto, Y., Ohta, K., Sasaki, Y., Segawa, S., and Tachibana, H. (1992). Efficient in vitro folding of the three-disulfide derivatives of hen lysozyme in the presence of glycerol. *FEBS Lett.* **303**, 11–14.

96 Michaelis, U., Rudolph, R., Jarsch, M., Kopetzki, E., Burtscher, H., and Schumacher, G. (1995). Process for the production and renaturation of recombinant, biologically active, eukaryotic alkaline phosphatase. *U.S. Patent* 5,434,067.

97 Chilson, O. P., and Chilson, A. E. (2003). Perturbation of folding and reassociation of lactate dehydrogenase by proline and trimethylamine oxide. *Eur. J. Biochem.* **270**, 4823–4834.

98 Kumar, T. K. S., Samuel, D., Jayaraman, G., Srimathi, T., and Yu, C. (1998). The role of proline in the prevention of aggregation during protein folding in vitro. *Biochem. Mol. Biol. Int.* **46**, 509–517.

99 Shiraki, K., Kudou, M., Fujiwara, S., Imanaka, T., and Takagi, M. (2002). Biophysical Effect of Amino Acids on the Prevention of Protein Aggregation. *J. Biochem. (Tokyo)* **132**, 591–595.

100 Buchner, J., Pastan, I., and Brinkmann, U. (1992a). A Method

for Increasing the Yield of Properly Folded Recombinant Fusion Proteins: Single-Chain Immunotoxins from Renaturation of Bacterial Inclusion Bodies. *Anal. Biochem.* **205**, 263–270.

101 Brinkmann, U., Buchner, J., Pastan, I. (1992). Independent domain folding of Pseudomonas exotoxin and single-chain immunotoxins: influence of interdomain connections. *Proc. Natl. Acad. Sci. USA* **89**, 3075–3079.

102 Arora, D., and Khanna, N. (1996). Method for increasing the yield of properly folded recombinant human gamma interferon from inclusion bodies. *J. Biotechnol.* **52**, 127–133.

103 Oneda, H., and Inouye, K. (1999). Refolding and recovery of recombinant human matrix metalloproteinase 7 (matrilysin) from inclusion bodies expressed by *Escherichia coli*. *J. Biochem.* **126**, 905–911.

104 Taneja, S., and Ahmad, F. (1994). Increased thermal stability of proteins in the presence of amino acids. *Biochem. J.* **303**, 147–153.

105 Arakawa, T., and Tsumoto, K. (2003). The effects of arginine on refolding of aggregated proteins: not facilitate refolding, but suppress aggregation. *Biochem. Biophys. Res. Comm.* **304**, 148–152.

106 Kita, Y., Arakawa, T., Lin, T.-Y., and Timasheff, S. N. (1994). Contribution of the Surface Free Energy Perturbation to Protein–Solvent Interactions. *Biochemistry* **33**, 15178–15189.

107 Rariy, R. V., and Klibanov, A. M. (1997). Correct protein folding in glycerol. *Proc. Natl. Acad. Sci. USA* **94**, 13520–13523.

108 Rariy, R. V., and Klibanov, A. M. (1999). Protein Refolding in Predominantly Organic Media Markedly Enhanced by Common Salts. *Biotechnol. Bioeng.* **62**, 704–710.

109 Summers, C. A., and Flowers, R. A. II (2000). Protein renaturation by the liquid organic salt ethylammonium nitrate. *Protein Sci.* **9**, 2001–2008.

110 Welton, T. (1999). Room-temperature ionic liquids. Solvents for synthesis and catalysis. *Chem. Rev.* **99**, 2071–2083.

111 Kragl, U., Eckstein, M., and Kaftzik, N. (2002). Enzyme catalysis in ionic liquids. *Curr. Opin. Biotechnol.* **13**, 565–571.

112 van Rantwijk, F., Madeira Lau, R., and Sheldon, R. A. (2003). Biocatalytic transformations in ionic liquids. *Trends Biotechnol.* **21**, 131–138.

113 Price, P. A., Stein, W. H., and Moore, S. (1969). Effect of Divalent Cations on the Reduction and Reformation of the Disulfide Bonds of Deoxyribonuclease. *J. Biol. Chem.* **244**, 929–932.

114 Smith, A. T., Santama, N., Dacey, S., Edwards, M., Bray, R. C., Thorneley, R. N. F., and Burke, J. F. (1990). Expression of a Synthetic Gene for Horseradish Peroxidase C in *Escherichia coli* and Folding and Activation of the Recombinant Enzyme with Ca^{2+} and Heme. *J. Biol. Chem.* **265**, 13335–13343.

115 Whitwam, R. E., Gazarian, I. G., and Tien, M. (1995). Expression of Fungal Mn Peroxidase in *E. coli* and Refolding to Yield Active Enzyme. *Biochem. Biophys. Res. Comm.* **216**, 1013–1017.

116 Doyle, W. A., and Smith, A. T. (1996). Expression of lignin peroxidase H8 in *Escherichia coli*: folding and activation of the recombinant enzyme with Ca^{2+} and Haem. *Biochem. J.* **315**, 15–19.

117 Rudolph, R., Gerschitz, J., and Jaenicke, R. (1978). Effect of Zinc(II) on the Refolding and Reactivation of Liver Alcohol Dehydrogenase. *Eur. J. Biochem.* **87**, 601–606.

118 Rudolph, R., Heider, I., and Jaenicke, R. (1977). Mechanism of Reactivation and Refolding of Glyceraldehyde-3-Phosphate Dehydrogenase form Yeast after Denaturation and Dissociation. *Eur. J. Biochem.* **81**, 563–570.

119 Bell, S., Hansen, S., and Buchner, J. (2002). Refolding and structural characterization of the human p53 tumor suppressor protein. *Biophys. Chem.* **96**, 243–257.

120 Wang, C.-C., and Tsou, C.-L. (1998).

Enzymes as chaperones and chaperones as enzymes. *FEBS Lett.* **425**, 382–384.

121 Schiene, C., and Fischer, G. (2000). Enzymes that catalyse the restructuring of proteins. *Curr. Opin. Struct. Biol.* **10**, 40–45.

122 Walter, S., and Buchner, J. (2002). Molecular Chaperones – Cellular Machines for Protein Folding. *Angew. Chem. Int. Ed.* **41**, 1098–1113.

123 Buchner, J., Brinkmann, U., and Pastan, I. (1992b). Renaturation of a Single-Chain Immunotoxin Facilitated by Chaperones and Protein Disulfide Isomerase. *Bio/Technology* **10**, 682–685.

124 Altamirano, M. M., Golbik, R., Zahn, R., Buckle, A. M., and Fersht, A. R. (1997) Refolding chromatography with immobilized mini-chaperones. *Proc. Natl. Acad. Sci. USA* **94**, 3576–3578.

125 Altamirano, M. M., García, C., Possani, L. D., and Fersht, A. R. (1999). Oxidative refolding chromatography: folding of the scorpion toxin Cn5. *Nature Biotechnol.* **17**, 187–191.

126 Cooper, A. (1992). Effect of Cyclodextrins on the Thermal Stability of Globular Proteins. *J. Am. Chem. Soc.* **114**, 9208–9209.

127 Sharma, L., and Sharma, A. (2001). Influence of cyclodextrin ring substituents on folding-related aggregation of bovine carbonic anhydrase. *Eur. J. Biochem.* **268**, 2456–2463.

128 Rozema, D., and Gellman, S. H. (1996a). Artificial Chaperone-assisted Refolding of Carbonic Anhydrase B. *J. Biol. Chem.* **271**, 3478–3487.

129 Rozema, D., and Gellman, S. H. (1996b). Artificial Chaperone-Assisted Refolding of Lysozyme: Competition between Renaturation and Aggregation. *Biochemistry* **35**, 15760–15771.

130 Daugherty, D. L., Rozema, D., Hanson, P. E., and Gellman, S. H. (1998) Artificial Chaperone-assisted Refolding of Citrate Synthase. *J. Biol. Chem.* **273**, 33961–33971.

131 Hanson, P. E., and Gellman, S. E. (1998). Mechanistic comparison of artificial-chaperone-assisted and unassisted refolding of urea-denatured carbonic anhydrase B. *Folding & Design* **3**, 457–468.

132 Mannen, T., Yamaguchi, S., Honda, J., Sugimoto, S., and Nagamune, T. (2001). Expanded-Bed Protein Refolding Using a Solid-Phase Artificial Chaperone. *J. Biosci. Bioeng.* **91**, 403–408.

133 Schade, B. C., Rudolph, R., Lüdemann, H.-D., and Jaenicke, R. (1980). Reversible High-Pressure Dissociation of Lactic Dehydrogenase from Pig Muscle. *Biochemistry* **19**, 1121–1126.

134 Hillson, N., Onuchic, J. N., and García, A. E. (1999). Pressure-induced protein-folding/unfolding kinetics. *Proc. Natl. Acad. Sci. USA* **96**, 14848–14853.

135 Gorovits, B. M., and Horowitz, P. M. (1998). High Hydrostatic Pressure Can Reverse Aggregation of Protein Folding Intermediates and Facilitate Acquisition of Native Structure. *Biochemistry* **37**, 6132–6135.

136 Foguel, D., Robinson, C. R., Caetano de Sousa, P. Jr., Silva, J. L., and Robinson, A. S. (1999). Hydrostatic Pressure Rescues Native Protein from Aggregates. *Biotechnol. Bioeng.* **63**, 552–558.

137 St. John, R. J., Carpenter, J. F., and Randolph, T. W. (1998) High pressure fosters protein refolding from aggregates at high concentrations. *Proc. Natl. Acad. Sci. USA* **23**, 13029–13033.

138 St. John, R. J., Carpenter, J. F., and Randolph, T. W. (2002) High-Pressure Refolding of Disulfide-Cross-Linked Lysozyme Aggregates: Thermodynamics and Optimization. *Biotechnol. Prog.* **18**, 565–571.

139 Meersman, F., and Heremans, K. (2003). High pressure induces the formation of aggregation-prone states of proteins under reducing conditions. *Biophys. Chem.* **104**, 297–304.

140 Xie, Y., and Wetlaufer, D. B. (1996) Control of aggregation in protein refolding: The temperature-leap tactic. *Protein Sci.* **5**, 517–523.

141 Lajmi, A. R., Wallace, T. R., and Shin, J. A. (2000). Short, Hydrophobic, Alanine-Based Proteins Based on the Basic Region/Leucine Zipper Protein Motif: Overcoming Inclusion Body Formation and Protein Aggregation during Overexpression, Purification, and Renaturation. *Protein Expr. Purif.* **18**, 394–403.

142 Hirose, M., Akuta, T., and Takahashi, N. (1989). Renaturation of Ovotransferrin Under Two-step Conditions Allowing Primary Folding of the Fully Reduced Form and the Subsequent Regeneration of the Intramolecular Disulfides. *J. Biol. Chem.* **264**, 16867–16872.

143 Rudolph, R., Zettlmeissl, G., and Jaenicke, R. (1979). Reconstitution of Lactic Dehydrogenase. Noncovalent Aggregation vs. Reactivation. 2. Reactivation of Irreversibly Denatured Aggregates. *Biochemistry* **18**, 5572–5575.

144 Williamson, R. A., Natalia, D., Gee, C. K., Murphy, G., Carr, M. D., and Freedman, R. B. (1996). Chemically and conformationally authentic active domain of human tissue inhibitor of metalloproteinases-2 refolded from bacterial inclusion bodies. *Eur. J. Biochem.* **241**, 476–483.

145 Creighton, T. E. (1985). Folding of proteins adsorbed reversibly to ion-exchange resins. *UCLA Symp. Mol. Cell. Biol.* 39, 249–258.

146 Creighton, T. E. (1990). Process for the production of a protein. *U.S. Patent* 4,977,248.

147 Geng, X., and Chang, X. (1992). High-performance hydrophobic interaction chromatography as a tool for protein refolding. *J. Chromatogr.* **599**, 185–194.

148 Walter, C., Höner zu Bentrup, K., and Schneider, E. (1992). Large Scale Purification, Nucleotide Binding Properties, and ATPase Activity of the MalK Subunit of *Salmonella typhimurium* Maltose Transport Complex. *J. Biol. Chem.* **267**, 8863–8869.

149 Stempfer, G., Höll-Neugebauer, B., and Rudolph, R. (1996). Improved Refolding of an Immobilized Fusion Protein. *Nature Biotechnol.* **14**, 329–334.

150 Bassuk, J. A., Braun, L. P., Motamed, K., Baneyx, F., and Sage, H. E. (1996). Renaturation of SPARC Expressed in *Escherichia coli* Requires Isomerization of Disulfide Bonds for Recovery of Biological Activity. *Int. J. Cell Biol.* **28**, 1031–1043.

151 Liñares, D., Echevarria, I., and Maña, P. (2004). Single-step purification and refolding of recombinant mouse and human myelin oligodendrocyte glycoprotein and induction of EAE in mice. *Protein Expr. Purif.* **34**, 249–256.

152 Franken, K. L. M. C., Hiemstra, H. S., van Meijgaarden, K. E., Subronto, Y., den Hartigh, J., Ottenhoff, T. H. M., and Drijfhout, J. W. (2000). Purification of His-Tagged Proteins by Immobilized Chelate Affinity Chromatography: The Benefits from the Use of Organic Solvents. *Protein Expr. Purif.* **18**, 95–99.

153 Rogl, H., Kosemund, K., Kühlbrandt, W., and Collinson, I. (1998) Refolding of *Escherichia coli* produced membrane protein inclusion bodies immobilised by nickel chelating chromatography. *FEBS Lett.* **432**, 21–26.

154 Berdichevsky, Y., Lamed, R., Frenkel, D., Gophna, U., Bayer, E. A., Yaron, S., Shoham, Y., and Benhar, I. (1999). Matrix-Assisted Refolding of Single-Chain Fv-Cellulose Binding Domain Fusion Proteins. *Protein Expr. Purif.* **17**, 249–259.

155 Shimizu, H., Fujimoto, K., and Kawaguchi, H. (2000). Improved refolding of denatured/reduced lysozyme using disulfide-carrying polymeric microspheres. *Colloids Surfaces B: Biointerfaces* **18**, 137–144.

156 Shalongo, W., Ledger, R., Jagannadham, M. V., and Stellwagen, E. (1987). Refolding of denatured thioredoxin observed by size-exclusion chromatography. *Biochemistry* **26**, 3135–3141.

157 Shalongo, W., Jagannadham, M. V.,

Flynn, C., and Stellwagen, E. (1989). Refolding of denatured ribonuclease observed by size-exclusion chromatography. *Biochemistry* **28**, 4820–4825.

158 Werner, M. H., Clore, G. M., Gronenborn, A. M., Kondoh, A., and Fisher, R. J. (1994). Refolding proteins by gel filtration chromatography. *FEBS Lett.* **345**, 125–130.

159 Gu, Z., Su, Z., and Janson, J.-C. (2001). Urea gradient size-exclusion chromatography enhanced the yield of lysozyme refolding. *J. Chromatogr.* A **918**, 311–318.

160 Gu, Z., Weidenhaupt, M., Ivanova, N., Pavlov, M., Xu, B., Su, Z.-G., and Janson, J.-C. (2002). Chromatographic Methods for the Isolation of, and Refolding of Proteins from, *Escherichia coli* inclusion Bodies. *Protein Expr. Purif.* **25**, 174–179.

161 Gu, Z., Zhu, X., Zhou, H., and Su, Z. (2003) Inhibition of aggregation by media selection, sample loading and elution in size exclusion chromatographic refolding of denatured bovine carbonic anhydrase B. *J. Biochem. Biophys. Methods* **56**, 165–175.

162 Tsumoto, K., Umetsu, M., Kumagai, I., Ejima, D., and Arakawa, T. (2003). Solubilization of active green fluorescent protein from insoluble particles by guanidine and arginine. *Biochem. Biophys. Res. Comm.* **312**, 1381–1386.

163 Burgess, R. R. (1996). Purification of Overproduced *Escherichia coli* RNA Polymerase σ Factors by Solubilizing Inclusion Bodies and Refolding from Sarkosyl. *Methods Enzymol.* **273**, 145–149.

164 Frankel, S., Sohn, R., and Leinwand, L. (1991). The use of sarkosyl in generating soluble protein after bacterial expression. *Proc. Natl. Acad. Sci. USA* **88**, 1192–1196.

165 Lowe, P. A., Marston, F. A. O., Sarojani, A., and Schoemaker, J. A. (1994). Process for the recovery of recombinantly produced protein from insoluble aggregate. *U.S. Patent* 5,340,926.

166 McCoy, K. M., and Frost, R. A. (1991). Method for solubilization and naturation of somatotropin. *U.S. Patent* 5,064,943.

167 Bradford, M. M. (1976). A rapid and sensitive method for the quantitation of microgram quantities of protein utilizing the principle of protein-dye binding. *Anal. Biochem.* **72**, 248–254.

39
Engineering Proteins for Stability and Efficient Folding

Bernhard Schimmele and Andreas Plückthun

39.1
Introduction

The industrial, biotechnological, and medical applications of proteins are often limited by an insufficient protein stability or related problems. Such applications commonly require that proteins be produced on a large scale and remain stable enough to fulfill their functions for a reasonable length of time, often under harsh conditions. However, natural proteins are typically only marginally stable, and it is thus a major challenge for protein engineers to optimize stability and folding efficiency. The approaches that have been successfully employed to achieve this goal are rational design, semi-rational strategies based on sequence comparisons, and the methods of directed protein evolution. Of course, these methods are not mutually exclusive and can be combined to solve practical problems. All studies employing these methods have revealed important rules for protein engineering and at the same time shed light on the principles and mechanisms responsible for the folding and stability of proteins. Recent advances in stability engineering have demonstrated that merely small changes in a given protein sequence can have profound effects on its biophysical properties. The major challenge is therefore to correctly identify and remedy these shortcomings. It is the goal of this chapter to summarize the biophysical principles and technological approaches useful in improving the biophysical properties of proteins through sequence modification.

Considering the enormous array of technologies involved in this endeavor, ranging from computer algorithms to selection technologies, it is not possible to give detailed experimental protocols in this chapter; instead, we will guide the reader to the cited literature.

39.2
Kinetic and Thermodynamic Aspects of Natural Proteins

39.2.1
The Stability of Natural Proteins

Evolution does not per se provide proteins with high stability. In fact, stability is just one of many evolutionary constraints on proteins. Proteins have to fold to a

Protein Folding Handbook. Part II. Edited by J. Buchner and T. Kiefhaber

ISBN: 3-527-30784-2

defined structure with adequate yield in a reasonable time and then have to be just stable enough to perform their function over a certain period. There is no evolutionary incentive to make a protein any "better" than what is needed to fulfill its cellular functions. In contrast, the use of a protein in a formulation at high concentration, its prolonged activity at 37 °C, and its large-scale expression and crystallization, just to name a few conditions, may put far higher demands on the protein than its natural environment. Thus, the natural sequence may not be able to provide these properties, but a mutant sequence may.

Proteins exist and have evolved in order to fulfill a given function, and evolution drives the structural properties of a protein mainly towards increased functionality [1]. In fact, most proteins are only marginally stable, with $\Delta G_{\mathrm{folding}}$ in the range of -20 to -60 kJ mol^{-1}. It is still a matter of debate whether this marginal stability is actually a "design feature," e.g., to allow degradation at a certain rate, whether it is caused by the selection pressure towards higher functionality that may not be compatible with high stability, or whether it is just a side effect of the lack of selection for high stability. Function is often linked to higher structural flexibility in certain regions of a protein. Lower stability as a result of this higher local structural flexibility might therefore simply represent an adaptation to increased functionality [2]. If this were generally true, however, stability engineering would fail in most cases, as it would not be able to reconcile stability with preserved protein function. An alternative, more optimistic view for the protein engineer is that marginal stability can be interpreted as a result of genetic drift [3]. In other words, lower stability is not intended; but it simply does not matter, provided that function is maintained. Random mutations occurring during evolution are more likely to destabilize the structure of a given parental protein sequence than to stabilize it or be neutral. However, as long as this stability decrease is not sufficient to render the protein nonfunctional, these destabilizing mutations are likely to accumulate in the sequence. This tendency has also been referred to as "sequence entropy" [4]. As a consequence, stability engineering could be interpreted as the art of identifying these unfavorable mutations in order to reestablish a more stable sequence.

The concept discussed above also sets the basis for the consensus approach to stability engineering, which is discussed in Section 39.3.1 Based on the physical principles of protein folding and a structure-based analysis of the interactions between amino acid residues, rational design can give hints as to which residues need to be altered to achieve a desired effect, and this is discussed in Section 39.3.2. The third focus will be set on the methods of directed evolution, which mimic the mechanisms of Darwinian evolution to evolve proteins with enhanced folding and stability properties (see Section 39.4).

39.2.2
Different Kinds of "Stability"

Before discussing in detail different strategies for rendering a protein more stable, it is worth taking a closer look at some basic features associated with protein stability and folding properties. The term "stability" itself is rather vague, and its precise meaning has often been adapted depending on the problem being addressed. This leads to different definitions of "stability." We will now briefly analyze the definitions and differences between thermodynamic, kinetic, and thermal stability, as well as folding efficiency, which is also sometimes discussed in this context. Even though these properties are interconnected, they are not equivalent. This has important implications for protein engineering, as it is difficult to predict in advance how a given mutation will influence each of these properties. These influences will in fact be different for any protein under investigation according to the free energies of the native and unfolded states and the folding intermediates, as well as the folding pathway and the respective rate constants.

39.2.2.1 Thermodynamic Stability

Thermodynamics describes the global unfolding behavior of a protein. The corresponding *thermodynamic stability* ΔG describes the differences between the free energies of the native (N) and the unfolded (U) states. Importantly, ΔG is an equilibrium property for a reaction involving two or more states. The simplest model of an unfolding reaction is the equilibrium of a two-state unfolding reaction:

$$\mathrm{N} \underset{k_{\mathrm{folding}}}{\overset{k_{\mathrm{unfolding}}}{\rightleftharpoons}} \mathrm{U}$$

where $k_{\mathrm{unfolding}}$ and k_{folding} are the rate constants of the respective unfolding and folding reactions. As a consequence, values for ΔG can be deduced only if the described process is fully reversible and no intermediate states are populated to a significant extent, unless they are explicitly known and measured. ΔG describes to what degree the two states are populated at a given temperature according to

$$\Delta G_{\mathrm{unfolding}} = -RT \ln K \tag{1}$$

where K is the equilibrium constant of the unfolding process. The treatment of experimental data is much simplified in such a model, as stability is affected only by the free energies of the folded and unfolded states. Because ΔG also provides an overall measure of all energetic contributions of interactions occurring upon protein folding, it is a convenient quantity for comparing the energetic effects of single amino acid replacements in a given folded structure. Because of the great scientific interest in deducing the principles of protein folding on a quantitative basis, many studies have been carried out with carefully selected model systems in which these strict requirements that allow the determination of ΔG are fulfilled. Moreover, ΔG represents an intrinsic quantity, at a given temperature in a given buffer, that is independent of the experimental setup and therefore reproducible in any case. The major problem in such studies, however, is confirmation that the observed transition can in fact be described as a fully reversible process, and denaturant-induced transition data alone are often not sufficient to allow reliable judgment. Only if data derived from measurements on the basis of different spectral probes (or better yet, additionally by differential scanning calorimetry, yielding the model-independent enthalpy change of unfolding) agree with each other are the deduced ΔG values likely to be correct. Nevertheless, many useful conclusions can be drawn about the effects of mutations even if these strict requirements cannot be completely fulfilled in all cases.

Another fact complicates the use of ΔG as a measure for protein stability. Mostly as a result of the fact that the hydrophobic effect is the major driving force of protein folding [5], ΔG itself is a characteristic, curved function of temperature. It is defined as

$$\Delta G(T) = G^U - G^N = \Delta H(T) - T\Delta S(T) = -RT \ln K \tag{2}$$

emphasizing especially that the enthalpy change ΔH is itself a function of temperature. This change of ΔH with T can be described by the heat capacity change

$$\Delta C_p = \left(\frac{\partial \Delta H(T)}{\partial T}\right)_p \tag{3}$$

Although often ignored, the temperature dependence of ΔG should thus not be left out of the account if one is dealing with the stability engineering of proteins. The large heat capacity change upon the transfer of nonpolar solutes to water, which is the basis of the hydrophobic effect, results in a curved function of ΔG versus temperature. By using the definition of $\Delta C_p, \Delta G$ can be approximated as a function of T, the melting temperature T_m, the enthalpy change at T_m $\Delta H(T_m)$, and ΔC_p.

$$\Delta G(T) = \left(1 - \frac{T}{T_m}\right)\Delta H(T_m) + (T - T_m)\Delta C_p - T\Delta C_p \ln \frac{T}{T_m} \tag{4}$$

If $\Delta G(T)$ is plotted as a function of T, the curve increases at low temperatures and decreases at high temperatures (Figure 39.1). The temperature at which folded and unfolded states are equally populated, and thus $\Delta G = 0$, is called the melting temperature T_m.

The respective curvature of ΔG versus temperature is strongly dependent on the change of heat capacity ΔC_p upon unfolding. A mutation may change $\Delta C_p, \Delta H(T_m)$, and T_m in any combination, thereby altering the shape and the position of this curve. Higher thermodynamic stability at a given temperature ($\Delta G(T)$) can thus be achieved, e.g., by an upshift of this curve with constant maximum. Right-shifting of the curve will decrease ΔG at lower temperatures but increase it at higher temperatures, which goes along with an increase of T_m. Flattening of the curve due to a lower change of heat capacity upon unfolding may cause lower thermodynamic stability at most temperatures, even though T_m is increased.

These considerations contain important implications for the analysis of engineered mutants. First, a measured decrease of ΔG at a certain temperature does not necessarily mean that at higher temperatures the thermodynamic stability might not have been increased. Second, by determining the change of ΔG, no conclusions can be drawn about a potential change of the melting temperature T_m. While they are typically related, a low ΔG at a given temperature does not necessar-

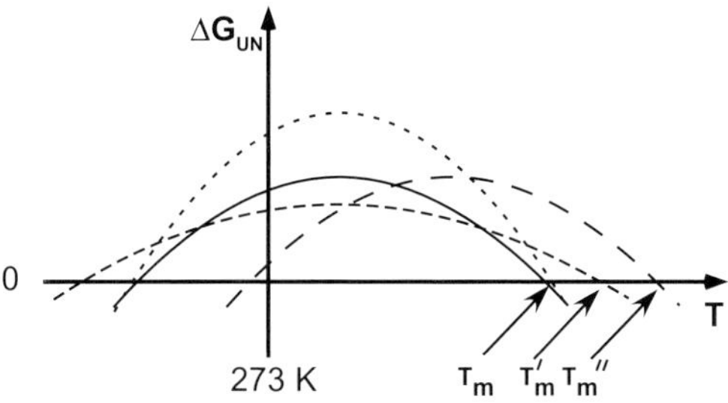

Fig. 39.1. The complex relationship between the melting temperature and the free energy of folding. Schematic representations of the free energy difference between the folded and unfolded states of a protein, ΔG_N, as a function of temperature, T. A typical protein is shown in curve 1 (————). The shape of this curve changes if a mutation affects ΔC_p, $\Delta H(T_m)$, and/or T_m of the protein. In the examples shown, not just one but several of these parameters are changed. Higher thermodynamic stability at a given temperature can phenomenologically be the result of a curve upshift (---------). Right-shifting the curved function (— — — —) results in a right-shift of the maximum of the ΔG_N function as well as in a shift of the melting temperature T_m towards higher temperatures ($T_m{}'$). If the ΔG_N function is flattened because of a small ΔC_p (----), a higher melting temperature ($T_m{}''$) can result, even though the thermodynamic stability is actually decreased over a broad temperature range. All depicted curves represent extreme cases, and typically a combination of the described alterations will occur upon mutating a protein.

ily mean that T_m will be decreased. T_m and ΔG must therefore be regarded as different properties. It should also be recalled that thermodynamic parameters, while easily reproduced, give no information about how long it will take until equilibrium is reached or, in the case of a nonreversible reaction, until a certain fraction of proteins is inactivated. Thus, thermodynamic stability does not necessarily provide information about whether a protein will meet the stability requirements for an intended application.

T_m itself often serves as a measure of protein stability, more precisely of *thermal stability*. In the literature, the expression "thermal stability" is again used with different meanings. In the described case, it represents the melting temperature for a reversible process. However, a complete thermodynamic analysis of the vast majority of proteins is not possible, because either intermediate folding states are populated or folding in the absence of denaturants that solubilize the unfolded state is not a fully reversible process. As will be discussed in the following section, thermal stability can nevertheless serve as a very practical means of describing protein stability, defined either as the transition temperature of an irreversible process or as the half-life of a protein under a given set of conditions.

39.2.2.2 **Kinetic Stability**

In most cases, outside the biophysical research lab, protein unfolding is an irreversible reaction. Initially, the aggregation of unfolded molecules or of folding intermediates prevent the back-reaction of folding, and this is followed, after prolonged times at high temperature, by chemical inactivation or, in impure samples, proteolysis. In a simple model, kinetic inactivation can be described as

$$\mathrm{N} \underset{k_{\text{folding}}}{\overset{k_{\text{unfolding}}}{\rightleftharpoons}} \mathrm{U} \xrightarrow{k_{\text{agg}}} \mathrm{X}$$

As discussed in the Introduction, one evolutionary constraint on proteins is that their three-dimensional structure remains viable for a certain period of time. In the case of non-equilibrium conditions and irreversible reactions, reaction kinetics becomes the important parameter. The folded state can resist high temperatures for a considerably long time if either the rate constant of inactivation or the rate constant of unfolding is sufficiently low. The rate constants are determined by the free energy of activation, e.g., the difference in energy between the folded state and the transition state of unfolding. Proteins can thus be kinetically stabilized by increasing this activation barrier.

This kinetic stability is different in some crucial aspects from the thermodynamic stability mentioned above. Engineering of proteins for enhanced stability has to deal with both aspects. The best mutations for enhancing kinetic stability will not necessarily be the best mutations for enhancing thermodynamic stability, and not every mutation that increases thermodynamic stability will automatically have a positive effect on protein half-life.

Kinetic stabilization is a common theme in nature, and there are several indications that many proteins from thermophilic organisms are indeed stabilized kinetically rather than thermodynamically [6, 7]. Another important example is that of proteins from the coats of viruses and phages that have to protect their genetic material under very adverse conditions [8]. In extreme cases the native state of a protein can even be less stable than its denatured state, but the native fold can still be kinetically trapped, and large kinetic unfolding barriers can provide the protein with an extremely long half-life [9].

There are different ways of describing and determining kinetic stability. Even if the reaction proceeds in an irreversible way, a practical "melting curve" can still be determined, and the observed transition is cooperative. The midpoint of this transition can serve as a practical means to compare the thermal resistance of different protein variants. Alternatively, one can use the half-life of the protein at a given temperature as an empirical means of stability. However, one has to be aware that this will not reflect equilibrium conditions. The observed values are actually kinetic values and thus are not independent of the exact experimental conditions, such as the protein concentration, the heating rate, etc. A more thorough analysis of kinetic stabilization must include kinetic measurements to determine the respective unfolding rates at different temperatures [10, 11].

For medical applications, engineered proteins usually have to fulfill defined stability requirements such as the absence of aggregation in the formulation used, long-term stability, and prolonged activity at 37 °C. The mere analysis of thermodynamic and kinetic parameters under defined reaction conditions in vitro is not always a reliable indicator of protein behavior under in vivo conditions. Mechanisms other than the intrinsic properties of the protein, such as proteolysis or aggregation with other proteins, can affect half-life. For practical utility, the half-life can therefore also be determined by measuring the percentage of molecules that retain their function after incubation under the respective conditions, e.g., in human serum at 37 °C for several days [12].

39.2.2.3 **Folding Efficiency**

A different issue of kinetics is related not to unfolding but to the folding of the protein in vivo. More precisely, the question is, which percentage of a protein will actually fold to the native state, as opposed to going to misfolded states, soluble aggregates, or inclusion bodies? Even though the folding efficiency in vivo is not directly related to stability, correlations can often be observed [13]. Additionally, the efficiency of protein folding in vivo is usually the predominant factor influencing the expression yield and is therefore also crucial for large-scale production of functionally intact proteins. Importantly, many mammalian proteins with a high potential for medical applications, especially those secreted or expressed on the cell surface, can rely on the complex folding machinery of the eukaryotic cell to reach their final native state, and the secretory quality-control system of eukaryotic cells allows discrimination of proteins by their folding behavior [14]. Moreover, they usually do not need to be expressed in high amounts in their native physiological context. The resulting lack of selection pressure on their efficiency of folding during evolution is likely to be one of the causes for the difficulties often observed when attempting their overexpression. Unfortunately, these tend to be the proteins of greatest pharmacological utility.

Despite the fact that thermodynamic stability underlies folding efficiency, the kinetic partitioning into productive folding or aggregation is influenced by many different factors. For illustrative purposes, we can again describe this by a very simple scheme:

$$\mathrm{U} \xrightarrow{k_{\mathrm{int}}} \mathrm{I} \begin{cases} \xrightarrow{k_{\mathrm{folding}}} \mathrm{N} \\ \xrightarrow{k_{\mathrm{agg}}} \mathrm{Agg} \end{cases}$$

Folding intermediates are often the source of aggregation, and the overall folding efficiency will therefore depend mainly on the nature of these intermediates. This includes the free energy of the folding intermediates themselves, as well as their half-lives and any efficient pathways to aggregation. In vivo, the situation becomes much more complex, as additional parameters such as interactions with cellular components and chaperones or degradation by host cell proteases come into play. The final output of a properly folded protein will therefore depend on all of these kinetic competitions [15]. Protein expression in the bacterial cytoplasm in many cases shows correlations between soluble expression yield and thermodynamic stability of the protein [16, 17]. Additional complications can arise from transport steps. For example, the expression of proteins in the periplasm of *E. coli* is dependent on the prior transport of the polypeptide chain through the inner membrane, and the folding yield is subsequently influenced by the folding and aggregation reaction in the periplasmic space as well as by interactions with periplasmic factors such as chaperones and proteases. In some cases, mutations that show positive effects on in vivo folding yield have no influence on the overall stability of the protein [18]. Conversely, mutations that strongly increase thermodynamic stability sometimes result in lower folding yields [19]. Nevertheless, many mutations act synergistically on both properties, because they are likely to reduce the free energy of the folded state as well as the free energy of folding intermediates and thereby lower the energetic activation barrier to folding [13].

39.3 The Engineering Approach

39.3.1 Consensus Strategies

39.3.1.1 Principles

As discussed in the preceding sections, marginal protein stability is likely to be a side effect of "sequence entropy" occurring during natural evolution, because the major driving force of evolution is positive selection towards an enhanced functional property, while stability has to be maintained at only a minimum level to secure function. Mutations are likely to occur in a random fashion during this process; the probability that a mutation will have a stabilizing effect on the protein is very low, whereas the probability that the mutation will have destabilizing effects is very high. However, as long as the remaining amino acid sequence is still able to fold into a given structure and the overall domain stability does not fall below a certain threshold, the resulting protein sequences will not be eliminated during the course of evolution [20]. Destabilizing mutations are therefore often selectively neutral and thus accumulate in a given parental sequence. The same should also be true for folding efficiencies. Most proteins are not needed at high concentrations or may even become harmful to the organism in such a case. Similar to stability, folding yield is selectively neutral, provided that the minimum level for cellular function is maintained.

This sets the basis for a semi-rational approach to protein stabilization, which is called the consensus approach [21] and is based on sequence statistics. Because mutations occur randomly, the distribution of amino acids at a given position in a set of homologous proteins can be described, in a very crude approximation, by Boltzmann's law. The consensus approach assumes that at a specific position in a sequence alignment of homologous proteins, the contribution of the respective consensus amino acid to the stability of the protein is on average higher than the contribution of any non-consensus amino acid. Replacement of all non-consensus amino acids in a sequence by the respective consensus amino acid should therefore increase the overall stability of the protein. Obvious advantages of the consensus approach are that it is comparatively simple and is not strictly dependent on structural information at high resolution.

The prerequisite for building a non-biased consensus is the availability of sequences homologous to the protein under investigation. The number of sequences should be large enough to make the sequence statistics reliable and to exclude bias in the resulting consensus sequences. Figure 39.2 shows an alignment of homologous sequences of single repeat modules, the smallest structural entity of a class of proteins known as leucine-rich repeat (LRR) proteins. Because the length of these modules varies among the different classes of LRR proteins – influencing their topology – only repeat modules of a length of 24 amino acid residues have been used for this alignment. The probability of each amino acid occurring at a given position is calculated to derive a consensus sequence, representing the most frequently occurring amino acid residue at each position. The distribution of residues at each position can provide information on structurally forbidden residues and allows weighing the consensus with respect to variability. In most cases, the consensus will contain the residues important for defining the structure of the proteins. In the case of an enzyme family, it will also include the "functional" residues, i.e., those of the active site. In the case of binding proteins, such as antibodies or repeat proteins, the "functional" residues (those involved in binding) are not conserved but are different for each individual molecule, which has to adapt to its target.

Although at first sight no sophisticated structural analysis seems to be required, this is true in only the simplest of cases, where a single family of related sequences can be represented by a single consensus. Frequently, multiple families have emerged that use mutually incompatible solutions of packing. A good example is that of antibodies for which subgroup-specific consensus building has been very fruitful [22]. Averaging over all families would simply yield the consensus of the most-represented family and, if they are equally represented, may result in mutually incompatible residues. Thus, structural analysis can be very helpful in deciding whether an "averaging" of different sequence families is permissible or not. Because of this problem of interacting residues, a simple averaging may lead to incompatible pairs; therefore, these residues should be changed only as groups. The danger of disrupting these interactions by substitutions with consensus residues is especially high in cases where a very broad set of sequences is used for the alignment. To minimize this risk, the sequence statistics can be extended by analysis of covariance in order to derive probabilities that describe the joint occurrence of amino acid residues at two defined positions [23]. As explained above, before deriving a consensus, the aligned sequences can be divided into subclasses, which are likely to contain interacting pairs or groups of residues. Certain variations of residues involved in salt bridges, distinct hydrogen-bonding patterns, or packing of the hydrophobic core are characterized by complementary changes between these subclasses, with mutations to a certain residue at one position being compensated by a mutation to a complementary residue at another position. The subclasses can be built either by being based on sequence homology alone or by including structural information if available. Even though the definition of these subclasses is always dependent on the homology cutoff set by the investigator, a simple dendrogram analysis can be used to group the complete set of sequences into distinct families. Additionally, by building the consensus sequence of each family separately, fol-

a sequence alignment

Position	1	2	3	4	5	6	7	8	9	10	11	12	13	14	15	16	17	18	19	20	21	22	23	24
Sequence 1	S	L	V	S	V	N	L	R	M	N	K	F	S	G	I	V	P	E	S	F	G	K	L	K
Sequence 2	K	L	A	R	L	D	L	T	S	N	R	L	Q	K	L	P	P	D	P	I	F	A	R	S
Sequence 3	S	L	Q	I	L	D	C	S	F	N	R	I	V	A	F	K	W	Q	E	L	Q	H	F	P
Sequence 4	A	L	T	T	L	D	A	S	F	N	E	L	V	A	L	S	P	G	T	L	D	G	L	S
...												...												
Sequence x	R	V	L	S	L	N	L	S	S	S	G	L	T	G	I	I	A	A	A	I	Q	N	L	T

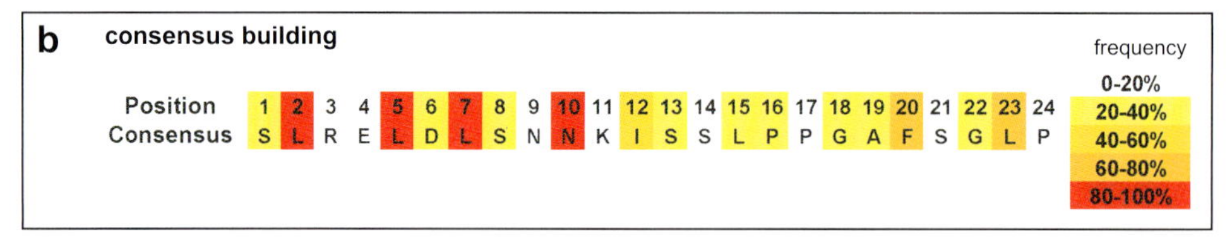

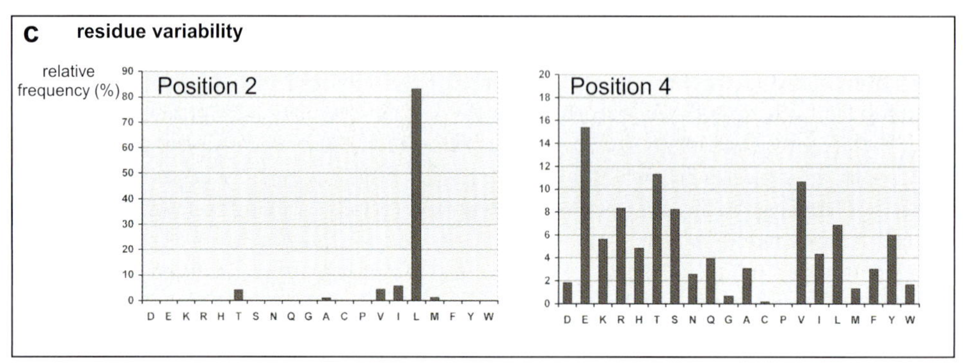

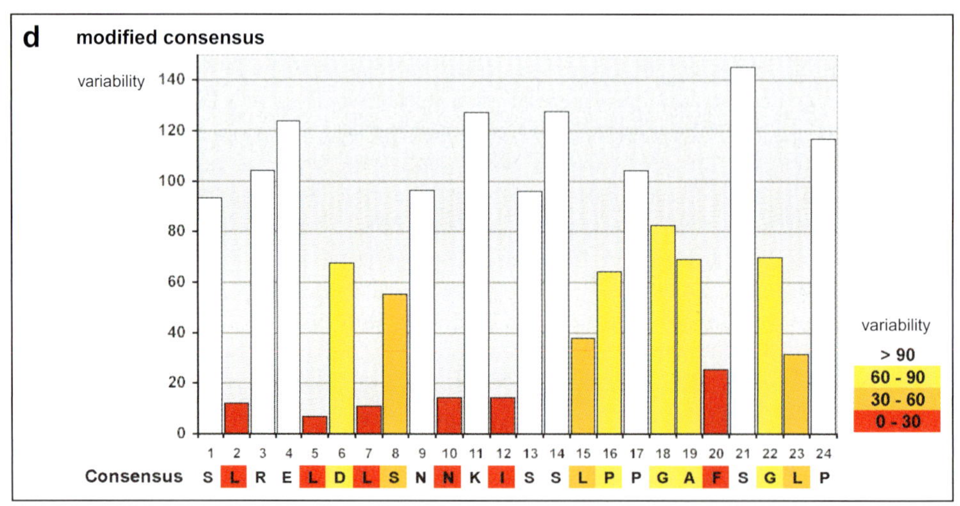

Fig. 39.2. Example of the analysis for deriving and analyzing a consensus sequence. (a) From a sequence alignment of 3077 sequences of 24 amino acid LRR motifs of the Pfam database (http://www.sanger.ac.uk/Software/Pfam/), the relative frequencies of amino acid residues at each position are calculated. (b) Based on the relative frequencies calculated from the alignment, a consensus sequence can be derived representing the most frequently

lowed by a comparison of these consensus sequences, distinct structural features of each group can be recognized in some cases.

The consensus concept has been applied successfully to a large variety of proteins and structural motifs to date. Important lessons for an effective application of the consensus approach can be learned from these studies, and we will therefore give a few examples to briefly discuss some of the advantages and limitations of the method.

39.3.1.2 Examples

There is always the concern that the stabilizing and destabilizing effects of introduced mutations will counterbalance each other and that the overall change in protein stability will be small. Steipe et al. [20] applied the method for the first time on immunoglobulin V_L domains and predicted 10 potentially stabilizing mutations. Six mutations were indeed stabilizing, three had no effect, and one was destabilizing. When applied to GroEL mini-chaperones, 34 predicted amino acid replacements were individually checked, out of which 13 were stabilizing, five showed no effect, and 16 were destabilizing [24]. Lehmann et al. [25] extended the approach to an entire protein. In a first set of experiments, 13 homologous sequences of a fungal phytase were used to build the consensus, and the resulting consensus enzyme showed an increase of 15–22 degrees in unfolding temperature and an increase of the temperature optimum for catalysis of 16–26 degrees compared with each of its parents. In a second set of experiments, additional mutations were predicted by simply adding more sequences to the alignment. By checking the effects of the individual mutations on thermal stability and combining mutations with positive effects, the unfolding temperature could be increased by an additional 21 degrees to 90 °C. No loss of catalytic activity of the enzyme was observed in any case. This work showed that the number of sequences used for the alignment is indeed an

occurring residue at each position. The color code is given on the right hand side of the panel. Residues occurring in more than 80% of all sequences at the respective position are colored in red. Based on these values, consensus sequences with a given homology threshold can be derived (not shown), i.e., with a higher threshold, more positions will be "undefined." (c) Preferences at a given position are reflected not only by the absolute frequency but also by the total number of different amino acid types occurring at each position (residue variability). As an example, the relative frequencies of each amino acid are shown for the highly conserved position 2 and the highly variable position 4. By plotting the relative frequencies of amino acids at a particular position, preferences for certain amino acid types as well as "forbidden" residues can be identified. At position 2, a strong preference for leucine can be observed, and the occurrence of other residues is restricted mostly to hydrophobic side chains. At position 4, all residues are "allowed" except for proline, which is "disallowed" due to secondary structure propensity reasons. (d) By normalizing the relative frequency of the consensus amino acid with the number of "allowed" residues at a given position, a modified consensus sequence can be deduced. The variability V is calculated according to $V = 100 \times N/F$, where N is the total number of different residue types occurring at each position and F is the frequency of the most frequent residue in percentage.

important factor because it might help to optimize ambiguous positions. In addition, the observed large increase in thermal stability could not be attributed to the effect of one single amino acid substitution but rather to the synergistic effects of many replacements.

Even though these and other studies show clearly that the consensus approach allows one to predict stabilizing mutations with a rather high success rate in a rapid way, the effect of each predicted mutation carries some uncertainty, and it is possible that some may contribute destabilizing effects that can counterbalance the stabilizing ones. However, the effects of stabilizing mutations were often found to be additive. Therefore, instead of examining each mutation individually, it is often useful to combine groups of "rather certain" mutations and others that are more speculative.

The application of the consensus concept to families of repeat proteins [23, 26–28] represents, in some respects, a special case due to a number of favorable features. It can nevertheless illustrate the importance of some principles of the approach. The non-globular fold of repeat proteins consists of repeated motifs of 20–40 amino acids. Several results indicate that consensus repeat proteins are indeed much more stable than natural repeat proteins [29, 30]. Repeat proteins might represent an extreme case in which the principles of the consensus approach become very apparent. The structural entities (the repeat modules) are small and thus each protein contributes several repeats to the databases; the number of available sequences consequently becomes very large compared with other proteins. Therefore, a consensus sequence for a single repeat module can easily be assigned and ambiguous positions will occur with lower frequency in the statistical output (Figure 39.2). In addition, interactions that are present within or between several repeats will add to the free energy of folding multiple times, while problem spots would equally be potentiated. Therefore, effects on stability are likely to be consecutively added by introducing additional modules to the array. Even though more-detailed analyses are still needed, initial results pinpoint some of the structural reasons for the stability gains observed upon building a consensus in each of these studies. The regular arrangement of structural motifs gives rise to a more regular H-bonding pattern with a higher number of inter- and intra-repeat H bonds [30]. Loop insertions in natural repeat proteins that are likely to result in more flexible local regions are removed, thereby eliminating local centers of unfolding. In the left-handed helical and disallowed regions of the Ramachandran plot, glycine residues are always present in the consensus proteins, while they are avoided in other places by this design, where their flexibility is not needed and may be harmful to stability.

39.3.2
Structure-based Engineering

Structure-based engineering relies on a detailed analysis of 3D structures, followed by site-directed mutagenesis. We avoid the term "rational" engineering, as it would elicit expectations of perfect predictability and implicitly suggest that all other approaches are free of logical reasoning. In structure-based engineering, positions have to be identified at which suboptimal amino acids in the original sequences lead to a loss of stability. Subsequently it needs to be specified which amino acids should be introduced as a replacement. The *ab initio* prediction of protein structure, however, is still not a feasible task due to the multitude of potential interactions within the protein and between protein and solvent, which leads to an extremely high number of possible conformations and intermediates of similar energy [31]. Hence, a prerequisite for structure-based engineering is, next to experimentally determined structures, usually the existence of a large experimental dataset within a group of structural homologues that can be used as a basis for predictions. High-resolution structures are necessary to allow the estimation of possible conformational, energetic, and steric influences upon replacement of particular amino acid side chains and can thus help to avoid unfavorable strain in the resulting mutants. Because of the present efforts in the field of structural genomics, these structure-based approaches are likely to become even more important in the future. With this structural information in hand, the goal of designing more stable variants is then to pinpoint particular regions and positions associated with possible stability defects and to subsequently find a better solution to the problem. In contrast, semi-rational approaches like the previously discussed consensus concept are rather crude methods for introducing stabilizing features. Structural and energetic analysis can be used to reexamine the changes proposed by the consensus approach and to fine-tune the system to reintroduce structural features that might have been lost in the averaging process.

Proteins of hyperthermophilic organisms have been of special interest for examining the structural mechanisms of thermostabilization and have been contemplated as guides for the engineering of "problem" proteins for better properties. From a phenomenological point of view, the basis of increased thermostability is frequently set by a flattened ΔG-versus-T curve that is due to a smaller change of heat capacity upon unfolding (Figure 39.1) or by a kinetic stabilization that is due to a strong decrease in the rate of unfolding. The crucial question is, however, what the molecular differences are that give these proteins their favorable properties. Genome-wide comparisons between hyperthermophilic and mesophilic organisms with respect to amino acid composition did not yield any obvious common rules of how these effects are achieved [32]. Therefore, hope was placed on the increasing number of pairwise high-resolution structure comparisons of thermophilic proteins with their mesophilic counterparts. While they provided a more differentiated picture, a "global" rule still could not be derived. A few highly specific mutations are often enough to provide considerably stabilizing effects, but the additive effect of many small contributions, none of them dramatic by itself, may be the usual case. Moreover, rather than relying on one universal strategy, nature utilizes a variety of strategies for the thermal adaptation of proteins [33]. In fact, the list of stabilizing structural features in hyperthermophilic proteins reflects the diverse principles of protein stability and folding that protein engineers try to exploit and that will be discussed in this section. High-resolution structures of thermophilic proteins can thus provide a detailed view of how nature implements these principles to create proteins of higher stability [34]. However, the lack of a unifying "rule" and the multitude of strategies nature uses provide an important lesson for protein engineers. Depending on the protein under investigation, the strategy of choice can be different, and even for a given protein there may be more than one optimal solution to the problem. Before choosing from the available set of strategies, the focus should therefore be on identifying potential "weak points" responsible for stability defects in a given protein structure.

We will now discuss some structural features associated with protein stability as well as strategies for altering these features towards more favorable biophysical properties. The given list of course lays no claim to completeness but should point out some important principles. Any replacement in a given sequence may have multiple effects on protein stability, and destabilizing effects can often outweigh the stabilizing ones. An assessment of potential destabilizing effects is therefore crucial. Wherever possible, references are made to the different forms of "stability" discussed in the Introduction. In the case study provided at the end of this section, an example will be given to demonstrate how consensus approaches and structural analysis can be combined to yield useful results.

39.3.2.1 Entropic Stabilization

An obvious strategy for increasing the free-energy difference between the folded and unfolded states, and thus the thermodynamic stability, is to decrease the entropy of the unfolded state. The underlying concept is to decrease the flexibility of the polypeptide chain, usually by introducing an additional intrachain linkage. Such entropic stabilization has become a common strategy for protein engineers. The prerequisite for success is that the mutations rendering the unfolded protein less flexible do not introduce unfavorable strain in the folded three-dimensional structure or result in any steric incompatibilities [35]. We now discuss several ways to achieve an entropic stabilization.

Introduction of Disulfide Bridges The introduction of additional disulfide bridges is a straightforward way of establishing an intrachain linkage to reduce the entropy of the unfolded state [36, 37]. The magnitude of the entropic effect is thought, as a crude approximation, to be proportional to the logarithm of the number of residues between the two bridged cysteines [38]. The spatial distance between the residues to be replaced with cysteines has to be evaluated with care in the model of the folded protein in order to prevent perturbations of the native structure upon formation of the disulfide bridge. It should be noted, however, that the energetic effect of additional disulfide bridges is not only entropic but also of a far more complex nature, giving rise to entropic as well as enthalpic contributions to the change in the free energy of folding [39, 40]. For example, an additional decrease in the free energy of folding can result from the reduced solvation energy of the unfolded state [40]. In contrast, a reduced solvation energy of the folded state would have the opposite effect, while residual structure in the denatured state would again push the equilibrium to the side of the folded protein. Because the disulfide bond itself is hydrophobic in nature, it is often engineered into the interior of the protein. This is not an easy task, as it can negatively affect core packing. Even though there are several examples of successful protein stabilization by introducing artificial disulfide bridges [41, 42], the complex energetic effects can also cause a destabilization of the protein [43]. Furthermore, the introduction of additional cysteines often results in a rather drastic decrease in folding efficiency, because incorrect and intermolecular disulfide formation can remove large portions of expressed protein by aggregation. Since disulfide formation does not occur in the cytoplasm, secretion to the bacterial periplasm or to the eukaryotic ER is required for functional expression, usually associated with lower yields than for the production of cytoplasmic proteins. Alternatively, if the protein is produced by refolding, redox conditions have to be adjusted, which can be difficult if the native protein also contains free cysteines.

Circularization An alternative approach with the same underlying concept is the circularization of proteins by fixing the loose N- and C-termini via a peptide bond. In addition to the entropic effect, the fixing of the loose ends can prevent local unfolding events occurring at the termini and thereby kinetically stabilize the native structure. With the discovery of inteins, which mediate protein-splicing reactions, a tool that allows the directed formation of peptide bonds between ends fused to different parts of the intein became available. Intein-mediated protein ligation has been used to covalently link the termini of β-lactamase, a protein that is especially amenable to this strategy due to the close proximity of its N- and C-termini [44]. In accordance with polymer theory, the thermal stability of the protein was enhanced by about 6 degrees, from 45 °C to 51 °C. For circularized DHFR [45] an increased half-life at elevated temperature was observed. The close proximity of the termini is of course a prerequisite for this procedure, and the stabilizing effect is likely to become marginal if the loose ends are linked via long unstructured loops. Similar to the situation upon introduction of artificial disulfide bridges, destabilizing enthalpic effects may negate the favorable entropic contribution [46]. In addition, low protein ligation efficiencies and difficulties in separating circular from linear forms of the protein often cause additional technical challenges. It remains to be seen whether this technology is robust enough for biotechnological or biomedical applications.

Shortening Solvent-exposed Loops Short, solvent-exposed loops are rather fixed in the native state, but a comparably large number of additional conformations become accessible in the unfolded state, while long loops have a large number of conformations also available in the native state. Thus, the shortening of loops should in principle lead to a relative decrease in the loss of conformational entropy upon folding. Conversely, increasing the loop lengths by insertion of glycine residues into the loops of the four-helix bundle protein Rop has indeed resulted in a strong and continuous decrease in thermodynamic stability [47]. In addition, loop shortening can have the effect of abolishing hot spots of local unfolding events and may result in kinetic stabilization. Even though it has become obvious that loop shortening or tying down of loops by external interactions is a common theme in thermostable proteins of thermophilic organisms [48, 49], the strategy is often hard to realize for a given protein target, as the danger of introducing additional strain in the native state is high, and solvent-exposed loops are often important with respect to function.

Reduction of Chain Entropies By considering the conformational entropies of amino acid side chains, another strategy for decreasing the entropy of the unfolded state becomes apparent. Because of the five-membered-ring nature of the proline side chain, it not only restricts the possible conformations of the preceding residue but also can adopt only a few conformations itself. It therefore has the lowest conformational entropy of all amino acids [50]. In contrast, glycine, which has no side chain, has the highest conformational entropy. Substitutions of non-glycine residues with proline or the replacement of glycines by other residues should therefore reduce the entropy of the unfolded state.

Positions that allow substitutions with proline are, however, very rare. Because proline is poorly compatible with α-helices and incompatible with β-strands, the position of a new proline must not be part of these secondary structure elements. At most positions in the native structure, the respective torsion angles will be incompatible, and the mutations are thus very much restricted to loop and turn regions. Again, care should be taken not to remove any favorable interactions of the replaced amino acid side chain [51]. In order to examine in advance whether the respective site is permissive for a substitution with proline, the dihedral angles of the site can be checked and should lie in the range of ϕ/ψ -50 to $-80/120$ to 180 or, alternatively, -50 to $-70/-10$ to -50.

Similar restrictions apply for the replacement of glycines with any other residue. In many cases this will create steric overlaps, and such negative structural crowding effects can outweigh the positive energetic benefits.

39.3.2.2 Hydrophobic Core Packing

Exposed residues are often directly involved in ligand or substrate binding and therefore often play a functional role. In contrast, the residues of a protein's interior usually play mostly a structural role, and the associated hydrophobic effect is thought to be the main driving force of protein folding and thermodynamic stability.

In known structures the core residues fill almost the entire interior space, provide many favorable van der Waals interactions, and maximize hydrophobic stabilization by exclusion of the solvent. In principle, an increase in thermodynamic stability of 4–8 kJ mol^{-1} can be achieved for each additional methylene group buried [52]. Paradoxically, the importance of the hydrophobic effect for folding and stability of proteins simultaneously limits its applicability for protein engineering. Because the hydrophobic core is already densely packed in almost all native proteins, most changes here will create over-packing or packing defects, causing an overall destabilization rather than an improvement in stability. In addition, even subtle changes of core residues can lead to a rearrangement of external residues and thereby alter the functional properties of the protein [53]. Care should also be taken to avoid the introduction of conformational strain by the mutation of core residues, as destabilizing effects from a strained conformation can sometimes compensate the energetic gain of an increase in buried hydrophobic volume [54]. Improvement of core packing must therefore be based on analyses made from high-resolution structural information in combination with sequence comparisons. This allows one to specifically look for cavities in the core that indicate imperfect packing. If the hydrophobic surface area around the cavity is large, additional van der Waals interactions can be provided by the introduction of sterically fitting alkyl or aryl groups from hydrophobic side chains, thereby decreasing the size of the cavity [55, 56].

39.3.2.3 Charge Interactions

Oppositely charged amino acid residues, if appropriately positioned, have the potential to form salt bridges, whereas like-charged residues lead to repulsions. The magnitude of the effect of charge-charge interactions on overall protein stability is still a matter of discussion [57]. In the case of ionic interactions between side chains buried in the hydrophobic core, the high energetic cost of transferring charged ions from aqueous solution to the low-dielectric interior of the protein also has to be taken into account. If a single charge were buried in a protein, which would be extremely rare in a natural protein, the design of an ion pair would be very attractive. However, if a hydrophobic pair were to be replaced by an ion pair, the resulting energy would have to be higher than the loss of the previous pair plus the cost of burying the charge. Nevertheless, the high contribution of buried salt bridges to the overall stability of the native protein structure underlines their potential for introducing additional stability [58]. The optimal spatial arrangement of the interacting side chains and their respective charges is, however, crucial. Moreover, buried charged side chains are often not only part of interacting charge pairs but also part of complex charge clusters built from many side chains, which are able to magnify the effect.

Similar rules apply for the interactions of charged residues on the protein surface. Only the perfect arrangement of charges seems to be able to make up for the desolvation penalty that has to be paid upon formation of a salt bridge. The effect on thermal stability, however, can be drastic. Increasing the free energy of unfolding by the changing of charges includes maximizing the number of salt bridges and, equally important, the removal of repulsive interactions [59], which are not uncommon in natural proteins. Predictions on a structural basis can be difficult due to the often higher flexibility of side chains on the protein surface, but simple models that allow predictions about potential stabilizing and destabilizing surface charges can be used [60]. The key is to consider not only nearest neighbors but also a whole network of charges that have to optimally interact and avoid repulsions. Because the surface charge distribution can have a huge impact on stability, but is defined by many residues at different positions, these residues are also a valuable target to be combined with selection techniques as discussed in Section 39.4.

A special case of electrostatic interaction is the "helix dipole." By reducing the net partial charges at the helical ends through placement of side-chains, which productively interact with the helix dipole, the helical structure is stabilized [61]. Introduction of negatively charged residues at the N-terminal end and positively charged residues at the C-terminal end leads to this stabilizing effect. The provided stability gain is, however, marginal (less than 4 kJ mol^{-1}).

39.3.2.4 Hydrogen Bonding

There was no initial reason to believe that intrachain hydrogen bonds in the native state would be more energetic than those of the unfolded chain to water [5]. By including terms of entropy change of the solvent and additional van der Waals interactions upon polar group burial, however, the positive contribution of hydrogen bonding to protein stability has become generally accepted [58, 62]. Despite this ongoing discussion, hydrogen-bonding patterns are a highly valuable target for the stability engineering of proteins. Because engineering deals with improving folded proteins, the major concern has to be how to satisfy the existing hydrogen-bonding network in a structural context. The basis for this endeavor is structural information of high resolution, and the most lucrative goal is to identify potential residues that represent buried but unsatisfied donors of hydrogen bonds. Site-directed mutagenesis of a nearby residue to provide a hydrogen-bonding acceptor can cause a stability gain in the range of 2–10 kJ mol^{-1}, depending on the geometry and other compensating effects [63, 64].

A special structural context that can provide significant additional stabilization by either hydrogen bonds or ionic interactions is the anchoring of relatively loose structural elements like loop structures or the N- and C-termini, thereby tightening "hot spots" of local unfolding [49, 65].

39.3.2.5 Disallowed Phi-Psi Angles

The stereochemistry of the polypeptide backbone can be defined by the dihedral angles ϕ and ψ, and any individual residue in a structure is defined by a single set of ϕ, ψ values. For conformational analysis of protein structures, the Ramachandran plot representing the dihedral angle space is an excellent starting point [66]. The ϕ, ψ values of amino acid residues in protein structures usually reside in three preferred or "allowed" regions of the Ramachandran plot, called right-handed helical, extended, and left-handed helical. The right-handed and extended conformations correspond to α-helix and β-strand secondary structures, respectively, and the vast majority of non-glycine residues lie within these two regions. The left-handed region corresponds to structural features at the termini of secondary structure elements and describes regions involved in the reversal of the polypeptide chain. There is a high preference in this region for glycine residues, as the β-carbons of non-glycine residues can sterically interact with the polypeptide backbone, resulting in unfavorable energies. In some cases, the substitution of non-glycine residues by glycines in the left-handed helical region increases thermodynamic stability (up to 8 kJ mol^{-1} in RNase H) [67]. Although the strict introduction of glycines in such positions is an important point to consider for de novo protein design, it does not represent a general rule for stabilizing the native states of proteins. In some cases, the energy penalty for the accommodation of unfavorable strain can be offset by lost unfavorable or new favorable local interactions, such as hydrogen bonding or hydrophobic interactions. Some replacements of this kind can therefore even lead to destabilization rather than stabilization [68].

The same rules apply in principle to the disallowed regions of the Ramachandran plot. Steric clashes result in a high energetic cost in the folded structure, especially for non-glycine residues in all disallowed regions. Stabilizing mutations to glycine have been introduced with energy gains of up to 18 kJ mol^{-1} [69]. It has been noted, however, that certain non-glycine residues also have propensities to occur in the disallowed regions – such as Asn and Ala in the type II′ turn region – and the energetic cost of their occurrences is often low [70]. Other residues found in the disallowed regions are small polar residues [71] that compensate for the energy cost by making additional hydrogen bonds. Unfavorable conformations often occur in very short loops [72], where the rest of the structure may constrain the loop efficiently, and interactions with the solvent may also offset the energy costs.

Conformational analysis by the Ramachandran plot therefore provides a convenient and fast way to assess possible conformational strain in the tertiary structure associated with particular target residues. However, a close inspection of possible side chain interactions is required, and the analysis should be extended to identify potential compensating features of the residues to be replaced.

39.3.2.6 Local Secondary Structure Propensities

Effects on the overall stability of a protein can also be influenced by the respective secondary structure propensities of amino acid residues for the α-helical and β-sheet conformations. However, these effects are usually marginal. Nevertheless, even at the expense of other favorable trends, such as avoiding the exposure of hydrophobic side chains to the solvent, a given residue is often favored at a certain position due to its secondary structure propensity [73]. If a particular secondary structure element does not form efficiently, many interactions between this element and other parts of the protein can be lost. The energetic consequences and the influences on folding efficiency are, however, hard to predict at this point.

39.3.2.7 Exposed Hydrophobic Side Chains

The removal of exposed hydrophobic side chains increases the polar surface of a protein. Such mutations are not likely to affect thermodynamic stability but presumably do affect the folding efficiency by influencing the rate of aggregation of intermediates. Interestingly, they also do not affect the solubility per se (the amount of native protein that can be dissolved in buffer) and seem to act mostly on folding intermediates [74]. Because of lateral interactions, e.g., between neighboring loops, partially exposed hydrophobic amino acids can even increase stability. It must be kept in mind however, that not only the aggregation pathway, but also the stabilities of folding intermediates are an important parameter for kinetic stability, which can possibly be influenced by the existence of exposed hydrophobic side chains. Moreover, hydrophobic cavities at the protein surface are often involved in specific binding functions of the protein, and the removal of these "functional hot spots" has to be avoided. Therefore, the hydrophobicity of the protein surface must be carefully balanced.

39.3.2.8 Inter-domain Interactions

In proteins consisting of more than one domain, additional principles come into play. The overall stability not only reflects the intrinsic stabilities of the single domains but is also influenced by the stability of the interface of the domains in cases where they are interacting with each other. Stabilization of this interface can be achieved mainly by increasing and optimizing the hydrophobic surface area of the interface. Because hydrophobic side chains at the interface are usually exposed during folding and transient opening of the domain interface, a tradeoff between interface stability and folding efficiency is often observed. Additional dramatic effects on stability can be observed in cases where the two domains exhibit very different intrinsic stabilities [75]. In such a scenario, the unfolding of the protein and the loss of function are strongly related to the unfolding of the less stable domain [76]. An important aspect of kinetic stabilization can be observed in two-domain proteins, where one domain can slow down the unfolding, and therefore the aggregation, of the other. Thus, the native state becomes kinetically stabilized in the assembly and a stable domain interface reduces the extent of its transient openings and, thereby, the resulting exposure of hydrophobic patches that would favor aggregation. Alternatively, covalent cross-links (e.g., disulfide bonds) between the interfaces of multimeric proteins can be introduced. For example, the introduction of disulfide bonds between the interfaces of *Lactobacillus* thymidylate synthase not only increases their thermal stability but also leads to reversible thermal unfolding [77].

39.3.3 Case Study: Combining Consensus Design and Rational Engineering to Yield Antibodies with Favorable Biophysical Properties

The following example illustrates how consensus approaches, rational design, and experimental data can be combined in a synergistic fashion to iteratively optimize biophysical properties.

The smallest form of an antibody able to bind the antigen in the same manner as the whole IgG consists of two domains, the variable domain of the heavy chain (V_H) and the variable domain of the light chain (V_L). Both domains interact with each other via an interfacial region of highly hydrophobic character. In single-chain Fv (scFv) antibody fragments, the two domains are covalently linked by a flexible linker region of typically 15–20 amino acids (Figure 39.3a). The binding site for the antigen usually involves three loop regions in each of the domains, named complementarity-determining regions (CDRs). Antibodies that are based on human antibody sequences possess great potential for many medical applications, either directly as an antibody fragment or after reconstructing an IgG [78]. Antibody fragments can be expressed in a convenient manner in *E. coli*, thereby providing rapid access to these proteins [79]. Nevertheless, there are often drastic differences between individual antibodies concerning their expression yield and their stabilities. Ideally, the recombinant antibodies would all provide favorable biophysical properties.

The starting point for such recombinant antibodies is a library. One fully synthetic library of this kind, the human combinatorial antibody library (HuCAL), was designed based on the consensus concept [22]. Based on human antibody germ-line sequences, several sequence families were created. Importantly, instead of averaging over all possible human antibody sequences, the consensus sequences were built for each family separately, resulting in seven consensus frameworks for V_H (V_H1a, V_H1b, V_H2, V_H3, V_H4, V_H5, and V_H6) and seven frameworks for V_L (V_κ1, V_κ2, V_κ3, V_κ4, V_λ1, V_λ2, and V_λ3). This diversity is important, as the use of different frameworks allows a variety of non-CDR contacts to the target, thereby greatly increasing the range of targets being recognized. By using this strategy, each human V_H and V_L subfamily that is frequently used during an immune response is represented by one consensus framework, and thus the immune response is closely mimicked. The consensus building was further restricted to the framework regions, while the CDRs were diversified in a manner guided by structure. Thereby, functional diversity is maintained in optimized sets of frameworks.

Instead of fully relying on the statistical output of the consensus building, structural modeling was employed in order to decrease the risk of disrupting interactions between certain amino acid residues that might be in contact with each other in the three-dimensional structure. Because many structures of antibody domains are available, the modeled structures of each framework family could be compared with the respective natural template structures. The models were checked according to several principles, which have been outlined in the preceding section. At this point the models were mostly checked for whether the interactions in natural antibody domains were correctly recreated. The consensus sequence models were inspected for any unfavorable strain in the structures, represented by unfavorable regions of the Ramachandran plot or any obvious cavities in the hydrophobic core. Moreover, the sequences were checked for the existence of residues already known to be involved in conserved intra-domain interaction patterns – such as conserved charge clusters and hydrogen-bonding patterns – as well as for exposed hydrophobic side chains known to decrease expression levels. Nevertheless, differences between the natural subclasses became apparent in the models.

Empirical results had suggested that certain natural framework types display more favorable stabilities and expression levels than others. These differences are already existing in the original natural human germ-line sequences. The intrinsic differences in terms of biophysical properties for each subtype were then experimentally explored in a systematic fashion, first on single domains, then on scFv fragments [13]. The experiments confirmed the observed trends, showing that V_H3 displays the highest thermodynamic stability and soluble expression level, when expressed as an individual domain, among all V_H subtypes. In contrast, V_H2, V_H4, and V_H6 display the least favorable properties in terms of stability, folding yield, and the tendency to aggregate. For the V_L domains, members of the V_κ

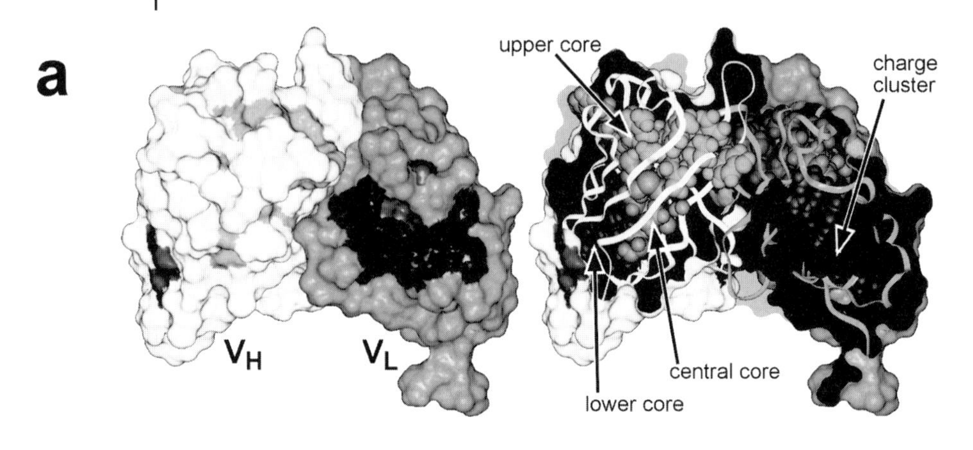

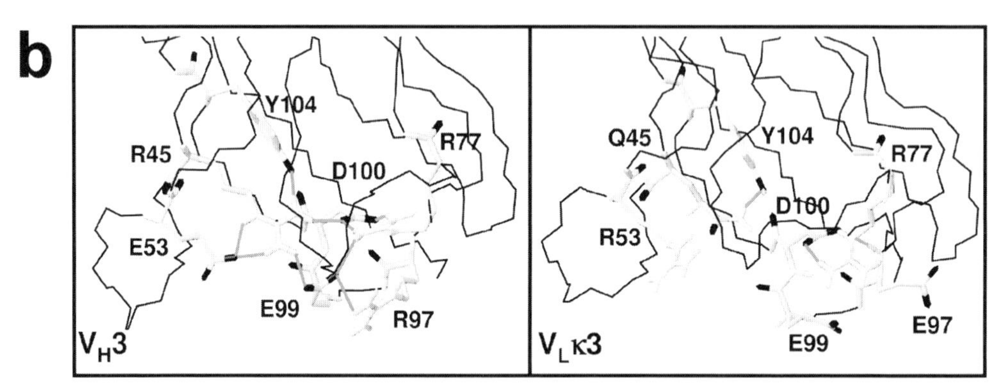

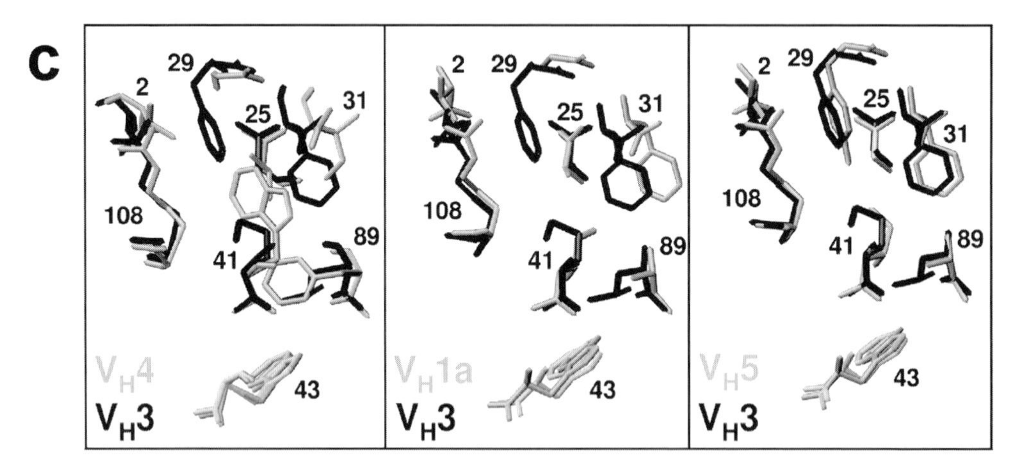

Fig. 39.3. (a) Structure of an antibody Fv fragment consisting of the variable heavy chain (V_H) and the variable light chain (V_L). Each domain V_H and V_L is characterized by three hydrophobic core regions (upper [green], central [yellow], and lower [orange] core) and a charge cluster at the base of each domain (red). Even though the residues defining these

subtypes showed slightly higher stabilities and expression yields than the V_λ subtypes, but the behavior was much more homogenous overall. In order to trace back these differences to the structural level, the model structures of the different subtypes were compared with each other. Several structural features can be invoked to explain the extraordinary stability of V_H3 domains in comparison with the even-numbered V_H subtypes.

First, differences in the hydrogen-bonding networks have an influence on the thermodynamic stabilities. Long-range interactions involving several residues are concentrated in a charge cluster at the base of V_H domains to establish a complex interaction network. In V_H3 domains the ionic and hydrogen-bonding interactions within this charge cluster are very well satisfied, whereas fewer interactions are observed at the analogous positions in other subtypes (Figure 39.3b). Corresponding to the subclass, different hydrogen-bonding networks are formed in the charge cluster of V_H domains. Some of these networks are less extended and contain a smaller number of interactions than in V_H3.

Additionally, based on the residues at three different positions in the first β-strand of the V_H domains, the domains can be classified into four different structural subtypes with respect to their conformations in the first framework region [80]. Mutations bringing together incompatible residues and thus "mixing" subtypes have previously been shown to have a large unfavorable effect on the stability of the whole scFv [81].

Also, clear differences can be observed for hydrophobic core packing of the family subtypes. The upper core region of V_H3 is densely packed, whereas cavities can be identified in V_H4, V_H1a, and V_H5 on the basis of structural alignments (Figure 39.3c). In the lower core, two of the stable domains have an aromatic residue, while the others do not.

Finally, a comparison of the Ramachandran plots showed additional non-glycine residues with positive ϕ-angles and residues with higher secondary structure pro-

regions are conserved within the same germ-line family, these sequence motifs differ between different germ-line families. (b) Arrangement of the residues defining the charge cluster of V_H3 and $V_\kappa3$ domains [13]. Importantly, the charge cluster consists of a network of buried charges and hydrogen bonds rather than pairwise interactions between individual residues. (c) Furthermore, subtype-dependent packing differences occur. As an example, the residues defining the upper hydrophobic core region of different human V_H subtypes are shown as structural superpositions. Structural alignments are shown of V_H4 (PDB entry 1DHV), V_H1a (1DHA), and V_H5 (1DHW) [22], with the most stable framework, V_H3 (1DHU) (left panel). While the upper core residues of V_H3 are densely packed, cavities occur for the other subtypes. In the least stable subtypes, V_H4 and V_H6, the bulky aromatic residues Phe29 and Phe31 are replaced by smaller residues, and the created space is only partly filled up by compensating residues Trp41 and Val25. The loss of the phenyl ring by replacement with Gly in V_H1a (middle panel) as well as the substitution of Leu89 by Ala are not compensated for by larger side chains at other positions, thereby creating hydrophobic cavities. The same is true for position 89 in V_H5 (right panel). Adapted from Ewert et al. [13].

pensity at certain positions for the even-numbered subtypes compared with the odd-numbered ones.

The immediate question was, therefore, whether the results of this structural trouble-shooting could be used to project favorable properties of V_H3 domains onto the less stable subtypes and thereby add another step to the optimization of antibody sequences while maintaining the structural diversity of the immune system. Instead of using the stable V_H3 framework exclusively, resulting in a loss of diversity, some point mutations might be enough to correct some of the shortcomings of the less stable domains.

Based on these comparisons, the mutation of six residues in a scFv containing a V_H6 framework led to an overall increase in thermodynamic stability of 20 kJ mol^{-1} and a fourfold increase in soluble expression yield [73], indeed bringing this framework to the level of V_H3. The effects of the single mutations on stability were almost fully additive, while the effects of folding efficiency (soluble expression yield) were only qualitatively additive. The most dramatic effects on thermodynamic stability were obtained by mutations removing an unsatisfied H-bond donor in the hydrophobic core and introducing glycines at positions with positive ϕ-angles. Individually, all mutations, except the one in the hydrophobic core, led to slight increases in soluble expression yield. Interestingly, one mutation that removed a hydrophilic, solvent-exposed glutamine residue by a replacement with hydrophobic valine on the basis of higher secondary structure propensity also significantly increased the expression yield, possibly since valine secures this stretch to be in β-sheet structure.

In antibodies, disulfide bond engineering has also been investigated. Optimized disulfide bonds engineered between V_H and V_L indeed significantly increased the half-life of an Fv fragment at 37 °C [82]. This strategy was subsequently extended to interchain disulfides in generic framework positions [83]. Even more stable proteins can be obtained by combining the single-chain Fv approach with the engineered disulfide [84, 85]. It should be pointed out, however, that the additional inter-domain disulfide significantly reduces the yield of folded protein when produced in the bacterial periplasm, and such proteins have to be prepared by in vitro refolding. Therefore, the additional disulfides are not ideal for antibody libraries; instead, optimized frameworks have shown the greatest promise for combining diversity with stability.

In summary, the consensus concept provides a convenient tool for proceeding with large steps in sequence space and a high probability of accumulating features that are favorable for higher stability. In many cases, structural analysis can serve as a trouble-shooting tool to identify shortcomings that might have been created by the consensus-building process or, as in the case of the antibodies described here, that are inherent to the natural sequence. In addition, it serves to rank the potential mutations identified by the consensus approach, keeping the mutational load on the target molecule to a minimum. Experimental data are important not only to validate these results but also to give important hints for future design approaches. Many entries in the table of experimental antibody stability data linked to mutations have come from directed evolution experiments. By combining semi-rational and rational approaches with experimental data, optimization of biophysical properties can be achieved in an iterative fashion. This interplay of experiment and structural analysis can therefore be an effective way to probe the vast sequence space in a systematic manner in order to find the valleys of free energy.

39.4 The Selection and Evolution Approach

39.4.1 Principles

The previous section may have led to the impression that mutations enhancing the biophysical properties of proteins can rapidly be identified. In the highlighted case of antibody domains, only the wealth of structural data and empirical measurements available allowed predictions with a high probability of success. Some of the empirical data used successfully for structure-based engineering have come, paradoxically, from directed evolution experiments (see Section 39.3.3). Despite the rapidly increasing amount of structural data and the better understanding of folding mechanisms of proteins, the effects of introduced mutations still cannot be predicted with a high degree of accuracy in most cases. The main reason is that even slight alterations in the primary sequence can lead to profound conformational changes in tertiary or quaternary structure; consequently, structural predictions have to be very accurate and usually must be backed up by empirical data.

Because rational engineering uses site-directed mutagenesis followed by biophysical investigation to probe the effects of specific amino acid substitutions, it becomes very labor-intensive if many mutants have to be checked individually and if no additional hints are available. Additionally, as soon as small synergistic effects of several mutations need to be checked, the combinatorial explosion rapidly exceeds the sample number that can be handled efficiently. More importantly, the restriction to certain target residues automatically excludes alternative solutions to a given problem that may not have been obvious by the initial analysis. For example, affinity improvement of a protein to its binding partner is often achieved more efficiently by slight spatial adjustments of residues directly involved in binding, rather than by substitution of these residues. This kind of spatial adjustment can be caused by mutation of residues whose location in the native structure is further away from the actual binding site (so-called "second-sphere mutations") [86]. Today, it is almost impossible to predict this kind of mutation by rational means.

When it comes to stability engineering of proteins, the problems associated with rational design procedures are even intensified. First, because the energetic contributions of single-site mutations are usually small, the need to sample multiple mutants – in which synergistic effects of several mutations are combined – is more acute. Second, the factors and principles responsible for the overall stability of a native protein are still far from being completely understood, and a multitude of different forces and interactions contribute to it. A complete analysis will have to consider not only the interaction network of the native protein but also the effects on the denatured state. Additionally, potential aggregation pathways will have to be considered. Even simple substitutions like the ones discussed in the previous sections often make contributions to the entropy as well as the enthalpy of the folded and unfolded states, including the solvent in either state. Third, because rational approaches always rely on the current theoretical knowledge, new principles underlying protein stability will rarely be uncovered. In any case, rational engineering requires a clear definition of the problem by the investigator in order to find a solution. Ironically, in the field of protein engineering, the exact definition of the problem is often the problem itself.

Thus, to overcome these limitations, an experimental setup is needed that allows the creation of a vast number of variants of a given protein and that subsequently can identify "superior" molecules that best fulfill a desired property. Nature samples the vast sequence space by the strategy of Darwinian evolution, a cyclic iteration of randomization and selection. Nature thereby adapts proteins to fulfill a function under the given environmental conditions. Recent developments in molecular biology have made it possible to mimic Darwinian evolution in a reasonable time in vitro. Not only has this "evolutionary approach" become the most powerful method to date for engineering proteins towards a desired property, but it also provides new insights into the mechanisms and principles that are responsible for this property. However, as has been illustrated in the case study on antibodies, such experiments can be used not only to solve a particular problem but also to gather information about which residues tend to become enriched in particular positions. This again provides a database for rational engineering.

Many different selection and evolution strategies have been developed in recent years, but all of them have several features in common that reflect the principles of Darwinian evolution. The starting point is the generation of a genetic library of mutants derived from the wild-type sequence of the protein under examination. Several methods exist to create sequence diversity. In error-prone PCR, the error rate of polymerases is increased by performing the PCR reaction in the presence of deoxynucleotide analogues or in the presence of other metal ions. By using bacterial mutator strains, which are characterized by deficient DNA repair systems, random mutations are introduced during DNA amplification in the bacterial cell, albeit usually at a lower frequency [87]. In DNA shuffling, which mimics the natural process of sexual recombination, genes are randomly fragmented by nuclease digestion and reassembled by a PCR reaction in which homologous fragments act as primers for each other [88] (Figure 39.4). The staggered extension process [89] is another possibility to obtain recombined genes in vitro. Here, the polymerase-catalyzed extension of template sequences is extremely abbreviated, and repeated cycles of denaturation and extension lead to several template switches – thereby recombining elements from different genes – before the extension finally yields full-length products. Additionally, techniques are available to focus the diversity to certain regions on the whole gene, such as the use of degenerate primers in PCR or the "doping" of a shuffling reaction with degenerate primers. Depending on the problem and the sequence under investigation, each of these methods has its

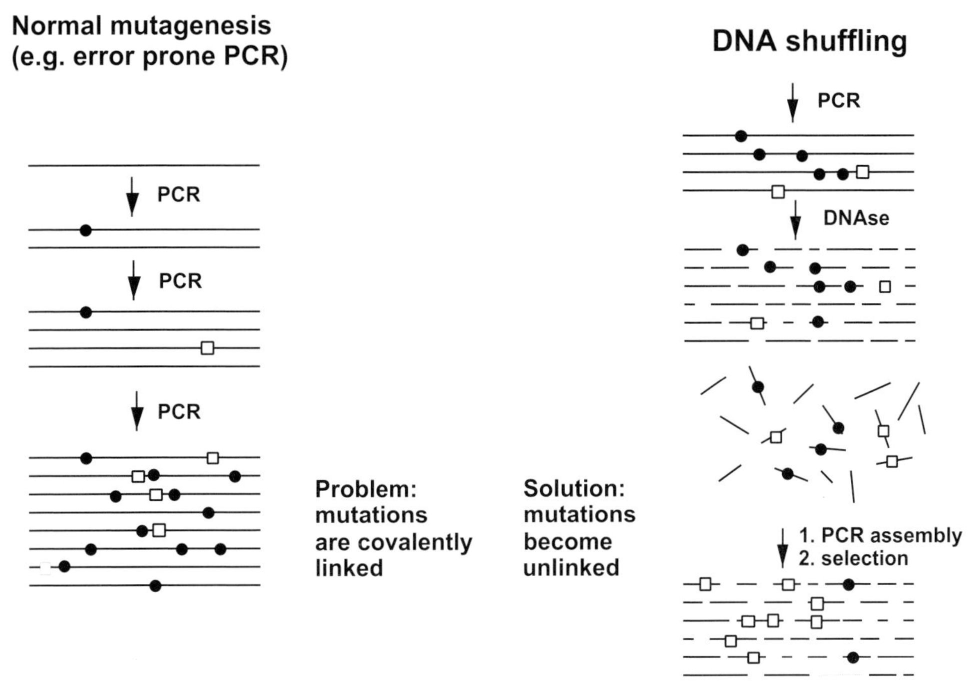

Fig. 39.4. Methods to create genetic diversity by biochemical means. On the left, error-prone PCR (see text) is depicted schematically. Two types of mutations are shown: favorable ones (open squares) and unfavorable ones (closed circles). In successive cycles of PCR, more mutations of each are introduced, and usually molecules will contain some of either type. Thus, the beneficial effect of the favorable mutations can be completely obscured by the presence of unfavorable ones, if the error rate is too high. Therefore, this method is most successful if it is not used at too high an error rate. On the right, DNA shuffling according to Stemmer [88] is shown. A short DNAse digestion breaks up the DNA into small pieces, and PCR is used to reassemble the gene. Thereby, mutations are "crossed" and genes with largely favorable mutations can be obtained that can be enriched by selection. Nevertheless, successful evolution experiments have been carried out with either method.

advantages and disadvantages. In any case, the generated library should obey two major criteria: it must be diverse enough to contain individual sequences with beneficial mutations, and it should be of high enough quality to reduce the experimental "noise" (such as sequences with stop codons or frameshifts) in the subsequent selection experiment [90].

A primary prerequisite of any selection system is the coupling of the genotype (the gene sequence) and phenotype (the respective protein displaying the properties) of any individual library member. Briefly, this can be done by two strategies. Either the gene and the protein need to be compartmentalized in cells or artificial compartments, such that the "improved" phenotype stays connected to the altered gene, or the protein has to be physically coupled to the gene, such that they can be isolated as a particle containing both gene and protein (Figure 39.5). By selecting

a Physical coupling

b Compartmentalization

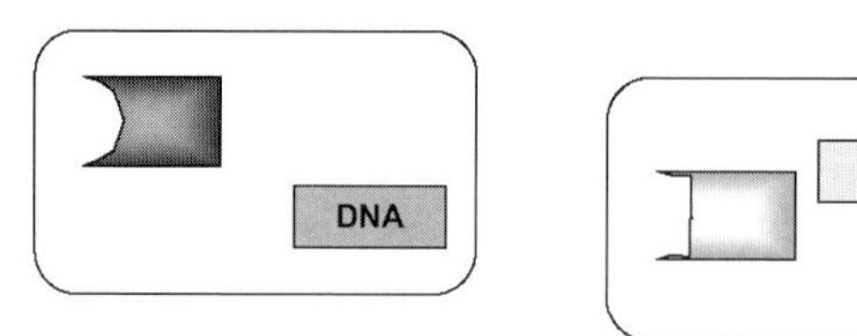

Fig. 39.5. Two principal strategies to link phenotype and genotype are depicted. The genetic material must be connected in a unique way to the protein, which defines the phenotype, such that the gene encoding the valuable mutation can be selectively amplified. A collection of two mutant proteins and mutant genes is shown (light gray and dark gray). (a) This connection can be realized through a direct link (e.g., in phage display or ribosome display), as shown on the top. In this case, all assemblies can freely diffuse in the same volume and the selection must filter out those proteins with the desired function, e.g., by binding to a ligand. (b) Alternatively, gene and protein must be in the same compartment. In nature, this is realized in cells, and microbial cells can be manipulated so that each takes up one variant of a mutant collection. The key is to identify the improved phenotype. This can be done by screening (individual assays on cells) or selection (giving cells with the desired property of the protein a growth advantage). Rather than natural cells, water-in-oil emulsions can be used to create artificial compartments of small water droplets in an oil phase. Usually, the emulsion must be broken and the proteins must be selected by binding to identify the one with the desired phenotype. For details, see text.

the proteins displaying the desired properties, the linked gene sequence can be inherited and amplified subsequently. The various selection technologies differ mainly in the way this physical linkage or compartmentalization is achieved. This will be discussed in more detail in the following section. The library must then be screened or subjected to selection for a certain function, and a defined selection pressure can be applied to direct the selection towards a molecular quantity of interest.

It is necessary at this point to discriminate between "screening" and "selection." Screening methods examine individual members of a library for a given property (e.g., catalytic activity or solubility). For certain properties, screening is often the only way to go. The number of mutants to be screened in a reasonable time depends on the versatility of the screening method. Despite constant progress in automation and miniaturization, even the best screens to date usually do not allow assessment of more than 10^6 variants in a reasonable time. In contrast, selection methods force the single library members to compete with each other, and members that best fulfill the specified criteria are enriched. Often, the selection is based on the binding of particular variants to an immobilized ligand. Note that the binding is only a "surrogate quantity" of the real property to be improved. The basis for a successful enrichment is an efficient counter-selection of variants that do not possess properties fulfilling these criteria. Especially during the selection of proteins for higher stability, an efficient counter-selection, in addition to the correct choice of the applied selection pressure, is one of the major experimental challenges [91].

The basic rule of screening and selection technology describes the importance of assigning the correct selection pressure. This has been succinctly phrased: "You only get what you screen for!" [92]. An analogous statement can be made for selection. Even though the rule sounds plausible, it is often difficult to translate this statement into an experimental setup, because an additional complication is introduced by the fact that an explicit selection pressure towards just one property is impossible to realize. Depending on the selection procedure used, more than one molecular property will have an influence on the enrichment process, including, for example, affinity or catalytic activity, thermodynamic stability, folding efficiency, and toxicity of the respective molecule. The outcome of the selection experiment will therefore always be a "compromise solution" with respect to the weighting of many properties in a given experimental setup. Assigning the right selection pressure therefore means biasing the selection towards a certain property, rather than exclusively altering this property. For these reasons we will discuss the application of the various selection technologies with an emphasis on the available selection pressures for stability engineering. Some of the described methods and examples will be based on mere screening of library members, but the main focus will be on the selection for favorable protein variants.

Choosing the appropriate selection pressure is just one side of the coin. The proper adjustment of its strength is another important factor. If the selection pressure on the system is too low, molecules with the desired properties will be lost in the background noise of the experiment – i.e., they are not enriched. On the other hand, if the selection pressure is too stringent, even the best variants might not pass the "survival threshold."

The genes of variants that survive the applied selection pressures are subsequently amplified and subjected to another so-called "round of selection." This either can be performed in the absence of further mutagenesis to simply enrich the best members of the initial library (which is constant) or, to completely mimic the Darwinian principle, the selected members can be subjected to alternating rounds of randomization or recombination before subsequent selection, leading to an adaptation of the library from round to round. However, as most of the mutations will be non-beneficial, care has to be taken that the mutation rate is low enough to allow successful enrichment of improved members and not to extinguish the whole population. We use the terms "combinatorial selection" for a process in the absence of such mutations and "evolutionary selection" for a process that includes such mutations.

It has been pointed out that the method of DNA family shuffling is in some respects similar to the consensus approach because it combines gene fragments

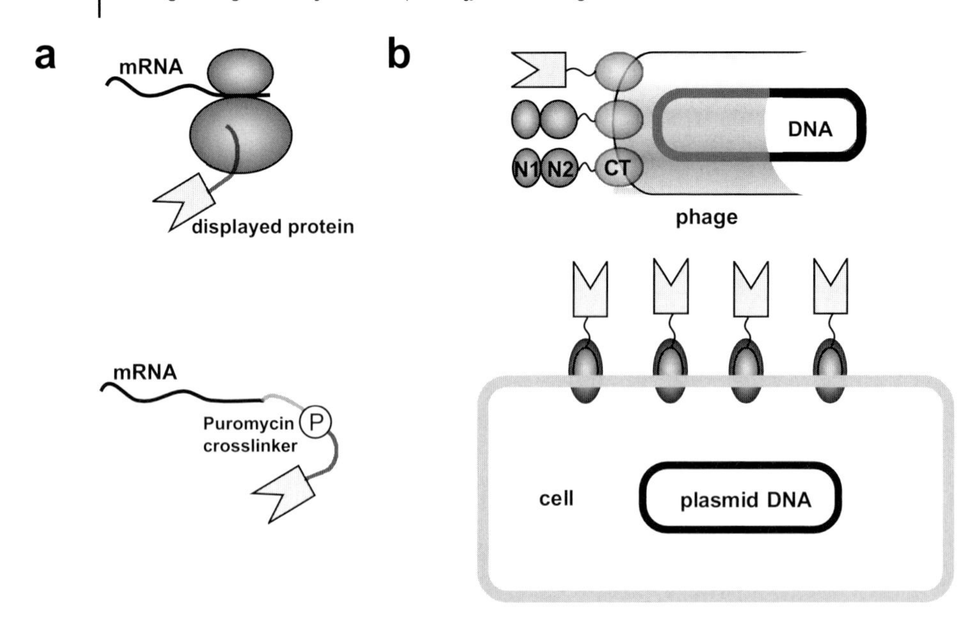

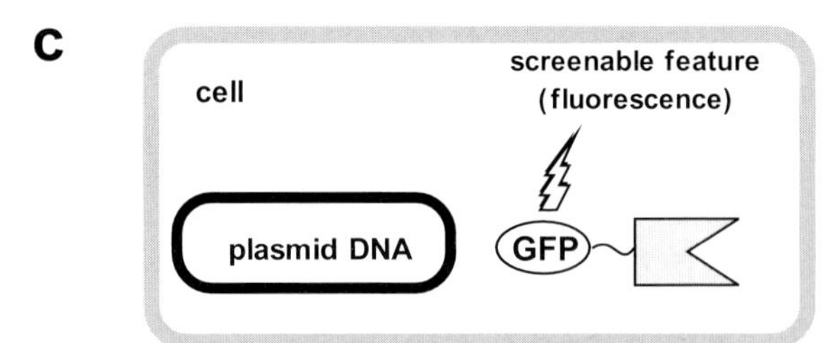

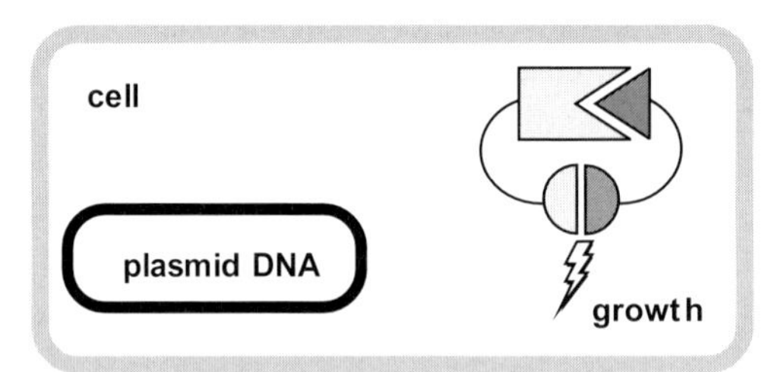

Fig. 39.6. Common display systems used for selection for enhanced biophysical properties. In in vitro display technologies (a), the proteins are produced by in vitro translation, and protein production does not rely on a host organism. The displayed proteins are linked to RNA either by stabilized ternary complexes (ribosome display, upper panel) or by a covalent puromycin cross-linker (mRNA display, lower panel), thereby establishing the genotype-phenotype linkage. (b) In contrast, partial in vitro display systems rely on cells to produce the displayed protein, but selections can be performed in vitro. In phage display (upper panel), the protein is displayed on the surface of filamentous phages, usually fused to

from homologous sources in a random fashion. For mere statistical reasons, the probability of replacing a residue with a consensus residue by gene shuffling is also higher than replacing it with a non-consensus residue [93]. There is, however, a crucial difference: when recombining genes in a random fashion, a selection or screening step is needed subsequently to identify the "fittest" members of the resulting collection. Rather than being based on theoretical assumptions, gene-shuffling methods mimic the natural process of evolution by identifying the members exhibiting a desired property by explicitly subjecting them to selective pressure for this desired property. In practice, however, even with the largest of libraries, a full "re-equilibration" of residues will not take place. In contrast, the consensus approach is based on the explicit assumption that the statistical preferences in a given (often limited or biased) set of sequences indeed reflect the energetic preferences. However, there may be other reasons that certain sequences are prominent.

39.4.2 Screening and Selection Technologies Available for Improving Biophysical Properties

As described above, any selection technology relies on four major steps: (1) generation of a genetic library, (2) establishment of the link between genotype and phenotype upon translation into protein, (3) subsequent screening or selection under defined conditions, and (4) re-amplification of selected members. Depending on how these steps are performed, selection technologies can be subdivided into in vitro methods, partial in vitro methods, and in vivo methods.

While in vivo systems rely on a host organism to express the protein and to carry the respective genetic information, all in vitro display technologies have in common that the protein production and the selection process are performed entirely in vitro, i.e., that the protein is obtained by cell-free translation. In the partial in vitro methods, the genetic information is introduced in cells, where protein production occurs, while the selection process is performed in vitro. All techniques differ in the way the physical linkage between the protein and its genetic information is established. The principles of linking genotype and phenotype are shown in Figure 39.6.

the CT domain of the minor coat protein g3p. The three domains (N1, N2, and CT) of the minor coat protein g3p are depicted. The phage particle also carries the gene encoding the displayed protein. In other partial in vitro technologies, the proteins are displayed on the surface of the expressing host cell itself, which are either bacterial cells (bacterial surface display) or yeast cells (yeast surface display) (lower panel) that also harbor the respective plasmid DNA. (c) In vivo screening or selection systems use the properties of fused reporter proteins for screening (upper panel) or split proteins for intracellular selection (lower panel) in which cellular growth and therefore amplification of the genetic material occurs only if the two protein halves are reconstituted upon interaction of the fused library protein with its target.

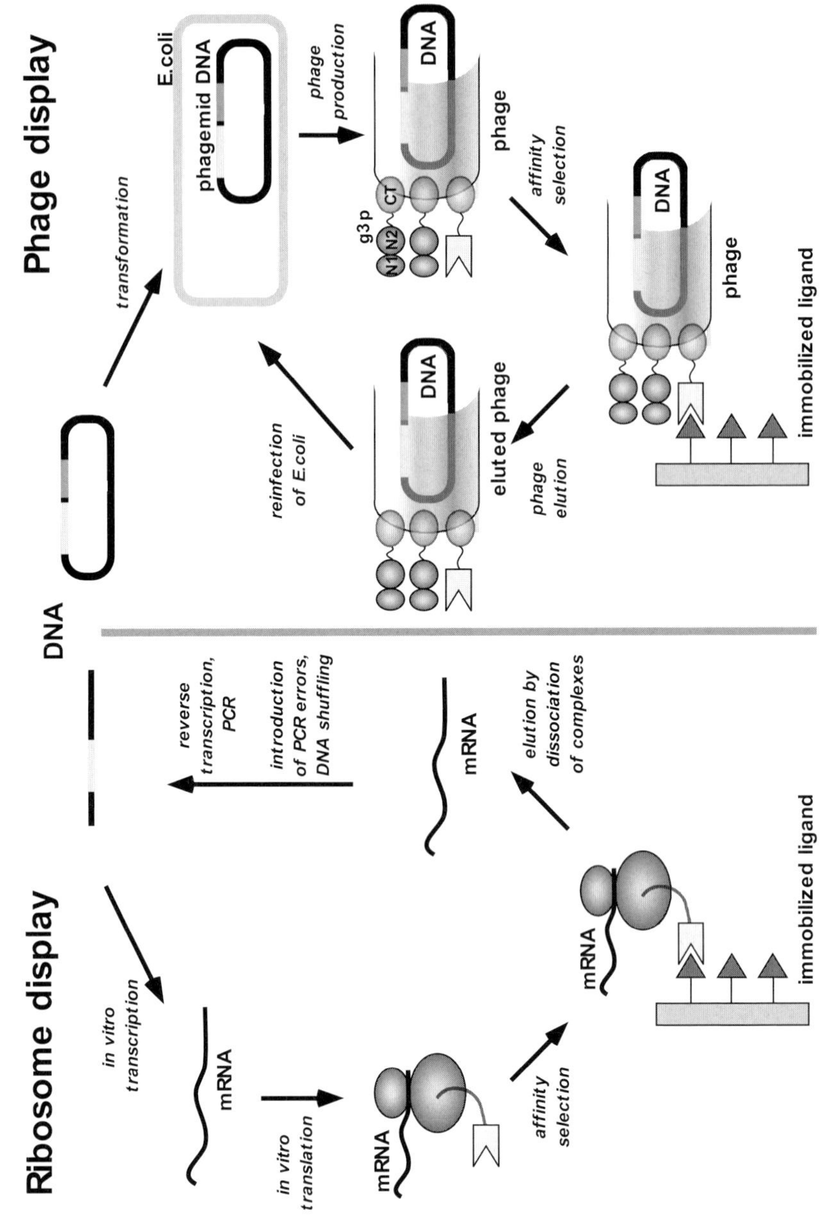

Phage display
transformation
E.coli
phagemid DNA
phage production
g3p
N1
N2
CT
DNA
phage
affinity selection
DNA
phage
immobilized ligand
phage elution
eluted phage
DNA
reinfection of E.coli
DNA
reverse transcription, PCR
introduction of PCR errors, DNA shuffling
mRNA
elution by dissociation of complexes
mRNA
immobilized ligand
Ribosome display
in vitro transcription
mRNA
in vitro translation
mRNA
affinity selection

Fig. 39.7

39.4.2.1 In Vitro Display Technologies

In vitro display technologies started off with the selection of peptides [94] and were then made efficient enough to select for functional proteins [95]. A number of technologies have been developed, which will be described briefly: ribosome display [95, 96], RNA-peptide fusions [97, 98], and water-in-oil emulsions [99, 100]. Water-in-oil emulsions constitute artificial compartments, rather than a physical link, and thereby create a phenotype-genotype coupling and can also use a selection step for interaction with a target.

In ribosome display, the genetic library, usually in the form of a PCR product, is directly used to produce mRNA by in vitro transcription. This mRNA contains a ribosome-binding site for the subsequent translation of the protein and several features that stabilize it against degradation. The encoded protein variants are expressed in a cell-free translation reaction by a stoichiometric number of ribosomes. The essential linkage of the translated proteins to their respective mRNA molecule is achieved by eliminating the stop codon and stabilizing the ribosomal complexes, which prevents the release of the translated protein from the ribosome. In the related method of mRNA-peptide fusions, additional steps are used after this stoichiometric translation to covalently couple a linker between the end of the mRNA and the protein. The ribosome is then removed and the complexes are purified. These complexes are then used to bind to an immobilized target. In all these methods, after the selection process the re-amplification of genetic material can be performed solely by biochemical means, using reverse transcriptase to first synthesize RNA-DNA hybrids, which are then subsequently amplified in a PCR reaction. The resulting DNA then serves as a template for the production of mRNA used for the next round of selection (Figure 39.7).

In vitro display technologies offer several advantages over in vivo selection systems. First, much larger library sizes are accessible because the creation of large molecular diversity on the genetic level does not pose a great challenge, and diversities up to 10^{13} are easily achieved. For in vivo systems, the critical step that limits the size of the displayed library is the transformation efficiency of the respective host organism. Only a small fraction of the initially generated pool will actually enter the cells, and, depending on the host organism, library sizes of about 10^{7} variants in yeast or up to 10^{11} variants in bacteria can be achieved. In contrast, for

Fig. 39.7. Selection cycle of ribosome display (left panel) and phage display (right panel). For ribosome display, all necessary steps are performed in vitro, whereas for phage display protein translation, folding, and phage assembly take place in the bacterial host cell. Other selection criteria can be applied to the protein of interest (see Figure 39.8) before selecting for binding to a target, thereby introducing selection pressures for higher stability. It has to be guaranteed, however, that the selection conditions are not too harsh to destroy the linkage of genotype and phenotype. Importantly, the genetic pool of library members in phage display remains constant in the course of several selection rounds, whereas in ribosome display, the genetic pool undergoes continuous modifications during the in vitro amplification steps by the limited accuracy of the polymerases, resulting in a true Darwinian evolution. In addition, this pool can deliberately and conveniently be altered by introducing additional mutations or recombination events such as DNA shuffling.

in vitro selection systems such a transformation step is not necessary and the diversity of the library is defined by the number of different RNA molecules added to the cell-free translation reaction or by the number of functional ribosomes, whichever number is smaller. For a library of 10^{14}, a 10-mL translation reaction is required. A second advantage is that, because translation and protein folding take place in vitro, reagents can be added during protein synthesis, which can either promote protein folding or minimize aggregation of the displayed proteins (e.g., chaperones), or substances that exert certain selection pressures can be added. In vitro translation encompasses an additional advantage: many proteins are not compatible with in vivo selection systems because the wild type or at least some of the mutants are toxic to the host organism, undergo severe degradation, or cannot be expressed functionally in the respective cellular environment. Last but not least, a major advantage results from the amplification process after the selection has been performed. Because all enzymes used to convert and amplify the genetic material possess an intrinsic error rate, the genetic pool is never constant and is continuously modified from round to round. Therefore, even without special measures to increase the error rate, usually some "evolution" is observed. The error rates can be additionally increased by performing error-prone PCR or by recombining favorable mutations by means of DNA shuffling as explained in the previous section.

39.4.2.2 Partial in Vitro Display Technologies

We define partial in vitro display technologies as those methods in which host organisms are employed to carry and amplify the genetic information and to produce the proteins from this genetic library, usually encoded on plasmids. The selection step, however, is performed in vitro. Therefore, selection pressures similar to those in the in vitro methods can be applied. The array of available techniques includes phage display [101], bacterial surface display [102], and surface display on yeast [103].

Phage display is still the most popular selection technique, and due to its robustness, most of the studies that aim for enhanced biophysical properties of proteins have utilized this technique. Originally, phage display was applied as a method for the identification and selection of peptides or proteins binding to a specific target [101]. More recently, several developments paved the way for its application as a tool for studying protein folding, and as a result, it is now used for the selection of proteins with improved biophysical properties.

We will briefly review the main principles of phage display in its most common format. Variants of the protein of interest are fused to the minor coat protein g3p of the filamentous bacteriophage, which consists of three domains (namely, N1, N2, and CT) that are connected by glycine-rich linkers. The fusion protein is usually encoded on a phagemid vector, expressed in the E. coli host and assembled. Typically, the other proteins necessary to produce an intact phage particle are provided by the helper phage. The assembled phages are secreted by the cell, which does not lyse, and the phages can be collected. While the protein is displayed on the exterior of the phage (typically as an N-terminal fusion to either N1 or CT), the gene of interest encoding a library member is packed into the phage particle upon its assembly in the bacterial host. The phages can then be subjected to the selection process. Phages compete with each other for binding to an immobilized target, using the displayed library members for interaction, and can thereby be captured. Optionally, before capturing, the phages can be subjected to harsh environmental conditions. The captured phages are subsequently amplified by re-infecting bacteria, which then produce phages for a new round of selection (Figure 39.7). In the ideal case, only phages that specifically recognize the target would be amplified. Due to reasons such as nonspecific "sticking" of phages to the surface, the efficiency of selection is in practice at most a 1000- to 10 000-fold enrichment over nonspecific molecules and can even be very much lower.

Several aspects are important when phage display is used as a means of evolving proteins with improved biophysical properties. The correct folding of the fusion protein during phage morphogenesis is an important parameter determining the frequency of incorporation of correctly folded proteins displayed on the phage. Folding intermediates that have a strong aggregation tendency lead to aggregation of a particular library member during the assembly process, and rather than the g3p fusion, the g3p wild-type protein from the helper phage will be incorporated. Despite this inherent selection pressure for the folding efficiency of displayed proteins, the enrichment of improved variants is slow [104], and the utility of phage display to study and improve folding of a given protein was initially not obvious [105]. Combined with a selection for protein functionality, it nevertheless sets the basis for selecting proteins according to their folding properties, as will be discussed in Section 39.4.3.2.

Since the selection procedure is performed in vitro, a large variety of external selection pressures can be applied to the phage particles and be combined with functional selection. If function is omitted as a direct indicator for the native state, other criteria that directly correlate with the native state have to be translated into a selectable feature. This opens the door to select for proteins, which do not bind a ligand, albeit with the caveat that it is not assured that the native state is being selected. Such general selection approaches for physical properties take advantage of the fact that unfolded proteins are more susceptible to proteolytic cleavage than are compactly folded ones [106]. One variation on this theme makes use of the modular nature of the g3p protein, thereby linking phage infectivity directly to the proteolytic susceptibility of the target protein. This system has been called "Proside" (protein stability increase by directed evolution) [107] and will be discussed in Section 39.4.4.4.

39.4.2.3 **In Vivo Selection Technologies**

In vivo selection technologies such as the yeast two-hybrid system [108, 109] or other split protein complementation assays [110, 111] are valuable tools for studying protein-protein interactions in living cells. They commonly employ reporter systems, in which a covalent link between the protein of interest and another so-called reporter protein is established. Cell growth or a colorimetric reaction is dependent on either a specific protein-protein interaction or the solubility of a critical component. The reporter protein transduces certain properties of the host protein, most importantly its native fold and its ability to interact and/or its solubility or its resistance to cellular proteases, to a screenable or selectable feature of the fused reporter protein itself [112]. Examples of reporter proteins are transcription factors [108], critical metabolic enzymes [111], or proteins that can easily be assayed [113].

Several facts, however, limit the applicability of in vivo systems for stability engineering. First, as the viability of the host organism has to be guaranteed, external selection pressures cannot easily be applied. Thermophilic organisms represent in some cases an interesting alternative, but conditions allowing selections are hard to establish. Additionally, the host environment as well as the fused reporter proteins possibly perturb the characteristics of the target protein. Furthermore, control experiments must ensure that growth is really dependent on the interaction of interest and that mutations in the host have not "short-circuited" the selection strategy.

39.4.3 Selection for Enhanced Biophysical Properties

As has been explained in the Introduction, desirable biophysical parameters include solubility, stability, and folding efficiency. Even though the various selection methods and conditions usually do not improve one of these properties exclusively, they are often biased towards one of them. We will therefore discuss different ways to exert selection pressure on the system with respect to the property that is likely to be changed.

39.4.3.1 Selection for Solubility

Solubility, correctly defined as the maximal concentration of the native protein that can be kept in solution, is usually not a property to be selected, as most proteins – with the exception of membrane proteins – are sufficiently soluble. The word "solubility" in the context of screening is often inaccurately used to refer to "soluble expression yield," and thus it usually mirrors the efficiency of folding in the cell. Even though soluble expression yield is not necessarily equivalent to the folding properties and even less to the stability of a protein, a correlation can often be observed, and soluble expression yield is comparatively easy to screen for. In some cases, a function for a given protein cannot be assigned, or it is difficult and laborious to screen for, and then this property becomes especially important.

The fluorescence yield of bacterial colonies expressing proteins or protein domains fused to the green fluorescent protein (GFP) correlates over a wide range with the soluble expression yield of the fused domain [114]. Screening for fluorescence intensity is a versatile method, and, combined with directed evolution, has the potential to select variants of proteins that are less aggregation-prone than their progenitors [115, 116]. In a similar setup, reporter proteins can be used as selection markers instead, such as by fusing the protein of interest to chloramphenicol acetyltransferase [117]. Assays exploiting protein-protein interactions can also be used for this purpose and they have the additional advantage that the fused reporter sequence can be much smaller and thereby minimize the risk of a perturbing influence of the reporter itself on the domain of interest [113]. To completely exclude such perturbing influences, reporter systems might possibly be established that do not need any fusion at all and exploit instead the stress response of the expression host cell. As the overexpression of proteins often activates particular stress response genes by the accumulation of insoluble aggregates, the respective gene promoters can be employed to activate reporter genes instead [118].

In special cases, reporter systems can also be combined with other selection methods to select for folding properties. Many intrinsically stable proteins contain permissive sites in loops, into which polypeptide chains of variable length can be inserted without loss of function of the host protein. The higher the number of residues inserted, the larger the entropic cost of ordering these residues will be. Consequently, the overall stability of the host protein should decrease. However, the entropic cost of inserting a folded sequence will be lower than that of an unfolded sequence of the same length. The probability of the host protein reaching the native (functional) state should thus correlate directly with the ability of the inserted sequence to fold into a compact structure. If host proteins with various inserted sequences are displayed on phages and if the host protein allows a selection for "foldedness" by means of binding to a target, folded sequence insertions are enriched. This system has been termed "loop entropy reduction phage display selection" [119]. While attractive in theory, it was found experimentally that mostly sequences that keep the hybrid protein soluble are enriched.

Protein solubility, as used here, can also be interpreted as a lower degree of exposed hydrophobic residues, and folded proteins usually display fewer hydrophobic residues on their surface than do unfolded proteins. Conversely, the exposure of hydrophobic residues is usually a sign of non-native states. Display technologies can thus, for example, use the interaction with hydrophobic surfaces to select against more hydrophobic, i.e., less "folded," proteins [120].

In summary, protein solubility can reflect in many cases the folding properties of proteins and thus be used as a selection criterion. Moreover, soluble expression in vivo is often a major requirement for the large-scale production and the convenient in vitro handling of proteins. In any case, solubility should never be confused with protein stability. Even though both properties might correlate in some cases, the governing principles are often of a different nature.

39.4.3.2 **Selection for Protein Display Rates**

The in vivo folding efficiency of proteins represents an intrinsic selection pressure when using phage display. Because correct folding is a prerequisite for both incorporation into the phage coat and binding to the target, the subsequent selection of phages that are able to bind to the target is partly influenced by the folding properties of the displayed molecule.

However, this does not automatically imply that the selection is driven towards superior folding properties. As mentioned above, the enrichment factors of proteins with superb folding behavior over the poorly folding members are low [104]. Moreover, even though the binding of a given protein to its target is a very direct way to monitor and screen for its proper folding, the selection criterion "folding" is not decoupled from the selection criterion "binding affinity." If the enrichment factor for one property (folding) is low, the selection is more likely to be driven towards the other (affinity). As a result, selected members will represent a compromise between folding properties, which only have to be sufficient under the given experimental conditions, and the binding affinity to their target.

By randomizing exclusively the residues that build the hydrophobic core of the IgG binding domain of peptostreptococcal protein L, which are not involved in ligand binding but determine the stability and folding kinetics of this small protein, Gu et al. [121] could unambiguously demonstrate the utility of phage display for studying the stability and folding efficiency of proteins. A more extensive randomization effort with subsequent characterization of the selected variants pointed out some important aspects of selection for folding [122]. First, it showed that folding kinetics is not the critical parameter for the selection but rather the overall stability of the mutants, a tendency which could be confirmed by more advanced selection approaches (see Section 39.4.4.2). Second, the authors demonstrated the utility of selection methods to identify certain residues that are important for the folding mechanism. Third, the selected pools were highly diverse and the thermodynamic stabilities of all variants were lower compared to the wild type. In fact, most of the selected proteins denatured just above room temperature.

These results illustrate the principle that any selection – be it natural or in the test tube – simply continues until the minimum requirements are met, in this case, functionality at the selection temperature. Consequently, this suggests that additional, more stringent selection pressures are necessary to really accomplish a directed evolution for improved properties.

39.4.3.3 Selection on the Basis of Cellular Quality Control

Additional selection pressure can be provided by the host organism itself. For example, the secretory quality-control system of eukaryotic cells discriminates proteins according to their folding behavior in an efficient way. This is based on mechanisms that lead to retention of misfolded proteins in the endoplasmic reticulum (ER), followed by degradation of these proteins. In yeast, the Golgi complex can reroute misfolded proteins, which have escaped ER retention, to the vacuole for degradation, thereby constituting an additional important quality-control pathway [14]. In combination with yeast surface display, these mechanisms can be employed to bias functional selections towards enhanced folding efficiency. Several studies have shown that the surface display rate of proteins strongly correlates with their thermostability and their soluble secretion efficiency [123, 124]. As an example, by making use of elevated temperatures during expression as an additional selection pressure, improved T-cell receptor fragments, whose thermostability exceeded by far the expression temperatures, could be obtained [125]. Even though the low transformation efficiency of yeast restricts the accessible library size, one obvious advantage of the method is the applicability to glycosylated eukaryotic proteins, which generally are not amenable to yeast two-hybrid or phage display methodologies [126].

39.4.4
Selection for Increased Stability

39.4.4.1 General Strategies

The key to all selections for stability is to introduce a threshold that separates the molecule with desired properties from the starting molecules. If the population initially lacks functionality, while only a few members are above the selection threshold, the selection system can distinguish between these slight energetic differences. This is the starting situation if the target protein is initially of very low stability and should be brought to "average" properties.

If, however, the starting protein is already of considerable stability, but should be brought to even higher stability, it can be advantageous to intentionally destabilize it prior to selection in order to find stabilizing mutations that reconstitute its functionality. This principle can be applied to very different kinds of proteins, provided that mutations are known that destabilize the protein to an extent that will subsequently allow the selection for alternative stabilizing mutations (Figure 39.8a). Because the effects of independent stabilizing mutations are often additive, the deliberately introduced destabilizing mutations can be reverted in the context of the additional newly selected ones, thereby rendering the molecule far more stable than the original one (Figure 39.8b).

In principle, all reagents and conditions known to destabilize proteins can decrease the number of library members populating the folded state and can therefore be used to exert increasing selection pressure on the system. Increased temperature and denaturing agents are obvious methods that are useful for selecting proteins of higher thermodynamic stability. The major problem is presented by the compatibility of the conditions with the selection method used.

39.4.4.2 Protein Destabilization

An example of the stabilization of a naturally unstable domain by using phage display was a selection performed with the prodomain of the protease subtilisin BPN′ [127]. At room temperature the prodomain folds into a stable conformation only upon binding to subtilisin. By randomizing positions that are not directly in contact with subtilisin and subsequent selection on subtilisin, a mutant was selected that showed an increase of $\Delta G_{\mathrm{unfolding}}$ by 25 kJ mol^{-1}, from -8 kJ mol^{-1} to 17 kJ mol^{-1}, despite the fact that the library size was comparatively small. Intriguingly, the predominant energetic contribution was mediated by a selected disulfide bond. Previously, a similar strategy had been used to select for thermodynamically favored β-turns of the B1 domain of protein G [128]. Based on considerations described above and on the fact that the replacement of amino acids on the surface of proteins would be predicted to have only moderate effects on stability, the authors reasoned that such a selection could be successful only for proteins of marginal stability. Most of the substitutions would then lead to a positive free energy of folding, and thus the molecule would fail to fold into a functional form at all. In fact, selected turn sequences showed clear sequence preferences only if less stable

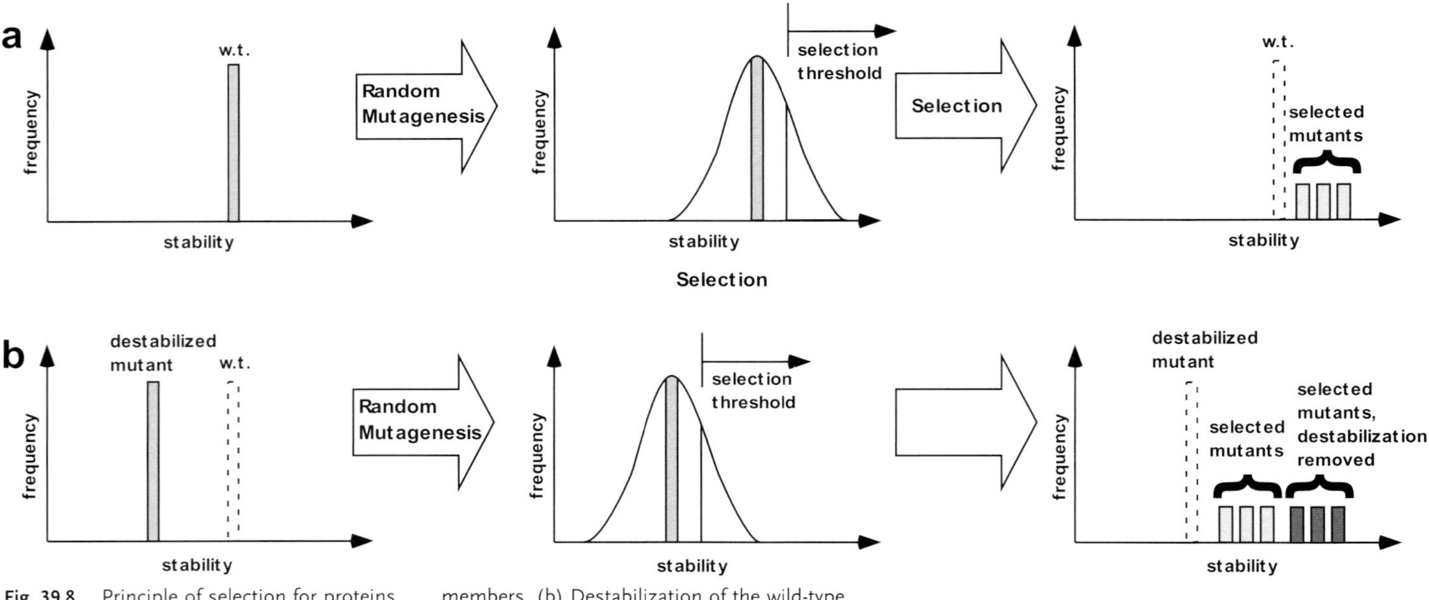

Fig. 39.8. Principle of selection for proteins with improved stability. (a) By random mutagenesis of a given sequence, many mutations are obtained. Some of these mutations are favorable for the stability of the native protein and others are unfavorable. As a result, a diverse pool of proteins with a distribution of different energies is created. Upon exposure of the pool to an external selection pressure, only mutants with stabilities exceeding the selection threshold can be recovered and amplified. However, the selection threshold has to be set high enough to allow an efficient selection of "improved" members. (b) Destabilization of the wild-type sequence prior to selection allows reducing the necessary selection threshold. Thereby, alternative mutations that stabilize the native fold of the protein can be identified and the selection process becomes more efficient. Moreover, additional stability gains can be achieved by removing the deliberately introduced destabilizing mutations after the selection. The initially lost energy is therefore regained, resulting in mutants of higher stability than the wild-type protein. Adapted from Wörn et al. [76].

host proteins were used to accept the turn. In this case, the selected sequences either resembled the wild-type turn or reflected the statistical preferences of turn sequences in the databases and stabilized the protein by 12–20 kJ mol^{-1}, compared to random sequences. Moreover, increased temperature was used as an additional selection pressure during the phage selection.

As discussed in the previous section, the inability of the wild-type target protein to fold is the prerequisite for performing an efficient positive selection for functionality by additional mutations. The internal disulfide bond of immunoglobulin domains significantly contributes to the stability of antibodies [129]. The dramatic loss of free energy of folding upon their removal usually renders antibodies nonfunctional. By first destabilizing a scFv antibody fragment through replacement of cysteines with other residues in both variable domains separately, a completely disulfide-free antibody was identified after several cycles of functional selection and recombination by DNA shuffling [17]. One globally stabilizing mutation was found to compensate for the initial stability loss. This study illustrates some important features of how several parameters exert influence on different stages of the selection process. Interestingly, the stabilizing mutation was already selected during the first rounds, showing that the loss of thermodynamic stability was the primary problem that had to be overcome. Because the stability gain of this mutation was large enough to shift the free energy of folding above the required threshold, all successive rounds did not affect protein stability but led to a fine-tuning with respect to the improvement of folding yield [17]. To illustrate the principle of additivity, reintroduction of the disulfide bridge into the selected variant yielded a scFv antibody of very high stability and superior expression yields [84].

39.4.4.3 **Selections Based on Elevated Temperature**

In vivo selection systems using thermophilic expression hosts have been employed for the stabilization of enzymes [130]. However, a rather large set of requirements must be met to apply such a selection system. In order to use stability of the enzyme at elevated temperatures as the selection criterion, its enzymatic activity must be vital to the thermophilic organism, and the corresponding gene of the thermophilic host has to be deleted. Randomized versions of a mesophilic enzyme can then be screened by means of metabolic selection. Although the approach is very powerful, the utility of these systems is restricted to special cases.

To evolve enzymes with altered thermal stability, conventional screening for enzymatic activity in vitro after randomization or recombination of the respective gene and subsequent expression of the enzyme is still the most widely used method. An activity screen rapid and sensitive enough to identify slightly improved members from a vast pool of mutants is the key feature of this directed evolution approach [131]. However, individual screens have to be developed for each specific class of enzymes. Even though screening limits the explorable sequence space considerably compared to selection, mesophilic enzymes such as subtilisin E [132] and *p*-nitrobenzyl esterase [133] could be converted into mutants functionally equivalent to thermophilic enzymes by only a few rounds of directed evolution. Similar to observations discussed before, only very few mutations were necessary to increase the melting temperatures by more than 14 degrees. An important feature of screening compared to selection is that the much smaller library size is partially compensated for by directly measuring the quantity of interest: enzymatic activity at elevated temperature. In contrast, most selection systems use a surrogate measure where "false positives" can lead to the phenotype by mechanisms different from the ones desired.

Elevated temperatures not only can be used in screening but also can be combined with selection technologies. While ribosome display is not an option – because low temperatures are essential to keep the ternary complexes of RNA, ribosome, and nascent polypeptide intact and thus ensure the coupling of genotype and phenotype – phage display has proved to be a very suitable method for harsh conditions due to the robustness of the phage particles. Nevertheless, future improvements of in vitro technologies may allow their application under more stringent conditions. As of today, in the case of in vitro technologies, it is vital to destabilize the protein first (see Section 39.4.4.2); then, very significant stability improvements can be selected [134].

The upper temperature limit for selections using filamentous M13 phages is approximately 60 °C [135]. Above this temperature, re-infection titers of the phages are severely decreased, presumably due to the irreversible heat denaturation of phage coat proteins. Phages displaying the protein of interest can be incubated up to this temperature, and proteins still able to function can be selected. It should be noted that the exposure of phages to higher temperature after phage assembly exerts a somewhat different stress on the proteins than in the methods described before. While in the previous examples the functionality of the proteins was influenced mainly by in vivo folding efficiency and the protein stability during the panning procedure, it is the irreversible unfolding reaction at a given temperature that now becomes an additional parameter. The fraction of unfolded molecules will therefore reflect the rate of unfolding and thus kinetic stability.

39.4.4.4 **Selections Based on Destabilizing Agents**

The use of protein destabilizing agents for biasing the selection pressure towards stability is limited to in vitro and partial in vitro display methods. Like temperature, the concentrations of denaturing agents can be controlled precisely and can be varied from round to round, allowing a gradual increase of stringency.

Even though phages are quite resistant to denaturing agents [136], one should be aware of possible general problems when selections are based on the chemical denaturation properties of the displayed proteins. However, if chemical denaturation is combined with a selection for binding, high concentrations of denaturant can prevent binding to the target, even if the protein is not yet unfolded: as the forces governing ligand binding are very similar to those responsible for protein stability, both are disturbed by chemical denaturants. Conversely, because chemical denaturation is in fact often reversible, removal or dilution of the denaturant prior to ligand binding will often result in refolding on the phage and thus release of selection pressure. Moreover, ionic denaturants such as guanidinium chloride weaken electrostatic interactions and strengthen hydrophobic ones. This might impair the selection of mutations that introduce additional ionic interactions on the protein surface [59] and may favor additional hydrophobic interactions, which is not necessarily desired.

If the protein is known to be stabilized by disulfide bridges, a strategy similar to the one used by Proba et al. [17] described above may be applicable. To identify globally stabilizing mutations, which compensate for the stability loss upon removal of disulfide bridges, selection can be performed in the presence of reducing agents such as DTT. Thus, the disulfide-forming cysteines do not have to be removed in advance. However, one should be aware of the fact that some disulfide bridges in proteins, once formed, are often hard to reduce, especially if they are buried within the protein core. Thus, it is advantageous to add the reducing agent at a time when the protein is not yet folded. Jermutus et al. [134] stabilized a scFv antibody fragment by using ribosome display. In contrast to the phage display method, the synthesis of the targeted protein occurs in vitro, which makes the polypeptide chain accessible to reagents during its synthesis on the ribosome and before folding has taken place. DTT was added during translation of the scFv, and its concentration was continuously increased from round to round. When using ribosome display, the population undergoes slight changes due to mutations occurring during PCR, resulting in an iterative adjustment of the selected variants towards tolerating the increasingly stringent conditions.

39.4.4.5 **Selection for Proteolytic Stability**

In each of the examples cited above, selection for stability was based on a functional selection. Therefore, only those proteins for which a specific function can be assigned could be targeted; in addition, this function has to be screenable or, better yet, selectable. It would be highly advantageous to completely uncouple function from stability in the selection process to extend the range of problems to which this can be applied. A more general approach would thus be an invaluable tool for engineering any given protein and for selecting stable folds from a pool of de novo designed proteins. For functional selections, an additional problem arises from the fact that after a certain stability threshold is reached, allowing enough proteins to populate the folded state, the selection is likely to run towards improved binding properties instead. On the other hand, certain mutations might be stabilizing but might result in slight structural rearrangements that affect functionality in a negative way. Thus, it will not be possible to recover these mutations [106].

A general approach that has proved to be powerful for stability engineering is based on the concept that compactly folded proteins are much more resistant to proteolysis than are partly folded or unfolded proteins [137]. Several variations on this theme have been applied to phage display selections. In principle, it is only necessary that the phages displaying proteins, which can be cleaved by the protease, can be efficiently separated from the phages displaying protease-resistant proteins. This can be achieved in a physical way by providing the displayed protein with an N-terminal tag sequence allowing capture of only phages with non-cleaved proteins [106]. Alternatively, the ability of phages to re-infect bacteria can be directly linked to the protease resistance of the protein of interest [107, 136]. This second alternative takes advantage of the modular structure of the protein domains responsible for phage infection. If the displayed domains are inserted between the carboxy-terminal domain of gp3 and the amino-terminal domains N1 and N2, which are required for phage infectivity, proteolytic cleavage of the displayed domain renders the phage noninfectious. This was inspired by the so-called selectively infective phage method (SIP), where it was shown that the g3p domains could be interrupted by additional domains and even an interacting pair [138]. By employing a phage system derived from the SIP technique, which lacks any wild-type g3p, it is assured that infectivity is completely abolished upon proteolytic cleavage [107] (Figure 39.9). Alternatively, phages can be engineered to make the remaining wild-type g3p itself susceptible to the proteolytic attack [136]. The principle of protease selection can also be applied to in vitro selection techniques in which the displayed protein is freely accessible [120].

Even though proteolysis seems at first glance to be less correlated to the stability of proteins than temperature, it is suitable for optimizing packing of the hydrophobic core [139] as well as for optimizing electrostatic interactions on the protein surface [59]. One reason for the strength of this approach is based on the fact that proteolytic cleavage is an irreversible reaction, which is not necessarily the case for denaturation by temperature or denaturing agents. Furthermore, protease resistance monitors the flexibility of the polypeptide chain rather than complete denaturation. It is therefore capable not only of selecting against completely unfolded variants but also of detecting local unfolding events. Because sites of local unfolding often initiate the global unfolding process, it can be advantageous to remove such sites. Nevertheless, the effective cleavage of flexible parts of the protein restricts the method to proteins that do not have extended flexible regions in the native state. Furthermore, a selection against the primary recognition sequence of the protease is clearly a possible outcome of such experiments.

None of the described methods can stand completely on its own. Many methods can easily be combined to increase the stringency of a selection. A combination of temperature stress and increasing amounts of denaturing agents may, for example, lead to higher flexibility in certain regions of the protein, thereby increasing the sensitivity of the subsequent proteolytic attack. Additionally, temperature and the concentration of denaturing agents allow a much tighter control of the selection pressure than do increasing concentrations of protease and thus open the possibility of a well-controlled gradual increase of selection stringency.

39.5
Conclusions and Perspectives

Several different approaches are now available for engineering proteins for enhanced biophysical properties. In many cases, few and specific mutations are sufficient to provide proteins of marginal stability with considerably stabilizing features. Thus, the challenge for protein engineering is to identify these positions in a given sequence among the vast number of possible changes and to correctly alter them. Each of the applied techniques has its own merits and bottlenecks. Instead

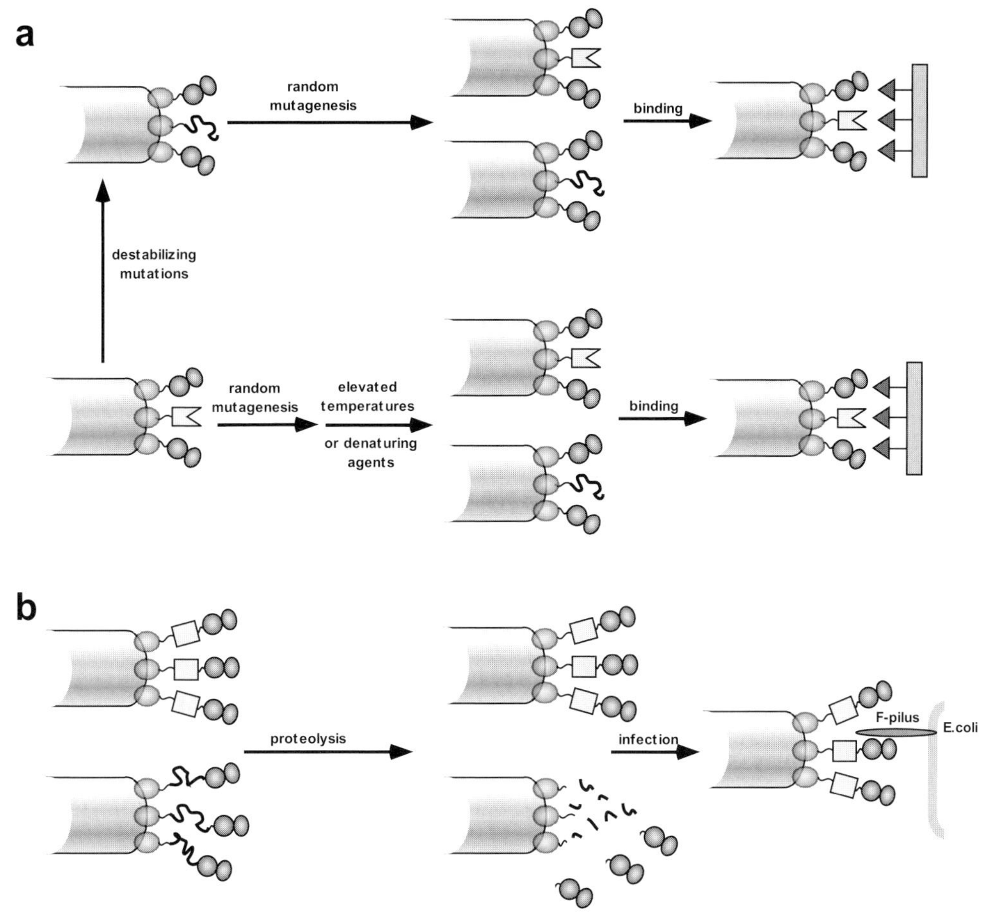

Fig. 39.9. Strategies for selecting for improved protein stability and folding by phage display. (a) Methods utilizing the binding to a given target (affinity selection) as a means for selecting members with improved folding behavior and higher stability from a protein library. Several types of selection pressure can be applied in order to recover mutants with enhanced stability. Destabilizing mutations are deliberately introduced (top) to render the protein "nonfunctional," and alternative stabilizing mutations are identified by selecting variants whose functionality is regained. Alternatively (bottom), elevated temperatures or denaturing agents can be used to render most of the protein variants "nonfunctional," allowing the selection of simply the "fittest" members. (b) The resistance of the target protein to a protease can be combined with the ability of phages to re-infect bacteria. The protein library is inserted between the C-terminal domain and the N-terminal domain in all copies of the g3p protein (cloned into the phage genome), which are needed for re-infection of bacterial cells. By proteolytic cleavage of the inserted protein, these domains are cut off and phage infectivity is lost. In alternative approaches, an N-terminal tag sequence is fused to the protein of interest instead of the N-terminal g3p domains. Upon proteolytic cleavage, the tag is lost. In contrast, phages presenting proteins that are resistant to proteolytic cleavage can subsequently be captured on an affinity matrix binding to the tag sequence. In order to further increase the selection threshold, the shown selection strategies can also be combined.

of playing them off against each other, the future challenge will be to find ways to synergistically use them to improve a given molecular property.

Evolutionary methods have the advantage of being much less biased by theoretical assumptions or working hypotheses. Additionally, proteins stable in new environments may be evolved, e.g., proteins that fulfill a given function in non-aqueous solutions or high concentrations of detergents. Because the biophysical principles in such environments are of a different nature, rational design has to rely on a much smaller empirical dataset. While rational approaches are likely to become more important as more structural and experimental data become available, notably also those from selection experiments, the large number of variations that are potentially able to improve the biophysical properties of a protein often still exceeds the experimentally accessible number. Moreover, because rational engineering has to rely on the available dataset, mutations that lie off the beaten track will rarely be identified. Nevertheless, selection experiments often identify the same "key" mutations in proteins. It is then useful to exploit this information and directly introduce such mutations. A "rational" analysis can also help to recombine important "key" mutations in selected clones and to reduce the effects of a selection-neutral genetic drift. Another combination of rational and combinatorial methods is the creation of "smart" libraries of variants. Library design that is based on such structural considerations and principles will therefore allow more accurate focusing of the selection on specific regions of interest and thereby increase the chances for success.

Acknowledgements

The authors thank Drs. Daniela Röthlisberger, Casim Sarkar and Annemarie Honegger for critical reading of the manuscript. B. S. was the recipient of a Kekulé fellowship from the Fonds der Chemischen Industrie.

References

1 Taverna, D. M. & Goldstein, R. A. (2002). Why are proteins marginally stable? *Proteins* 46, 105–9.
2 Zavodszky, P., Kardos, J., Svingor & Petsko, G. A. (1998). Adjustment of conformational flexibility is a key event in the thermal adaptation of proteins. *Proc. Natl. Acad. Sci. U.S.A.* 95, 7406–11.
3 Arnold, F. H., Giver, L., Gershenson, A., Zhao, H. & Miyazaki, K. (1999). Directed evolution of mesophilic enzymes into their thermophilic counterparts. *Ann. N.Y. Acad. Sci.* 870, 400–3.
4 Govindarajan, S. & Goldstein, R. A. (1995). Searching for Foldable Protein Structures Using Optimized Energy Functions. *Biopolymers* 36, 43–51.
5 Dill, K. A. (1990). Dominant forces in protein folding. *Biochemistry* 29, 7133–55.
6 Pappenberger, G., Schurig, H. & Jaenicke, R. (1997). Disruption of an ionic network leads to accelerated thermal denaturation of D-glyceraldehyde-3-phosphate dehydrogenase from the hyperthermophilic bacterium Thermotoga maritima. *J. Mol. Biol.* 274, 676–83.
7 Sterner, R. & Liebl, W. (2001).

Thermophilic adaptation of proteins. *Crit. Rev. Biochem. Mol. Biol.* 36, 39–106.

8 Zlotnick, A. & Stray, S. J. (2003). How does your virus grow? Understanding and interfering with virus assembly. *Trends Biotechnol.* 21, 536–42.

9 Jaswal, S. S., Sohl, J. L., Davis, J. H. & Agard, D. A. (2002). Energetic landscape of alpha-lytic protease optimizes longevity through kinetic stability. *Nature* 415, 343–6.

10 Fersht, A. R., Bycroft, M., Horovitz, A., Kellis, J. T., Jr., Matouschek, A. & Serrano, L. (1991). Pathway and stability of protein folding. *Philos. Trans. R. Soc. Lond. B. Biol. Sci.* 332, 171–6.

11 Jäger, M. & Plückthun, A. (1999). Domain interactions in antibody Fv and scFv fragments: effects on unfolding kinetics and equilibria. *FEBS Lett.* 462, 307–12.

12 Willuda, J., Honegger, A., Waibel, R., Schubiger, P. A., Stahel, R., Zangemeister-Wittke, U. & Plückthun, A. (1999). High thermal stability is essential for tumor targeting of antibody fragments: engineering of a humanized anti-epithelial glycoprotein-2 (epithelial cell adhesion molecule) single-chain Fv fragment. *Cancer Res.* 59, 5758–67.

13 Ewert, S., Huber, T., Honegger, A. & Plückthun, A. (2003). Biophysical properties of human antibody variable domains. *J. Mol. Biol.* 325, 531–53.

14 Ellgaard, L., Molinari, M. & Helenius, A. (1999). Setting the standards: quality control in the secretory pathway. *Science* 286, 1882–8.

15 Martineau, P. & Betton, J. M. (1999). In vitro folding and thermodynamic stability of an antibody fragment selected in vivo for high expression levels in Escherichia coli cytoplasm. *J. Mol. Biol.* 292, 921–9.

16 Schuler, B. & Seckler, R. (1998). P22 tailspike folding mutants revisited: effects on the thermodynamic stability of the isolated beta-helix domain. *J. Mol. Biol.* 281, 227–34.

17 Proba, K., Wörn, A., Honegger, A. & Plückthun, A. (1998). Antibody scFv fragments without disulfide bonds made by molecular evolution. *J. Mol. Biol.* 275, 245–53.

18 Knappik, A. & Plückthun, A. (1995). Engineered turns of a recombinant antibody improve its in vivo folding. *Protein Eng.* 8, 81–9.

19 Wörn, A. & Plückthun, A. (1999). Different equilibrium stability behavior of ScFv fragments: identification, classification, and improvement by protein engineering. *Biochemistry* 38, 8739–50.

20 Steipe, B., Schiller, B., Plückthun, A. & Steinbacher, S. (1994). Sequence statistics reliably predict stabilizing mutations in a protein domain. *J. Mol. Biol.* 240, 188–92.

21 Lehmann, M. & Wyss, M. (2001). Engineering proteins for thermostability: the use of sequence alignments versus rational design and directed evolution. *Curr. Opin. Biotechnol.* 12, 371–5.

22 Knappik, A., Ge, L., Honegger, A., Pack, P., Fischer, M., Wellnhofer, G., Hoess, A., Wolle, J., Plückthun, A. & Virnekäs, B. (2000). Fully synthetic human combinatorial antibody libraries (HuCAL) based on modular consensus frameworks and CDRs randomized with trinucleotides. *J. Mol. Biol.* 296, 57–86.

23 Mosavi, L. K., Minor, D. L., Jr. & Peng, Z. Y. (2002). Consensus-derived structural determinants of the ankyrin repeat motif. *Proc. Natl. Acad. Sci. U.S.A.* 99, 16029–34.

24 Wang, Q., Buckle, A. M., Foster, N. W., Johnson, C. M. & Fersht, A. R. (1999). Design of highly stable functional GroEL minichaperones. *Protein Sci.* 8, 2186–93.

25 Lehmann, M., Loch, C., Middendorf, A., Studer, D., Lassen, S. F., Pasamontes, L., van Loon, A. P. & Wyss, M. (2002). The consensus concept for thermostability engineering of proteins: further proof of concept. *Protein Eng.* 15, 403–11.

26 Binz, H. K., Stumpp, M. T., Forrer, P., Amstutz, P. & Plückthun, A.

(2003). Designing repeat proteins: well-expressed, soluble and stable proteins from combinatorial libraries of consensus ankyrin repeat proteins. *J. Mol. Biol.* 332, 489–503.

27 Main, E. R., Xiong, Y., Cocco, M. J., D'Andrea, L. & Regan, L. (2003). Design of stable alpha-helical arrays from an idealized TPR motif. *Structure (Camb)* 11, 497–508.

28 Stumpp, M. T., Forrer, P., Binz, H. K. & Plückthun, A. (2003). Designing repeat proteins: modular leucine-rich repeat protein libraries based on the mammalian ribonuclease inhibitor family. *J. Mol. Biol.* 332, 471–87.

29 Forrer, P., Binz, H. K., Stumpp, M. T. & Plückthun, A. (2004). Consensus design of repeat proteins. *Chembiochem* 5, 183–9.

30 Kohl, A., Binz, H. K., Forrer, P., Stumpp, M. T., Plückthun, A. & Grütter, M. G. (2003). Designed to be stable: crystal structure of a consensus ankyrin repeat protein. *Proc. Natl. Acad. Sci. U.S.A.* 100, 1700–5.

31 Levitt, M., Gerstein, M., Huang, E., Subbiah, S. & Tsai, J. (1997). Protein folding: the endgame. *Annu. Rev. Biochem.* 66, 549–79.

32 Bohm, G. & Jaenicke, R. (1994). Relevance of sequence statistics for the properties of extremophilic proteins. *Int. J. Pept. Protein Res.* 43, 97–106.

33 Jaenicke, R. & Bohm, G. (1998). The stability of proteins in extreme environments. *Curr. Opin. Struct. Biol.* 8, 738–48.

34 Vieille, C. & Zeikus, G. J. (2001). Hyperthermophilic enzymes: sources, uses, and molecular mechanisms for thermostability. *Microbiol. Mol. Biol. Rev.* 65, 1–43.

35 van den Burg, B. & Eijsink, V. G. (2002). Selection of mutations for increased protein stability. *Curr. Opin. Biotechnol.* 13, 333–7.

36 Matsumura, M., Signor, G. & Matthews, B. W. (1989). Substantial increase of protein stability by multiple disulphide bonds. *Nature* 342, 291–3.

37 Perry, L. J. & Wetzel, R. (1984). Disulfide bond engineered into T4 lysozyme: stabilization of the protein toward thermal inactivation. *Science* 226, 555–7.

38 Pace, C. N., Grimsley, G. R., Thomson, J. A. & Barnett, B. J. (1988). Conformational stability and activity of ribonuclease T1 with zero, one, and two intact disulfide bonds. *J. Biol. Chem.* 263, 11820–5.

39 Zhang, T., Bertelsen, E. & Alber, T. (1994). Entropic effects of disulphide bonds on protein stability. *Nat. Struct. Biol.* 1, 434–8.

40 Doig, A. J. & Williams, D. H. (1991). Is the hydrophobic effect stabilizing or destabilizing in proteins? The contribution of disulphide bonds to protein stability. *J. Mol. Biol.* 217, 389–98.

41 Martensson, L. G., Karlsson, M. & Carlsson, U. (2002). Dramatic stabilization of the native state of human carbonic anhydrase II by an engineered disulfide bond. *Biochemistry* 41, 15867–75.

42 Ivens, A., Mayans, O., Szadkowski, H., Jurgens, C., Wilmanns, M. & Kirschner, K. (2002). Stabilization of a $(\beta\alpha)_8$-barrel protein by an engineered disulfide bridge. *Eur. J. Biochem.* 269, 1145–53.

43 Betz, S. F. & Pielak, G. J. (1992). Introduction of a disulfide bond into cytochrome c stabilizes a compact denatured state. *Biochemistry* 31, 12337–44.

44 Iwai, H. & Plückthun, A. (1999). Circular beta-lactamase: stability enhancement by cyclizing the backbone. *FEBS Lett.* 459, 166–72.

45 Scott, C. P., Abel-Santos, E., Wall, M., Wahnon, D. C. & Benkovic, S. J. (1999). Production of cyclic peptides and proteins in vivo. *Proc. Natl. Acad. Sci. U.S.A.* 96, 13638–43.

46 Camarero, J. A., Fushman, D., Sato, S., Giriat, I., Cowburn, D., Raleigh, D. P. & Muir, T. W. (2001). Rescuing a destabilized protein fold through backbone cyclization. *J. Mol. Biol.* 308, 1045–62.

47 Nagi, A. D. & Regan, L. (1997). An inverse correlation between loop

length and stability in a four-helix-bundle protein. *Fold. Des.* 2, 67–75.

48 Russell, R. J., Ferguson, J. M., Hough, D. W., Danson, M. J. & Taylor, G. L. (1997). The crystal structure of citrate synthase from the hyperthermophilic archaeon pyrococcus furiosus at 1.9 Å resolution. *Biochemistry* 36, 9983–94.

49 Macedo-Ribeiro, S., Darimont, B., Sterner, R. & Huber, R. (1996). Small structural changes account for the high thermostability of 1[4Fe-4S] ferredoxin from the hyperthermophilic bacterium Thermotoga maritima. *Structure* 4, 1291–301.

50 Matthews, B. W., Nicholson, H. & Becktel, W. J. (1987). Enhanced protein thermostability from site-directed mutations that decrease the entropy of unfolding. *Proc. Natl. Acad. Sci. U.S.A.* 84, 6663–7.

51 Sriprapundh, D., Vieille, C. & Zeikus, J. G. (2000). Molecular determinants of xylose isomerase thermal stability and activity: analysis of thermozymes by site-directed mutagenesis. *Protein Eng.* 13, 259–65.

52 Pace, C. N. (1992). Contribution of the hydrophobic effect to globular protein stability. *J. Mol. Biol.* 226, 29–35.

53 Lim, W. A., Hodel, A., Sauer, R. T. & Richards, F. M. (1994). The crystal structure of a mutant protein with altered but improved hydrophobic core packing. *Proc. Natl. Acad. Sci. U.S.A.* 91, 423–7.

54 Ventura, S., Vega, M. C., Lacroix, E., Angrand, I., Spagnolo, L. & Serrano, L. (2002). Conformational strain in the hydrophobic core and its implications for protein folding and design. *Nat. Struct. Biol.* 9, 485–93.

55 Ohmura, T., Ueda, T., Ootsuka, K., Saito, M. & Imoto, T. (2001). Stabilization of hen egg white lysozyme by a cavity-filling mutation. *Protein Sci.* 10, 313–20.

56 Chen, J. & Stites, W. E. (2001). Higher-order packing interactions in triple and quadruple mutants of staphylococcal nuclease. *Biochemistry* 40, 14012–9.

57 Hendsch, Z. S. & Tidor, B. (1994). Do salt bridges stabilize proteins? A continuum electrostatic analysis. *Protein Sci.* 3, 211–26.

58 Pace, C. N. (2001). Polar group burial contributes more to protein stability than nonpolar group burial. *Biochemistry* 40, 310–3.

59 Martin, A., Sieber, V. & Schmid, F. X. (2001). In-vitro selection of highly stabilized protein variants with optimized surface. *J. Mol. Biol.* 309, 717–26.

60 Loladze, V. V., Ibarra-Molero, B., Sanchez-Ruiz, J. M. & Makhatadze, G. I. (1999). Engineering a thermostable protein via optimization of charge-charge interactions on the protein surface. *Biochemistry* 38, 16419–23.

61 Shoemaker, K. R., Kim, P. S., York, E. J., Stewart, J. M. & Baldwin, R. L. (1987). Tests of the helix dipole model for stabilization of alpha-helices. *Nature* 326, 563–7.

62 Pace, C. N., Shirley, B. A., McNutt, M. & Gajiwala, K. (1996). Forces contributing to the conformational stability of proteins. *FASEB J.* 10, 75–83.

63 Loladze, V. V., Ermolenko, D. N. & Makhatadze, G. I. (2002). Thermodynamic consequences of burial of polar and non-polar amino acid residues in the protein interior. *J. Mol. Biol.* 320, 343–57.

64 McDonald, I. K. & Thornton, J. M. (1994). Satisfying hydrogen bonding potential in proteins. *J. Mol. Biol.* 238, 777–93.

65 Hennig, M., Darimont, B., Sterner, R., Kirschner, K. & Jansonius, J. N. (1995). 2.0 Å structure of indole-3-glycerol phosphate synthase from the hyperthermophile Sulfolobus solfataricus: possible determinants of protein stability. *Structure* 3, 1295–306.

66 Ramachandran, G. N., Ramakrishnan, C. & Sasisekharan, V. (1963). Stereochemistry of polypeptide chain configurations. *J. Mol. Biol.* 7, 95–9.

67 Kimura, S., Kanaya, S. & Nakamura, H. (1992). Thermostabilization of

Escherichia coli ribonuclease HI by replacing left-handed helical Lys95 with Gly or Asn. *J. Biol. Chem.* 267, 22014–7.

68 Takano, K., Yamagata, Y. & Yutani, K. (2001). Role of amino acid residues in left-handed helical conformation for the conformational stability of a protein. *Proteins* 45, 274–80.

69 Stites, W. E., Meeker, A. K. & Shortle, D. (1994). Evidence for strained interactions between side-chains and the polypeptide backbone. *J. Mol. Biol.* 235, 27–32.

70 Vega, M. C., Martinez, J. C. & Serrano, L. (2000). Thermodynamic and structural characterization of Asn and Ala residues in the disallowed II′ region of the Ramachandran plot. *Protein Sci.* 9, 2322–8.

71 Gunasekaran, K., Ramakrishnan, C. & Balaram, P. (1996). Disallowed Ramachandran conformations of amino acid residues in protein structures. *J. Mol. Biol.* 264, 191–8.

72 Pal, D. & Chakrabarti, P. (2002). On residues in the disallowed region of the Ramachandran map. *Biopolymers* 63, 195–206.

73 Ewert, S., Honegger, A. & Plückthun, A. (2003). Structure-based improvement of the biophysical properties of immunoglobulin V_H domains with a generalizable approach. *Biochemistry* 42, 1517–28.

74 Nieba, L., Honegger, A., Krebber, C. & Plückthun, A. (1997). Disrupting the hydrophobic patches at the antibody variable/constant domain interface: improved in vivo folding and physical characterization of an engineered scFv fragment. *Protein Eng.* 10, 435–44.

75 Brandts, J. F., Hu, C. Q., Lin, L. N. & Mos, M. T. (1989). A simple model for proteins with interacting domains. Applications to scanning calorimetry data. *Biochemistry* 28, 8588–96.

76 Wörn, A. & Plückthun, A. (2001). Stability engineering of antibody single-chain Fv fragments. *J. Mol. Biol.* 305, 989–1010.

77 Gokhale, R. S., Agarwalla, S., Francis, V. S., Santi, D. V. & Balaram, P. (1994). Thermal stabilization of thymidylate synthase by engineering two disulfide bridges across the dimer interface. *J. Mol. Biol.* 235, 89–94.

78 Hudson, P. J. (1999). Recombinant antibody constructs in cancer therapy. *Curr. Opin. Immunol.* 11, 548–57.

79 Plückthun, A., Krebber, A., Krebber, C., Horn, U., Knüpfer, U., Wenderoth, R., Nieba, L., Proba, K. & Riesenberg, D. (1996). Producing antibodies in *Escherichia coli*: From PCR to fermentation. In *Antibody Engineering* (McCafferty, J., Hoogenboom, H. R. & Chiswell, D. J., eds.), pp. 203–252. IRL Press, Oxford.

80 Honegger, A. & Plückthun, A. (2001). The influence of the buried glutamine or glutamate residue in position 6 on the structure of immunoglobulin variable domains. *J. Mol. Biol.* 309, 687–99.

81 Jung, S., Spinelli, S., Schimmele, B., Honegger, A., Pugliese, L., Cambillau, C. & Plückthun, A. (2001). The importance of framework residues H6, H7 and H10 in antibody heavy chains: experimental evidence for a new structural subclassification of antibody V_H domains. *J. Mol. Biol.* 309, 701–16.

82 Glockshuber, R., Malia, M., Pfitzinger, I. & Plückthun, A. (1990). A comparison of strategies to stabilize immunoglobulin Fv-fragments. *Biochemistry* 29, 1362–7.

83 Jung, S. H., Pastan, I. & Lee, B. (1994). Design of interchain disulfide bonds in the framework region of the Fv fragment of the monoclonal antibody B3. *Proteins* 19, 35–47.

84 Wörn, A. & Plückthun, A. (1998). Mutual stabilization of V_L and V_H in single-chain antibody fragments, investigated with mutants engineered for stability. *Biochemistry* 37, 13120–7.

85 Young, N. M., MacKenzie, C. R., Narang, S. A., Oomen, R. P. & Baenziger, J. E. (1995). Thermal stabilization of a single-chain Fv antibody fragment by introduction of a disulphide bond. *FEBS Lett.* 377, 135–9.

86 Arkin, M. R. & Wells, J. A. (1998). Probing the importance of second sphere residues in an esterolytic antibody by phage display. *J. Mol. Biol.* 284, 1083–94.

87 Low, N. M., Holliger, P. H. & Winter, G. (1996). Mimicking somatic hypermutation: affinity maturation of antibodies displayed on bacteriophage using a bacterial mutator strain. *J. Mol. Biol.* 260, 359–68.

88 Stemmer, W. P. (1994). Rapid evolution of a protein in vitro by DNA shuffling. *Nature* 370, 389–91.

89 Zhao, H., Giver, L., Shao, Z., Affholter, J. A. & Arnold, F. H. (1998). Molecular evolution by staggered extension process (StEP) in vitro recombination. *Nat. Biotechnol.* 16, 258–61.

90 Amstutz, P., Forrer, P., Zahnd, C. & Plückthun, A. (2001). In vitro display technologies: novel developments and applications. *Curr. Opin. Biotechnol.* 12, 400–5.

91 Forrer, P., Jung, S. & Plückthun, A. (1999). Beyond binding: using phage display to select for structure, folding and enzymatic activity in proteins. *Curr. Opin. Struct. Biol.* 9, 514–20.

92 Zhao, H. & Arnold, F. H. (1997). Combinatorial protein design: strategies for screening protein libraries. *Curr. Opin. Struct. Biol.* 7, 480–5.

93 Lehmann, M., Pasamontes, L., Lassen, S. F. & Wyss, M. (2000). The consensus concept for thermostability engineering of proteins. *Biochim. Biophys. Acta* 1543, 408–415.

94 Mattheakis, L. C., Bhatt, R. R. & Dower, W. J. (1994). An in vitro polysome display system for identifying ligands from very large peptide libraries. *Proc. Natl. Acad. Sci. U.S.A.* 91, 9022–6.

95 Hanes, J. & Plückthun, A. (1997). In vitro selection and evolution of functional proteins by using ribosome display. *Proc. Natl. Acad. Sci. U.S.A.* 94, 4937–4942.

96 He, M. & Taussig, M. J. (1997). Antibody-ribosome-mRNA (ARM) complexes as efficient selection particles for in vitro display and evolution of antibody combining sites. *Nucleic Acids Res* 25, 5132–4.

97 Roberts, R. W. & Szostak, J. W. (1997). RNA-peptide fusions for the in vitro selection of peptides and proteins. *Proc. Natl. Acad. Sci. U.S.A.* 94, 12297–302.

98 Kurz, M., Gu, K., Al-Gawari, A. & Lohse, P. A. (2001). cDNA – protein fusions: covalent protein–gene conjugates for the in vitro selection of peptides and proteins. *ChemBioChem* 2, 666–72.

99 Tawfik, D. S. & Griffiths, A. D. (1998). Man-made cell-like compartments for molecular evolution. *Nat. Biotechnol.* 16, 652–6.

100 Doi, N. & Yanagawa, H. (1999). STABLE: protein-DNA fusion system for screening of combinatorial protein libraries in vitro. *FEBS Lett.* 457, 227–30.

101 Smith, G. P. (1985). Filamentous fusion phage: novel expression vectors that display cloned antigens on the virion surface. *Science* 228, 1315–7.

102 Georgiou, G., Stathopoulos, C., Daugherty, P. S., Nayak, A. R., Iverson, B. L. & Curtiss, R., 3rd. (1997). Display of heterologous proteins on the surface of microorganisms: from the screening of combinatorial libraries to live recombinant vaccines. *Nat. Biotechnol.* 15, 29–34.

103 Boder, E. T. & Wittrup, K. D. (1997). Yeast surface display for screening combinatorial polypeptide libraries. *Nat. Biotechnol.* 15, 553–7.

104 Jung, S. & Plückthun, A. (1997). Improving in vivo folding and stability of a single-chain Fv antibody fragment by loop grafting. *Protein Eng.* 10, 959–66.

105 Hoess, R. H. (2001). Protein design and phage display. *Chem. Rev.* 101, 3205–18.

106 Finucane, M. D., Tuna, M., Lees, J. H. & Woolfson, D. N. (1999). Core-directed protein design. I. An experimental method for selecting

stable proteins from combinatorial libraries. *Biochemistry* 38, 11604–12.

107 Sieber, V., Plückthun, A. & Schmid, F. X. (1998). Selecting proteins with improved stability by a phage-based method. *Nat. Biotechnol.* 16, 955–60.

108 Fields, S. & Song, O. (1989). A novel genetic system to detect protein-protein interactions. *Nature* 340, 245–6.

109 Bai, C. & Elledge, S. J. (1996). Gene identification using the yeast two-hybrid system. *Methods Enzymol.* 273, 331–47.

110 Johnsson, N. & Varshavsky, A. (1994). Split ubiquitin as a sensor of protein interactions in vivo. *Proc. Natl. Acad. Sci. U.S.A.* 91, 10340–4.

111 Michnick, S. W. (2001). Exploring protein interactions by interaction-induced folding of proteins from complementary peptide fragments. *Curr. Opin. Struct. Biol.* 11, 472–7.

112 Waldo, G. S. (2003). Genetic screens and directed evolution for protein solubility. *Curr. Opin. Chem. Biol.* 7, 33–8.

113 Wigley, W. C., Stidham, R. D., Smith, N. M., Hunt, J. F. & Thomas, P. J. (2001). Protein solubility and folding monitored in vivo by structural complementation of a genetic marker protein. *Nat. Biotechnol.* 19, 131–6.

114 Waldo, G. S., Standish, B. M., Berendzen, J. & Terwilliger, T. C. (1999). Rapid protein-folding assay using green fluorescent protein. *Nat. Biotechnol.* 17, 691–5.

115 Pedelacq, J. D., Piltch, E., Liong, E. C., Berendzen, J., Kim, C. Y., Rho, B. S., Park, M. S., Terwilliger, T. C. & Waldo, G. S. (2002). Engineering soluble proteins for structural genomics. *Nat. Biotechnol.* 20, 927–32.

116 Wurth, C., Guimard, N. K. & Hecht, M. H. (2002). Mutations that reduce aggregation of the Alzheimer's Aβ42 peptide: an unbiased search for the sequence determinants of Aβ amyloidogenesis. *J. Mol. Biol.* 319, 1279–90.

117 Maxwell, K. L., Mittermaier, A. K., Forman-Kay, J. D. & Davidson, A. R. (1999). A simple in vivo assay for increased protein solubility. *Protein Sci.* 8, 1908–11.

118 Lesley, S. A., Graziano, J., Cho, C. Y., Knuth, M. W. & Klock, H. E. (2002). Gene expression response to misfolded protein as a screen for soluble recombinant protein. *Protein Eng.* 15, 153–60.

119 Minard, P., Scalley-Kim, M., Watters, A. & Baker, D. (2001). A "loop entropy reduction" phage-display selection for folded amino acid sequences. *Protein Sci.* 10, 129–34.

120 Matsuura, T. & Plückthun, A. (2003). Selection based on the folding properties of proteins with ribosome display. *FEBS Lett.* 539, 24–8.

121 Gu, H., Yi, Q., Bray, S. T., Riddle, D. S., Shiau, A. K. & Baker, D. (1995). A phage display system for studying the sequence determinants of protein folding. *Protein Sci.* 4, 1108–17.

122 Kim, D. E., Gu, H. & Baker, D. (1998). The sequences of small proteins are not extensively optimized for rapid folding by natural selection. *Proc. Natl. Acad. Sci. U.S.A.* 95, 4982–6.

123 Kowalski, J. M., Parekh, R. N., Mao, J. & Wittrup, K. D. (1998). Protein folding stability can determine the efficiency of escape from endoplasmic reticulum quality control. *J. Biol. Chem.* 273, 19453–8.

124 Shusta, E. V., Kieke, M. C., Parke, E., Kranz, D. M. & Wittrup, K. D. (1999). Yeast polypeptide fusion surface display levels predict thermal stability and soluble secretion efficiency. *J. Mol. Biol.* 292, 949–56.

125 Shusta, E. V., Holler, P. D., Kieke, M. C., Kranz, D. M. & Wittrup, K. D. (2000). Directed evolution of a stable scaffold for T-cell receptor engineering. *Nat. Biotechnol.* 18, 754–9.

126 Boder, E. T. & Wittrup, K. D. (2000). Yeast surface display for directed evolution of protein expression, affinity, and stability. *Methods Enzymol.* 328, 430–44.

127 Ruan, B., Hoskins, J., Wang, L. & Bryan, P. N. (1998). Stabilizing the subtilisin BPN′ pro-domain by phage display selection: how restrictive is the amino acid code for maximum protein stability? *Protein Sci.* 7, 2345–53.

128 Zhou, H. X., Hoess, R. H. & DeGrado, W. F. (1996). In vitro evolution of thermodynamically stable turns. *Nat. Struct. Biol.* 3, 446–51.

129 Goto, Y. & Hamaguchi, K. (1979). The role of the intrachain disulfide bond in the conformation and stability of the constant fragment of the immunoglobulin light chain. *J. Biochem. (Tokyo)*. 86, 1433–41.

130 Akanuma, S., Yamagishi, A., Tanaka, N. & Oshima, T. (1999). Further improvement of the thermal stability of a partially stabilized Bacillus subtilis 3-isopropylmalate dehydrogenase variant by random and site-directed mutagenesis. *Eur. J. Biochem.* 260, 499–504.

131 Farinas, E. T., Bulter, T. & Arnold, F. H. (2001). Directed enzyme evolution. *Curr. Opin. Biotechnol.* 12, 545–51.

132 Zhao, H. & Arnold, F. H. (1999). Directed evolution converts subtilisin E into a functional equivalent of thermitase. *Protein Eng.* 12, 47–53.

133 Giver, L., Gershenson, A., Freskgard, P. O. & Arnold, F. H. (1998). Directed evolution of a thermostable esterase. *Proc. Natl. Acad. Sci. U.S.A.* 95, 12809–13.

134 Jermutus, L., Honegger, A., Schwesinger, F., Hanes, J. & Plückthun, A. (2001). Tailoring in vitro evolution for protein affinity or stability. *Proc. Natl. Acad. Sci. U.S.A.* 98, 75–80.

135 Jung, S., Honegger, A. & Plückthun, A. (1999). Selection for improved protein stability by phage display. *J. Mol. Biol.* 294, 163–80.

136 Kristensen, P. & Winter, G. (1998). Proteolytic selection for protein folding using filamentous bacteriophages. *Fold. Des.* 3, 321–8.

137 Fontana, A., Polverino de Laureto, P., De Filippis, V., Scaramella, E. & Zambonin, M. (1997). Probing the partly folded states of proteins by limited proteolysis. *Fold. Des.* 2, R17–26.

138 Krebber, C., Spada, S., Desplancq, D. & Plückthun, A. (1995). Co-selection of cognate antibody-antigen pairs by selectively-infective phages. *FEBS Lett.* 377, 227–31.

139 Finucane, M. D. & Woolfson, D. N. (1999). Core-directed protein design. II. Rescue of a multiply mutated and destabilized variant of ubiquitin. *Biochemistry* 38, 11613–23.

Index

Note on page numbers: Roman numbers indicate parts, Arabic numbers indicate pages.

a

A_{220} weight-concentration extinction coefficient **II**/ 1235
AAA proteases **II**/ 1033
AAA proteins **II**/ 442, 1031
AAA superfamily, oligomerization **II**/ 111
ab initio prediction **II**/ 1293
ABC transporter **II**/ 925
absorbance spectroscopy **I**/ 23
ACBP *see* acyl-CoA-binding protein
accessible surface area **I**/ 130, 383, 415
– *see also* solvent accessible surface area
acetylation **II**/ 444–445
N-α-acetyltransferases **II**/ 434, 444
acidic phospholipase **II**/ 687
aconitase **II**/ 709
acrylodan **II**/ 219
actin **II**/ 200 ff, 732 ff, 756 ff, 761, 763, 1267
– binding assay **II**/ 765
– folding assay **II**/ 764–765
activation free energy **I**/ 411
acyl CoA-binding protein **I**/ 499, 620
acyl phosphatase **I**/ 898, **II**/ 387, 1101, 1148
adenylate kinase **I**/ 612, 623
adhesive pili **II**/ 965
AffiGel Blue **II**/ 711, 713
affinity chromatography, ATPases **II**/ 109
AFM *see* atomic force microscopy
AGADIR **I**/ 358
aggregates **II**/ 78, 265, 572, 575, 626, 945, 948, 1057, 1095, 1098, 1118, 1180, 1246, 1264, 1317
aggregation **I**/ 368, 388, 543, 680, 692, **II**/ 6 ff, 16 ff, 54 ff, 162 ff, 177, 182 ff, 388, 436, 461, 469, 492 ff, 516, 626, 659, 680 ff, 702 ff, 756, 884, 894, 921, 925, 942, 971, 975, 1005, 1030, 1033, 1059, 1069, 1073, 1093, 1103 ff, 1176, 1185, 1200, 1245 ff, 1287, 1299
– assay **II**/ 174, 179 ff
– diseases **II**/ 1105
– kinetics **I**/ 694
– mature aggregates **II**/ 1211
– modifiers **II**/ 1189
– propensities **I**/ 368
– role of ClpB in **II**/ 111
– thermal **II**/ 180
aggresome system **II**/ 1105, 1228
aging **II**/ 1187 ff
Aha1 **II**/ 794–795
– function **II**/ 795
– structure **II**/ 795
α-helix *see* helix
AK *see* adenylate kinase
alamethicin **II**/ 895
alcohol dehydrogenase **II**/ 868, 1260
aldolase **II**/ 52
alkaline phosphatase **II**/ 360, 1258
alkylating agent **II**/ 82
ALP *see* α-lytic protease
Alzheimer's disease **I**/ 694, 739, **II**/ 198, 398, 1098, 1103 ff, 1115, 1148, 1175
– amyloid peptide Aβ **II**/ 1101 ff, 1208
– amyloid precursor protein (APP) **II**/ 1106
– secretase **II**/ 1106

Protein Folding Handbook, Part II. Edited by J. Buchner and T. Kiefhaber

ISBN: 3-527-30784-2

amide hydrogen exchange I/ 13, 398, 502, 634
– competition experiments I/ 652
– equilibrium I/ 661
– intrinsic exchange rate constant I/ 638
– mechanism I/ 642
– pulse labeling experiments I/ 656
amide proton exchange I/ 294
D-amino acids I/ 317, II/ 1208
AMP-PNP II/ 730, 731, 734
AMS II/ 690
amylin (IAPP) II/ 226, 1101, 1107
amyloid II/ 198 ff, 1096, 1148, 1200
– amlyoidogenic II/ 198
– amyloid epitope II/ 1225
– amyloidoses II/ 198
– diffraction pattern II/ 198
– structure II/ 1179
amyloid disease II/ 1096, 1106, 1115, 1148
– amyloid precursors II/ 1116
– familial II/ 198
– senile systemic amyloidosis (SSA) II/ 1116
– serum amyloid P (SAP) II/ 1107
– systemic II/ 1106
– transthyretin (TTR) II/ 1116
– treatment II/ 1108
amyloid fibers II/ 1150, 1152
amyloid fibrils II/ 1097, 1105, 1210
– assembly II/ 1098
– seeds II/ 1155
– therapeutic strategy II/ 1106
– toxicity II/ 1104
amyloid β-peptide II/ 1101 ff, 1180 ff, 1208
amyloidosis *see* amyloid disease
analytical ultracentrifugation II/ 3, 864
– determination of the dissociation constant II/ 63
Anfinsen cage II/ 734
1-anilino-8-naphthalensulfonate (ANS) II/219 ff
– fluorescence I/ 498
– binding I/ 858
animal models II/ 1204
anti-Hammond behavior I/ 424
antibody II/ 495, 563, 970
– C_H1 II/ 566
– C_H3 I/ 922, II/ 47, 59
– combinational library II/ 1301
– conformation-specific II/ 77
– disulfide bond II/ 1321
– domain II/ 41
– engineering II/ 1300–1305
– Fab fragment II/ 56 ff, 1258
– – folding II/ 57
– – structure II/ 56
– fragments I/ 955, II/ 54
– frameworks II/ 1301
– Fv structure II/ 1302
– heavy chains II/ 566, 572
– light chains II/ 565, 566, 572
– single-chain fragments II/ 1256 ff, 1265
– single-chain Fv fragment II/ 1265, 1300 ff, 1321
– stability II/ 1321
– V_L domains II/ 1291, 1302
α1-antritrypsin II/ 46
APIase II/ 415
Apj1p II/ 443
apomyoglobin I/ 216, 459, 464, 471, 660, 859, 863, 889, 899, 900
arc repressor I/ 971
arginine II/ 20, 1258, 1267
Aristotle II/ 21
aromatic interactions I/ 349
ASA *see* accessible surface area
– *see* solvent accessible surface area
aspartate transcarbamylase I/ 984, II/ 42
assembly factor II/ 970
assembly of Fe/S centers II/ 499
assembly-competent II/ 943
association II/ 16 ff
– phage II/ 8
ataxia 1 II/ 1184
ataxin II/ 493
ataxin-1 II/ 1181, 1184
ATCase *see* aspartate transcarbamylase
ATF6 II/ 574
atomic force microscopy I/ 1112, II/ 39, 41, 225 ff, 234, 265
– contact mode II/ 225
– tapping mode II/ 225

ATP II/ 187
– binding II/ 168 ff
– depletion II/ 85
– hydrolysis II/ 163, 167, 168 ff, 176, 188
– – assay II/ 596
– – cooperativity II/ 114
ATP-agarose II/ 108
ATP-dependent proteases
– assembly II/ 263
– unfolding II/ 263
ATPase II/ 169, 171, 176
– affinity chromatography II/ 109
– assays II/ 124 ff, 188
– enzymatic activity II/ 108, 495
autosomal dominant II/ 1201
auxilin II/ 500 ff

b

B-domain of staphylococcal protein A I/ 552
B1 domain of protein G, folding mechanism I/ 505
– C-terminal hairpin I/ 333
Ba proteins, mechanism II/ 536
backbone
– chain entropy I/ 574
– dynamics, ^{15}N relaxation rates I/ 786
– solvation I/ 147, 726
bacteriophage
– fd (m13) II/ 205
– P22 II/ 22
– *Salmonella* phage P22 II/ 206
– T4 II/ 206
bacteriorhodopsin I/ 1009, II/ 878, 892, 894
Bae II/ 951
Bag proteins II/ 520, 536 ff, 740
– Bag-1 II/ 571
– – purification II/ 547–548
– domain structure II/ 536
– family members II/ 537
Bap II/ 520, 538, 572
barnase II/ 41, 255–256, 264 ff, 994, 998
β-barrel domain II/ 939
barstar I/ 456, 545, 611, 621, 683, II/ 256, 921
basic pancreatic trypsin inhibitor II/ 10, 26
BFP II/ 49
binding equilibria
– analytical solutions II/ 133 ff
– iterative solutions II/ 138 ff
bioluminescence resonance energy transfer II/ 901
biotin II/ 1208, 1219
biotin-avidin system I/ 617
BiP II/ 78, 85, 494 ff, 504, 521, 523, 564 ff, 572 ff, 650, 661
– binding site II/ 566
– binding specificity II/ 565
– *see also* immunoglobulin heavy-chain binding protein
birefringency II/ 198 ff, 214
Bis-ANS binding II/ 1079
BMP-1 II/ 651
bone morphogenetic protein *see* BMP
bovine carbonic anhydrase II/ 469
bovine pancreatic trypsin inhibitor I/ 11, 607, 1179, II/ 255, 366, 921 ff, 950
bovine seminal ribonuclease II/ 46
bovine serum albumin II/ 1007
BPTI *see* bovine pancreatic trypsin inhibitor
Brønsted plot I/ 417
BRET *see* bioluminescence resonance energy transfer
Brownian motion II/ 222, 995
Brownian ratchet II/ 1000
BSE II/ 1105, 1119, 1124
bulk-phase monomer II/ 1213
burst-phase reaction I/ 388, 394

c

C. elegans II/ 1207
c-cpn *see* CCT
CA *see* carbonic anhydrase
CaBP1 II/ 572
cadherins II/ 890
calbindin I/ 929
calcineurin II/ 323, 389, 1063
– 3-D structure II/ 279
– autoinhibitory peptide II/ 279
– phosphatase II/ 278
calcium II/ 565, 620
calcium depletion II/ 84

calmegin **II**/ 575
calmodulin **I**/ 457, **II**/ 278, 323
calnexin **II**/ 78 ff, 564, 572 ff, 617, 622 ff
calreticulin **II**/ 78 ff, 564, 572 ff, 617, 622 ff
calsequestrin **II**/ 335
carbonic anhydrase **I**/ 215, 225, 922, **II**/ 1263, 1266
carbonyl cyanide 3-chlorophanylhydrazone **II**/ 85
carboxypeptidase **II**/ 626
casein **II**/ 1057, 1058
casein kinase **II**/ 782
caseinolytic protease *see* Clp
castanospermine **II**/ 86
cataracts **II**/ 835
CATH classification **II**/ 33 ff
CATH database **II**/ 34
cation-π-interactions **I**/ 349
cavities **I**/ 216
cavity formation **I**/ 132, 352
– MD simulations **I**/ 133
cAMP receptor protein **I**/ 980
CCT **II**/ 725 ff, 756 ff, 791
– ATPase **II**/ 731, 742
– folding assays **II**/ 742
– mechanism **II**/ 731–736
– purification **II**/ 742
– structure **II**/ 726–727
– substrate proteins **II**/ 736–738
CCV *see* clathrin-coated vesicles
CD *see* circular dichroism
Cdc37 **II**/ 791
– chaperone activity **II**/ 793
– influence on Hsp90 **II**/ 791
– structure **II**/ 792–793
Cdc42Hs **II**/ 324
cell dystrophy **II**/ 1231
cellulose-binding domain **II**/ 1265
Cer1 **II**/ 494, 567
CFTR *see* cystic fibrosis transmembrane conductance regulator
chain conformation, polymer models **I**/ 809
chain dimension **I**/ 721
chain entropy **I**/ 729
chain stiffness **I**/ 810
chaperone **I**/ 935, 1100, **II**/ 6 ff, 8, 699 ff
– assays **II**/ 174 ff, 693, 841 ff
– ATPase activity **II**/ 164
– binding sites **II**/ 184, 502, 602
– chemical **II**/ 170
– cofactors **II**/ 168 ff
– complexes **II**/ 175 ff, 183, 185
– function **II**/ 162, 176 ff
– intramolecular **II**/ 170
– model substrates **II**/ 177
– promiscuity **II**/ 164, 166, 172
– regulation **II**/ 168
– specific **II**/ 171
– substrates **II**/ 164 ff, 172 ff, 183, 185
chaperone-like function **II**/ 394
chaperonins **II**/ 699 ff, 725, 1029, 1052–1056
– chloroplasts **II**/ 1055
– group I **II**/ 726
– group II **II**/ 726
– structure **II**/ 726, 726–727
characteristic ratio **I**/ 811
charge-charge interactions **II**/ 1297
charge-dipole interactions **I**/ 348
chelating agents **II**/ 84
chemical denaturation *see* solvent denaturation
chemical shift index **I**/ 746
chevron plot **I**/ 382, 393, 415, 427, **II**/ 37
– curvature **I**/ 385
CHIP **II**/ 79, 542, 794, 796
– purification **II**/ 549–550
chloramphenicol acetyltransferase **II**/ 1316
chloramphenicol transacetylase **II**/ 943
chloroplast **II**/ 79, 1047
– 14-3-3 proteins **II**/ 1065, 1068
– ClpC **II**/ 1066, 1069
– Com70 **II**/ 1065, 1068, 1069
– Cpn60 **II**/ 1066, 1069, 1071
– FKBP13 **II**/ 1073
– guidance complex **II**/ 1068
– Hsp100 **II**/ 1070
– Hsp70 **II**/ 1065 ff, 1070–1071
– prolyl isomerases **II**/ 1073
– protein translocation **II**/ 1065, 1066
– S87 **II**/ 1070
– Tic22 **II**/ 1069

– Tic40 II/ 1070
– Tic110 II/ 1069
– Toc receptors II/ 1066
– Toc12 II/ 1069
– Toc34 II/ 1066
– Toc64 II/ 1069
– Toc64 receptor II/ 1068
– Toc75 II/ 1066
– Toc159 II/ 1066
– translocon II/ 1066, 1069
– unfolding II/ 259
cholera toxin II/ 687
chorion II/ 198
chymotrypsin II/ 76, 385
CI2 I/ 431
circular dichroism I/ 38, 287, 324, 331, 464, 502, II/ 18, 36, 213 ff, 297 ff, 1212
– membrane proteins I/ 1020
circular permutation II/ 268
circularization II/ 1295
citrate synthase I/ 693, II/ 8, 173 ff, 180 ff, 366, 680 ff, 782, 841 ff, 869, 942, 1053, 1059, 1074
– assay II/ 842
CJD *see* Creutzfeld-Jakob disease
clathrin II/ 500, 501
clathrin-coated vesicles II/ 500
ClbA II/ 502
Clp proteins II/ 1056 ff
– ATP dependence II/ 262–263
– ClpA II/ 260, 1033, 1050, 1057 ff
– ClpAP II/ 260 ff, 501
– ClpB II/ 503, 1027, 1032 ff
– – ATPase cycle II/ 111 ff
– – dissociation constant ATP binding II/ 112
– – NBD mutants II/ 112
– – nucleotide-binding II/ 111 ff
– – nucleotide-binding domains II/ 111
– $ClpB_{Tth}$ II/ 113
– – NBD mutants II/ 115
– – nucleotide binding II/ 128
– – oligomerization II/ 113
– ClpC II/ 1057 ff, 1069 ff
– ClpD II/ 1057 ff
– ClpP II/ 261, 1033, 1058, 1072
– ClpX II/ 261
– ClpXP II/ 260 ff
– Erd1 II/ 1057
– FtsH II/ 260
– HslUV II/ 260
– Lon II/ 260
– Mcx1 II/ 1033
– structure II/ 261, 262
Clp/Hsp100 II/ 503, 948, 1033
clusterin II/ 1103
Cns1 II/ 788, 794, 796
CNX II/ 623
co-chaperone II/ 171
co-immunoprecipitation II/ 600
coagulation II/ 1212
cofactors II/ 176, 187
coil library I/ 729
coiled coils I/ 974, II/ 657, 757, 761, 887, 889
cold denaturation I/ 100, 209, 332, 455, 683
cold denatured proteins I/ 209
cold shock protein I/ 456, 544, 551, 553, 615, 621
– CspB I/ 189
– peptide B-1 II/ 1230
colicin A I/ 1013, II/ 947
collagen I/ 935, II/ 78, 199 ff, 380, 942
– biosynthesis I/ 1060, II/ 649
– chain association II/ 657
– cross-linked I/ 1089, 1097
– cruciform-shaped structure II/ 657
– disulfide bond II/ 652, 653, 659
– fibrillar collagen II/ 649
– folding
– – mechanism I/ 546
– – model I/ 1085 ff
– glycosylation II/ 651
– heterotrimer assembly II/ 659
– hydrogen bonds I/ 1064
– hydroxylase II/ 652
– hydroxylation II/ 658
– model peptides I/ 1066, 1097
– procollagen II/ 650
– proline modifications I/ 1078
– prolyl-4-hydroxylase II/ 659
– propeptide domains II/ 649 ff
– C-proteinase II/ 650, 653, 656

– sequence variations I/ 1075
– stability II/ 658
– telopeptides II/ 650
– thermal denaturation I/ 1069
– triple helix II/ 650
– unfolded chain I/ 1077
colloidal assembly II/ 1212–1217
combinatorial libraries II/ 892
compressibility coefficient I/ 108
confocal microscopy II/ 1229
conformational fluctuations I/ 575, 588, 612
conformational strain I/ 349
conformational substates I/ 608, 648
conformational switch I/ 929
Congo red II/ 198 ff, 1179, 1181, 1222
contact order I/ 869, II/ 330
continuous-flow I/ 689
– absorbance spectroscopy I/ 501
– FRET I/ 620
Cos-7 cells II/ 1229
counter ion binding I/ 906
covalent fluorescent labeling II/ 217
Cp7 II/ 796
Cpn10 II/ 1053, 1056
Cpn20 II/ 1053–1056
Cpn60 II/ 1071
– αCpn60 II/ 1053, 1054
– βCpn60 II/ 1053–1056
– concentration in the chloroplast II/ 1053
– Cpn60 binding assay II/ 1075–1076
– purification II/ 1076–1077
Cpr3 II/ 1021, 1027
Cpr6 II/ 788, 791
Cpr7 II/ 788, 791
Cpx II/ 951 ff
CREB-binding protein II/ 1207
Creutzfeld-Jakob disease (CJD) II/ 1106, 1116, 1124, 1144
δ Cro repressor I/ 682
cross-β structure II/ 198 ff
cross-interaction parameter I/ 420, 423
cross-linking II/ 47, 63, 423, 440, 460, 468, 572, 600, 622, 662, 924
– dithio-*bis*-succinimidylpropionate II/ 85
– DSS II/ 481
– EDC II/ 481
– glutaraldehyde II/ 485
– *bis*-maleimidohexane II/ 85
CRT II/ 623
cryo-electron microscopy II/ 232, 860
crystallin
– α-crystallin I/ 544, II/ 830 ff, 858 ff
– – alpha II/ 858
– – binding site II/ 862 ff
– – chaperone function II/ 859, 862 ff
– – function II/ 863
– – modifications II/ 860 ff
– – molecular weight distribution II/ 859
– – purification II/ 863
– – substrate binding II/ 861
– – subunit exchange II/ 859
– β-crystallin II/ 14, 45
– $\beta\gamma$-crystallin II/ 44
– γ-crystallin II/ 4, 37 ff, 45
– ζ-crystallin II/ 683
α-crystallin domain II/ 814
crystallization II/ 1213
CS *see* citrate synthase
curli II/ 965
cyclic β-hairpin I/ 331
cyclin II/ 736
cyclin D-dependent kinase inhibitor I/ 540
cyclodextrin II/ 1261
cyclophilin I/ 923, 930, II/ 78, 381, 382, 572, 661, 942, 1030, 1050, 1063
cyclosporin A II/ 382
Cyp40 II/ 786
cysteine II/ 358
– determination of free thiols II/ 370
– knot I/ 956
cysteine string proteins II/ 501
cystic fibrosis transmembrane conductance regulator II/ 79, 884
cystine II/ 4, 14, 939
– knot I/ 1081, 1089, 1097
cytochrome b II/ 256
cytochrome b_2 II/ 999, 1011
cytochrome bc1 II/ 892
cytochrome c I/ 110, 458, 471, 493, 501, 620, 659, 824, 835, 860, 864, 889, 899, 900, 905, II/ 10
– folding mechanism I/ 512

cytochrome c oxidase II/ 893
cytoskeleton II/ 763
cytotoxicity II/ 1228

d

D. melanogaster II/ 1207
Daf II/ 440
Daf-16 II/ 1188, 1189
Debye-Hückel screening I/ 904
DegP II/ 946 ff
DegQ II/ 946
degradation tag I/ 367
DegS II/ 946 ff
denaturant binding I/ 831
– GdmCl I/ 228
– urea I/ 223
denaturation II/ 177, 1033
denatured proteins
– MD simulations I/ 1148
– residual structure I/ 218, 220, 225, 344, 434, 472
detergents II/ 1257
Dexter mechanism I/ 821
DHFR *see* dihydrofolate reductase
diamide II/ 77, 91
dielectric constant I/ 165, 173
diffusion coefficient I/ 673, II/ 284
– intramolecular I/ 592
– NMR I/ 765
diffusion-controlled reactions, test for – I/ 847
digitonin II/ 92, 662, 664
dihydrofolate reductase I/ 865, 985, II/ 12, 13, 261 ff, 419, 994 ff, 1030, 1295
– import into mitochondria II/ 1009–1011
– methotrexate II/ 255
– stabilization by methotrexate II/ 1009
dimeric protein
– kinetic data analysis I/ 969
– two-state folding I/ 970
diphtheria toxin I/ 1013, II/ 256
dipole-dipole interaction I/ 348
dipyridyl sulfide II/ 690
disaggregation
– by chaperones II/ 1056 ff
– by hexafluoroisopropanol II/ 1209
– by trifluoroacetic acid II/ 1209
disease *see* human folding disease
displacement titrations II/ 113
distance determination I/ 586
distance distribution function I/ 590
disulfide *see* cystine
disulfide bond I/ 946, II/ 79 ff, 359, 620 ff, 679, 889, 969, 970, 1247 ff
– air oxidation II/ 1251, 1253
– analysis II/ 601
– artificial disulfide bridges II/ 1295
– catalysis II/ 358
– catalytic II/ 678
– entropic effect II/ 1294
– formation I/ 11, II/ 80, 81, 90, 358, 369, 677
– isomerization II/ 365
– mixed disulfide II/ 366
– mutation II/ 87
– oxidation II/ 267
– possible combinations of 2n cysteine II/ 1254
– reduction II/ 369
– regulatory II/ 678, 679
– reshuffling II/ 1254
– structural II/ 678
– trapping I/ 947
disulfide isomerase II/ 365, 1254, 1261
– chaperone activity II/ 366
– pathway II/ 368
– structure II/ 366
disulfide oxidase, assay II/ 370
disulfide-bonded proteins II/ 1247
dithio-*bis*-succinimidylpropionate II/ 85
dithioerythritol (DTE) II/ 1251
dithionitrobenzoate II/ 370
dithiothreitol (DTT) II/ 77, 83, 90, 128 ff, 177, 179, 360 ff, 620, 1251
DjC7 II/ 520
DNA triplet expansion II/ 1202
DnaJ II/ 488, 499 ff, 520, 528, 532, 565, 567, 684 ff, 1027, 1028, 1050, 1051, 1231
– affinity for substrates II/ 533

DnaK **II**/ 123, 415 ff, 436, 459 ff, 493, 499 ff, 519, 525, 571, 572, 684, 686, 688, 756, 921, 1002, 1003, 1034, 1051, 1052, 1067, 1185, 1248, 1261
– binding motif **II**/ 523–524
– structure with peptide **II**/ 524
– substrate proteins **II**/ 492
DNAse **II**/ 1260
DnJ/Hsp40 **II**/ 1157
domain architecture **II**/ 34
domain interactions **II**/ 1300
domain shuffling **II**/ 33
domain swapping **I**/ 7, **II**/ 14 ff, 44, 45
double-jump experiments **I**/ 10, 386, 394, 920, 936, 1053
downhill folding **I**/ 438
4-DPS **II**/ 77
drk **I**/ 555
– structure **II**/ 286, 288
DsbA **II**/ 359 ff, 678, 951, 969, 970, 1261
– reaction cycle **II**/ 360
DsbB **II**/ 359 ff, 361 ff
– quinone reductase activity **II**/ 364
– reaction cycle **II**/ 363
DsbC **II**/ 359 ff, 365 ff, 946
DsbD **II**/ 359 ff, 365 ff
DsbG **II**/ 359 ff, 365 ff, 946
– reaction cycle **II**/ 365
DSS **II**/ 481
DTE *see* dithioerythritol
DTNB *see* dithionitrobenzoate
DTT *see* dithiothreitol
dynamic quenching, acrylamide **II**/ 302
– oxygen **II**/ 302
– Stern-Volmer equation **II**/ 302
dynamin **II**/ 501

e

E-cadherin **II**/ 902
ECCE database **II**/ 939
ECM *see* extracellular matrix
EDC **II**/ 481
EDEM **II**/ 576
EGF *see* epidermal growth factor
elasticity **II**/ 41
elastin **II**/ 204
electron microscopy **II**/ 18, 1222
– image processing **II**/ 742–747
electron paramagnetic resonance spectroscopy **II**/ 110, 882
electrostatic attraction **II**/ 1229
electrostatic interactions **II**/ 880
– long-ranged **I**/ 348
electrostatic models
– all-atom models **I**/ 166
– continuum (Poisson-Boltzmann) **I**/ 171
– dipolar lattice models **I**/ 168
– generalized Born (GB) model **I**/ 172
– PDLD-approach **I**/ 168
electrostatic theories **I**/ 163
Ellman's reagent **II**/ 370, 630, 690
EM *see* linear extrapolation method
N-end rule **II**/ 443
end-to-end contact formation rate constants, table **I**/ 817
end-to-end diffusion **I**/ 834
end-to-end distance **I**/ 811
end-to-end distribution function **I**/ 813
endoglycosidase H **II**/ 81, 83
– digestion **II**/ 593
– resistance **II**/ 96
endoplasmic reticulum **II**/ 7 ff, 259, 494, 651
– calcium depletion **II**/ 84
– multi-protein complex **II**/ 572
– protein degradation *see* ERAD
– quality control **II**/ 564
– retrotranslocation **II**/ 575
– signal peptide **II**/ 79
– thapsigargin **II**/ 84
– translocation **II**/ 432, 576
– translocon **II**/ 563
endoplasmin **II**/ 84
energy landscape **I**/ 424, 716, 1139
– roughness **I**/ 469
– theory **I**/ 466
engineering
– antibody engineering **II**/ 1300–1305
– chain entropy **II**/ 1296
– display technologies **II**/ 1310–1315
– disulfide bond **II**/ 1304
– DNA shuffling **II**/ 1306, 1311

- evolutionary strategies II/ 1305–1315
- immunoglobulin V_L domains II/ 1291
- libraries II/ 1307
- phage display II/ 1310 ff
- proteolytic stability II/ 1323–1325
- ribosome display II/ 1310 ff
- stability II/ 1250, 1319–1325
- staggered extension process II/ 1306
- structure-based II/ 1292–1300
- yeast surface display II/ 1318
engrailed homeodomain I/ 1150
enthalpy of unfolding I/ 81
enthalpy-entropy compensation I/ 79, 139, 344, 346, 729
entropy, conformational I/ 351, II/ 286
entropy of unfolding I/ 83
- convergence temperature I/ 140
environmentally sensitive fluorophores II/ 217
epidermal growth factor I/ 953
EPR *see* electron paramagnetic resonance spectroscopy
equilibrium unfolding transitions
- analysis of I/ 872
- multistate transitions I/ 84
ERAD II/ 575, 629, 631
- pathway II/ 82
Erdj 1–5 II/ 568 ff
Ero1p II/ 82, 678, 686, 688
ERp57 II/ 78, 622, 633
ERp72 II/ 572
Erv proteins II/ 82
N-ethyl maleimide II/ 82, 667, 998
europium II/ 1219
EX1 exchange mechanism I/ 644
EX2 exchange mechanism I/ 644
excess heat capacity I/ 73, 84, 85
excluded volume I/ 719
excluded volume effect I/ 812
extracellular deposits II/ 198
extracellular matrix II/ 660
extremophiles II/ 23 ff
eye lens II/ 830, 858 ff

f

F1 antigen II/ 965
F_{ab} fragment *see* antibody
farnesylation II/ 527
fast kinetics I/ 454
- simulation approaches I/ 472
fatal familial insomnia II/ 1116
Fes1 II/ 520, 538
Fes1p II/ 571
FFF *see* field flow fractionation
fiber II/ 197 ff, 265, 965, 1097, 1116, 1175
- assembly II/ 974
- common conformational II/ 1181
- diffraction II/ 1119
- filaments II/ 207
- formation I/ 544, 693
- lag phase II/ 1099
- oligomeric intermediates II/ 1181
- protofilaments II/ 206
- surface II/ 229
fibrillin II/ 204
fibrin II/ 203 ff
fibrinogen II/ 203 ff
fibritin II/ 657
fibronectin II/ 39, 220
field-flow fractionation II/ 221
Fim proteins II/ 965–976
first-order phase transition I/ 101
FK506-binding proteins II/ 78, 381, 464, 791, 1050
FKBP *see* FK506-binding proteins
FkpA II/ 942, 946
flagella
- bacterial flagellum II/ 204
- flagellin II/ 205
- FlgI II/ 358
- FlgM II/ 321
- - structure II/ 283, 287
Flory's isolated-pair hypothesis I/ 265, 722
fluorescein II/ 1208, 1229
fluorescence
- anisotropy II/ 525
- energy transfer I/ 31
- intrinsic II/ 215 f
- ANS fluorescence II/ 305

– dynamic quenching II/ 302
– fluorescence anisotropy II/ 303
– fluorescence resonance energy transfer II/ 303
– time-resolved methods I/ 588
– tryptophan fluorescence II/ 301
– microscopy II/ 1221
– nucleotide analogues, structures II/ 107
– quantum yield I/ 586
– quenching I/ 31
– spectroscopy I/ 465, 495
fluorescence resonance energy transfer (FRET) I/ 499, 557, 573 ff, 600, 819, II/ 48, 217, 264, 900, 1180
– data analysis I/ 603
– donor/acceptor pairs II/ 900
– green fluorescent protein II/ 900
– orientation factor I/ 583
– time-resolved measurements I/ 601
– transfer efficiency I/ 600
– transient intermediates I/ 619
fluorescent labels II/ 525
– nucleotide analogues II/ 116 ff
foci II/ 1230
Förster critical distance I/ 579, 585
Förster equation I/ 580
Förster transfer *see* fluorescence resonance energy transfer (FRET)
folding funnel I/ 466, 716, 866
folding models I/ 867
– analytical solutions I/ 402, II/142 ff
– dimeric proteins I/ 978
– elucidation I/ 394
– pro-domain assisted folding I/ 1043
foldon domain I/ 1098, II/ 657
Fourier transform infrared spectroscopy (FTIR) I/ 295, 1022, II/ 54, 239
FRAP II/ 1180, 1182, 1187, 1190–1192
free energy
– of solvation I/ 131
– of transfer I/ 129
freely jointed chain I/ 810
freeze concentration II/ 1210
FRET *see* fluorescence resonance energy transfer
FtsH II/ 501, 502
FtsY II/ 921
FtsZ II/ 211
fusin proteins II/ 1212

g

β-galactosidase II/ 460, 538, 1247, 1267
GALLEX II/ 902
gamma interferon II/ 1258
GAPDH II/ 469
GCN4 I/ 420, 437, 974
– structure II/ 283, 287
GdmCl *see* guanidinium chloride
gel exclusion chromatography II/ 295
gel filtration II/ 48, 1208
geldanamycin II/ 110
– structure II/ 774
gene-3-protein I/ 922, 939
GFP *see* green fluorescent protein
GHKL-ATPases II/ 775
Gibbs-Duhem relation I/ 104
Gibbs-Helmholtz equation I/ 975, 991
GimC II/ 439, 442
Gln/Asn-rich proteins II/ 1153
glucagon-like peptide 1 receptor II/ 1251
glucocorticoid receptor II/ 1061
ξ-glucosidase II/ 7
α-glucosidase II/ 173, 948, 1265
glucosidase II II/ 576
glutamine repeats II/ 1175 ff
glutaraldehyde II/ 47, 63
glutaredoxin II/ 684
glutathione II/ 369, 680, 686 ff, 1254, 1265
– ratio GSW:GSSG in vivo II/ 82
glutathione S-transferase (GST) II/ 898, 1206
– pull-down II/ 599
N-glycans II/ 617
– analysis II/ 636
glyceraldehyde-3-phosphate dehydrogenase II/ 42, 366, 687, 1260
glycophorin A I/ 1011, II/ 887 ff, 900, 902
glycosyl transferase, assay II/ 637
glycosylation II/ 16, 80–81
– castanospermine II/ 86
– glucosyltransferase II/ 619 ff
– glycan processing II/ 619

- kifunensine II/ 86
- N-glycan structure II/ 618
- N-glycosylation II/ 592, 617 ff
- – consensus sequences II/ 618
- – endo H digestion II/ 593
- – manipulation of glycosylation sites II/ 594
- – PNCase F digestion II/ 594
- – tunicamycin treatment II/ 593
- O-glycosylation II/ 83, 651
- tunicamycin II/ 86

gold clusters II/ 266, 995
gp41 envelope protein I/ 984
gp55-P II/ 890
GpE, mechanism II/ 535
granulocyte colony-stimulating factor II/ 1252
Greek key topology II/ 37, 44
green fluorescent protein (GFP) II/ 49, 83, 1190, 1207, 1267, 1316
GroE II/ 62, 699 ff, 1248
- assisted folding II/ 706 ff, 717 ff
- catalyst II/ 709
- efficiency II/ 708 ff
- folding kinetics II/ 708 ff
- reaction cycle II/ 706

GroEL II/ 62, 167, 169, 181 ff, 264, 429, 459, 684, 686, 699 ff, 725 ff, 761–762, 948, 951, 1029, 1050 ff, 1186, 1261
- allostery II/ 701, 704 ff, 709, 710
- Anfinsen cage II/ 708
- apical domains II/ 700 ff, 703, 705
- ATP binding II/ 701, 704, 707, 708
- binding and hydrolysis of ATP II/ 704 ff
- central cavity II/ 700 ff
- cooperativity II/ 704, 707, 716
- fluorescent label II/ 704
- mechanism II/ 707
- peptide-binding II/ 702
- purifcation II/ 710 ff
- reaction cycle II/ 705 ff
- single ring mutant II/ 708, 710, 716
- structure II/ 699 ff, 700 ff
- substrates II/ 702–710
- Trp-containing contaminations II/ 711 ff

GroES II/ 169, 186, 187, 684, 686, 699, 701 ff, 726 ff, 756 ff, 1029, 1050 ff, 1261
- interaction with GroEL II/ 701 ff, 705 ff, 716
- mobile loop II/ 701
- purification II/ 713 ff
- structure II/ 701 ff

ground state effects I/ 413, 428
Grp78/Bip II/ 84, 622 ff, 633
Grp94 II/ 572, 622, 769 ff, 782–783
Grp170 II/ 495, 567, 571, 572
GrpE II/ 499 ff, 520, 536, 565, 571, 572, 684, 686, 1024, 1050 ff
- purification II/ 546–547
- structure II/ 534

GSH *see* glutathione
GSSG *see* glutathione
GTPases II/ 1066
guanidinium chloride *see* solvent denaturation
gyrase B II/ 775, 778
- structure II/ 779, 780

h

Hac1 II/ 573
β-hairpin I/ 315, II/ 757
- design I/ 317
- nucleation I/ 316
- relaxation kinetics I/ 461
- stability I/ 332, 334, 359

Hammond behavior I/ 413, 424, 428
HasA II/ 925
HCGβ II/ 572
Hch1 II/ 794–795
Hdj1 II/ 504, 1185, 1186, 1231
Hdj2 II/ 493, 504, 1185, 1186
heat capacity I/ 81, 100, 137
heat shock II/ 162 ff, 172, 177
- granules II/ 834, 837
- proteins II/ 946
- – *see also* Hsp
- response
- – rpoH II/ 501
- – transcription factor II/ 501

HEDJ II/ 568, 569

helix
– 3_{10}-helix **I**/ 251
– π-helix **I**/ 251
– α-helix **II**/ 198 ff
– – capping **I**/ 248
– – design **I**/ 252
– – dipole **I**/ 185, 253, 267, 348, **II**/ 1297
– – end effects **I**/ 254
– – MD simulations **I**/ 1146
– – nomenclature **I**/ 248
– – nucleation **I**/ 265
– – packing **I**/ 1010
– – peptide models **I**/ 146, 252
– – phosphorylation **I**/ 276
– – programs **I**/ 270
– – propensity **I**/ 352, **II**/ 271, 878
– – Ramanchandran plot **I**/ 320
– – relaxation kinetics **I**/ 461
– – side chains **I**/ 255
– – single sequence approximation **I**/ 266
– – stability **I**/ 145, 270
– – structure **I**/ 247
helix-coil
– model **I**/ 265, 358
– parameters **I**/ 257
– theory **I**/ 257, 261
– transition enthalpy **I**/ 82
hemagglutinin **II**/ 77, 623, 657
heme-binding domain **II**/ 1011
hemoglobin **II**/ 3, 1213, 1253
– sickle hemoglobin **II**/ 1215
heparinase **II**/ 1248
heptad repeats **II**/ 202, 757
HEWL *see* lysozyme
1,1,1,3,3,3-hexafluoro-2-propanol **I**/ 887
hexokinase **II**/ 46, 52
Hip **II**/ 520, 541–542, 571
– purification **II**/ 548–549
hirudin **II**/ 372
HIV envelope **II**/ 81
HIV-1 capsid **II**/ 393
Hoff enthalpy **I**/ 1064
Hofmeister series **I**/ 904, **II**/ 1258
homeodomain **I**/ 457
homodimeric proteins **II**/ 43
Hop **II**/ 542, 740, 786, 1070
– purification **II**/ 549
– structure **II**/ 789
HPr **I**/ 544
Hsc20 **II**/ 499
Hsc66 **II**/ 499
Hsc70 **II**/ 500 ff, 740
HscA **II**/ 521
HscB **II**/ 528, 530
HscC **II**/ 521 ff
hSec63 **II**/ 568, 569
Hsf **II**/ 772
Hsf-1 **II**/ 1189
HslVU **II**/ 501
Hsp10 **II**/ 1021, 1027, 1029
Hsp12 **II**/ 831
Hsp16.3 **II**/ 838
Hsp16.5 **II**/ 837
Hsp16.9 **II**/ 837, 859
Hsp17 **II**/ 837
Hsp21 **II**/ 1050, 1074, 1075
– purification **II**/ 1077–1078
Hsp25 **II**/ 835
– purification **II**/ 841
Hsp26 **II**/ 838
– purification **II**/ 840
Hsp27 **II**/ 832, 859, 1074
Hsp33 **II**/ 46, 677, 679, 680, 686
– activation **II**/ 681
– chaperone network **II**/ 684, 686
– mechanism **II**/ 684
– structure **II**/ 682
– substrate proteins **II**/ 683
– zinc **II**/ 681
Hsp40 **II**/ 171, 431, 436, 740
– J-domain **II**/ 437
Hsp42 **II**/ 831
Hsp47 **II**/ 171, 651, 652, 660, 661
Hsp60 **II**/ 165, 699 ff, 948, 993, 1021, 1026 ff
Hsp70 **II**/ 79, 165 ff, 184, 415, 431, 490, 499 ff, 563, 564, 571, 756, 763, 764, 786, 1021, 1022, 1034, 1050, 1061, 1066 ff, 1157, 1182 ff
– affinity for peptides **II**/ 525–526
– ATPase activities **II**/ 495
– ATPase cycle **II**/ 519, 565

– binding site II/ 1067
– chaperone cycle II/ 785
– chloroplasts II/ 1048–1051
– Css1 II/ 1067
– DNA replication II/ 502
– general functions II/ 516
– mitochondrial II/ 257 ff
– protein translocation II/ 496
– purification II/ 544–545
– ribosome association II/ 436, 496
– structure II/ 517
– substrate interaction II/ 516, 522
– TPR containing co-chaperones II/ 540
– uncoating II/ 500
Hsp78 II/ 503, 1021, 1027, 1032 ff
Hsp80, chloroplast Hsp90 II/ 1061–1062
Hsp82 II/ 772
Hsp90 II/ 79, 165, 169, 171, 181, 496, 520, 540, 740, 1050, 1061, 1062
– analysis of quaternary structure II/ 804–805
– ATP binding II/ 110, 775
– ATPase activities II/ 782
– ATPase cycle II/ 109, 780 ff
– chaperone assays II/ 801, 802–803
– chaperone cycle II/ 784–787, 785
– clients II/ 774–778
– co-chaperones II/ 784–787
– cross-linking assay II/ 807–808
– dissociation constant II/ 110
– domain structure II/ 769–770
– drug target II/ 773
– evolution II/ 768–770
– Grp94 II/ 769, 771
– HtpG II/ 769, 771
– in vivo assays in yeast II/ 797–798, 801
– inhibitors II/ 773–774
– isothermal titration calorimetry II/ 807
– large PPIases II/ 790–791
– middle domain structure II/ 778, 779
– mutants II/ 799
– nucleotide analogues II/ 110
– partner proteins II/ 784–787
– peptide-binding II/ 783
– regulation II/ 772–773
– steroid hormone receptors II/ 784, 787
– Sti1 II/ 786
– structure of FKBP51 II/ 790
– substrate proteins II/ 774–778
– substrate-binding site II/ 783
– N-terminal dimerization II/ 781
– N-terminal domain structure II/ 775, 779
– X-ray structure II/ 110
Hsp100 II/ 165, 516, 1056
Hsp100/Clp II/ 1033
Hsp104 II/ 113, 181, 442, 443, 1033, 1157, 1185
– ATPase enzymatic parameters II/ 115
– cooperativity II/ 115 ff
– nucleotide binding II/ 115
Hsp104p II/ 1158
Hsp110 II/ 517
Hsp170 II/ 517
HspA1L II/ 1185, 1186
HspB2 II/ 840
HspB3 II/ 840
HspBP1 II/ 520, 538, 571, 572
HtpG II/ 769, 771, 772, 782
HtrA II/ 946, 947
human folding disease II/ 198, 1200
– Alzheimer's disease II/ 1206
– androgen receptor II/ 1201
– ataxias II/ 1201
– dentatorubral-pallidoluysian atrophy II/ 1202
– Huntington's disease II/ 1201
– Machado-Joseph disease II/ 1202
– motor disorder II/ 1201
– neurodegenerative diseases II/ 1233
– neurologic symptoms II/ 1203
– psychiatric symptoms II/ 1201
– spinal and bulbar muscular atrophy (SBMA) II/ 1201
– spinocerebellar ataxia 17 II/ 1201
huntingtin II/ 493, 1177, 1181 ff, 1205
– htt exon II/ 1206
Huntington's disease II/ 198, 265, 1104, 1148, 1151, 1175 ff
hydration I/ 81, 220, 688
– enthalpy I/ 82
– entropy I/ 83, 106

– shell I/ 122
– – contributions to δS_C and δV I/ 106
hydrodynamic dimensions I/ 687
hydrodynamic radius I/ 543, 673, 678, 680, II/ 284
– NMR I/ 765
hydrodynamic volume I/ 222
hydrogen bond I/ 145, 348, 351, 420, II/ 881
– burial I/ 153
– collagen I/ 1064
– effect of pressure I/ 121
– enthalpy I/ 82
– inventory I/ 149
hydrogen bonding II/ 1298
hydrogen exchange I/ 634 ff, II/ 309
hydrogen peroxide II/ 690
hydrophobic collapse I/ 388, 394, 471, 505, 542, 575, 621
hydrophobic core I/ 144
– solvation, free energy I/ 142
hydrophobic effect I/ 83, 100, 128, 350, II/ 882, 1284, 1296
hydrophobic free energy I/ 128, 136, 137
hydrophobic interactions II/ 166, 175, 703
hydrophobic staple I/ 249
hydrophobic thickness of membranes I/ 1006
hygromycin B II/ 438
hyperthermophilic proteins II/ 1293
hysteresis I/ 1099

i

IAC *see* iodoacetic acid
IANBD amide II/ 219
IAM *see* iodoacetamide
IAPP (amylin) II/ 1107, 1225
im14 II/ 1026
immunity protein Im7 I/ 518, 865
immunofluorescence staining II/ 589
immunoglobulin *see also* antibody
– fold II/ 39
– – Caf1M-Caf1 complex II/ 972
– heavy-chain binding protein II/ 564
– light chain II/ 1225
immunoprecipitation II/ 75, 79, 81, 82, 460, 482, 623, 954, 1036
– protocol II/ 93
immunotoxin II/ 1258, 1261
in vitro translation II/ 662, 898
inclusion body II/ 6 ff, 16, 22, 54, 59, 898, 942, 1105, 1118, 1120, 1175, 1176, 1245 ff, 1267
– formation II/ 1247
– isolation II/ 1250, 1268
– solubilization II/ 1250, 1269
inclusion body binding proteins II/ 1231, 1246
influenza hemagglutinin II/ 572, 897
influenza M2 protein II/ 888
infrared spectroscopy II/ 299 ff
insulin I/ 929, II/ 173, 182, 226, 842, 1097, 1268
– aggregation II/ 846, 861
– assay II/ 867
– receptor II/ 626
insulin-like growth factor II/ 1252
inteins II/ 1295
inter-subunit interfaces, hydrophobicity II/ 43
interleukin I/ 929
interleukin-1β I/ 865
interleukin 4 II/ 1255
interleukin 21 II/ 1255, 1258
intermediate filaments II/ 202 ff
intermediates I/ 871
– burst phase I/ 471, 492, 863, 876, 891
– dimeric I/ 978
– disulfide bonds I/ 948
– equilibrium I/ 221, 686
– high-energy – I/ 393
– intramolecular distances I/ 619
– kinetic I/ 7, 380, 384, 388, 394, 517, 540, 550, 619, 651, 659, 686, 862, 865, 898
– molten-globule I/ 858
– obligatory I/ 396
– prolyl isomerization I/ 928
internal friction I/ 458
internal hydration I/ 216
interrupted refolding experiments I/ 387, 391, 395, 1053

intersystem crossing I/ 821
intrachain diffusion I/ 459, 815, 816, 826
intraneuronal nuclear inclusions II/ 1204
intrinsic tryptophan fluorescence II/ 130
intrinsic viscosity I/ 719
intrinsically unstructured proteins *see* natively disordered proteins
invertase II/ 16, 51
iodoacetamide II/ 82, 218
iodoacetic acid II/ 82, 179
Ire1/Ern1 II/ 573
islet amyloid polypeptide II/ 220
iterative annealing II/ 708

j

in-cell NMR II/ 321
J-domain II/ 441, 500, 567, 569, 570, 1004, 1026, 1028, 1052, 1069
J-domain proteins II/ 492 ff, 526–527, 786, 1005, 1051
– binding to Hsp70 II/ 530–532
– domain structure II/ 527
– purification II/ 545–546
– structures II/ 528
– subfamilies II/ 527
– substrate specificity II/ 532
– substrate-binding domain II/ 530
– Zn 2^+-binding domain II/ 527 ff, 533
Jac1 II/ 499, 500, 1021, 1027, 1028
Jacobsen-Stockmeyer theory I/ 821
Jem1 II/ 504
Jem1p II/ 567, 568
jigsaw puzzle II/ 25
JPOI II/ 568

k

Kar2 II/ 494, 495, 504, 564 ff
keratin II/ 202
kifunensine II/ 86
kinetic mechanism I/ 390
– lag-phase I/ 396, II/ 1213
kinetic nucleus II/ 1213
kinetic partitioning I/ 398
– model II/ 923
knowledge-based potentials I/ 357
Kramer's theory I/ 470, 840
Kratky plot II/ 293
Kuhn length I/ 811
Kuhn segments I/ 715

l

labeling II/ 218, 623
– of proteins I/ 595
α-lactalbumin I/ 213, 225, 542, 690, 859, 899, II/ 36, 173, 541, 921
– bound Ca^{2+} II/ 182
– pH sensitivity II/ 867
– reduced carboxymethylated II/ 178 ff
β-lactamase I/ 899, II/ 419, 943, 948, 1071, 1262, 1295
lactate dehydrogenase II/ 7, 17, 20, 26, 52, 53, 1258, 1264
– release II/ 1230
β-lactoglobulin I/ 219, 228, 660, 888, 890
– cold denaturation I/ 210
LamB II/ 943, 944, 945, 946
lamins II/ 202
Langevin equation I/ 813
laser photolysis I/ 457
LDH *see* lactate dehydrogenase
lectin II/ 621, 622, 967
Leffler plot I/ 415, 418, 425, 427
– *see also* α-value
– curved I/ 421
lens II/ 863
– aging II/ 863
– transparency II/ 863
leucin-rich repeat (LRR) proteins, sequence statistics II/ 1289
leucine zipper II/ 283, 889, 890, 1206
– AB I/ 974
leucocin I/ 887
Levinthal's paradox I/ 5, 856
Lewy bodies II/ 1100
Lhs1 II/ 494
Lhs1p II/ 567, 571
Lifson-Roig model I/ 262
ligand binding I/ 88

light microscopy II/ 197
– birefringence II/ 1223
– pathological stain for amyloid II/ 1223
– polarized light microscopy II/ 1223
light scattering I/ 462, 673 ff, 674, II/ 62, 175, 179, 180, 220 ff, 693, 1213
– aggregation II/ 866
– classical II/ 221
– dynamic I/ 673, II/ 221, 295
– interpretation II/ 869
– multi-angle II/ 223
– multi-angle laser II/ 864 ff
– multi-angle laser photometers II/ 221
– Rayleigh II/ 1225
– static I/ 673
limited proteolysis II/ 76
Linderstrøm-Lang equation I/ 643
linear extrapolation method I/ 48, 973
linear lattice effect II/ 1204
lipid bilayer I/ 999
liposome encapsulation II/ 1229
local interactions I/ 887
LolA II/ 945
Lomize-Mosberg model I/ 269
Lon II/ 501
loop formation kinetics I/ 826 ff
low-pressure denaturation I/ 108
luciferase I/ 982, II/ 367, 421, 493, 520, 533, 538, 543, 680, 683, 693, 736, 740, 783, 901, 1158
– bacterial II/ 10, 53
– *photinus pyralis* II/ 173
lysozyme I/ 225, 389, 544, 660, 682, 689, 888, II/ 10, 34, 36, 54, 55, 469, 765, 868, 1106, 1250, 1258 ff
– aggregation II/ 861
– folding mechanism I/ 394, 397
lysyl hydroxylase II/ 651, 658
α-lytic protease I/ 1037, II/ 1252
lyticase II/ 1175
LZ16A I/ 984

m

M13 major coat protein II/ 880
M2 proton channel II/ 889
MABA-ADP/ATP II/ 106, 116 ff
macromolecular crowding II/ 24 ff
macroscopic cellular inclusions II/ 1228
macroscopic order II/ 1224
magnetic relaxation dispersion I/ 202
major histocompatibility complex II/ 623
– class I II/ 573, 738
– heavy-chain II/ 626
malate dehydrogenase II/ 21, 173, 178, 183, 708, 717 ff, 869
– cytoplasmic II/ 18
– enzyme assay II/ 718 ff
– mitochondrial II/ 18
MalE II/ 902
maleimides II/ 218
bis-maleimidohexane II/ 85
MalS II/ 946, 947
maltose-binding protein II/ 173, 901, 902, 919, 942 ff
MANT-ADP/ATP II/ 118–121
Map1p II/ 434, 438
mass spectrometry I/ 959, II/ 866
mastoparan I/ 959 f, II/ 309, 866
MBP *see* maltose binding protein
Mcx1 II/ 1021
MD *see* molecular dynamics
Mdg1 II/ 568, 569
MDH *see* malate dehydrogenase
Mdj1 II/ 494, 503, 504, 1005, 1021, 1027, 1032
Mdj2 II/ 1028
2-ME *see* 2-mercaptoethanol
mechanical unfolding I/ 1112, 1118, II/ 1000
melittin I/ 906, II/ 895
membrane
– composition I/ 1007
– polarity II/ 882
– thickness I/ 1006, 1016
membrane helices
– G xxx G II/ 884
– proline II/ 884

membrane proteins II/ 1265
– β-barrel I/ 1003
– bitopic II/ 874, 886
– chemical synthesis II/ 898
– electrophoresis II/ 898
– equilibrium stability I/ 1005
– folding kinetics I/ 1012
– folding mechanism I/ 1004
– GALLEX II/ 902
– α-helical I/ 1003
– helix-helix interactions II/ 878
– hydrodynamic methods II/ 899–900
– insertion I/ 1012 f
– integration II/ 876
– lateral interactions I/ 1010
– lipid-protein interactions II/ 893–894
– NMR II/ 888
– polytopic proteins II/ 876
– POSSYCCAT II/ 902
– recombinant expression II/ 897–898
– stability I/ 1009
– structure II/ 877, 883, 886
– TOXCAT system II/ 901
– TOXR system II/ 901
MepA II/ 365, 366
2-mercaptoethanol II/ 77, 1251
metal binding, to α-helix I/ 250
metastable states II/ 381
methionine aminopeptidase II/ 434, 443
Mge1 II/ 494, 500, 503, 993, 1005, 1021 ff
MHC *see* major histocompatibility complex
β-2-microglobulin II/ 324, 623, 1105, 1225
microtubules II/ 203 ff, 760
– α-tubulin II/ 202
– β-tubulin II/ 203
– tubulins II/ 203
misfolding II/ 61, 1093
mitochondria II/ 79, 494, 1020
– AAA proteases II/ 1031
– aggregation versus degradation II/ 1033
– analysis of aggregation II/ 1036
– β-barrel proteins II/ 988
– Brownian ratchet II/ 992 ff, 1025
– chaperone network II/ 1027
– chaperonins II/ 1029
– degradation assay II/ 1036
– DnaJ II/ 991
– folding II/ 1026–1029
– – assay II/ 1036
– genome II/ 987
– import into the matrix II/ 255–256, 990
– import mechanism II/ 991 ff, 1001, 1024–1026
– import motor II/ 988, 991 ff, 1001 ff, 1023–1025
– imported proteins II/ 1026–1029
– matrix-processing peptidase II/ 993
– membrane potential II/ 258, 267, 990, 1001, 1022
– mtHsp70 II/ 991
– power stroke II/ 992 ff
– protein import II/ 989, 1023
– – assay II/ 1006–1008, 1035
– protein synthesis II/ 1030
– protein unfolding II/ 1022
– quality-control system II/ 1032
– TIM complex II/ 257
– TIM23 complex II/ 988 ff, 1002, 1023
– Tim44 II/ 991, 1002 ff, 1023
– TOM complex II/ 257, 988 ff, 1023
– translocase II/ 257, 987, 990, 991, 1022
– translocation channel II/ 258, 994–995, 1001–1003
– – dimensions II/ 266
– translocation machineries II/ 257–258, 989, 1023
– unfolding II/ 256 ff, 988 ff
model substrates II/ 177, 179
molar concentration II/ 1237
molecular chaperone II/ 1218
– ATPase cycles II/ 105 ff, 109
molecular crowding II/ 566, 1218
molecular dynamics simulations I/ 1144 ff
– biased unfolding I/ 1150
– continuum solvent models I/ 1185
– forced unfolding I/ 1151
– high-temperature simulations I/ 1150
– long-range electrostatics I/ 1177
– photo-switch I/ 1187
– polarizability I/ 1175, 1179
– protein electrostatics I/ 1172
– protein-solvent systems I/ 1174

– real artifacts I/ 1170
– relaxation times and spatial scales I/ 1172
– solvent I/ 1173, 1181
– techniques and protocols I/ 1155
– time scales I/ 1184
– transition state I/ 1153
molecular mass determination I/ 677, 680, II/ 864, 1237
molecular mechanics I/ 1170
molecular switch II/ 378
molten globule I/ 14, 100, 109, 685, 686, 857, 889, 903, II/ 620, 626, 861, 921
– characterization I/ 858
– hydration I/ 213, 215
– packing I/ 215
MRD *see* magnetic relaxation dispersion
mtHsp70 II/ 506, 993, 996, 1000 ff, 1025 ff
Mtj1 II/ 568
Mtj1p II/ 498
MTT reduction II/ 1230
multi-domain proteins
– association II/ 32 ff
– folding II/ 32 ff
multichannel detection techniques I/ 461
MUP-I II/ 324
muscular atrophy II/ 1175
MutL II/ 775, 778, 779
m-value I/ 55, 415, 431, 648, 831, 840, 1006
– equilibrium I/ 384
– kinetic – I/ 383
MyoD II/ 783
myoglobin I/ 458, II/ 1097, 1128
myosin II/ 200 ff

n

NAC II/ 432 ff, 445, 446
– purification II/ 446 ff
nascent polypeptide chain II/ 423, 430, 497, 939
nascent polypeptide-associated complex II/ 423
NatA II/ 432, 438, 445, 446
native state
– dynamics I/ 81
– fluctuation I/ 550
– hydrogen exchange I/ 648
natively disordered proteins I/ 739, II/ 43, 275 ff, 1121
– amino acid compositions II/ 314
– assembly/dissambly functions II/ 325
– characterization II/ 275, 282, 288, 293
– – by NMR spectroscopy II/ 281 ff
– – by X-ray crystallography II/ 288
– definition II/ 278
– evolution II/ 315 ff
– functions II/ 275, 322
– highly entropic chains II/ 325
– hydrodynamic volume II/ 295 ff
– in databases II/ 292
– phosphorylation II/ 327
– scoring matrix II/ 320
– secondary structure II/ 282, 299
– sequence alignment II/ 317 ff, 320
– structure in vivo II/ 321
NBD *see* nucleotide binding domain
near-field scanning optical microscopy II/ 197
NEM *see* N-ethyl maleimide
nerve growth factor I/ 956, II/ 1252
neurodegenerative diseases II/ 1175
neurofibrillary tangles II/ 207
neurofilaments II/ 202
neuronal dysfunction II/ 1228
neurotoxicity II/ 1219
neurotoxin α62 I/ 952
neutron diffraction II/ 882
NGV *see* nerve growth factor
NM domain of Sup35 I/ 544
N8-MABA-AMP, synthesis II/ 119
N8-MABA-ATP, NMR II/ 120
– synthesis II/ 119
NMR spectroscopy I/ 292, 536, 737 ff
– dynamic I/ 550
– experimental protocols I/ 556
– in-cell II/ 321, 332
– limitations I/ 552
– line shape analysis I/ 547, 551
– photo-CIDNP I/ 544, 767
– pulsed field gradient methods II/ 284
– real-time I/ 538, 863, 921
– relaxation experiments II/ 332 ff

– resolution I/ 552
– secondary structure II/ 332
– solid-state II/ 197
– spin relaxation I/ 550
– translational diffusion coefficient II/ 332
non-Arrhenius behavior I/ 1147
nonnative proteins II/ 166, 174 ff, 177, 183
nonnatural amino acids II/ 1208
nonprolyl isomerization I/ 386, 930
nuclear lamina II/ 202 ff
nuclear localization signal (NLS) II/ 1207, 1229
nuclear magnetic resonance *see* NMR
nucleation I/ 8, 1048
– β-hairpins I/ 316
– mechanism I/ 6
nucleation-condensation model I/ 419
nucleation-dependent polymerization I/ 698, II/ 1215
nucleation-propagation mechanism I/ 546
nucleotides
– depletion
– – DnaK II/ 123
– – methods II/ 123
– fluorescent modifications II/ 106 ff
– purity II/ 121 ff
nucleotide binding II/ 108 ff, 127 ff
– [α-^{32}P]ADP II/ 129
– active site titration II/ 131
– assays II/ 594 f
– affinity II/ 128
– competition experiments II/ 132
– equilibrium dialysis II/ 129
– filter binding II/ 129
– fluorescence titration II/ 130
– fluorescent analogues II/ 131
– isothermal titration calorimetry II/ 127
– stoichiometry II/ 128, 131
– time-resolved measurements II/ 141
– tryptophan fluorescence II/ 130
nucleotide-binding proteins II/ 106
nucleotide-exchange factors II/ 534, 539, 571
nucleotide-releasing factors II/ 565
nucleus II/ 1215

o

oligomeric proteins
– association II/ 32 ff
– dissociation constant II/ 43, 45 ff
– equilibrium stability I/ 966, II/ 46
– equilibrium transitions I/ 988
– folding kinetics I/ 968, II/ 32 ff
– kinetic data analysis I/ 992
– reconstitution II/ 49 ff
oligomerization I/ 680, 693, II/ 1201
oligosaccharyl transferase II/ 80, 563
OmpA I/ 1015, II/ 360, 459, 463, 921, 924, 943 ff
OmpC II/ 944
OmpF II/ 943, 944
OmpT II/ 948, 950
organelle-specific markers II/ 589
ornithine transcarbamylase II/ 493
ORP150 II/ 567, 571
osmolytes II/ 1258
OST *see* oligosaccharyl transferase
oxidation
– cysteine, reduced and oxidized II/ 1255
– proteins II/ 690
oxidative folding I/ 949, II/ 1120, 1136, 1253, 1247, 1257–1260
– low-molecular-weight thiol compounds II/ 1254
– volume yield II/ 1256
oxidoreductase, catalytic motif II/ 360
– mixed disulfide II/ 361
– redox potential II/ 360

p

P body II/ 1160
P-glycoprotein II/ 79
P22 tail spike protein II/ 42, 52, 59, 1262
p23 II/ 786, 789, 794
– binding to Hsp90 II/ 789
– hormology to sHsps II/ 789
– structure II/ 789–790, 790
P300 II/ 1221
p53 I/ 985, II/ 325, 398, 791, 1261
PAB1 II/ 437

packing I/ 144
packing defects I/ 99
Pam16 II/ 1021, 1026
Pam18 II/ 495, 504, 1021, 1026, 1028
pan-amyloid antibodies II/ 1225
Pap oroteins II/ 967–975
PAR-PMPS assay zinc II/ 690
parallel pathways I/ 413, 425, 429, 437
Parkinson's disease II/ 198, 265, 1100, 1107, 1115, 1148, 1175
paromomycin II/ 438
partially unfolded forms I/ 648 ff
particle growth II/ 220 ff
parvalbumin EF-loop I/ 835
parvulin II/ 78, 382, 943
PDC *see* pyruvate decarboxylase
PDI *see* protein disulfide isomerase
PDZ domain II/ 946 ff
peptide bond
– APIase activity II/ 420
– *cis-trans* isomerization II/ 378
– – assay II/ 416–418, 424–426
– conformation II/ 415
– formation II/ 429
– imidic bond II/ 415
– rotation II/ 377, 415–416
peptide library II/ 463, 922
peptidyl-prolyl isomerase I/ 11, 386, 920, 1100, II/ 7 ff, 57, 382,399–401, 415, 421, 434, 460 ff, 661, 993, 1030
– affinity for substrate II/ 384
– assays II/ 383 ff, 384, 385, 399, 941
– chaperone-like function II/ 393
– Cpr6 II/ 786
– cyclophilin II/ 78, 1063
– cyclosporin A II/ 389, 942
– Cyp18 II/ 389, 390, 391, 392, 393
– Cyp40 II/ 786, 790
– CypA II/ 390
– drug complex II/ 389
– drug targets II/ 383
– families II/ 381
– FK506 II/ 389, 394, 942
– FK506-binding proteins (FKBPs) II/ 78, 394, 941, 1063
– – FKB12 II/ 394
– – FKB13 II/ 1064
– – FKBP51 II/ 790
– – FKBP52 II/ 790
– – structure of FKBP51 II/ 790
– FkpA II/ 941
– immunosuppression II/ 388
– in vivo function I/ 935
– kinetic mechanism II/ 389
– large PPIases II/ 790–791
– multi-domain PPIases II/ 388
– Par10 II/ 394
– parvulin II/ 78, 389, 397
– physiological II/ 388
– Pin1 II/ 398
– PpiA II/ 941
– PpiD II/ 941
– presenter-function II/ 389
– protease-coupled assay II/ 385 f
– protein folding assay II/ 387
– rapamycin II/ 389
– reduction induced assay II/ 386
– Roc4 II/ 1064
– single domain cyclophilis II/ 390
– single domain PPIases II/ 388
– SurA II/ 941
– TLP20 II/ 1064
Per100p II/ 571
periodic boundary conditions (PBCs) I/ 1177
periplasm II/ 359, 938, 965, 1247
– periplasmic chaperone II/ 968
– proteases II/ 946
– protein composition II/ 939
peroxidase II/ 1260
persistence lengths I/ 715, 811
PEST sequences I/ 368
Pfam database II/ 1290
pH jumps I/ 458
pH titrations I/ 298
phage display II/ 1312, 1318, 1322, 1323
– peptide display II/ 522–523
phase boundary I/ 100
phase diagrams I/ 99, 114
– re-entrant I/ 102
PhoA II/ 946
PhoE II/ 877, 943, 944

phosphoglycerate kinase **I**/ 471, 612, 620, 683, 695
phospholamban **II**/ 888, 889, 900
photolysis **I**/ 459
phylogenetic trees **II**/ 797
phytase **II**/ 1291
pili **II**/ 204
– 3-D structure **II**/ 972
– agglutination assay **II**/ 977
– assembly **II**/ 970
– assembly platform **II**/ 970
– biogenesis **II**/ 953
– Caf1-Caf1M structure **II**/ 974
– chaperone **II**/ 965
– complementation **II**/ 972
– donor strand **II**/ 972
– expression of subunits **II**/ 978
– FaeD **II**/ 976
– helical rod **II**/ 965
– P pili **II**/ 970, 975
– periplasm **II**/ 975
– pilins **II**/ 205
– rod **II**/ 967
– Sec YEG **II**/ 969
– stability **II**/ 975
– structure **II**/ 970
– subunit interactions **II**/ 972
– subunit purification **II**/ 979
– type 1 pili **II**/ 967 ff
– type P **II**/ 967
– usher **II**/ 970 ff
Pim1 **II**/ 1021, 1032, 1034
PIN1 **I**/ 930
PIP_3 kinase **II**/ 1148
pK_a-value
– calculation **I**/ 179
– in α-helices **I**/ 280
– shift **I**/ 177
PNGase F **II**/ 81
Υ-point **I**/ 716, 719
point mutation, effect of unfolded state **I**/ 345
polar zipper **II**/ 1206
poly(A)-binding proteins **II**/ 498
polyglutamine repeats
– aggregates **II**/ 1151, 1177 ff, 1183 ff
– CAG **II**/ 1200
– diseases **II**/ 1175, 1200
– length **II**/ 1177
– peptides **II**/ 1178
– β-structure **II**/ 1179
– toxicity **II**/ 1177, 1181
polylysine **II**/ 1097
polymer dynamics, theory of **I**/ 813
polyproline stretch **II**/ 381
– type II helix **I**/ 265, **II**/ 1205
– – conformation **I**/ 725
polythreonine **II**/ 1097
porins **I**/ 1002, **II**/ 943
POSSYCCAT **II**/ 902
posttranslational modifications, hydroxylation of proline **I**/ 1063
potassium channel **II**/ 326, 877
power-stroke model **II**/ 996, 1000
Pp5 **II**/ 791
PPI *see* peptidyl-prolyl isomerase
PpiA **II**/ 942
PpiD **II**/ 943
Ppt1 **II**/ 788
Prc **II**/ 948
pre-equilibria **I**/ 392
pre-equilibrium constant **II**/ 1217
pre-exponential factor **I**/ 383, 470, 840
prefoldin **II**/ 739, 756 ff
– complementation **II**/ 760–761
– cooperation with CCT **II**/ 761
– evolution **II**/ 757–758
– role in actin folding **II**/ 763
– structure **II**/ 757, 759
preprolactin **II**/ 460
pressure
– denaturation **I**/ 99, 108, 121
– effect on tolding **I**/ 421, 455
pressure-jump-experiments **I**/ 455
pressure-temperature phase diagram **I**/ 107
prion disease **II**/ 1114
– generation of infectious prions **II**/ 1131 ff
– heterodimer model **II**/ 1115
– mechanism **II**/ 1116
– nucleation-polymerization model **II**/ 1115, 1133
– prion hypothesis **II**/ 1115

– prion strains II/ 1149 ff
– role of chaperone II/ 1158
– role of chaperones II/ 1157
– species barrier II/ 1149 ff
– strain phenomenon II/ 1132 ff
– template assistance II/ 1115
prion protein II/ 220, 442 ff, 1120
– artificial prion II/ 1153, 1156
– bacterial production II/ 1120
– cellular prion protein (PrP C) II/ 1114
– copper binding II/ 1121
– CPEB II/ 1154
– crystal-like oligomer II/ 1134
– disulfide bond II/ 1120 ff
– Doppel protein II/ 1123
– fiber growth assay II/ 1162, 1164
– flexible tail II/ 1121
– folding II/ 1126, 1151
– GFP fusion II/ 1148
– glycosylation II/ 1121, 1134
– NMR structure II/ 1122, 1123
– oligopeptide repeat II/ 1116, 1119, 1156 ff
– oxidative refolding II/ 1136
– PIN^+ II/ 1145
– point mutation II/ 1129, 1130
– posttranslational modification II/ 1122
– prion infection protocol II/ 1164
– propagons II/ 1159
– protease resistance II/ 1118, 1131
– protein-only hypothesis II/ 1114, 1144
– protein X II/ 1134
– PSI^+ II/ 442, 1146
– purification II/ 1136
– Rnq1 II/ 1145, 1154, 1158
– scrapie isoform (PrP S c) II/ 1114
– seeds II/ 1155
– species barrier II/ 1124
– stability II/ 1125
– strain phenomenon II/ 1116
– structure II/ 1121
– Sup35p II/ 442, 1145, 1146
– thioflavin T assay II/ 1161
– two state folding II/ 1125
– Ure2p II/ 1145, 1147
– yeast prions II/ 1144
pro-domain I/ 1032
– function in folding I/ 1042, 1046
pro-peptide I/ 956
prochymosin II/ 1267
procollagen *see* collagen
productive collision II/ 1212
progress heat capacity I/ 73
proinsulin II/ 687, 1254, 1256
proline I/ 921, II/ 55, 1258
– D-prolines II/ 1226
– proline-limited folding, test for I/ 932
– prolyl bond II/ 379 ff
proline/prolyl isomerase *see* peptidyl-prolyl isomerase
prolyl hydroxylase II/ 651, 654, 658
– C-propeptide II/ 688
prolyl isomerization I/ 386, 916, 1045, II/ 57, 59, 379 ff, 380
– acivation energies II/ 941
– bioactivity II/ 381
– effect on folding kinetics I/ 918
– isomer-specific enzymes II/ 381
– isomers II/ 381
– native-state I/ 929
– phosphorylation-dependent I/ 930
– test for I/ 919
propidium iodide exclusion II/ 1230
protease
– digestion II/ 76
– – intermediates for II/ 313
– protease III II/ 948
– protection assay II/ 590
– resistance II/ 95, 999
– sensitivity II/ 311
proteasomal degradation II/ 629
proteasome II/ 265, 439, 564, 948, 1187
– ATP dependence II/ 262–263
– structure II/ 260, 262
– unfolding II/ 261
protection factor I/ 641, 657, 889
protein
– design II/ 1252
– determination of the nucleotide content II/ 122
– flexibility II/ 284 ff
– – NH-NOE II/ 284

- nonnative II/ 166
- purification, his-tagged proteins II/ 578 ff
- self-organization II/ 3 ff

protein disulfide isomerase I/ 12, 952, II/ 7 ff, 78, 570, 572, 622, 627, 651, 654, 659, 677, 678, 684 ff
- chaperone function II/ 687
- RB60 II/ 1062

protein expression
- bacterial expression II/ 578, 580
- baculovirus II/ 581
- cytosolic expression II/ 578
- mammalian cells II/ 583
- periplasmic expression II/ 580
- yeast expression II/ 580

protein folding
- assisted II/ 163
- classical view I/ 856
- code II/ 11 ff
- diseases II/ 1093, 1095
- in vivo II/ 21, 73
- jigsaw puzzle II/ 9
- mechanism I/ 857, 870
- new view I/ 856
- speed limit I/ 839
- two-state I/ 380, 868, 970, II/ 15

protein G II/ 1319

protein stability I/ 343, 346, 381, II/ 276, 1281
- engineering II/ 1284, 1288
- in vivo I/ 366
- prediction I/ 353
- proteases I/ 1039
- salt effects I/ 906
- stability curve I/ 77, 976

protein structure
- higher-order structure II/ 276
- prediction from sequence II/ 313
- structural state II/ 277
- thermodynamic hypothesis II/ 276
- thermodynamic state II/ 277

protein structure database (PDB) I/ 357

protein-protein interactions I/ 184
- assays II/ 599

proteinase II/ 76

proteinase K II/ 1118, 1119

proteoliposome II/ 975

protofilaments II/ 1097, 1210

PrP C I/ 695, II/ 1117, 1118, 1119

PrP Sc II/ 1118, 1119

psbd41 I/ 552

pseudo-first-order kinetics II/ 1219

PSI^+ II/ 1144

PUFs *see* partially unfolded forms

pulse chase experiment II/ 74, 82, 623, 954
- basic protocol II/ 88
- cells in suspension II/ 91
- in vitro assay II/ 92

pulse labeling II/ 919

puromycin II/ 460

pyruvate decarboxylase II/ 42, 43

q

quality control II/ 23, 617, 624, 1032, 1094
- reaction cycle II/ 625

quenched-flow I/ 502, 658

quinone II/ 361 f, 364, 369

r

rabbit reticulocyte lysate II/ 1007

RAC *see* ribosome-associated complex

radicicol II/ 773, 783

radius of gyration I/ 677, 689, 714, 719, 720

rafts II/ 895

Ramachandran plot I/ 319, II/ 12, 1298–1299

Raman scattering I/ 464

Raman spectroscopy I/ 296

random coil I/ 712, 768, II/ 1205
- model I/ 718
- theory I/ 713

rapid mixing experiments I/ 494
- NMR I/ 539
- technical instrumentation I/ 521

rate constants
- microscopic I/ 380
- observable I/ 380

rate equilibrium free energy relationships (REFER) I/ 411
- linear I/ 414
- nonlinear I/ 420

Rayleigh ratio I/ 677
Rayleigh scattering I/ 720
reaction coordinate I/ 381, 412, 424, 426, 428, 430, 469
reactive oxygen species II/ 679, 681, 1033
receptor dimerization II/ 891
recruitment kinetics II/ 1221
redox regulation II/ 677, 680
reducing agents II/ 177
reduction of proteins II/ 689–690
REFER *see* rate equilibrium free energy relationships
refolding II/ 186, 975, 1245, 1252
– L-Arg II/ 1258
– assay II/ 175, 186
– cofactors II/ 1260
– cyclodextrin II/ 1262
– matrix-assisted II/ 1264
– mixed disulfide II/ 1255
– pressure-induced II/ 1262
– temperature-leap II/ 1263
– volume yield II/ 1256
refractive index I/ 700, II/ 863
relaxation kinetics I/ 454
RepA II/ 503
repeat proteins II/ 1292
γ-repressor I/ 550
residual dipolar couplings, conformational dependence I/ 760
resilin II/ 204
reticulocyte lysate II/ 268, 656
reverse transcriptase II/ 783, 796
reverse-phase HPLC II/ 1212, 1234
rhodanese II/ 173, 177, 186, 187, 543, 687, 765, 783, 1262
ribonuclease A I/ 208, 225, 386, 517, 684, 691, 918, 953, II/ 4, 7, 12, 26, 366, 419, 921, 1254 ff
– fast-folding species I/ 9
– prolyl isomerization I/ 924
– reversible folding I/ 3, 924
– slow-folding I/ 9
– thermal unfolding I/ 205
ribonuclease B II/ 620
ribonuclease T1 I/ 540, 545, 549, 681, 682, 684, 930, II/ 8, 361 ff, 387, 460 ff
– prolyl isomerization I/ 926
ribosomal protein L9 I/ 552
ribosome II/ 429, 429 ff, 460, 465, 496, 939
– arrested II/ 480
– associated chaperone II/ 459
– exit site II/ 471, 497
– exit tunnel II/ 459
– nascent chain complexes (RNCs) II/ 480
– nascent chains II/ 505
– ribosomal proteins II/ 430
– ribosome-associated proteins II/ 434
– tunnel II/ 423, 429, 439, 470, 497
– tunnel exit II/ 445
– yeast II/ 434
ribosome display II/ 1312, 1313, 1323
ribosome-associated complex II/ 432, 438 ff, 497, 505
– purification II/ 446 ff
Rim44 II/ 1021
Rlg1 II/ 573
RNase A *see* ribonuclease A
RNase H II/ 1298
RNase I II/ 365, 366
RNase T1 *see* ribonuclease T1
Rop II/ 1295
ROS *see* reactive oxygen species
rotational correlation time II/ 285, 286
Rouse model I/ 814
RPA70 II/ 318
Rubisco II/ 173, 709, 1053 ff, 1067 ff
– binding protein II/ 1071
rubulin II/ 733

s

salt bridge I/ 347, II/ 1297
– networks I/ 348
SAXS *see* small angle X-ray scattering
SBT *see* subtilisin
scaled particle theory I/ 136
scanning electron microscopy II/ 228
– critical point drying II/ 229
– microscopy II/ 197

scanning transmission electron microscopy II/ 197
Scel II/ 506
Scj1 II/ 504, 568
Scj1p II/ 567, 568
scorpion toxin II/ 951
– Lqh-8/6 I/ 929
scrapie II/ 1144
ScSls1p II/ 571
SDF2-L1 II/ 572
SDS-PAGE II/ 97
– gel shift assay I/ 1004, 1016
– two-dimensional II/ 98
Sec II/ 939
Sec systemia II/ 259
Sec61 channel II/ 632
Sec61p II/ 651, 877
Sec63 II/ 495, 531
Sec63p II/ 567, 568
SecA I/ 980, II/ 921 ff
SecB II/ 919
– affinity for substrates II/ 921, 924
– binding motif II/ 922
– catalytic cycle II/ 930
– interaction with SecA II/ 927 ff
– structure II/ 922, 923, 927
second virial coefficient I/ 677, 719
secondary structure
– induction by alcohols I/ 886
– propensity I/ 357
SecY II/ 925
SecYEG II/ 877, 921
seeds II/ 1211
self-assembly II/ 241
self-interaction parameter I/ 420
SEM *see* scanning electron microscopy
semi-permeabilized cell II/ 92, 662, 663
serpin II/ 206
SH2 domain I/ 929
SH3 domain I/ 431, 433, II/ 1148
– structure II/ 286
β-sheet I/ 314 ff, II/ 198 ff, 1205
– cooperativity I/ 336
– cross-β structure II/ 1206
– H-bonding II/ 1225
– MD simulations I/ 1144
– model peptides I/ 332
– multistranded I/ 334
– native sequences I/ 316
– peptide models I/ 315
– propensity I/ 319, 352
– Ramanchandran plot I/ 320
Shiga toxin II/ 569
SHR *see* steroid hormon receptor
sHsps II/ 830 ff, 1058 ff, 1074 f
– alpha-crystallin domain II/ 858 ff
– assay for Hsp21 II/ 1078
– cellular functions II/ 831 ff
– diseases II/ 836
– dynamics II/ 839, 846
– heat stress granules II/ 1059
– homology to p23 II/ 789
– Hsp16.9 II/ 1059
– Hsp21 II/ 1058, 1059, 1060
– posttranslational modifications II/ 839
– purification II/ 840
– quaternary structure II/ 837, 1059
– release of substrate II/ 835
– substrate II/ 832, 834, 932
– subunit exchange II/ 840
side-chain conformation, J coupling constants I/ 684
sigma 32 II/ 772
sigma E II/ 947, 951
– anti-sigma factor II/ 952
– RpoE II/ 951
– RseA II/ 952
– RseB II/ 952
sigma factor II/ 943, 1267
σ^{32} II/ 502
– degradation II/ 501, 502
– half-life II/ 501
signal peptide II/ 79
signal recognition particle II/ 432 ff, 445, 460, 470, 498, 569, 877, 921, 924
– eucaryotic II/ 80
– receptor Fts-Y II/ 435
– structure II/ 435
signal sequence II/ 922
– *E. coli* II/ 939
– in bacteria II/ 926
Sil1p II/ 571

silk II/ 199 ff
– cross-β II/ 208
– major ampullate (dragline) II/ 208
– β-silks II/ 207
– Tussah II/ 208
single molecule experiments I/ 1113
– FRET I/ 613
single sequence approximation I/ 473
singular value decomposition I/ 480
siRNA II/ 88
Sis1 II/ 498, 504, 520, 528, 571
Sis1p II/ 437, 438, 439, 443, 572
– temperature sensitive variant II/ 437
site-directed mutagenesis I/ 417
six-helix bundle I/ 984
size-exclusion chromatography II/ 184, 222, 846, 864
Skp II/ 940, 944, 945, 950, 954
Sls1 II/ 571
Sls1p II/ 572
small angle X-ray scattering (SAXS) I/ 462, 502, 720, II/ 293
– Kratky plot II/ 293
small heat shock proteins *see* sHsps
snake toxins I/ 951
SNARE II/ 887, 889, 896, 902
SNase *see* staphylococcal nuclease
solid phase peptide synthesis I/ 281
solubility II/ 1316–1317
solvation I/ 350, 831
– free energy I/ 150, 178, 346
– insertion-model I/ 132
– shell I/ 136
– small molecules I/ 178
θ-solvent I/ 715
solvent accessible surface area (ASA) I/ 81, 141
solvent denaturation I/ 218, 383, 684
– acid I/ 899, 901
– alcohols I/ 884, 888
– denaturant binding model I/ 54
– GdmCl I/ 225, II/ 176 ff
– linear extrapolation method I/ 224, 973
– mechanism I/ 45, 219
– salt I/ 904
– Tanford's model I/ 54
– 2,2,2-trifluoroethanol (TFE) I/ 228
– urea I/ 219, II/ 176 ff, 183
– m-value I/ 55
somatotropin II/ 1267
spectral density function II/ 286
spectroscopically silent reactions, detection I/ 938
spin labels I/ 766
spinocerebellar ataxia II/ 1203
split-ubiquitin system II/ 903
SPPS *see* solid phase peptide synthesis
SPR *see* surface plasmon resonance
Src kinase II/ 791, 795, 800
SRP *see* signal recognition particle
Ss1p II/ 440
Ssa II/ 493, 494, 498, 522, 1158
Ssb II/ 436, 498, 505, 756, 763, 764
Ssc II/ 494 ff, 500 ff, 525, 1003 f, 1021 f, 1027 f
Sse II/ 494
Ssf1 II/ 794
Ssi1 II/ 494, 567
Ssq1 II/ 499 f, 1021, 1027 f
SSS theory *see* Szabo-Schulten-Schulten theory
Ssz II/ 436, 441, 442, 497, 505, 506, 519,
staphylococcal nuclease I/ 496, 861, 865, 929, II/ 419
STCH II/ 519
steady-state concentration II/ 1218
steric repulsion I/ 349
steric restriction I/ 730
steroid hormone receptor II/ 783, 787, 789
– glucocorticoid receptor II/ 785
Sti1 II/ 520, 542, 788, 793
Sti1p II/ 443, 1070
Stokes radius *see* hydrodynamic radius
Stokes-Einstein equation I/ 678
stopped-flow I/ 391, 689
– FRET I/ 620
streptavidin II/ 1219
stretched exponential I/ 401, 470, 471
structural water molecules I/ 206, 361
structure-function paradigm II/ 275 ff, 280
subcellular fractionation II/ 589
subcellular localizaton II/ 588

substrate binding II/ 166 ff
substrate proteins II/ 172
subtilisin I/ 1033, II/ 1252, 1319, 1321
– experimental protocols I/ 1049 ff
– stability I/ 1040
subunits, association II/ 13 ff
succinate dehydrogenase complex II/ 1031
sucrose density-gradient II/ 865 ff
Sup35
– preparation of fibers II/ 1161
– purification II/ 1160
superoxide dismutase II/ 1180
SurA II/ 943, 944, 945, 946
– structure II/ 944
surface hydration I/ 207, 222
surface plasmon resonance spectroscopy II/ 223 ff, 526, 808–809, 1219
– carboxymethyl dextran II/ 224
– self-assembled monolayers II/ 224
– sensor chips II/ 224
SV40 large T-antigen II/ 531
SVD *see* single value decomposition
Swa2 II/ 504
synaptobrevin II/ 887, 888
synaptophysin II/ 887
syntaxin II/ 887
synthetic lethality II/ 460, 950
α-synuclein II/ 1101, 1107, 1180, 1181
– ANS binding II/ 306
Szabo-Schulten-Schulten theory I/ 816

t

T cell receptor II/ 891, 892
– subunits II/ 565
T4 phage fibritin I/ 1098
tail spike endorhamnosidase II/ 22 ff
tail spike proteins II/ 59, 948
– folding mechanism II/ 61
– structure II/ 60
tandem repeats I/ 1125
TAP II/ 623
TAT II/ 939
TAT translocase II/ 988
TATA-bindig protein II/ 1201
tau II/ 1101
– tauopathies II/ 207
TB6 domain of human fibrillin-1 I/ 929
TBP II/ 1207
TCEP II/ 861, 866 ff
Tcm62 II/ 1031
TCP-1 *see* CCT
temperature-jump experiments I/ 7, 456
temperature-sensitive mutants II/ 59, 1034
tendamistat I/ 315, 386, 393, 421, 431, 887, 932, II/ 387, 419, 420
C-terminal rule I/ 367
N-terminal rule I/ 366
tetrameric proteins I/ 983
tetratrico peptide repeats II/ 444, 496, 527, 788
TF *see* trigger factor
TFE *see* 2,2,2-trifluoroethanol
thermal denaturation I/ 52, 99, 177, 184, 121, 208, 455, 682
– collagen I/ 1064
– transition temperature I/ 73, II/ 1284
– two-state I/ 71
thermal stability I/ 141
– electrostatic contributions I/ 188
– mechanism I/ 77
– two-state I/ 71
thermodynamic cycle I/ 448
thermophiles II/ 24
thermosome II/ 725, 726, 727, 731
thioflavin S II/ 227
thioflavin T II/ 226, 1179, 1222
thiol trapping II/ 690
thioredoxin II/ 360, 365, 367, 368, 684
thioredoxin reductase II/ 367, 684, 1247
thioxo dipeptides II/ 417
Thornton's reacting bond rule I/ 426
thrombin II/ 203
ThT fluorescence II/ 1213
thyroglobulin II/ 572
thyroxine II/ 1106
Tim16 II/ 1026
TIM23 II/ 997, 1026
Tim44 II/ 495, 993, 1022 ff
time-resolved fluorescence II/ 1219
– quenching I/ 1018

– tryptophan quenching **I**/ 1015
tissue-type plasminogen activator **II**/ 1249, 1255
titin **II**/ 39 ff, 264, 997, 1000, 1134
– thermodynamic stability **II**/ 41
TLP40 **II**/ 1065
TM-helix *see* transmembrane helix
Tom70 **II**/ 496, 788, 793–794
topomer search model **II**/ 330
TOXCAT system **II**/ 901
toxicity **II**/ 1201
ToxR **II**/ 901 ff
TPR domain **II**/ 540, 788, 790, 794, 796, 1070
TPR protein *see* tetratrico peptide repeats
transcription factors **II**/ 1206
transcription/translation reaction **II**/ 479
transcription/translation system **II**/ 475
transfer model
– gas-liquid **I**/ 154
– liquid-liquid **I**/ 128
transferrin **II**/ 78
transition state **I**/ 1136
– characterization **I**/ 411
– ensemble **I**/ 1154
– movement **I**/ 413, 424, 435
transition state theory **I**/ 383
– application to protein folding **I**/ 717
translation **II**/ 429, 665
translocase **II**/ 921
– *E. coli* **II**/ 925
translocation **II**/ 79, 495, 498
translocation channel **II**/ 996
translocon **II**/ 497, 939
transmembrane helix
– amino acid sequences **II**/ 878–879
– basis for interaction **II**/ 881–882
– Cys residues **II**/ 885
– entropic factors **II**/ 882
– free energy of helix associaton **II**/ 894
– G xxx G motif **II**/ 886, 887, 893
– heptad repeat motif **II**/ 883, 886, 891, 893
– homodimers **II**/ 887
– in thermophilic organisms **II**/ 885
– interactions **II**/ 880
– polar clamp **II**/ 885
– prediction **II**/ 878–879
– sequence **II**/ 883–885
– serine zipper **II**/ 885
– stability **II**/ 878–879
– structures of bitopic proteins **II**/ 888
transmissible encephalopathies **II**/ 1103, 1114, 1144
– bovine spongiform encephalopathy **II**/ 1116
– familial CJD **II**/ 1116
– Kuru **II**/ 1116
– variant CJD **II**/ 1116
transmission electron microscopy **II**/ 230
– metal shadowing **II**/ 230
– negative staining **II**/ 231
transthyretin **II**/ 1106, 1225
trap mutants
– BiP **II**/ 87
– GroEL **II**/ 87
TRAP1 **II**/ 782
TriC **II**/ 442, 498
tricorn **II**/ 948
2,2,2-trifluoroethanol **I**/ 279, 687, 887
– effect on folding kinetics **I**/ 896
trigger factor **I**/ 934, **II**/ 423, 438, 445, 459, 461, 463, 464, 470, 756, 924
– activity assay **II**/ 474
– binding to ribosomes **II**/ 470
– chaperone activity **II**/ 469
– domain structure **II**/ 464, 465
– PPIase activity **II**/ 462
– purification **II**/ 473
– substrate proteins **II**/ 461, 462, 493
trimeric proteins **I**/ 983
triple helix **I**/ 1061
– folding kinetics **I**/ 1081
– helix – coil transition **I**/ 1066
– stability **I**/ 1062
– structure **I**/ 1076
triplet quenching **I**/ 821
triplet-triplet energy transfer **I**/ 577, 821
tropomyosin **II**/ 200 ff
troponin C **I**/ 611
TRP domain, structure **II**/ 541
Trp-cage **I**/ 1149
TRP2 **II**/ 520
Trx *see* thioredoxin

trypsin II/ 76
trypsinogen II/ 329
tryptophan repressor protein I/ 981
tryptophan zipper I/ 324
TSE disease II/ 1115
TSEs II/ 1114
tsp *see* tail spike protein
TTET *see* triplet-triplet energy transfer
tubulin II/ 732 ff, 761 ff
tunicamycin II/ 86, 593
β-turn I/ 316
two-hybrid screen II/ 586
two-state folding I/ 380, 868
– dimeric proteins I/ 970
two-state model
– calorimetric test I/ 75
– kinetic test I/ 382
type 2 diabetes II/ 1105, 1107

u

ubiquitin I/ 471, II/ 261, 267, 575, 1177
– folding mechanism I/ 508
– ubiquitin-proteasome system I/ 366, II/ 1228
UDP-glycoprotein glucosyltransferase II/ 572, 576, 619
unfolded protein response II/ 573, 574, 938, 953
unfolded protein response element II/ 573
unfolded proteins I/ 710
– NMR assignment I/ 740
– NMR parameters I/ 744
– residual structure I/ 727, 840
– steric restrictions I/ 730
unfolded state, dimensions I/ 687
unfolding II/ 168, 177, 254, 708, 995, 1005, 1022, 1033
– analysis in vivo II/ 1010–1011
UPR *see* unfolded protein response
UPRE *see* unfolded protein response element
Ure2 I/ 922
urea II/ 176 ff, 183
urokinase II/ 366

v

α-value I/ 411, 414, 419, 840
– *see also* Leffler plot
Φ-value analysis I/ 417, 433, 445, 472, 840
– frictional I/ 449
– mechanical unfolding I/ 1132
van der Waals interactions I/ 128, 144, 350, II/ 881
van't Hoff Enthalpy I/ 990
van't Hoff equation I/ 975
vesicular stomatitis virus G-protein II/ 565, 896
V_H II/ 1301–1303
vimentin II/ 202
virus
– adenovirus II/ 206
– reovirus II/ 206
– tobacco mosaic virus II/ 205
– viral fibers II/ 208
– virion II/ 205
viscosity I/ 470
volume changes I/ 99
von Hippel-Lindau tumor suppressor protein II/ 737

w

Walker sequence motifs II/ 111
water condensation analysis II/ 1217
water-filled nanotube II/ 1227
WD40 proteins II/ 738, 742, 764
words of wisdom I/ 594, 721
WW domain I/ 339, 457

x

X-ray crystallography II/ 211
– disordered regions II/ 291
– isomorphous replacement II/ 289
– molecular replacement II/ 290
– multi-wavelength anomalous dispersion II/ 290
– phase problem II/ 289
X-ray diffraction II/ 198 ff
Xap2 II/ 796

y

Yael II/ 952
YaeL protease II/ 945
Ydj1 II/ 443, 497, 504, 527, 528, 529, 1158
yeast prion II/ 198
yeast two-hybrid system II/ 1315–1316
YEP II/ 1182
YFP II/ 1187

z

Zimm-Bragg model I/ 261
Zimm model I/ 814
Zimm plot I/ 678
zinc II/ 679, 681, 926, 929
– binding constants II/ 692
zinc center II/ 679
zinc finger proteins II/ 679
zinc-coordinating thiols II/ 679
Zipper model, collagen folding I/ 1092
Zuo1 II/ 497, 498, 505
zuotin II/ 437, 440, 441, 442